THE FOOD ADDITIVES BOOK

THE FOOD ADDITIVES BOOK

Nicholas Freydberg, Ph.D.,
and Willis A. Gortner, Ph.D.

BANTAM BOOKS
Toronto • New York • London • Sydney

THE FOOD ADDITIVES BOOK
A Bantam Book / June 1982

Published simultaneously in hardcover and trade paperback

Grateful acknowledgment is made for permission to include the following copyrighted material:

Information on sugar content in cereals reprinted from *Brand Name Guide to Sugar* by Ira L. Shannon. Copyright © 1977 by Ira L. Shannon. Used by permission of Nelson-Hall Publishers, Chicago.

Information from table 2-5 of "Gas-Liquid Chromatographic Analysis of Sugars in Ready-to-Eat Breakfast Cereals", *Journal of Food Science*, Vol. 45, No. 1, pp. 138-141. Reprinted by permission. Copyright © 1980 by Institute of Food Technologists.

Library of Congress Cataloging in Publication Data

Freydberg, Nicholas.
The food additives book.

Includes index.
1. Food additives. I. Gortner, Willis Alway, 1913-
II. Title.
TX553.A3F73 664'.06 81-15044
ISBN 0-553-01376-9 (pbk.) AACR2

Published simultaneously in the United States and Canada

Bantam Books are published by Bantam Books, Inc. Its trademark, consisting of the words "Bantam Books" and the portrayal of a rooster, is Registered in U.S. Patent and Trademark Office and in other countries. Marca Registrada. Bantam Books, Inc., 666 Fifth Avenue, New York, New York 10103.

PRINTED IN THE UNITED STATES OF AMERICA

0 9 8 7 6 5 4 3 2 1

ACKNOWLEDGMENTS

Many people have helped us in preparing this book. Our Dedication of the book acknowledges our great debt to the many scientists in the FASEB Select Committee on GRAS Substances and the Life Sciences Research Office. Miss Christine Aylor has helped keep us abreast of the scientific reports on additives issuing from that office.

The staff of the Bureau of Foods of the Food and Drug Administration, and especially Drs. Sanford A. Miller, F. Edward Scarbrough, and Corbin Miles; and of the Human Nutrition Center of the U.S. Department of Agriculture, particularly Drs. Frank Hepburn, Wayne R. Wolf, and John L. Weihrauch, have been most helpful and supportive in tracing some of the scientific details important in our assessment of various of the food additives.

Drs. Horace L. Sipple, Byron T. Webb, Mildred Rodriguez, and Steven R. Tannenbaum provided valuable assistance as scientific consultants.

Our research associates, Barbara Schlesinger, Elizabeth Vizza, and Margaret Curtin, bore the major share of the formidable task of purchasing the thousands of brands of food that were required, traveling to distant parts of the country for that purpose, and extracting the information from their labels needed for the Inventory-of-Brands Section and preparing it for publication. It would be difficult to duplicate their enthusiasm and standards of accuracy during the many months it took to accomplish this work. The authors, besides, are grateful for their many valuable suggestions which have contributed materially to making this volume more useful for the reader.

Acknowledgment of our debt to others would be incomplete without expressing our indebtedness to Judy Knipe, the editor assigned to us by our publisher. Her perceptive queries, insightful challenges, her rewritings and rearrangements have been of inestimable value toward achieving a clarity of presentation necessary for our objective—a readily understandable, helpful, and accurate guide to additives and branded foods for the food shopper.

DEDICATION

This work is dedicated to the distinguished scientists, researchers, and scholars, who, as the FASEB Select Committee on GRAS Substances (SCOGS) or as the staff of the FASEB Life Sciences Research Office (LSRO), devoted their expertise and time for a decade (1971–81) to the collection and the assessment of the bulk of the information used by the authors.

In a true sense, we, as the authors-of-record, regard these individuals as "co-authors," and justly so, considering that the data and judgments contained in this volume would have been inconceivable to collect and present here without the benefit of their efforts.

Now, at last, in large measure because of them, the American public has available to it the relevant knowledge it needs in an area of vital significance, that of food additives, in an understandable and reliable form, and easy to put into practice.

Richard G. Allison, Ph.D. (LSRO).
Aaron M. Altschul, Ph.D. (SCOGS), Professor of Community Medicine and International Health, Georgetown University School of Medicine.
Sue Ann Anderson, Ph.D. (LSRO).
Joseph F. Borzelleca, Ph.D. (SCOGS), Professor of Pharmacology, Virginia Commonwealth University Medical College.
C. Jelleff Carr, Ph.D. (Director, LSRO 1972–77).
Herman I. Chinn, Ph.D. (LSRO).
Harry G. Day, Sc.D. (SCOGS), Professor Emeritus of Chemistry, Indiana University.
Samuel B. Detwiler, M.A. (LSRO).
Kenneth D. Fisher, Ph.D. (Director, LSRO 1977–).
Samuel J. Fomon, M.D. (SCOGS), Professor of Pediatrics, College of Medicine, University of Iowa.
Andrew F. Freeman, B.S. (LSRO).
George W. Irving, Jr., Ph.D. (Chairman, SCOGS).
Bert N. La Du, Jr., M.D. (SCOGS), Chairman, Department of Pharmacology, University of Michigan.
John R. McCoy, V.M.D. (SCOGS), Professor of Comparative Pathology, College of Medicine and Dentistry, Jersey-Rutgers Medical School.
Sanford A. Miller, Ph.D. (SCOGS), Professor of Nutritional Biochemistry, Massachusetts Institute of Technology.
Gabriel L. Plaa, Ph.D. (SCOGS), Chairman, Department of Pharmacology, University of Montreal.
Frederic R. Senti, Ph.D. (Associate Director, LSRO).
Michael B. Shimkin, M.D. (SCOGS), Professor of Community Medicine and Oncology, University of California School of Medicine at San Diego.
R. G. H. Siu, Ph.D. (SCOGS), Consultant.
Marian E. Swendseid, Ph.D. (SCOGS), Professor of Nutrition and Biological Chemistry, University of California School of Public Health at Los Angeles.
John M. Talbot, M.D. (LSRO).
Michael J. Wade, Ph.D. (LSRO).
John L. Wood, Ph.D. (SCOGS), Distinguished Service Professor of Biochemistry, University of Tennessee Center for Health Sciences.

CONTENTS

WHAT THIS BOOK IS ALL ABOUT

Some of us have become confused about and fearful of a wide range of chemicals that are being added to our foods. Our distress has been accentuated by the continuing stream of *new research findings,* some adverse, others exaggerated by the media, which have fueled the concerns of consumers about the safety of foods available in the marketplace.

Our authoritative sources of information are not always of much help; they too have contributed to the uncertainty with conflicting advice. In 1980 the Departments of Health, Welfare & Education and Agriculture set dietary guidelines for us, only to have these disputed by the Food and Nutrition Board of the National Research Council. It didn't end there, for nutrition experts in the government, in turn, challenged these revisions. Our *experts* have been at loggerheads about saccharin, disputing whether this synthetic sweetener, which has caused cancer in laboratory animals, is a sufficient hazard to bar it in human food. They differ as well on whether excessive intake of cholesterol contributes to the occurrence of heart attacks.

As an increasingly urbanized society, we are becoming less and less able as individuals to grow our own food in sufficient quantity and variety. Some people have questioned whether the wide range of additives employed for many purposes is really needed. It is difficult to quarrel with the necessity of additives to preserve processed foods during their manufacture and shipment and while stored on grocery shelves for a reasonable period of time. Nor can we contest their "usefulness" if they enhance flavor, appearance, create satisfactory and appetizing textures, and so on, *as long as they are safe.*

What is difficult for a food consumer to be able to determine, within the limits of current knowledge, is which additives are safe, which are not, and which are questionable as to safety. This volume, the authors believe, will provide you with the necessary scientific and technological information to make those determinations for the additives you will find listed on the labels of branded food items. It should enable you to put this knowledge into practice with ease when you shop for food.

The safety ratings which this volume provides reflect the risks a consumer may face from including in his diet, at times frequently, certain food additives. They do *not* consider the possible benefits. The safety of ferric salts, for example, is questioned here because of insufficient information and poor biological availability. They are used as iron supplements and have nutritional benefits, as does calcium caseinate, whose use is also cautioned because of the possible presence of a substance which may (it is not certain) have an adverse effect on humans. The *risks* can be judged objectively; experimental studies can provide information on the nature of a possible hazard and the likelihood of it. By contrast, the *benefits* will vary for each person, depending on individual needs. With this in mind, the book tells why the additive has been used in formulating the food, and enables the reader to make an informed choice based on its value to him for a wholesome, satisfying, and safe diet.

A relevant question then is: Why hasn't the information contained in this volume been made available to the general public earlier? A Presidential Directive in 1969, tacitly acknowledging the need, ordered this done; and commencing in 1971, the massive task of evaluating the safety of hundreds of food additives was undertaken by a group of distinguished scientists. Collection of the worldwide experimental and clinical information has been ongoing for a decade, as have the assessments of safety.

You may have noted earlier that this volume is dedicated to this remarkable group of scientists (referred to as SCOGS—Select Committee on GRAS Substances). They were selected from the variety of disciplines concerned with food: biochemistry, organic chemistry, pathology, physiology, nutrition, food science, oncology, pharmacology, human and veterinary medicine. Their effort may have no parallel in science. The project has been conducted under the authority and direction of the Federation of American Societies for Experimental Biology (FASEB), a consortium of six scientific associations who bore the responsibility for the selection of the participants and provided their own research facility, the Life Sciences Research Office (LSRO), as a functioning staff.

The authors, on their part, have devoted over two years to winnowing the vast amounts of material provided in the reports issued by SCOGS, in order to make the information readily comprehensible to the layman. These reports provide a major part of the material contained in the second half of this volume. As the SCOGS reports did not cover all of the substances permitted as additives in food—those omitted were principally ones which the FDA had not classified "generally recognized as safe" (GRAS)*—the additional additives have been reviewed by the authors, or commissioned to qualified scientists for that purpose; and these too are included here.

For the most part, the safety assessments made by the authors have conformed with the judgments arrived at by the Select Committee. In some instances they have not, and, when in disagreement, SCOGS has arrived at a somewhat more severe rating in twice the number of instances as *The Food Additives Book*. (For this comparison, see p. 715.) The authors, of course, acknowledge their full responsibility for the judgments made in the book, irrespective of whether or not they are in agreement with SCOGS.

With the intention of making it easy to put the knowledge contained here into practice, an Inventory-of-Brands has been provided. It covers over 100 food categories, lists more than 6000 branded foods, and alongside each you will find any additives contained in it which are regarded in this volume as of possible concern to health. An explanatory introduction to this section follows immediately. It can be used as a basic shopping list and, at the least, as a means whereby ingredients to avoid in a specific category of food can be quickly identified.

---

*The 1958 Food Additives Amendments to the Food, Drug and Cosmetic Act of 1938 exempted from prior approval by the FDA substances which were "generally recognized as safe" by experts. These became known as GRAS substances. The amendments did require prior approval by the FDA for new commercially added food ingredients based on proof of safety.

THE INVENTORY OF BRANDS

THE PURPOSE OF THE INVENTORY OF BRANDS

The purpose of this section is to make it easy to apply the information on additives supplied in the dictionary section of this book when you go to market: to select foods both wisely and safely.

It can enable you to determine beforehand whether a specific food item included here contains an additive that may be of concern to your health; and if it does, to look for similar items which do not.

However, you are not likely to find every brand of food you prefer contained in the inventory section. To do so for all of the readers of this book would have required listing almost everything carried on grocery store shelves, possibly 25,000 items in all. We've compromised on about a quarter of this number, mostly best-sellers; otherwise the size of the book would make it unmanageable and its cost prohibitive.

If a food item of interest to you is not here, the inventory nevertheless should enable you to know what additives-of-concern to look for in that item. For example, if you fail to find a preferred soup included, you can easily determine the additives-of-concern to watch out for when shopping for it; they are likely to be the ones present with some frequency in the soups that are listed. By means of a breakdown of items into over 100 product categories, you have available sufficient representation in each to gain an excellent idea of the additives to avoid. And if you do come across an additive still different from these, you always can turn to the dictionary portion of this book to find what you'll need to know about it. The INVENTORY OF BRANDS covers:

- nationally distributed brands
- regional brands
- supermarket private brands
- brands in health food stores

All of these have representation here and are identified in each product category, the idea being to enable you to make comparisons and find purchasing alternatives. The national brands usually are best-sellers. Brands of regional origin (indicated by "R" on the left margin), usually perishable foods, are of significance in categories such as baked goods and dairy products. Again, space restrictions permitted representation of a limited number of supermarket chains, but these are among the largest in the country. Items sold in health food stores are becoming an increasingly important source for many shoppers, in part because of concern about additives. It has not been possible to determine whether these are regional or national in distribution.

It is of relevance to note here that all of the foods listed in this inventory were purchased by the authors at retail outlets, and none was obtained by other means; their selection was determined solely on their popularity or the requirements of a well-rounded representation.

Be certain to take note of this column heading, which is present in each product category:

ADDITIVES-OF-CONCERN
BOLD —CONCERN TO ALL
LIGHT —CONCERN TO SOME.

It has been possible to differentiate, based on research and clinical experience, between additives that are of concern to all of us, and those which are of concern to some people and not to others. If any additives-of-concern are contained in an item, you'll find them listed alongside in this column.

It may startle you, at first, to find that so many food items contain a number of additives-of-concern. But note that the ones in light type are of relevance only to limited groups of people; for instance, to those allergic to certain ingredients in grains (gluten), or to dairy products (albumin, whey); or to women who are pregnant or lactating; or to anyone afflicted with high blood pressure. There should be little reason for others to avoid these additives.

When you find an additive-of-concern listed alongside a brand, it is suggested that you refer to the review of the additive in the dictionary section, where you will find the scientific evidence for the assessment of caution, and the identification of those individuals who should avoid a particular additive and why.

THE EXTENT OF THE PRESENCE OF ADDITIVES-OF-CONCERN IN THE FOOD ITEMS LISTED IN THE INVENTORY OF BRANDS*

The Inventory-of-Brands contains 6,212 items of branded, packaged foods that were purchased in grocery and health food stores. They provide representation in over 100 food product categories.

Half of these (50 percent) contain one or more ingredients regarded in this volume to be additives-of-concern to the health of all consumers.† Another 11 percent contain ingredients regarded here as additives-of-concern to some (but not all) consumers. In all, 6 out of 10 of these branded items contained substances of health concern to some or all of the U.S. public.

A significant difference exists in this respect between food items distributed through the usual food outlets and those distributed through health food stores. Sixty-five percent of the items that were purchased in grocery stores contained additives-of-concern (A-O-C), compared to 31 percent in health food stores, or more than double proportionally.

TYPE OF RETAIL OUTLET		TOTAL ITEMS	W/A-O-C FOR ALL	W/A-O-C FOR SOME	W/NO A-O-C
Grocery	Number	5420	2938	582	1900
	Percent	100%	54%	11%	35%
Health Food	Number	792	155	89	548
	Percent	100%	20%	11%	69%
Total Both	Number	6212	3093	671	2448
	Percent	100%	50%	11%	39%

These totals obscure substantial differences in the incidence of additives-of-concern between product categories. For in-

*The claim is not made here that the items listed in the Inventory-of-Brands section are a representative cross section of all of the packaged foods to be found on retailers' shelves. The objectives that guided our selection were inclusion of the more popular, nationally distributed brands, and, in addition, examples of regional and supermarket private brands which would enable the reader to make comparisons. Because of this bias in selection, the extent of the presence of additives-of-concern found in these items cannot be assumed to be an accurate reflection of all processed foods available to the U.S. consumer.

†These items in some instances also contained additives-of-concern to some (not all) consumers.

stance, relatively few sweet spreads, vegetables, baking ingredients, fruit and fruit juices, and drinks sold in grocery outlets contain additives-of-concern; on the average less than two out of ten do. In contrast, eight out of ten packaged dinners, baked goods, soups, and low-calorie foods and beverages, on the average, contain additives of concern; these foods are not commonly found in health food stores.

The table which follows provides this information for each of twenty-three major food groups contained in the Inventory of Brands section for both grocery and health food outlets.

Presence or Absence of Additives-of-Concern in items sold in grocery and health food stores by major food product categories (arranged in descending order of absence of additives-of-concern in grocery store items)

	GROCERY STORES National, Regional, and Private Brands				HEALTH FOOD STORES Items sold in these outlets			
Major Categories	TOTAL NUMBER OF ITEMS	% WITH A-O-C FOR ALL	% WITH A-O-C FOR SOME	% WITH NO A-O-C	TOTAL NUMBER OF ITEMS	% WITH A-O-C FOR ALL	% WITH A-O-C FOR SOME	% WITH NO A-O-C
Jellies & Other Sweet Spreads; Nut & Seed Butters	200	6%	1%	93%	65	-%	-%	100%
Vegetable Juices & Vegetables	434	17	1	82	48	-	10	90
Baking Ingredients	99	20	-	80	31	3	3	94
Fruit, Fruit Drinks, Fruit Juices	336	22	1	77	82	5	7	88
Pasta; Potatoes (Instant); Rice	118	53	3	44	35	6	23	71
Baby Foods	202	44	14	42	2	*	*	*
Beverage Mixes & Beverages	256	53	9	38	24	4	13	83
Snack Items	187	60	6	34	46	20	4	76
Bread & Bread Products; Crackers	580	38	29	33	61	8	21	71
Fish & Shellfish	139	58	10	32	12	*	*	*
Cereal, Cereal Bars (Breakfast & Snack)	110	65	5	30	27	30	11	59
Dairy Products & Substitutes	598	55	15	30	88	50	9	41
Gravies, Sauces, & Seasonings	138	65	7	28	28	7	21	71
Candy & Gum	144	65	13	22	35	20	6	74
One-Course Dinners	179	75	7	18	44	23	9	68
Pickles, Salad Dressings, & Other Condiments	147	73	13	14	45	40	18	42
Baked Sweet Goods, Not Frozen or Refrigerated	507	74	14	12	19	58	21	21
Low-Calorie Beverages & Foods	202	84	6	10	-	-	-	-
Baking Mixes, Dessert Toppings, Gelatin, Pudding	279	89	2	9	17	23	12	65
Soups	164	59	32	9	40	5	17	78
Meat & Poultry & Substitutes	163	85	7	8	29	62	3	35

	GROCERY STORES National, Regional, and Private Brands				HEALTH FOOD STORES Items sold in these outlets			
Major Categories	TOTAL NUMBER OF ITEMS	% WITH A-O-C FOR ALL	% WITH A-O-C FOR SOME	% WITH NO A-O-C	TOTAL NUMBER OF ITEMS	% WITH A-O-C FOR ALL	% WITH A-O-C FOR SOME	% WITH NO A-O-C
Baked Goods, Frozen & Refrigerated	138	83	11	6	7	*	*	*
Frozen Dinners, Pizza, Pot Pies	100	91	5	4	7	*	*	*
Total All Categories	5420	54%	11%	35%	792	20%	11%	69%

*Insufficient representation for percentages to be meaningful

THE LIMITATIONS OF FOOD LABELS FOR IDENTIFYING ADDITIVES-OF-CONCERN

The Food and Drug legislation presently permits exemptions in the requirement that specific additives be listed on a label when contained in some foods. In such instances, a means has been devised in the inventory to alert you when an additive-of-concern may be present.

• Artificial color and some natural colors.

When a certified (FD & C) artificial color is used, the actual color used need not be identified; only "artificial color" or some variant of the term must appear on the label, unless evidence exists that a specific one used is a proven hazard. In that case, it must be listed. FD & C Yellow No. 5 (Tartrazine) is the only artificial color at present subjected to this requirement, as it causes an allergic reaction in some people. Disagreement, however, exists concerning the possibility of harmful effects caused by some other artificial colors; and to caution about it, artificial color when noted on a label will appear in the additives-of-concern column. A rule-of-thumb method proposed here for assessing whether disputed artificial colors may be present is by the color of the food. If the color of the food is red, orange, or violet, these disputed food colors may be in it. Whatever the colors, children, especially hyperactive ones, are cautioned to avoid artificial color additives until questions concerning their effects are resolved. A detailed discussion is available under Food Colors in the dictionary section (see p. 534).

Dairy products like cheese and frozen desserts are exempted

by law from the necessity of even declaring that they contain artificial color, except in such instances as FD & C Yellow No. 5. Some manufacturers now voluntarily state the presence of artificial color on the label, or its absence. However, if not listed, or unless specifically stated as not being present, it remains uncertain whether artificial color has been used in these foods. A notation has been placed at the beginning of these product categories to alert the reader to the possibility of the presence of artificial color when it does not appear in the additives-of-concern column.

A few natural colors, too, can be of concern to health, such as carmine or saffron. When "natural color," but no *specific* natural color, appears on an ingredient label, you'll find it noted in the additives-of-concern column. Unfortunately, no method has been devised here to help determine whether or not it refers to a possibly harmful natural color.

- Artificial flavor and natural flavor.

The flavors used in foods may be derived from natural or synthetic substances; mostly they consist of complicated combinations of these, perhaps a dozen or more. The flavor chemist can choose from almost 2000 compounds for this purpose; by law, an individual listing of these substances is not required, and it is most unlikely to be revealed, for these combinations are regarded as valuable trade secrets. With no way to determine which components have been used from the formidable number that are available, it has not been possible to provide safety assessments for flavors identified as "artificial" or "natural." It is the only instance in this volume where we have been unable to do so. When a single flavor is reported as present, such as ethyl methyl phenylglycidate or ionone, an assessment is possible, and has been noted in the additives-of-concern column whenever the evidence indicates it to be warranted.

- Fats and oils.

A manufacturer is permitted by FDA regulations to identify in the ingredient list the range of fats and oils which may have been used, but one cannot be certain of the ones that actually are in the food. For example, the label on Dolly Madison Cinnamon Rolls lists these fats and oils that may be present:

> . . . vegetable and/or animal shortening (may contain one or more of the following: soya oil, beef fat, lard, partially

hydrogenated [soya oil, cottonseed oil, palm oil, coconut oil, palm kernel oil, beef fat and/or lard]) . . .

You will find this reduced list of animal fats and vegetable oils appearing in the additives-of-concern column alongside this Dolly Madison breakfast pastry:

(BEEF FAT, LARD; COCONUT, PALM KERNEL OILS)

The procedure followed in these instances is to extract the ones that are highly saturated, either naturally or by processing (hydrogenation), and in sufficient quantity if present in the food to be regarded as of possible health concern. An explanation of the method for determining quantity will be found in both Animal Fats (see p. 479) and Vegetable Oils (see p. 660) in the dictionary section.

- Inadequate identification of ingredients.

At times, a general rather than a specific term is used for an ingredient on a label, such as shortening instead of coconut oil, or natural color rather than saffron. This general term can refer to any of a number of ingredients and poses a problem for its safety evaluation when some of these possibilities are of concern and others are not. Thirteen general terms found on labels to which this applies are:

Animal fats; cellulose derivatives; coloring, food colors; natural colors, vegetable colors; seaweed; shortening, vegetable oils; softeners, stabilizers, thickeners, vegetable gums

The inability to determine the specific ingredient that may be present has led, for reasons of caution, to the placement of these terms in the additives-of-concern column when they appear on a label. A listing of the specific ingredients they can have reference to follows, separated into those regarded as additives-of-concern and those that are not.

Additives-of-Concern	*Not Additives-of-Concern*
ANIMAL FATS	
Beef fat (tallow); butter (butter fat); lard (pork fat); mutton (fat); stearic acid; calcium stearate	Marine (fish) oil; poultry fat and skin

Additives-of-Concern	*Not Additives-of-Concern*
CELLULOSE DERIVATIVES	
Hydroxypropyl cellulose; methyl ethyl cellulose	Cellulose derivatives *other than* hydroxypropyl cellulose, methyl ethyl cellulose
COLORING, FOOD COLORS	
Artificial color; certified color; FD & C colors; cochineal and carmine; paprika, turmeric, and their oleoresins; saffron	Natural color *other than* cochineal and carmine; paprika, turmeric, and their oleoresins; saffron
NATURAL COLORS, VEGETABLE COLORS	
Cochineal and carmine; paprika, turmeric, and their oleoresins; saffron	Natural color *other than* cochineal and carmine; paprika, turmeric and their oleoresins; saffron
SEAWEED	
Alginates; carrageenan and furcelleran; dulse and kelp	Agar-agar
SHORTENING, VEGETABLE OILS	
Coconut oil; hydrogenated vegetable oils; palm kernel oil; rapeseed oil	Non- and partially hydrogenated vegetable oils *other than* coconut, palm kernel, and rapeseed oils
SOFTENERS, STABILIZERS, THICKENERS, VEGETABLE GUMS	
Alginates; carrageenan and furcelleran; dulse and kelp; glycerol esters of wood rosin; gum arabic; gum tragacanth; guar gum; hydroxypropyl cellulose; locust bean gum; methyl ethyl cellulose; modified starch	Agar-agar; cellulose derivatives *other than* hydroxypropyl cellulose, methyl ethyl cellulose; unmodified or gelatinized starch; xanthan gum

Evidence regarding the safety of each of the ingredients listed in both columns is presented in the dictionary section of this volume.

Another situation of inadequate identification arises when an ingredient specified on a label contains a component which, if listed by itself, would have been identified as an additive-of-concern. This is true of a number of substances extracted from milk (buttermilk solids, milk derivatives, milk protein, milk solids, sour milk or cream solids) which may contain casein, lactalbumin, or whey, all viewed here as additives-of-concern.* A similar circumstance occurs in "high protein flour" which can contain albumin, casein, or gluten, all warranting caution for some people. These milk extractives and the flour will appear in the additives-of-concern column when present in a food.

- Absence of ingredient lists on labels.

Some manufacturers state on the label only that their product is "pure" fruit juice or "100%" coffee, or whatever, and do not supply a formal list of ingredients. They may make no reference whatever to ingredients, assuming perhaps that the content of foods like frozen fish fillets, simple frozen vegetables, and, on occasion, frozen and refrigerated meats, is self-evident. Pastas, nuts, dried and frozen fruits not infrequently will omit an ingredient list. It has made common sense to us to consider these products as additive-free, and they have been identified in the inventory as containing no additives-of-concern.

THE OMISSION OF SODIUM CHLORIDE (TABLE SALT) FROM THE ADDITIVES-OF-CONCERN COLUMN

Sodium chloride is regarded as an additive-of-concern in the review of this substance in the dictionary section, because anyone with high blood pressure or a family history of this tendency, or normal infants (in their prepared food) should avoid an excess. In spite of this, you will not find salt in the additives-of-concern column alongside items that contain it. It is present in so many foods that, regrettably, its constant repetition might serve little purpose except to obscure the crucial fact that it is the amount and not its mere presence which is important.

We have no way of determining from the label the amount of sodium chloride there is in a food item. Its order in the list of ingredients does provide a coarse measurement, as the largest

---

*Milk itself is considered a food, not an additive. It is not included as an additive-of-concern in this book, although some people may have a milk intolerance.

quantity of a substance in a food will come first in the list, the next in amount second, and so on.

In comparison with the ingredients which precede it on the list, the relatively small amount of salt used in foods would usually place it at, or near, the bottom. For this reason, the position of salt on the ingredients list makes it a doubtful means for determining the actual amount present or for discriminating between similar food items. To demonstrate the difficulty, a comparison is available between two well-known white breads, each of which lists salt as 6th in the ingredient order. Two slices of one of these is reported as containing 335 milligrams, while two slices of the other brand contains less than a quarter of this amount, 81 milligrams.

Perhaps more important for this book's purposes, although salt may be well down in the order of ingredients, there is no assurance that the amount is not a hazard to vulnerable consumers. It is eighth on the list of a popular frozen macaroni and cheese dish; yet a small serving of 6 ounces contains 780 milligrams, a substantial amount, and especially so if one is tempted to take another helping.

Guidance on the amount of sodium in common foods is available from the booklet "The Sodium Content of Your Food," prepared under the auspices of the U.S. Department of Agriculture and available from Superintendent of Documents, U.S. Printing Office, Washington, D.C. 20402. It lists the sodium content of an extensive range of common foods, not by brands but by food categories (carbonated beverages, coffee, fruit drinks and juices, specific cheeses, other dairy products, and so on).

• A caution: check the food label

THE INGREDIENTS MAY HAVE CHANGED.

The items listed in this inventory were purchased from September 1979 to February 1981. Food processors do change ingredients, and for certainty inspect the list on the label of an item you may wish to purchase to determine whether any additives regarded here as of concern have been removed or added in the interim.

GUIDE TO STYLE OF PRESENTATION IN THE INVENTORY OF BRANDS

In order to provide accurately and concisely the information on additives of concern which may be present in any of the thousands of items listed in this inventory, it has been necessary to adopt a terse styling procedure which is explained below:

FOOD BRANDS AND ITEMS

PRODUCER AND/OR BRAND	TYPE FACE
SWANSON	**BOLD FACE**
RELATED ITEMS SAME BRAND	
GERBER	
—HIGH MEAT DINNER	LIGHT FACE
BEEF W/ VEGETABLES; CHICKEN W/ VEGETABLES	ITALICS, LIGHT FACE
SINGLE ITEMS	
SWANSON	
FRENCH TOAST W/ SAUSAGES	LIGHT FACE
SUPERMARKET	
STOP & SHOP	**BOLD FACE**
SUPERMARKET PRIVATE BRAND—PRIVATE LABEL	
STOP & SHOP	**BOLD FACE**
—OUR FAMOUS	**BOLD FACE**

ADDITIVES-OF-CONCERN

ADDITIVES-OF-CONCERN TO ALL	
CARRAGEENAN	**BOLD FACE**
SODIUM NITRATE	
ADDITIVES-OF-CONCERN TO SOME PEOPLE	
GLUTEN	LIGHT FACE
SODIUM ALGINATE	

WORDS IN PARENTHESES AFTER AN ADDITIVE-OF-CONCERN

Some additives appear on labels in a number of ways. An attempt has been made to standardize these by placing the basic term first, followed in parentheses by any others appearing with it on the ingredient list.

GLUTEN (WHEAT)
WHEY (SWEET DAIRY)
BUTTER (GRADE AA)

GLUTEN (VITAL WHEAT)
WHEY (DRIED OR MODIFIED)
BUTTER (SWEET CREAM)

WORDS IN BRACKETS AFTER AN ADDITIVE-OF-CONCERN

These have been added to complete the proper name of an additive incompletely stated on the label, or to clarify what the substance is.

LOCUST BEAN [GUM]
BEEF OIL [FAT]

SPELLING AND PUNCTUATION OF BRANDS AND ITEMS

These are presented *exactly* as they appear on the label, even if they appear to have been mispelled or seem to be inconsistent.

Super-Stix
Cocoa Chp 'N Nut
Oatmeal N' Molasses

Hi-Proteen
Parmagian
Parmigiana (in another instance)

FD&C PRECEDING A COLOR ADDITIVE

FD & C preceding a color indicates that it is permitted in food, drugs and cosmetics. On occasions it is omitted in ingredients lists, but any artificial color permitted in food *must* be an FD & C color. To avoid confusion, it has been added in those instances when it had been absent.

TWO SPECIAL CASES

ARTIFICIAL COLORS

Artificial color without specification of all of the colors that are used is regarded in this volume as an additive of general concern. It appears in a variety of ways in an ingredient list; at times by itself, at times along with one or more of the colors that have been used but not necessarily limited to these. The examples below demonstrate what had been listed on the label and what has appeared in this book.

AS LISTED ON THE LABEL	AS LISTED IN THIS BOOK
Artificial Color (including FD & C Yellow No. 5)	**ARTIFICIAL COLOR** W/ FD & C YELLOW NO. 5
Artificial Color, FD & C Yellow No. 5	**ARTIFICIAL COLOR,** FD & C YELLOW NO. 5

Another variation apparently indicates that the associated color is the *sole* artificial color that has been used. This color in the example shown below is of concern to some people but not others. Artificial color, therefore, appears in very light capital letters, as does the specified color.

ON LABEL	IN BOOK
Artificial Color (FD & C Blue No. 1)	ARTIFICIAL COLOR FD & C BLUE NO. 1

ANIMAL FATS AND VEGETABLE OILS

Highly saturated fats and oils, either alone or as part of a blend (such as a shortening), when present in sufficient quantity in a food are regarded as additives of concern for all consumers.

If one of these fats or oils is listed by itself, it is certain to be in the food.

SATURATED FATS OR OILS LISTED SEPARATELY
TALLOW
PALM KERNEL OIL
BEEF FAT

But when one or more of these are part of a blend, at this writing the producer is permitted to list the range of fats and oils which *may* have been used, but one cannot be certain of the ones that actually are in the food. Considering the objectives of this volume and the possibility that these substances can be present, it is believed that a caution is needed.

SATURATED FATS AND/OR OILS AS PART OF A BLEND

Blend as listed on a label	Fats and oils in blend listed in book
Animal and/or vegetable shortening (contains one or more of the following: beef fat, partially hydrogenated soybean, coconut, palm kernel and palm oils)	**(BEEF FAT; COCONUT, PALM KERNEL OILS)**

TABLE OF CONTENTS

INVENTORY OF BRANDS

I. BABY FOODS

INFANT BISCUITS, COOKIES, TOAST

BRANDS	ADDITIVES-OF-CONCERN **BOLD—CONCERN TO ALL** LIGHT—CONCERN TO SOME
GERBER TEETHING BISCUITS; "ANIMAL SHAPED" COOKIES	WHEY (DAIRY)
ZWIEBACK TOAST	**MACE; NUTMEG**
NABISCO NATIONAL ARROWROOT BISCUIT	CASEIN
ZWIEBACK TOAST	EGG WHITES; LACTALBUMIN; **MACE; NUTMEG**

INFANT CEREALS

BEECH-NUT HONEY MIXED; MIXED; RICE	NO ADDITIVES OF CONCERN
FAMILIA SWISS BABY FOOD	NO ADDITIVES OF CONCERN
GERBER HIGH PROTEIN; MIXED; MIXED CEREAL W/ BANANA; OATMEAL; OATMEAL W/ BANANA; RICE	NO ADDITIVES OF CONCERN

BRANDS	ADDITIVES-OF-CONCERN BOLD—CONCERN TO ALL LIGHT—CONCERN TO SOME
ITEMS IN HEALTH FOOD STORES HEALTH VALLEY NATURAL BABY FOOD BROWN RICE CEREAL W/ FRUIT	NO ADDITIVES OF CONCERN

INFANT FRUIT JUICE

BEECH-NUT APPLE; MIXED FRUIT; ORANGE; ORANGE PINEAPPLE	NO ADDITIVES OF CONCERN
GERBER APPLE; APPLE-PLUM; MIXED FRUIT; ORANGE; ORANGE-APPLE-BANANA	NO ADDITIVES OF CONCERN
HEINZ APPLE; APPLE-PRUNE; MIXED FRUIT; ORANGE-APPLE-BANANA	NO ADDITIVES OF CONCERN

INFANT FORMULA

ENFAMIL CONCENTRATED LIQUID; READY TO	**CARRAGEENAN; COCONUT OIL**

(continues)

BRANDS	ADDITIVES-OF-CONCERN BOLD—CONCERN TO ALL LIGHT—CONCERN TO SOME
USE; W/ IRON CONCENTRATED LIQUID; W/ IRON READY TO USE	
PRO SOBEE	
MILK-FREE FORMULA CONCENTRATED LIQUID	**CARRAGEENAN**
MILK-FREE FORMULA READY TO USE	**CARRAGEENAN**; GUAR GUM
SIMILAC	
CONCENTRATED LIQUID; READY TO FEED; W/ IRON CONCENTRATED LIQUID; W/ IRON READY TO FEED	**CARRAGEENAN; COCONUT OIL**
ISOMIL CONCENTRATED LIQUID	**CARRAGEENAN; MODIFIED CORN STARCH**
ISOMIL LIQUID READY TO FEED	**CARRAGEENAN**
SMA	
CONCENTRATED LIQUID; READY TO FEED	**CALCIUM CARRAGEENAN;** WHEY (ELECTRODIALIZED DEMINERALIZED)

BRANDS	ADDITIVES-OF-CONCERN BOLD—CONCERN TO ALL LIGHT—CONCERN TO SOME

INFANT STRAINED CEREALS, MEATS, VEGETABLES

BEECH-NUT	
BEEF & EGG NOODLE DINNER; MACARONI, TOMATO & BEEF DINNER; RICE CEREAL W/ APPLESAUCE & BANANAS; TURKEY RICE DINNER	**MODIFIED CORN STARCH**
CARROTS; GARDEN VEGETABLES; GREEN BEANS; VEAL; VEGETABLE LAMB DINNER	NO ADDITIVES OF CONCERN
COTTAGE CHEESE W/ PINEAPPLE JUICE	**MODIFIED TAPIOCA STARCH**
SWEET POTATOES IN BUTTER SAUCE	**BUTTER**
VEGETABLE BACON DINNER	**MODIFIED CORN STARCH; SMOKED BACON;* SMOKED YEAST***
VEGETABLE HAM DINNER	**MODIFIED CORN STARCH; SMOKED YEAST***
GERBER	
BEEF; CARROTS; GREEN BEANS; HAM; *(continues)*	NO ADDITIVES OF CONCERN

*Although "smoked" does not refer to a specific additive, it represents a process utilizing wood smoke, as "smoke flavoring" does, and therefore the food may contain some cancer-causing benzopyrene chemicals. For this reason, when a food or any of its constituents have been smoked, this has been noted in the additives-of-concern column.

BRANDS	ADDITIVES-OF-CONCERN **BOLD—CONCERN TO ALL** LIGHT—CONCERN TO SOME
LAMB; MIXED CEREAL W/ APPLESAUCE & BANANAS; MIXED VEGETABLES; PEAS; PORK; RICE CEREAL W/ APPLESAUCE & BANANAS; SWEET POTATOES; TURKEY; VEAL	
BEEF & EGG NOODLES W/ VEGETABLES; CHICKEN & NOODLES; CREAMED CORN; MACARONI-TOMATO W/ BEEF; TURKEY & RICE W/ VEGETABLES; VEGETABLES & BEEF; VEGETABLES & TURKEY	**MODIFIED CORN STARCH**
CREAMED SPINACH	MILK SOLIDS (WHOLE)
VEGETABLES & BACON	**NATURAL SMOKE FLAVORING**
—HIGH MEAT DINNER	
BEEF W/ VEGETABLES; CHICKEN W/ VEGETABLES	NO ADDITIVES OF CONCERN
HEINZ	
BEEF; CARROTS; CHICKEN; GREEN BEANS; LAMB; *(continues)*	NO ADDITIVES OF CONCERN

BRANDS	ADDITIVES-OF-CONCERN BOLD—CONCERN TO ALL LIGHT—CONCERN TO SOME
HEINZ *(Continued)*	
SQUASH; SWEET POTATOES	NO ADDITIVES OF CONCERN
BEEF & EGG NOODLES; CHICKEN NOODLE DINNER; CREAMED PEAS; TURKEY RICE DINNER W/ VEGETABLES; VEGETABLES & BACON; VEGETABLES, EGG NOODLES, & TURKEY; VEGETABLES & HAM; VEGETABLES & LAMB	**MODIFIED CORN STARCH**
CREAMED CORN; MACARONI, TOMATOES, & BEEF	MILK SOLIDS (WHOLE); **MODIFIED CORN STARCH**
OATMEAL W/ APPLES & BANANAS; RICE CEREAL W/ APPLES & BANANAS	**FERRIC ORTHOPHOSPHATE**
VEGETABLES, DUMPLINGS, & BEEF	**MODIFIED CORN STARCH; PAPRIKA; SMOKED YEAST;* TURMERIC**

*Although "smoked" does not refer to a specific additive, it represents a process utilizing wood smoke, as "smoke flavoring" does, and therefore the food may contain some cancer-causing benzopyrene chemicals. For this reason, when a food or any of its constituents have been smoked, this has been noted in the additives-of-concern column.

BRANDS	ADDITIVES-OF-CONCERN **BOLD—CONCERN TO ALL** LIGHT—CONCERN TO SOME

INFANT STRAINED DESSERTS, FRUITS

BEECH-NUT APPLES & APRICOTS; APPLESAUCE; APPLESAUCE & BANANAS; PEACHES; PEARS	NO ADDITIVES OF CONCERN
APRICOTS W/ TAPIOCA & APPLE JUICE; BANANAS W/ TAPIOCA & APPLE JUICE; FRUIT DESSERT	**MODIFIED TAPIOCA STARCH**
GERBER APPLE BLUEBERRY; APPLESAUCE; PEACHES; PEARS	NO ADDITIVES OF CONCERN
APRICOTS W/ TAPIOCA; BANANAS W/ PINEAPPLE & TAPIOCA; PLUMS W/ TAPIOCA	**MODIFIED TAPIOCA STARCH**
CHOCOLATE CUSTARD PUDDING; DUTCH APPLE DESSERT; FRUIT DESSERT; PEACH COBBLER; VANILLA CUSTARD PUDDING	**MODIFIED CORN STARCH**
HEINZ APPLESAUCE; PEACHES; PEARS; PEARS & PINEAPPLE	NO ADDITIVES OF CONCERN

BRANDS	ADDITIVES-OF-CONCERN BOLD—CONCERN TO ALL LIGHT—CONCERN TO SOME
HEINZ *(Continued)*	
BANANAS & PINEAPPLE & TAPIOCA; PINEAPPLE ORANGE	**MODIFIED CORN & TAPIOCA STARCHES**
CUSTARD PUDDING; TUTTI FRUTTI	**MODIFIED CORN STARCH**
DUTCH APPLE DESSERT; PEACH COBBLER	**MODIFIED TAPIOCA STARCH**

JUNIOR & TODDLER CEREALS, MEATS, VEGETABLES

BEECH-NUT *JUNIOR*	
CHICKEN NOODLE DINNER; TURKEY RICE DINNER; VEGETABLE BEEF DINNER	**MODIFIED CORN STARCH**
CARROTS; CHICKEN; GREEN BEANS; LAMB; SWEET POTATOES	NO ADDITIVES OF CONCERN
SPAGHETTI, TOMATO, & BEEF DINNER	**MODIFIED CORN STARCH;** WHEY (DRY)
SPLIT PEAS & HAM DINNER	**MODIFIED CORN STARCH; SMOKED YEAST***

BRANDS	ADDITIVES-OF-CONCERN **BOLD—CONCERN TO ALL** LIGHT—CONCERN TO SOME
VEGETABLE BACON DINNER	**MODIFIED CORN STARCH; SMOKED BACON;* SMOKED YEAST***
GERBER *—JUNIOR* BEEF; CARROTS; CHICKEN; LAMB; OATMEAL W/ APPLESAUCE & BANANAS; SWEET POTATOES; TURKEY & RICE W/ VEGETABLES; VEAL; VEGETABLES & BEEF	NO ADDITIVES OF CONCERN
BEEF & EGG NOODLES W/ VEGETABLES; MACARONI-TOMATO W/ BEEF; VEGETABLES & TURKEY	**MODIFIED CORN STARCH**
RICE CEREAL W/ MIXED FRUIT	**MODIFIED TAPIOCA STARCH**
VEGETABLES & BACON	**NATURAL SMOKE FLAVORING**
VEGETABLES & CHICKEN	**TURMERIC**

*Although "smoked" does not refer to a specific additive, it represents a process utilizing wood smoke, as "smoke flavoring" does, and therefore the food may contain some cancer-causing benzopyrene chemicals. For this reason, when a food or any of its constituents have been smoked, this has been noted in the additives-of-concern column.

BRANDS	ADDITIVES-OF-CONCERN BOLD—CONCERN TO ALL LIGHT—CONCERN TO SOME
GERBER *(Continued)*	
—HIGH MEAT DINNER BEEF W/ VEGETABLES; HAM W/ VEGETABLES; TURKEY W/ VEGETABLES	NO ADDITIVES OF CONCERN
—TODDLER MEAL BEEF STEW; CHICKEN STEW	**MODIFIED CORN STARCH**
HEINZ *JUNIOR* CARROTS; LAMB; SWEET POTATOES	NO ADDITIVES OF CONCERN
CEREAL & EGGS; CHICKEN NOODLE DINNER; MACARONI, TOMATOES, & BEEF; CREAMED PEAS; TURKEY RICE DINNER W/ VEGETABLES; VEGETABLES & BEEF	**MODIFIED CORN STARCH**
CREAMED CORN	MILK SOLIDS (WHOLE); **MODIFIED CORN STARCH**
CREAMED GREEN BEANS	MILK SOLIDS (WHOLE); **MODIFIED CORN STARCH; SMOKED YEAST***
VEGETABLES & BACON	**MODIFIED CORN STARCH; SMOKED YEAST***

*Although "smoked" does not refer to a specific additive, it represents a process utilizing wood smoke, as "smoke flavoring" does, and therefore the food may contain some cancer-causing benzopyrene chemicals. For this reason, when a food or any of its constituents have been smoked, this has been noted in the additives-of-concern column.

BRANDS	ADDITIVES-OF-CONCERN **BOLD—CONCERN TO ALL** LIGHT—CONCERN TO SOME

JUNIOR DESSERTS, FRUITS

BEECH-NUT	
APPLESAUCE & RASPBERRIES; PEACHES; PEARS	NO ADDITIVES OF CONCERN
APRICOTS W/ TAPIOCA; PLUMS W/ TAPIOCA & APPLE JUICE	**MODIFIED TAPIOCA STARCH**
BANANA DESSERT & APPLE JUICE; CUSTARD PUDDING	**MODIFIED CORN STARCH**
GERBER	
APPLE BLUEBERRY; APPLESAUCE; PEACHES; PEARS	NO ADDITIVES OF CONCERN
APPLE RASPBERRY	**MACE**
APRICOTS W/ TAPIOCA	**MODIFIED CORN STARCH; MODIFIED TAPIOCA STARCH**
BANANAS W/ PINEAPPLE & TAPIOCA; PLUMS W/ TAPIOCA	**MODIFIED TAPIOCA STARCH**
FRUIT DESSERT; RASPBERRY DESSERT W/ YOGURT; VANILLA CUSTARD PUDDING	**MODIFIED CORN STARCH**

BRANDS	ADDITIVES-OF-CONCERN **BOLD—CONCERN TO ALL** LIGHT—CONCERN TO SOME
HEINZ APPLES & CRANBERRIES W/ TAPIOCA; APRICOTS W/ TAPIOCA; BANANAS & PINEAPPLE W/ TAPIOCA	**MODIFIED CORN STARCH; MODIFIED TAPIOCA STARCH**
APPLES & PEARS; APPLESAUCE & APRICOTS; PEACHES; PEARS	NO ADDITIVES OF CONCERN
CUSTARD PUDDING; FRUIT DESSERT; PINEAPPLE ORANGE	**MODIFIED CORN STARCH**
DUTCH APPLE DESSERT; PEACH COBBLER	**MODIFIED TAPIOCA STARCH**

II. BAKED GOODS, FROZEN & REFRIGERATED

FROZEN BAGELS, BREAD DOUGH, MUFFINS, ROLLS

BRANDS	ADDITIVES-OF-CONCERN BOLD—CONCERN TO ALL LIGHT—CONCERN TO SOME
HOWARD JOHNSON'S *TOASTEES* BLUEBERRY	**BHA**
CORN	NO ADDITIVES OF CONCERN
LENDER'S *BAGELS* EGG FLAVORED; ONION; PLAIN; RAISIN 'N HONEY; WHEAT 'N HONEY	GLUTEN (VITAL WHEAT GLUTEN OR HI-GLUTEN FLOUR OR UNBLEACHED GLUTEN FLOUR)
PEPPERIDGE FARM *MUFFINS* BLUEBERRY; BRAN W/ RAISINS; CORN	**MODIFIED FOOD STARCH**
RHODES FROZEN BREAD DOUGH/WHITE	WHEY
RICH'S WHITE BREAD DOUGH	WHEY (DRIED)
SARA LEE PARKER HOUSE ROLLS	CASEIN; WHEY

BRANDS	ADDITIVES-OF-CONCERN **BOLD—CONCERN TO ALL** LIGHT—CONCERN TO SOME
STOUFFER'S PARTY GARLIC BREAD	**HYDROGENATED SOYBEAN OIL;** MONOSODIUM GLUTAMATE
SUPERMARKET PRIVATE BRANDS	
A & P	
—ANN PAGE BREAD DOUGH	NO ADDITIVES OF CONCERN

FROZEN BREAKFAST FOODS (EGG SUBSTITUTES, FRENCH TOAST, PANCAKES, WAFFLES)

AUNT JEMIMA	
—FRENCH TOAST	
CINNAMON SWIRL	**ARTIFICIAL COLOR FD & C** BLUE NO. 1, **RED NO. 3,** YELLOW NO. 5 & NO. 6; **BHA; BHT;** WHEY
PLAIN	**BHA; BHT;** WHEY
—PANCAKE BATTER	
BLUEBERRY; BUTTERMILK; PLAIN	NO ADDITIVES OF CONCERN
—WAFFLES	
BLUEBERRY	ARTIFICIAL COLOR FD & C BLUE NO. 2, YELLOW NO. 5 & NO. 6; MONOSODIUM GLUTAMATE; WHEY
ORIGINAL	ARTIFICIAL COLOR FD & C YELLOW NO. 5 & NO. 6; MONOSODIUM GLUTAMATE; WHEY

BRANDS	ADDITIVES-OF-CONCERN BOLD—CONCERN TO ALL LIGHT—CONCERN TO SOME
DOWNYFLAKE	
PANCAKES	**ARTIFICIAL COLOR** W/ FD & C YELLOW NO. 5; **(HYDROGENATED COTTONSEED, PALM OILS);*** WHEY SOLIDS
WAFFLES	**ARTIFICIAL COLOR** W/ FD & C YELLOW NO. 5; WHEY
EGGO ***WAFFLES***	
BRAN	WHEY (DRIED DAIRY)
PLAIN; W/ IMITATION BLUEBERRIES	**ARTIFICIAL COLORING;** WHEY
FLEISCHMANN'S EGG BEATERS	**ARTIFICIAL COLORS; CALCIUM & SODIUM CASEINATES;** EGG WHITES; **FERRIC ORTHOPHOSPHATE;** WHEY
MORNINGSTAR FARMS SCRAMBLERS	**ARTIFICIAL COLORS; CALCIUM CASEINATE;** CAROB BEAN GUM; EGG WHITES; **FERRIC ORTHOPHOSPHATE;** GUAR GUM; **MODIFIED CORN STARCH**
ROMAN MEAL	
WAFFLES	WHEY SOLIDS
SWANSON	
FRENCH TOAST W/ SAUSAGES	**CALCIUM CASEINATE;** WHEY

*Blend may contain one or more of these saturated oils.

BRANDS	ADDITIVES-OF-CONCERN **BOLD—CONCERN TO ALL** LIGHT—CONCERN TO SOME
SWANSON *(Continued)*	
PANCAKES W/ SAUSAGES	**CALCIUM CASEINATE; SODIUM CASEINATE;** WHEY
SUPERMARKET PRIVATE BRANDS	
A & P	
—ANN PAGE WAFFLES	**ARTIFICIAL COLOR** W/ FD & C YELLOW NO. 5; WHEY
FIRST NATIONAL	
—EDWARDS-FINAST WAFFLES	**ARTIFICIAL COLOR** W/ FD & C YELLOW NO. 5; WHEY
—FINAST WAFFLES	**ARTIFICIAL COLOR** W/ FD & C YELLOW NO. 5; WHEY
SAFEWAY	
—BEL-AIR *WAFFLES*	
BUTTERMILK ROUND	**ARTIFICIAL COLOR** W/ FD & C YELLOW NO. 5; **(HYDROGENATED COTTONSEED, PALM OILS)***
PLAIN	**ARTIFICIAL COLOR;** WHEY
—LUCERNE	
BREAKFAST TREAT EGG SUBSTITUTE	**ARTIFICIAL COLOR; CALCIUM CARRAGEENAN;** EGG WHITES; MONOSODIUM GLUTAMATE
STOP & SHOP	
—STOP & SHOP WAFFLES	**ARTIFICIAL COLOR** W/ FD & C YELLOW NO. 5; WHEY

BRANDS	ADDITIVES-OF-CONCERN **BOLD—CONCERN TO ALL** LIGHT—CONCERN TO SOME

FROZEN CAKES, DOUGHNUTS, PASTRIES

CHOCK FULL O' NUTS *CAKE* MARBLE; POUND	EGG WHITES
DRESSEL'S *CAKE* SPECIAL PARTY FUDGE; WHIPPED CREAM BIRTHDAY FUDGE	**ARTIFICIAL COLOR; CONFECTIONER'S GLAZE;** GUM ARABIC; **MODIFIED FOOD STARCH;** WHEY
MORTON *DONUT SHOP* HONEY BUNS	**ARTIFICIAL COLOR; (BEEF FAT, LARD);*** GUAR GUM; GUM ACACIA
JELLY DONUTS	**ARTIFICIAL COLOR; MODIFIED FOOD STARCH;** SODIUM ALGINATE
SUGAR 'N SPICE MINI-DONUTS	**(BEEF FAT, LARD);*** CASEIN; **MODIFIED FOOD STARCH;** WHEY
PEPPERIDGE FARM *—CAKE SUPREME* BANANA	**MODIFIED FOOD STARCH;** WHEY
BOSTON CREME; CHOCOLATE; WALNUT	**MODIFIED FOOD STARCH; (PALM KERNEL OIL)***

*Blend may contain one or more of these saturated fats and/or oils.

BRANDS	**ADDITIVES-OF-CONCERN** BOLD—CONCERN TO ALL LIGHT—CONCERN TO SOME
PEPPERIDGE FARM—*CAKE SUPREME (Continued)*	
LEMON COCONUT	EGG WHITES; **MODIFIED FOOD STARCH**
—LAYER CAKE	
DEVIL'S FOOD; GERMAN CHOCOLATE; GOLDEN; VANILLA	**CALCIUM CASEINATE; (PALM KERNEL OIL);* MODIFIED FOOD STARCH; SODIUM CASEINATE;** WHEY
—MISCELLANEOUS CAKES	
APPLE-WALNUT W/ RAISINS	**MODIFIED FOOD STARCH; NUTMEG**
CARROT	**NUTMEG**
POUND	**BUTTER (SWEET CREAM)**
RICH'S CHOCOLATE ECLAIRS	**ARTIFICIAL COLOR; (COCONUT, PALM KERNEL OILS);* SODIUM CASEINATE**
SARA LEE	
—CHEESE CAKE	
CHERRY CREAM	**CALCIUM CARRAGEENAN;** LOCUST BEAN GUM; **MODIFIED FOOD STARCH;** PROPYLENE GLYCOL ALGINATE; **U.S. CERTIFIED FOOD COLOR**
FRENCH	GUAR GUM; **MODIFIED FOOD STARCH; TURMERIC EXTRACT**

*Blend may contain one or more of these saturated oils.

BRANDS	ADDITIVES-OF-CONCERN **BOLD—CONCERN TO ALL** LIGHT—CONCERN TO SOME
ORIGINAL CREAM	GUAR GUM; GUM ARABIC; LOCUST BEAN GUM; **MODIFIED FOOD STARCH**
—ORIGINAL BUTTER RECIPE CAKE	
APPLE WALNUT; CARROT; COCONUT	**BUTTER (GRADE AA); MODIFIED FOOD STARCH**
CHOCOLATE BROWNIES; FRESH ORANGE	NO ADDITIVES OF CONCERN
FRESH BANANA	GUM ARABIC
—MISCELLANEOUS CAKES	
BLACK FOREST; STRAWBERRY SHORT	**CALCIUM CARRAGEENAN;** GUAR GUM; GUM TRAGACANTH; LOCUST BEAN GUM; **MODIFIED FOOD STARCH;** PROPYLENE GLYCOL ALGINATE; WHEY
CHOCOLATE BAVARIAN	EGG WHITES; **MODIFIED FOOD STARCH**
ORIGINAL POUND	NO ADDITIVES OF CONCERN
RAISIN POUND	**MODIFIED FOOD STARCH**
—MISCELLANEOUS COFFEE CAKES, DANISH, ROLLS	
ALMOND LIGHT COFFEE RING; APPLE CRUNCH ROLLS;	CASEIN; GUAR GUM; **MODIFIED FOOD STARCH**

(continues)

BRANDS	ADDITIVES-OF-CONCERN BOLD—CONCERN TO ALL LIGHT—CONCERN TO SOME
SARA LEE *(Continued)*	
BLUEBERRY LIGHT COFFEE RING; RASPBERRY LIGHT COFFEE RING	CASEIN; GUAR GUM; **MODIFIED FOOD STARCH**
APPLE DANISH	**CALCIUM CARRAGEENAN;** LOCUST BEAN GUM; **MODIFIED FOOD STARCH**
BUTTER STREUSEL COFFEE CAKE	**BUTTER (GRADE AA)**
HONEY ROLLS	CASEIN; GUAR GUM; GUM ARABIC; **MODIFIED FOOD STARCH**
STOUFFER'S	
BOSTON CREAM CUPCAKES; YELLOW CUPCAKES	**ARTIFICIAL COLOR; MODIFIED CORN STARCH**
CHOCOLATE CHIP CRUMB CAKES; POUND CAKE	**ARTIFICIAL COLOR**
CREAM FILLED CUPCAKES	**ARTIFICIAL COLOR; (COCONUT, PALM KERNEL OILS);* CONFECTIONER'S GLAZE; MODIFIED CORN STARCH**
SUPERMARKET PRIVATE BRANDS	
SAFEWAY	
—BEL-AIR *DONUTS*	
GLAZED	**(BEEF TALLOW, COCONUT OIL, LARD);* MODIFIED FOOD STARCH; SODIUM CASEINATE;** WHEY

BRANDS	ADDITIVES-OF-CONCERN BOLD—CONCERN TO ALL LIGHT—CONCERN TO SOME
JELLY	**ARTIFICIAL COLOR; (BEEF TALLOW, COCONUT OIL, LARD);* SODIUM CASEINATE;** WHEY
ITEMS IN HEALTH FOOD STORES	
BARBARA'S	
CAROB MACAROON; COCONUT MACAROON	EGG WHITES
CHOCOLATE BROWNIE	NO ADDITIVES OF CONCERN
FOOD FOR LIFE *CAKE*	
BANANA; CARROT; POUND	NO ADDITIVES OF CONCERN
DATE-NUT	**NUTMEG**

FROZEN PIE CRUST, PIES, TARTS, TURNOVERS

MORTON *GREAT LITTLE DESSERTS PIE*	
APPLE	**(COCONUT OIL, LARD);* MODIFIED FOOD STARCH**

*Blend may contain one or more of these saturated fats and/or oils.

BRANDS	**ADDITIVES-OF-CONCERN** **BOLD—CONCERN TO ALL** LIGHT—CONCERN TO SOME
MORTON *GREAT LITTLE DESSERTS PIE (Continued)*	
BANANA CREAM; CHOCOLATE CREAM	**ARTIFICIAL COLOR; (BEEF FAT, LARD; HYDROGENATED COCONUT, COTTONSEED, PALM KERNEL, SOYBEAN OILS);* HYDROXYPROPYL CELLULOSE;** GUAR GUM; **MODIFIED FOOD STARCH; SODIUM CASEINATE**
BLUEBERRY	**MODIFIED FOOD STARCH**
MRS. SMITH'S	
BOSTON CREAM PIE	**ARTIFICIAL COLOR; CALCIUM CARRAGEENAN; MODIFIED FOOD STARCH; SODIUM CASEINATE**
LEMON MERINGUE PIE	**ARTIFICIAL COLOR;** EGG WHITES; LOCUST BEAN GUM; **MODIFIED FOOD STARCH**
—DEEP DISH PIE	
APPLE; GOLDEN DELUXE APPLE; BLUEBERRY; DUTCH APPLE CRUMB	**ARTIFICIAL COLOR; MODIFIED FOOD STARCH**
PUMPKIN CUSTARD	**ARTIFICIAL COLOR; CALCIUM CARRAGEENAN;** LOCUST BEAN GUM; **MODIFIED FOOD STARCH**
WALNUT	**ARTIFICIAL COLOR**
ORONOQUE ORCHARDS PIE CRUSTS	NO ADDITIVES OF CONCERN

BRANDS	ADDITIVES-OF-CONCERN **BOLD—CONCERN TO ALL** LIGHT—CONCERN TO SOME
PEPPERIDGE FARM	
APPLE CRISS-CROSS PASTRY; APPLE PIE TARTS; APPLE TURNOVERS; BLUEBERRY TURNOVERS; RASPBERRY TURNOVERS	**MODIFIED FOOD STARCH**
APPLE STRUDEL	**(COCONUT OIL);*** EGG WHITES; **MODIFIED FOOD STARCH**
PATTY SHELLS	NO ADDITIVES OF CONCERN
SARA LEE *PIE*	
APPLE	**MODIFIED FOOD STARCH**
PUMPKIN	**CALCIUM CARRAGEENAN;** LOCUST BEAN GUM; **MODIFIED FOOD STARCH;** WHEY
SUPERMARKET PRIVATE BRANDS	
SAFEWAY	
—BEL-AIR	
BLUEBERRY PIE; BOYSENBERRY PIE; CHERRY PIE	**ARTIFICIAL COLOR; (LARD);* MODIFIED STARCH;** WHEY (SWEET)
DEEP DISH PIE CRUST SHELLS	**ARTIFICIAL COLOR; (BEEF FAT, LARD; HYDROGENATED PALM, SOYBEAN OILS);* BHA; BHT;** WHEY SOLIDS

*Blend may contain one or more of these saturated fats and/or oils.

BRANDS	ADDITIVES-OF-CONCERN **BOLD—CONCERN TO ALL** LIGHT—CONCERN TO SOME
—BEL-AIR *(Continued)*	
DUTCH APPLE PIE	**ARTIFICIAL COLOR; MODIFIED STARCH;** WHEY (SWEET)
WINN DIXIE	
—DIXIANA PIE CRUST SHELLS	**ARTIFICIAL COLOR** W/ FD & C YELLOW NO. 5; **(BEEF FAT, LARD; HYDROGENATED PALM, SOYBEAN OILS);* BHA; BHT;** WHEY SOLIDS

REFRIGERATED DOUGH PRODUCTS (BISCUITS, COOKIES, DANISH, ROLLS)

BRANDS	ADDITIVES-OF-CONCERN
PILLSBURY	
—BISCUITS	
BUTTERMILK; COUNTRY STYLE	**BHA**
1869 BRAND BAKING POWDER; HUNGRY JACK BUTTERMILK FLAKY	**(BEEF FAT; HYDROGENATED PALM, SOYBEAN OILS);* BHA**
1869 BRAND BUTTERMILK	**(HYDROGENATED PALM, SOYBEAN OILS)***
HUNGRY JACK BUTTER TASTIN'	**ARTIFICIAL COLOR; (BEEF FAT; HYDROGENATED PALM, SOYBEAN OILS);* BHA**

BRANDS	ADDITIVES-OF-CONCERN BOLD—CONCERN TO ALL LIGHT—CONCERN TO SOME
—DINNER ROLLS	
BUTTERFLAKE	**(BEEF FAT; HYDROGENATED PALM, SOYBEAN OILS);* BHA; BUTTER**
CRESCENT	**(BEEF FAT; HYDROGENATED PALM, SOYBEAN OILS);* BHA;** GLUTEN (VITAL WHEAT)
—SLICE 'N BAKE COOKIES	
CHOCOLATE CHIP; OATMEAL RAISIN; SUGAR	**(BEEF FAT; HYDROGENATED COTTONSEED, PALM, SOYBEAN OILS);* BHA; CARRAGEENAN**
—DANISH & TURNOVERS	
APPLE FLAKY TURNOVER PIES	**ARTIFICIAL COLOR; (BEEF FAT; HYDROGENATED PALM, SOYBEAN OILS);* BHA; MODIFIED CORN STARCH; NUTMEG;** PROPYLENE GLYCOL ALGINATE
CARAMEL DANISH; ORANGE DANISH	**ARTIFICIAL COLOR; (BEEF FAT; HYDROGENATED PALM, SOYBEAN OILS);* BHA**

*Blend may contain one or more of these saturated fats and/or oils.

BRANDS	ADDITIVES-OF-CONCERN **BOLD—CONCERN TO ALL** LIGHT—CONCERN TO SOME
PILLSBURY *(Continued)*	
CINNAMON RAISIN DANISH	**ARTIFICIAL COLOR; (BEEF FAT; HYDROGENATED PALM, SOYBEAN OILS);* BHA; CARRAGEENAN;** MILK PROTEIN; **SODIUM CASEINATE;** WHEY
SUPERMARKET PRIVATE BRANDS	
FIRST NATIONAL	
—EDWARDS CRESCENT DINNER ROLLS	**BEEF FAT; (HYDROGENATED COTTONSEED, PALM, SOYBEAN OILS);* BHA; BHT**
HOMESTYLE BISCUITS; TEXAS STYLE OLD FASHIONED BISCUITS	**BEEF FAT; BHA; BHT;** WHEY SOLIDS
—FINAST BUTTERMILK BISCUITS	**BEEF FAT; BHA; BHT;** WHEY SOLIDS
LUCKY	
—LADY LEE BUTTERMILK BISCUITS	**BEEF TALLOW** OR **LARD;** BUTTERMILK SOLIDS; **CARRAGEENAN;** GUAR GUM; WHEY
HOMESTYLE BISCUITS	**BEEF TALLOW** OR **LARD; CARRAGEENAN;** GUAR GUM; WHEY
SAFEWAY	
—MRS. WRIGHT'S BUTTERMILK BISCUITS	**BEEF TALLOW** OR **LARD;** BUTTERMILK SOLIDS; **CARRAGEENAN;** GUAR GUM; WHEY

BRANDS	**ADDITIVES-OF-CONCERN** **BOLD—CONCERN TO ALL** LIGHT—CONCERN TO SOME
CRESCENT DINNER ROLLS	**BEEF FAT; BHA; BHT (HYDROGENATED COTTONSEED, PALM, SOYBEAN OILS)***
HOMESTYLE BISCUITS	**BEEF TALLOW** OR **LARD; CARRAGEENAN;** GUAR GUM; WHEY

*Blend may contain one or more of these saturated fats and/or oils.

IV. BAKED SWEET GOODS, NOT FROZEN OR REFRIGERATED

R alongside a brand indicates regional (not national) distribution, based on shopping experience.

BISCUITS & COOKIES

BRANDS	ADDITIVES-OF-CONCERN BOLD—CONCERN TO ALL LIGHT—CONCERN TO SOME
ARCHWAY —*FAMILY STYLE* COCONUT CHOCOLATE CHIP; OATMEAL N' NUT; PECAN CRUNCH	NO ADDITIVES OF CONCERN
—*HOME STYLE* APRICOT FILLED; BLUEBERRY FILLED; RASPBERRY FILLED; STRAWBERRY FILLED	**ARTIFICIAL COLOR** OR **COLORS; (HYDROGENATED COTTONSEED, PALM OILS);* MODIFIED FOOD STARCH**
CHERRY NOUGAT; ICED MOLASSES; MERRY MINTS; NEW ORLEANS CAKE	**ARTIFICIAL COLOR**
CHOCOLATE CHIP DROP; CHOCOLATE CHIP SUPREME; COOKIE JAR HERMIT;	NO ADDITIVES OF CONCERN

(continues)

BRANDS	ADDITIVES-OF-CONCERN BOLD—CONCERN TO ALL LIGHT—CONCERN TO SOME
DATE FILLED OATMEAL; DATE 'N NUT; DUTCH COCOA; FUDGE NUT BAR; GRANDMA'S MOLASSES; PEANUT BUTTER 'N CHIPS; RUTH'S GOLDEN OATMEAL	
COCOA SPRINKLES	**ARTIFICIAL COLOR; CONFECTIONER'S SHELLAC; VEGETABLE GUM**
FROSTED FINGERS	**ARTIFICIAL COLOR; CONFECTIONER'S SHELLAC;** EGG WHITES; **VEGETABLE GUM**
HOLIDAY SUGAR	FD & C BLUE COLOR NO. 1, **RED NO. 3 & NO. 40,** YELLOW NO. 5 & NO. 6
BURRY	
BURRY'S BEST CHOCOLATE CHIP; BURRY'S BEST SUGAR FUDGE	**(HYDROGENATED PALM OIL);*** WHEY (DAIRY OR DRIED DAIRY)
BURRY'S BEST COCONUT MACAROON	WHEY (DRIED DAIRY)
BURRY'S BEST WALNUT FUDGE;	**(HYDROGENATED PALM OIL)***

(continues)

*Blend may contain one or more of these saturated oils.

BRANDS	**ADDITIVES-OF-CONCERN** BOLD—CONCERN TO ALL LIGHT—CONCERN TO SOME
BURRY *(Continued)*	
OXFORD VANILLA SANDWICH CREMES	**(HYDROGENATED PALM OIL)***
BUTTER FLAVORED	ARTIFICIAL COLOR FD & C YELLOW NO. 5 & NO. 6; **(HYDROGENATED PALM OIL)***
CINNAMON SHORTBREAD; SUGAR SHORTBREAD	EGG WHITES (DRIED); **(HYDROGENATED PALM OIL);*** WHEY (DRIED DAIRY)
FUDGETOWN (CHOCOLATE)	**(HYDROGENATED PALM OIL, LARD)***
FUDGETOWN (VANILLA); HAPPYNIKS	**ARTIFICIAL COLOR; (HYDROGENATED PALM OIL, LARD)***
GAUCHO	EGG WHITES (DRIED); WHEY (DRIED DAIRY)
MR CHIPS	**ARTIFICIAL COLOR; (PALM KERNEL, HYDROGENATED PALM OILS);*** WHEY (DRIED DAIRY)
SCOOTER-PIE DEVIL'S FOOD; SCOOTER-PIE ORIGINAL	MILK PROTEIN (HYDROLYZED)
CARR'S	
ASSORTED BISCUITS FOR CHEESE; WHEATMEAL BISCUITS	**(COCONUT, PALM KERNEL OILS)***

BRANDS	**ADDITIVES-OF-CONCERN** **BOLD—CONCERN TO ALL** LIGHT—CONCERN TO SOME
R **DUTCH TWINS**	
SUGAR WAFERS	**ARTIFICIAL COLORS; (COCONUT OIL)***
ENTENMANN'S	
BUTTER (APRICOT JELLY)	**BUTTER; MODIFIED CORN STARCH**
CHOCOLATE CHIP	NO ADDITIVES OF CONCERN
FFV	
FANCY TART	**ARTIFICIAL COLOR; CERTIFIED COLOR**
MINT SANDWICH	**ARTIFICIAL COLOR; (BEEF FAT);*** WHEY POWDER
PEANUT BUTTER SANDWICH	**(BEEF FAT)***
SPRINKLED MALLOWS	**ARTIFICIAL COLOR;** WHEY
STRIPED MALLOWS	WHEY
FIRESIDE	
CHOCOLATE CHIP	**ARTIFICIAL COLOR; (BEEF FAT, LARD; COCONUT, PALM KERNEL OILS);*** WHEY
GÖTEBORGS	
SWEDISH GINGER SNAPS	NO ADDITIVES OF CONCERN

*Blend may contain one or more of these saturated fats and/or oils.

BRANDS	ADDITIVES-OF-CONCERN BOLD—CONCERN TO ALL LIGHT—CONCERN TO SOME
JACOBS BISCUITS FOR CHEESE	**BEEF FAT, HYDROGENATED COCONUT, MARINE, PALM KERNEL OILS; BHA;** WHEY POWDER
KEEBLER	
CHOCOLATE FUDGE SANDWICH	**ARTIFICIAL COLORING; (BEEF FAT, COCONUT OIL, LARD);*** WHEY
DELUXE GRAHAMS; VANILLA CREMES	**ARTIFICIAL COLORING; (LARD; COCONUT, PALM KERNEL OILS);*** WHEY
DOUBLE NUTTY; ICED RAISIN BAR; PITTER PATTER	WHEY
FUDGE CREMES	**(COCONUT OIL, LARD);*** WHEY
FUDGE MARSHMALLOW	**(BEEF FAT, LARD; COCONUT, PALM KERNEL OILS)***
FUDGE STICKS	**ARTIFICIAL COLORING; (COCONUT, PALM KERNEL OILS);*** WHEY
PECAN SANDIES OLD FASHION OATMEAL;	NO ADDITIVES OF CONCERN
RICH 'N CHIPS	**ARTIFICIAL COLOR** W/ FD & C YELLOW NO. 5; **(BEEF FAT; COCONUT, PALM KERNEL OILS);*** WHEY

BRANDS	ADDITIVES-OF-CONCERN **BOLD—CONCERN TO ALL** LIGHT—CONCERN TO SOME
VANILLA WAFERS	**(BEEF FAT);*** WHEY
LITTLE BROWNIE	
ASSORTED CREMES; LEMON CREMES; PEANUT BUTTER CREMES	**ARTIFICIAL COLOR; (BEEF OIL [FAT])***
GINGER SNAPS	**(BEEF OIL [FAT])***
LITTLE DEBBIE	
NUTTY BAR; PEANUT BARS; SUPER-STIX	WHEY
MASTER SUGAR CINNAMON TOAST	NO ADDITIVES OF CONCERN
R **MAURICE LENELL**	
BUTTERSCOTCH OATMEAL	**NUTMEG**
BUTTERSCOTCH; PEANUT BUTTER STARS	**VEGETABLE SHORTENING**
CHINESE ALMOND	NO ADDITIVES OF CONCERN
FAMILY ASSORTMENT	**U.S. CERTIFIED COLOR WHEN USED**
MC VITIES	
WHEATOLO BISCUITS	**(BEEF FAT; COCONUT, PALM KERNEL OILS)***

*Blend may contain one or more of these saturated fats and/or oils.

	BRANDS	ADDITIVES-OF-CONCERN **BOLD—CONCERN TO ALL** LIGHT—CONCERN TO SOME
R	**MOTHERS**	
	CHOCOLATE CHIP ANGEL; FUDGE 'N CHIPS; ICED OATMEAL; OATMEAL	NO ADDITIVES OF CONCERN
	CIRCUS ANIMAL; ICED RAISIN	**ARTIFICIAL COLOR; (COCONUT, PALM KERNEL OILS)***
	CLASSIC; COOKIE PARADE ASSORTMENT; FLAKY FLIX FUDGE CREME WAFERS; FLAKY FLIX VANILLA CREME WAFERS	**ARTIFICIAL COLOR; (COCONUT, PALM KERNEL OILS);*** WHEY
	PEANUT BUTTER; SUGAR	WHEY
	STRIPED SHORTBREAD	ARTIFICIAL COLOR FD & C YELLOW NO. 5; WHEY
	WHOLE WHEAT FIG BARS	**ARTIFICIAL COLOR;** EGG WHITES; WHEY
	—*BAKERY WAGON* DATE FILLED OATMEAL; HONEY FRUIT BARS	NO ADDITIVES OF CONCERN
	FROSTED LEMON	**ARTIFICIAL COLOR;** WHEY
	RANGER BARS	**ARTIFICIAL COLOR; BHA**
	RASPBERRY FILLED	**ARTIFICIAL COLOR; MODIFIED CORN STARCH**

BRANDS	ADDITIVES-OF-CONCERN **BOLD—CONCERN TO ALL** LIGHT—CONCERN TO SOME
R MURRAY	
BIG BOY ASSORTED	**ARTIFICIAL COLOR; (BEEF TALLOW, PALM KERNEL OIL);* BHA**
BIG BOY DUPLEX; BIG BOY VANILLA	**(BEEF TALLOW, PALM KERNEL OIL);* BHA; BHT**
CHOCOLATE CHIP; COCONUT MACAROON; FUDGE NUT; PEANUT BUTTER CREMES; VANILLA CREMES	**(BEEF TALLOW, PALM KERNEL OIL);* BHA; BHT;** WHEY
FIG BARS	**BHA; BHT;** WHEY
PEANUT BUTTER JOYS	**ARTIFICIAL COLOR** W/ FD & C YELLOW NO. 5; **(BEEF TALLOW, PALM KERNEL OIL);* BHA; BHT;** WHEY
NABISCO	
BISCOS SUGAR WAFERS; BISCOS TRIPLE DECKER; BUTTER FLAVORED; CHIPS AHOY!; COCOANUT BARS; FIG NEWTONS; IDEAL CHOCOLATE PEANUT BARS; MALLOMARS; PEANUT CREME; RAISIN FRUIT BISCUIT	WHEY
BISCOS WAFFLE CREMES	**ARTIFICIAL COLOR;** WHEY

*Blend may contain one or more of these saturated fats and/or oils.

BRANDS	ADDITIVES-OF-CONCERN BOLD—CONCERN TO ALL LIGHT—CONCERN TO SOME
NABISCO *(Continued)*	
BROWN EDGE WAFERS; FAMOUS CHOCOLATE WAFERS; FUDGE FUDGE; NILLA WAFERS; OREO	**(LARD);*** WHEY
CAMEO	**(LARD);* SODIUM CASEINATE;** WHEY
CHOCOLATE CHIP SNAPS; CHOCOLATE CHOCOLATE CHIP; DEVIL'S FOOD CAKES; GINGER SNAPS	NO ADDITIVES OF CONCERN
COCOANUT CHOCOLATE CHIP; OATMEAL; PECAN SHORTBREAD; SUGAR RINGS	**(LARD)***
FAMOUS COOKIE ASSORTMENT	**(BEEF FAT, LARD);*** EGG WHITES; WHEY
FANCY DIP GRAHAMS; MYSTIC MINT;	**(LARD, PALM KERNEL OIL);*** WHEY
LORNA DOONE; SOCIAL TEA BISCUITS	**(BEEF FAT, LARD);*** WHEY
NATIONAL ARROWROOT BISCUIT	CASEIN
NUTTER BUTTER	**(HYDROGENATED PEANUT OIL, LARD);*** WHEY

BRANDS	**ADDITIVES-OF-CONCERN** BOLD—CONCERN TO ALL LIGHT—CONCERN TO SOME
UNEEDA BISCUIT	**(HYDROGENATED COTTONSEED, PALM OILS)***
VANILLA CREME SANDWICH	**(LARD);* TURMERIC OLEORESIN**
PEEK FREANS	
ARROWROOT BISCUITS; GINGER CRISP BISCUITS	NO ADDITIVES OF CONCERN
FRUIT CREME BISCUITS	**ARTIFICIAL COLOR; (BEEF FAT)***
LEMON PUFF	**ARTIFICIAL COLOR; PALM KERNEL OIL;** WHEY POWDER
"NICE" BISCUITS	**ARTIFICIAL COLOR**
PETIT BEURRE	**BUTTER**
RICH TEA BISCUITS; SHORTCAKE BISCUITS; SWEET MEAL BISCUITS	**(BEEF FAT)***
PEPPERIDGE FARM	
BORDEAUX; CHAMPAGNE; FAVORITES W/ CHIPS & NUTS; PIROUETTES CHOCOLATE LACED; SEVILLE; SHORTBREAD; SOUTHPORT	**(COCONUT OIL);*** EGG WHITES

*Blend may contain one or more of these saturated fats and/or oils.

BRANDS	ADDITIVES-OF-CONCERN BOLD—CONCERN TO ALL LIGHT—CONCERN TO SOME
PEPPERIDGE FARM *(Continued)*	
BROWNIE CHOCOLATE NUT; CAPRI; CINNAMON SUGAR; FAVORITES W/ SUGAR & SPICE; LIDO; MOLASSES CRISPS; OATMEAL RAISIN; SUGAR	**(COCONUT OIL)***
BRUSSELS; CHOCOLATE CHIP; MILANO; ORLEANS SANDWICH	EGG WHITES
BUTTER; CHESSMEN; KITCHEN HEARTH DATE-NUT GRANOLA; ST. MORITZ	**BUTTER**
GINGER-MAN; NASSAU; PEANUT; TAHITI; ZANZIBAR	NO ADDITIVES OF CONCERN
IRISH OATMEAL	**(COCONUT OIL);* NUTMEG**
MINT MILANO	**(COCONUT, PALM KERNEL OILS);*** EGG WHITES
R RIPPIN' GOOD	
CHIP-CHIP	**(BEEF FAT, COCONUT, PALM KERNEL OILS)***
PICADILLY JELLY	ALGIN; **STABILIZERS; U.S. CERTIFIED COLOR;** WHEY (DAIRY)
SANDWICH	**(BEEF FAT);* U.S. CERTIFIED COLOR;** WHEY (DAIRY)

*Blend may contain one or more of these saturated fats and/or oils.

BRANDS	**ADDITIVES-OF-CONCERN** **BOLD—CONCERN TO ALL** LIGHT—CONCERN TO SOME
SALERNO GINGER SNAPS	NO ADDITIVES OF CONCERN
STELLA D'ORO	
ALMOND TOAST (MANDEL); ANGELICA GOODIES; ANGINETTI; ANISETTE SPONGE; ANISETTE TOAST; BREAKFAST TREATS; FRUIT CRESCENTS; GOLDEN BARS; THE HOSTESS W/ THE MOSTEST ASSORTMENT; LADY STELLA ASSORTMENT; MARGHERITE COMBINATION; SESAME (REGINA)	**ARTIFICIAL COLOR** OR **COLORS**
CHINESE DESSERT; EGG JUMBO; ROMAN EGG BISCUITS (RUM & BRANDY); ROMAN EGG BISCUITS (VANILLA); SPICE DROPS PFEFFERNUSSE; SUGARED EGG BISCUITS	NO ADDITIVES OF CONCERN
COMO DELIGHT	**ARTIFICIAL COLOR; MACE**
SUNSHINE	
BUTTER FLAVORED; CHIP A ROOS; GINGER SNAPS; GOLDEN FRUIT; TOY; VANILLA WAFERS	WHEY OR DAIRY WHEY

BRANDS	ADDITIVES-OF-CONCERN BOLD—CONCERN TO ALL LIGHT—CONCERN TO SOME
SUNSHINE *(Continued)*	
CHOCOLATE FUDGE; HYDROX; VANILLA HYDROX; VIENNA FINGERS	**(COCONUT OIL);*** WHEY
CHOCOLATE NUGGETS; OATMEAL W/ GROUND RAISINS; OATMEAL PEANUT SANDWICH	NO ADDITIVES OF CONCERN
FIG BARS; PEANUT BUTTER WAFERS	**ARTIFICIAL COLOR;** WHEY
HYDE PARK ASSORTMENT	**ARTIFICIAL COLORS; (COCONUT, PALM KERNEL OILS);*** WHEY
MARSHMALLOW BARS	**ARTIFICIAL COLOR; MODIFIED FOOD STARCH**
SPRINKLES	**ARTIFICIAL COLOR; FOOD SHELLAC**
SUGAR WAFERS	**ARTIFICIAL COLOR; (COCONUT OIL);*** WHEY
SWISSTYLE GRAHAMS	**(COCONUT, PALM KERNEL OILS; HYDROGENATED COTTONSEED, PALM, SOYBEAN OILS);*** WHEY
R TOGGENBURGER	
CHOCOLATE CREAM FILLED SNACK WAFERS	**VEGETABLE SHORTENING**

*Blend may contain one or more of these saturated oils.

BRANDS	ADDITIVES-OF-CONCERN **BOLD—CONCERN TO ALL** LIGHT—CONCERN TO SOME
R VAN DE KAMP'S	
ALMOND ICE BOX; DUTCH CRISP; DUTCH SHORTBREAD BARS; LEMON SNACK; MAPLE PECAN; SPRITZ SHORTBREAD	MILK PROTEIN
COCONUT MACAROONS	EGG WHITES; GUM TRAGACANTH
R WALKERS	
SHORTBREAD FINGERS	**BUTTER**
SUPERMARKET PRIVATE BRANDS	
A & P	
—A & P	
BUTTER FLAVORED; CHOCOLATE CHIP; VANILLA WAFERS	WHEY
—ANN PAGE	
BUTTER PECAN	WHEY (SWEET DAIRY)
CHOCOLATE SANDWICH CREME; VANILLA SANDWICH CREME	**ARTIFICIAL COLORS**
OLD FASHIONED CHOCOLATE CHIP; OLD FASHIONED FUDGE SUGAR; OLD FASHIONED OATMEAL; OLD FASHIONED SUGAR	NO ADDITIVES OF CONCERN

BRANDS	ADDITIVES-OF-CONCERN BOLD—CONCERN TO ALL LIGHT—CONCERN TO SOME
FIRST NATIONAL **—FINAST** FIG BARS	NO ADDITIVES OF CONCERN
SAFEWAY **—BUSY BAKER** ANIMAL; COCONUT; VANILLA SNAPS	**(HYDROGENATED SOYBEAN OIL);*** WHEY OR WHEY SOLIDS
CHOCOLATE SNAPS; FAMILY ASSORTMENT; VANILLA WAFERS	**ARTIFICIAL COLOR** OR **COLORS; (HYDROGENATED SOYBEAN OIL);*** WHEY
RAISIN BRAN FRUIT BARS	WHEY
—BUSY BAKER *GOOD OLD FASHIONED* CHOCOLATE CHIP; OATMEAL	**(HYDROGENATED SOYBEAN OIL)***
COCOA CHIPS; PINK SUGAR WAFERS	**ARTIFICIAL COLOR; (BEEF FAT);*** WHEY
DUTCH APPLE BARS; FIG BARS; MARSHMALLOW PUFFS	**ARTIFICIAL COLOR;** WHEY
LEMON CREMES	**ARTIFICIAL COLOR; (BEEF FAT)***
PEANUT BUTTER CREME; WHOLE WHEAT FIG BARS	WHEY

BRANDS	ADDITIVES-OF-CONCERN **BOLD—CONCERN TO ALL** LIGHT—CONCERN TO SOME
—SCOTCH BUY	
COCOA DROP	**ARTIFICIAL COLOR; (COCONUT, PALM KERNEL OILS)***
STOP & SHOP	
—STOP & SHOP	
CHOCO CHIP; FIG BARS; FIG SQUARES; REGULAR HERMITS	NO ADDITIVES OF CONCERN
—SUN GLORY	
BUTTER PECAN	WHEY (SWEET DAIRY)
CHOCOLATE CHIP; CHOCOLATE SQUARES; COCONUT BARS; FUDGE SUGAR; LEMON CUSTARD SQUARES; OATMEAL; PEANUT BUTTER SQUARES	**ARTIFICIAL COLOR**
ICED FUDGIES	**ARTIFICIAL COLOR;** EGG WHITES; WHEY (SWEET DAIRY)
ICED OATMEAL 'N RAISIN	EGG WHITES
ICED SPICE; SUGAR	**ARTIFICIAL COLOR; NUTMEG**
VANILLA SQUARES	NO ADDITIVES OF CONCERN
WINN DIXIE	
—CRACKIN GOOD	
ACE ASSORTMENT	**ARTIFICIAL COLOR; (BEEF FAT, LARD);* BHA;** WHEY (DAIRY)

*Blend may contain one or more of these saturated fats and/or oils.

BRANDS	ADDITIVES-OF-CONCERN **BOLD—CONCERN TO ALL** LIGHT—CONCERN TO SOME
WINN DIXIE—CRACKIN GOOD *(Continued)*	
ALMOND CRUNCH; ANIMAL; BUTTER; CHOCOLATE CHIP; COCONUT CREMES; FROSTED FRUIT; FUDGE RIPPLES; LEMON COCONUT ROUNDS; OATMEAL; VANILLA WAFERS	**(BEEF FAT, LARD);* BHA;** WHEY (DAIRY)
ASSORTED CREMES; BIG 60 SANDWICH CREMES; LEMON CREMES	**ARTIFICIAL COLOR** W/ FD & C YELLOW NO. 5; **(BEEF FAT, LARD);* BHA;** WHEY (DAIRY)
COCOA CHIP 'N NUT; FIG-BARS; GINGER CRISPS; VANILLA KREMO	**BHA;** WHEY (DAIRY OR POWDERED)
DEVILSFOOD CAKES	WHEY
ICED APPLESAUCE; PEANUT BUTTER; PECAN JOY	**(BEEF FAT, LARD);* BHA**
ICED SUGAR CRISPS	**(BEEF FAT, LARD);* BHA; MACE;** WHEY (DAIRY)
PEANUT BUTTER WAFERS	**ARTIFICIAL COLOR;** WHEY
ITEMS IN HEALTH FOOD STORES	
DONNA'S	
GOURMET NATURAL HONEY SNACK HAWAIIAN	NO ADDITIVES OF CONCERN

BRANDS	**ADDITIVES-OF-CONCERN** **BOLD—CONCERN TO ALL** LIGHT—CONCERN TO SOME
—NATURAL HONEY COOKIE BROWNIE; RAISIN NUT	NO ADDITIVES OF CONCERN
EL MOLINO MILLS HONEY ANIMAL	NO ADDITIVES OF CONCERN
HEALTH VALLEY OATMEAL; PEANUT BUTTER; WHEAT GERM & MOLASSES	**PURE CREAMERY BUTTER;** WHEY POWDER
RAISIN BRAN	**PURE CREAMERY BUTTER**
—NATURAL SNAPS CAROB & HONEY; COCONUT & HONEY; GINGER & HONEY; LEMON & HONEY; YOGURT & HONEY	**PURE CREAMERY BUTTER;** WHEY POWDER
HOFFMAN'S ENERGY-PLUS; HI-PROTEEN	**SODIUM CASEINATE**
SOVEX COCONUT; MOLASSES; PEANUT BUTTER	WHEY
GRANOLA	EGG WHITES; WHEY

*Blend may contain one or more of these saturated fats.

BRANDS	ADDITIVES-OF-CONCERN **BOLD—CONCERN TO ALL** LIGHT—CONCERN TO SOME

BREAKFAST PASTRIES, BROWNIES, CAKES, SNACK CAKES

	BRANDS	ADDITIVES-OF-CONCERN
R	BUTTERMAID FARMCREST ECLAIRS	**ARTIFICIAL COLOR;** GUAR GUM; **(LARD);*** LOCUST BEAN GUM; **MODIFIED FOOD STARCH**
	COUNTRY GOOD DANISH BEAR CLAW	**ARTIFICIAL COLORINGS;** LOCUST BEAN GUM
R	DOLLY MADISON ANGEL FOOD CAKE	EGG WHITES
	CINNAMON ROLLS	**ARTIFICIAL COLOR; (BEEF FAT, LARD; COCONUT, PALM KERNEL OILS);*** LOCUST BEAN GUM; **MODIFIED CORN STARCH;** WHEY
	CREME CAKES	**ARTIFICIAL COLOR; CALCIUM CASEINATE;** GUAR GUM; **MODIFIED CORN STARCH;** SODIUM ALGINATE; **SODIUM CASEINATE;** WHEY
	DANISH ROLLS RASPBERRY	**ARTIFICIAL COLOR; (COCONUT, PALM KERNEL OILS; HYDROGENATED COTTONSEED, SOYBEAN OILS);*** LOCUST BEAN GUM; **MODIFIED CORN STARCH;** SODIUM ALGINATE; WHEY

BRANDS	ADDITIVES-OF-CONCERN BOLD—CONCERN TO ALL LIGHT—CONCERN TO SOME
DESSERT ROLL	**ARTIFICIAL COLOR; CALCIUM CASEINATE;** SODIUM ALGINATE; **SODIUM CASEINATE;** WHEY
POUND CAKE	**ARTIFICIAL COLOR; (BEEF FAT, LARD);* CALCIUM CASEINATE;** EGG WHITES; **SODIUM CASEINATE;** WHEY
ZINGERS (VANILLA)	**ARTIFICIAL COLOR; CALCIUM CASEINATE;** EGG WHITES; GUAR GUM; LOCUST BEAN GUM; **MODIFIED CORN STARCH;** SODIUM ALGINATE; **SODIUM CASEINATE;** WHEY
DRAKES	
COFFEE CAKE JR.	**ARTIFICIALLY COLORED WITH VEGETABLE COLORS; CALCIUM CASEINATE;** EGG WHITES; **SODIUM CASEINATE;** WHEY
CREME FINGERS	**ARTIFICIALLY COLORED WITH VEGETABLE COLORS; CALCIUM CASEINATE;** EGG WHITES; **MODIFIED CORN STARCH; SODIUM CASEINATE;** WHEY

*Blend may contain one or more of these saturated fats and/or oils.

BRANDS	ADDITIVES-OF-CONCERN BOLD—CONCERN TO ALL LIGHT—CONCERN TO SOME
DRAKES *(Continued)*	
DEVIL DOGS	**CALCIUM CASEINATE; SODIUM CASEINATE;** WHEY
MARBLE POUND CAKE JR.; POUND CAKE JR.	**CALCIUM CASEINATE; SODIUM CASEINATE;** WHEY
RING DING; SWISS ROLL; YODELS	**CALCIUM CASEINATE;** EGG WHITES; **(PALM KERNEL OIL);* SODIUM CASEINATE;** WHEY
ENTENMANN'S	
ALL BUTTER POUND CAKE; APPLE PUFFS; BANANA CRUNCH LOAF; CHEESE BUNS; CHOCOLATE CHIP CRUMB LOAF; FUDGE CAKE; MARSHMALLOW FUDGE SQUARE	**MODIFIED CORN STARCH**
APRICOT CRUMB SQUARE; RASPBERRY DANISH TWIST; STRAWBERRY CHEESE DANISH; WALNUT DANISH TWIST	**ARTIFICIAL COLOR; COLORING**

	BRANDS	ADDITIVES-OF-CONCERN **BOLD—CONCERN TO ALL** LIGHT—CONCERN TO SOME
	CHEESE COFFEE CAKE; CINNAMON DANISH TWIST; CINNAMON FILBERT; CRUMB COFFEE CAKE; PECAN RAISIN LOAF; WALNUT DANISH RING	**COLORING**
	MARBLE LOAF; PUMPKIN WALNUT CAKE	NO ADDITIVES OF CONCERN
R	**HOLSUM**	
	ANGELFOOD CAKE	EGG WHITES
	APPLE STREUDEL	**ARTIFICIAL COLOR;** FD & C YELLOW NO. 5; **(BEEF FAT);*** CASEIN; **MODIFIED FOOD STARCH**
	CINNAMON ROLLS (PECAN, PLAIN, RAISIN)	**ARTIFICIAL COLOR;** GUAR GUM; WHEY
	COCONUT LAYER CAKE	**ARTIFICIAL COLOR;** EGG WHITES; LOCUST BEAN GUM; WHEY
	DEVIL'S DELITE	**ARTIFICIAL COLOR;** EGG WHITES; **MODIFIED FOOD STARCH;** WHEY

*Blend may contain one or more of these saturated fats and/or oils.

BRANDS	ADDITIVES-OF-CONCERN **BOLD—CONCERN TO ALL** LIGHT—CONCERN TO SOME
HOLSUM *(Continued)*	
DEVIL'S FOOD LAYER CAKE	EGG WHITES; LOCUST BEAN GUM; WHEY
FUDGE BAR	**(BEEF TALLOW, LARD);*** EGG WHITES
PECAN SPINS	**ARTIFICIAL COLOR; (BEEF FAT, LARD);*** WHEY
POUND CAKE	**ARTIFICIAL COLOR;** EGG WHITES; WHEY
HOSTESS	
BIG WHEELS	**(BEEF FAT, LARD);* MODIFIED FOOD STARCH; SODIUM CASEINATE;** WHEY
BREAKFAST BAKE SHOP ICED HONEY BUN	**ARTIFICIAL COLOR; (BEEF FAT, LARD);*** LOCUST BEAN GUM
FILLED CUP CAKES	**MODIFIED FOOD STARCH; SODIUM CASEINATE;** WHEY
SUZY Q'S	**SODIUM CASEINATE;** WHEY
TWINKIES	**ARTIFICIAL COLOR; MODIFIED FOOD STARCH; SODIUM CASEINATE;** WHEY
R KING'S FUDGE FROSTEES	**ARTIFICIAL COLORS;** EGG WHITES; WHEY

	BRANDS	**ADDITIVES-OF-CONCERN** BOLD—CONCERN TO ALL LIGHT—CONCERN TO SOME
R	**KITCHEN PRIDE**	
	APPLE SANDWICH CAKES	**ARTIFICIAL COLOR; MODIFIED CORN STARCH;** WHEY SOLIDS
	LITTLE DEBBIE	
	APPLE DELIGHTS; NUTTY BARS	WHEY
	APPLE SPICE CAKES	**ARTIFICIAL COLORS;** EGG WHITES
	FUDGE ROUNDS CAKE; SWISS CAKE ROLLS	**ARTIFICIAL COLOR** OR **COLORS;** WHEY
	JEL-CREME ROLLS	**ARTIFICIAL COLORS;** EGG WHITES; **MODIFIED FOOD STARCH;** WHEY
	PEANUT BUTTER BROWNIES	EGG WHITES; WHEY
R	**MICKEY CAKES**	
	APPLE WALNUT	**(BEEF TALLOW, LARD)***
	CARROT CAKE	LOCUST BEAN GUM
	FUDGE BAR	LOCUST BEAN GUM; **(PALM KERNEL OIL)***
R	**NISSEN**	
	COCOA SNAX	WHEY SOLIDS
	COFFEE CAKE	**ARTIFICIAL COLOR;** EGG WHITES; GLUTEN (WHEAT); GUAR GUM; WHEY

*Blend may contain one or more of these saturated fats and/or oils.

	BRANDS	ADDITIVES-OF-CONCERN BOLD—CONCERN TO ALL LIGHT—CONCERN TO SOME
	NISSEN *(Continued)*	
	CREME HORNS	**ARTIFICIAL COLOR;** LOCUST BEAN [GUM]; **TALLOW**
	FUDGE CAKE	**ARTIFICIAL COLOR;** EGG WHITES; GLUTEN (WHEAT); GUAR GUM; LOCUST BEAN GUM; WHEY
R	**REGENT** FUDGE BROWNIE	EGG WHITES; **(SATURATED COTTONSEED, PALM OILS);*** WHEY POWDER
	STELLA D'ORO LADY FINGERS	NO ADDITIVES OF CONCERN
R	**SVENHARD'S**	
	APPLE ROLLS; BEAR CLAWS; BREAKFAST CLAWS; BREAKFAST HORNS; BUTTER CRUNCH ROLLS; CINNAMON HORNS; FUDGE-ETTES; HORNS A PLENTY; LEMON ROLLS; RAISIN-ETTES; RAISIN SNAILS; WALNUT ROLLS	**ARTIFICIAL COLOR** OR **COLORS**
	APPLE STREUDEL	**MODIFIED CORN STARCH**
R	**TABLE TALK**	
	GOLDEN FUDGE CAKE	**ARTIFICIAL COLORS;** EGG WHITES; GUAR GUM; **MODIFIED FOOD STARCH**
	SOUR CREAM POUND CAKE	**MODIFIED FOOD STARCH**

	BRANDS	ADDITIVES-OF-CONCERN **BOLD—CONCERN TO ALL** LIGHT—CONCERN TO SOME
R	VACHON	
	HALF MOON SPONGE CAKES W/ CREAMED FILLING	**ARTIFICIAL COLOR; SODIUM CASEINATE**
	POPOVER PUFF W/ STRAWBERRY FILLING	**ARTIFICIAL COLOR; (BEEF TALLOW, LARD);*** SODIUM ALGINATE
R	VAN DE KAMP'S	
	ANGEL FOOD LOAF	EGG WHITES
	APPLE MUFFINS; BANANA NUT LOAF; HONEY BRAN MUFFINS; LADY FINGERS	NO ADDITIVES OF CONCERN
	APPLE PUFFS	**ARTIFICIAL COLOR; MODIFIED FOOD STARCH; PARTIALLY HYDROGENATED VEGETABLE SHORTENING**
	APPLESAUCE CAKE	**ARTIFICIAL COLOR**
	BLUEBERRY MUFFINS	**SODIUM CASEINATE**
	BUTTERHORNS; PINEAPPLE SWIRLS	**ARTIFICIAL COLOR;** GUM TRAGACANTH
	CARROT CAKE	**CARRAGEENAN; PARTIALLY HYDROGENATED VEGETABLE SHORTENING**
	CHOCOLATE ICED ANGEL FOOD CAKE	**CARRAGEENAN;** EGG WHITES; **PARTIALLY HYDROGENATED VEGETABLE SHORTENING**

*Blend may contain one or more of these saturated fats and/or oils.

BRANDS	ADDITIVES-OF-CONCERN BOLD—CONCERN TO ALL LIGHT—CONCERN TO SOME
VAN DE KAMP'S *(Continued)*	
CINNAMON ROLLS	EGG WHITES; **ENZYME OF ASPERGILLUS ORYZAE;** GUM TRAGACANTH; WHEY
DATE NUT LOAF	**VEGETABLE SHORTENING**
GOLDEN CHOCOLATE CAKE	**ARTIFICIAL COLOR;** EGG WHITES; **MODIFIED FOOD STARCH; PARTIALLY HYDROGENATED VEGETABLE SHORTENING;** WHEY
MILK CHOCOLATE CUPCAKES	**MODIFIED FOOD STARCH;** WHEY
RASPBERRY ROLLS	**ARTIFICIAL COLOR; ENZYME OF ASPERGILLUS ORYZAE;** GUM TRAGACANTH
WALNUT SWIRLS	**ARTIFICIAL COLOR;** EGG WHITES; GUM TRAGACANTH; WHEY
SUPERMARKET PRIVATE BRANDS	
PUBLIX	
—**PUBLIX** FRUIT CAKE	**ARTIFICIAL COLOR**
STOP & SHOP	
—(NO BRAND NAME) APPLE SWEET ROLL; PECAN COFFEE RING	ALGIN; **ARTIFICIAL COLOR;** LOCUST BEAN GUM; **MODIFIED FOOD STARCH;** WHEY

BRANDS	ADDITIVES-OF-CONCERN BOLD—CONCERN TO ALL LIGHT—CONCERN TO SOME
CINNAMON RAISIN COFFEE RING; CINNAMON RAISIN ROLL	**ARTIFICIAL COLOR;** LOCUST BEAN GUM; WHEY
LADY FINGERS	NO ADDITIVES OF CONCERN
PECAN TWIRLS	**ARTIFICIAL COLOR;** FD & C YELLOW NO. 5
—OUR FAMOUS FUDGE CAKE	LOCUST BEAN GUM; WHEY
—OUR OWN ANGEL CAKE; SPONGE CAKE FEATHERLIGHT	EGG WHITES
MAPLE WALNUT CAKE	LOCUST BEAN GUM; WHEY
MARBLE PARTY BAR; MARBLE POUND CAKE	**ARTIFICIAL COLOR;** EGG WHITES; **MODIFIED FOOD STARCH;** WHEY
RASPBERRY FILLED FUN-BALLS	ALGIN; **ARTIFICIAL COLOR; CARRAGEENAN; MODIFIED FOOD STARCH;** WHEY
SPONGE CAKE	WHEY
—STOP & SHOP CHOCOLATE CHIP BUTTERSCOTCH BROWNIES; CREME FILLED COCOA ROLL	WHEY OR WHEY SOLIDS

BRANDS	**ADDITIVES-OF-CONCERN** **BOLD—CONCERN TO ALL** LIGHT—CONCERN TO SOME
—STOP & SHOP *(Continued)*	
GOLDEN CIRCLE CAKE; MARBLE CIRCLE CAKE	**ARTIFICIAL COLOR;** EGG WHITES; **MODIFIED FOOD STARCH;** WHEY
JELLY ROLL	**ARTIFICIAL COLORS; VEGETABLE GUM;** WHEY SOLIDS
WALNUT CHOCOLATE BROWNIES	GUAR GUM
WINN DIXIE	
—CRACKIN GOOD	
OATMEAL SNACK CAKES	**BHA;** EGG WHITES (DRIED); SODIUM ALGINATE; WHEY (DAIRY)
RAISIN SNACK CAKES	**ARTIFICIAL COLOR;** WHEY (DRIED)
ROYAL STRIPED MALLOWS	**ARTIFICIAL COLORS** W/ FD & C YELLOW NO. 5; **(BEEF FAT, LARD);* BHA;** SODIUM ALGINATE; WHEY (DAIRY)
—DIXIE DARLING	
ANGEL FOOD CAKE	EGG WHITES
PECAN TWIRLS	**ARTIFICIAL COLOR;** FD & C YELLOW NO. 5

DOUGHNUTS

ENTENMANN'S	
COUNTRY STYLE; CRUMB	**ARTIFICIAL COLOR; COLORING**

BRANDS	ADDITIVES-OF-CONCERN **BOLD—CONCERN TO ALL** LIGHT—CONCERN TO SOME
R HOLSUM	
DONUTS	**ARTIFICIAL COLOR; BHA; BHT; SHORTENING; SODIUM CASEINATE**
CHOCOLATE COATED	**ARTIFICIAL COLOR; (BEEF TALLOW, LARD; COCONUT, PALM KERNEL OILS);* MACE; NUTMEG;** WHEY
SUGAR	**ARTIFICIAL COLOR; MACE; NUTMEG**
HOSTESS	
DONETTE GEMS; FROSTED DONETTE GEMS	**ARTIFICIAL COLOR; (BEEF FAT, LARD);* SODIUM CASEINATE**
—*BREAKFAST BAKE SHOP*	
FINER DONUTS ASSORTMENT; FINER DONUTS FAMILY PAK; FROSTED	**ARTIFICIAL COLOR; (BEEF FAT, LARD);*** GUAR & KARAYA GUMS
R MICKEY *MINI-DONUTS*	
CHOCOLATE COATED	**ARTIFICIAL COLOR; (BEEF TALLOW, LARD; COCONUT, PALM KERNEL OILS);* MACE; NUTMEG;** WHEY
CRUNCH; SUGAR COATED	**ARTIFICIAL COLOR; MACE; NUTMEG**

*Blend may contain one or more of these saturated fats and/or oils.

	BRANDS	ADDITIVES-OF-CONCERN **BOLD—CONCERN TO ALL** LIGHT—CONCERN TO SOME
R	MOTHER PARKER'S OLD FASHIONED	NO ADDITIVES OF CONCERN
R	NISSEN ASSORTED; CINNAMON; PLAIN	**ARTIFICIAL COLOR;** GUAR & KARAYA GUMS
	CHOCOLATE COCOANUT	**ARTIFICIAL COLOR**
R	VAN DE KAMP'S CHOCOLATE ICED OLD FASHIONED	**VEGETABLE GUMS;** WHEY
	GLAZED OLD FASHIONED	**MODIFIED FOOD STARCH; VEGETABLE GUMS;** WHEY
	OLD FASHIONED	**ARTIFICIAL COLOR; BHA; BHT**
	RAISED	**ARTIFICIAL COLOR;** GLUTEN (WHEAT)
	SUPERMARKET PRIVATE BRANDS LUCKY —HARVEST DAY CHOCOLATE COATED	**ARTIFICIAL COLOR; (COCONUT, PALM KERNEL OILS);* MACE; NUTMEG;** WHEY
	PLAIN; POWDERED SUGAR	**ARTIFICIAL COLOR; MACE; NUTMEG**
	SAFEWAY —MRS. WRIGHT'S CHOCOLATE FLAVOR FROSTED	**ARTIFICIAL COLOR; (BEEF FAT, LARD);*** GUAR GUM; WHEY

BRANDS	**ADDITIVES-OF-CONCERN** BOLD—CONCERN TO ALL LIGHT—CONCERN TO SOME
DONUTS (FROSTED); DONUTS (PLAIN)	**ARTIFICIAL COLOR; (BEEF FAT, LARD);*** GUAR GUM; **MODIFIED FOOD STARCH;** WHEY
STOP & SHOP	
—STOP & SHOP BAKERY	
SNACK PACK (PLAIN)	EGG WHITES; GUAR GUM; **PAPRIKA; SODIUM CASEINATE; TURMERIC;** WHEY
SOUR CREME OLD FASHION	**ARTIFICIAL COLOR; MODIFIED WHEAT STARCH;** WHEY
—DAISY DONUTS	
CHOCOLATE COVERED; W/ CINNAMON; W/ POWDERED SUGAR; FROSTED	EGG WHITES; GUAR GUM; **PAPRIKA; SODIUM CASEINATE; TURMERIC;** WHEY
COUNTRY STYLE (W/ CINNAMON); COUNTRY STYLE (W/ POWDERED SUGAR)	**PAPRIKA; TURMERIC;** WHEY
GLAZED	**ARTIFICIAL COLOR; (BEEF FAT);* SODIUM CASEINATE;** WHEY
WINN DIXIE	
—WINN DIXIE GLAZED	**ARTIFICIAL COLOR; FUNGAL PROTEASE OBTAINED FROM ASPERGILLUS ORYZAE;** GUAR GUM; WHEY

*Blend may contain one or more of these saturated fats and/or oils.

BRANDS	ADDITIVES-OF-CONCERN **BOLD—CONCERN TO ALL** LIGHT—CONCERN TO SOME

PIES

DOLLY MADISON CHOCOLATE SNACK PIE	**ARTIFICIAL COLOR; (BEEF FAT, LARD; COCONUT, PALM KERNEL OILS);* CALCIUM CASEINATE;** LOCUST BEAN GUM; **MODIFIED CORN STARCH;** SODIUM ALGINATE; **SODIUM CASEINATE;** WHEY
—FRUIT PIE APPLE	**(BEEF FAT, LARD; COCONUT, PALM KERNEL OILS);* CALCIUM CASEINATE;** LOCUST BEAN GUM; **MODIFIED CORN STARCH; SODIUM CASEINATE;** WHEY
LEMON	**ARTIFICIAL COLOR; (BEEF FAT, LARD; COCONUT, PALM KERNEL OILS);* CALCIUM CASEINATE;** LOCUST BEAN GUM; **MODIFIED CORN STARCH; SODIUM CASEINATE; SODIUM NITRATE;** WHEY
DRAKE'S APPLE	CAROB BEAN GUM; EGG ALBUMIN; **MODIFIED CORN STARCH;** SODIUM ALGINATE; WHEY

	BRANDS	ADDITIVES-OF-CONCERN **BOLD—CONCERN TO ALL** LIGHT—CONCERN TO SOME
	ENTENMANN'S	
	APPLE; APPLE CRUMB	**MODIFIED CORN STARCH**
	HOSTESS *FRUIT PIE*	
	APPLE	**ARTIFICIAL COLOR; (BEEF FAT, LARD);* MODIFIED FOOD STARCH; NUTMEG**
	BLUEBERRY; CHERRY; LEMON	**ARTIFICIAL COLOR; (BEEF FAT, LARD);* MODIFIED FOOD STARCH**
R	TABLE TALK	
	BLUEBERRY	EGG WHITES; **LARD; MODIFIED FOOD STARCH; SODIUM CASEINATE;** WHEY
	BOSTON CREAM	**ARTIFICIAL COLOR; CARRAGEENAN; (COCONUT, PALM KERNEL OILS);*** LOCUST BEAN GUM; WHEY
	CHOCOLATE CREAM	**ARTIFICIAL COLOR; METHYL ETHYL CELLULOSE; MODIFIED FOOD STARCH; SODIUM CASEINATE;** WHEY
	LEMON MERINGUE	**ARTIFICIAL COLORS;** EGG WHITES; LOCUST BEAN GUM; **MODIFIED FOOD STARCH;** WHEY
	SQUASH	**ARTIFICIAL COLOR;** EGG WHITES; **SODIUM CASEINATE;** WHEY

*Blend may contain one or more of these saturated fats and/or oils.

BRANDS	ADDITIVES-OF-CONCERN BOLD—CONCERN TO ALL LIGHT—CONCERN TO SOME
R TABLE TALK *(Continued)*	
—JUNIOR PIES	
APPLE	**ARTIFICIAL COLORS; LARD; MODIFIED FOOD STARCH;** WHEY
BLUEBERRY	**LARD; MODIFIED FOOD STARCH;** WHEY
CHOCOLATE ECLAIR	**ARTIFICIAL COLORS;** GUAR GUM; **LARD;** LOCUST BEAN GUM; **MODIFIED FOOD STARCH;** WHEY
R VAN DE KAMP'S	
APPLE; BERRY	**ARTIFICIAL COLOR;** GUAR GUM; **MODIFIED STARCH**
CHERRY; LEMON	**ARTIFICIAL COLOR; MODIFIED FOOD STARCH**
HIGH TOP APPLE	LOCUST BEAN GUM; WHEY
LEMON MERINGUE	**ARTIFICIAL COLOR;** EGG WHITES; LOCUST BEAN GUM; **MODIFIED FOOD STARCHES;** WHEY
PEACH	LOCUST BEAN GUM; **MODIFIED FOOD STARCH;** WHEY
SUPERMARKET PRIVATE BRANDS	
STOP & SHOP	
—OUR OWN	
APPLE; BLUEBERRY; CHERRY; PINEAPPLE	**MODIFIED FOOD STARCH**
CUSTARD	**ARTIFICIAL COLOR; PAPRIKA; TURMERIC**

BRANDS	ADDITIVES-OF-CONCERN **BOLD—CONCERN TO ALL** LIGHT—CONCERN TO SOME
ECLAIR	LOCUST BEAN GUM
HOLIDAY MINCE	NO ADDITIVES OF CONCERN
LEMON	**ARTIFICIAL COLOR; MODIFIED FOOD STARCH;** SODIUM ALGINATE
SQUASH	**CARRAGEENAN;** LOCUST BEAN GUM; **MODIFIED FOOD STARCH; NUTMEG**
—STOP & SHOP FRUIT PIE LEMON	**ARTIFICIAL COLOR;** GUM ARABIC; **MODIFIED FOOD STARCH;** SODIUM ALGINATE; **SODIUM CASEINATE**
WINN DIXIE	
—CRACKIN GOOD	
BANANA; DEVIL FOOD	**ARTIFICIAL COLOR** OR **COLORS** W/ FD & C YELLOW NO. 5; **(BEEF FAT, LARD);* BHA;** SODIUM ALGINATE; WHEY (DAIRY)
CHOCOLATE; COCONUT	**(BEEF FAT, LARD);* BHA;** SODIUM ALGINATE; WHEY (DAIRY)
OATMEAL CREME	**ARTIFICIAL COLOR;** MILK SOLIDS; WHEY SOLIDS
WILD CHERRY	**ARTIFICIAL COLOR; (BEEF FAT, LARD);* BHA;** SODIUM ALGINATE; WHEY (DAIRY)

*Blend may contain one or more of these saturated fats.

IV. BAKING INGREDIENTS

ALMOND PASTE, BAKING CHOCOLATE, CAROB POWDER, COCONUT, FLAVORED CHIPS

BRANDS	ADDITIVES-OF-CONCERN BOLD—CONCERN TO ALL LIGHT—CONCERN TO SOME
BAKER'S	
ANGEL FLAKE COCONUT; GERMAN'S SWEET CHOCOLATE; SEMI-SWEET CHOCOLATE; UNSWEETENED CHOCOLATE	NO ADDITIVES OF CONCERN
CHOCOLATE FLAVOR BAKING CHIPS	MILK DERIVATIVE; **PALM KERNEL OIL**
HERSHEY'S	
BAKING CHOCOLATE UNSWEETENED; COCOA; REAL CHOCOLATE MINI CHIPS	NO ADDITIVES OF CONCERN
NESTLÉ	
BUTTERSCOTCH ARTIFICIAL FLAVORED MORSELS	**ARTIFICIAL COLORS; PALM KERNEL OIL**
CHOCO BAKE	**BHA; HYDROGENATED COTTONSEED OIL**

BRANDS	ADDITIVES-OF-CONCERN BOLD—CONCERN TO ALL LIGHT—CONCERN TO SOME
MILK CHOCOLATE MORSELS; SEMI-SWEET CHOCOLATE TOLL HOUSE MORSELS	NO ADDITIVES OF CONCERN
ODENSE ALMOND PASTE	NO ADDITIVES OF CONCERN
REESE'S PEANUT BUTTER FLAVORED CHIPS	**(PALM KERNEL OIL)***
SUPERMARKET PRIVATE BRANDS	
A & P	
—ANN PAGE	
BUTTERSCOTCH CHIPS	**ARTIFICIAL COLOR**
SEMI-SWEET CHOCOLATE CHIPS	**PALM KERNEL OIL**
SAFEWAY	
—TOWN HOUSE	
SEMI-SWEET CHOCOLATE CHIPS; SWEETENED FANCY SHRED COCONUT; SWEETENED FLAKED COCONUT	NO ADDITIVES OF CONCERN
ITEMS IN HEALTH FOOD STORES	
CARA COA CAROB POWDER	NO ADDITIVES OF CONCERN
NIBLACK CAROB POWDER	NO ADDITIVES OF CONCERN

*Blend may contain this saturated oil.

BRANDS	ADDITIVES-OF-CONCERN BOLD—CONCERN TO ALL LIGHT—CONCERN TO SOME

EXTRACTS, FLAVORS, & FOOD COLORS

DURKEE	
IMITATION BUTTER FLAVOR	**CERTIFIED FOOD COLOR; VEGETABLE GUM**
CHOCOLATE FLAVOR; IMITATION PEPPERMINT EXTRACT	**CERTIFIED FOOD COLOR**
IMITATION COCONUT FLAVOR	**VEGETABLE GUM**
IMITATION MAPLE FLAVOR; IMITATION RUM FLAVOR	NO ADDITIVES OF CONCERN
McCORMICK	
PURE ALMOND EXTRACT; PURE ANISE EXTRACT; PURE MINT & PEPPERMINT EXTRACT; PURE ORANGE EXTRACT	NO ADDITIVES OF CONCERN
IMITATION RUM EXTRACT	**FD & C RED NO. 40** & YELLOW NO. 5
IMITATION STRAWBERRY EXTRACT	**ETHYL METHYL PHENYL GLYCIDATE; FD & C RED NO. 40;** IONONE

BRANDS	ADDITIVES-OF-CONCERN **BOLD—CONCERN TO ALL** LIGHT—CONCERN TO SOME
SUPERMARKET PRIVATE BRANDS **A & P** **—ANN PAGE** PURE ALMOND EXTRACT; PURE LEMON EXTRACT; PURE VANILLA EXTRACT	NO ADDITIVES OF CONCERN
FIRST NATIONAL **—EDWARDS** PURE VANILLA EXTRACT	NO ADDITIVES OF CONCERN
LUCKY **—LADY LEE** PURE VANILLA EXTRACT	NO ADDITIVES OF CONCERN
SAFEWAY **—CROWN COLONY** ALMOND EXTRACT; LEMON EXTRACT; PEPPERMINT EXTRACT; VANILLA EXTRACT	NO ADDITIVES OF CONCERN
ARTIFICIAL BANANA FLAVOR; ARTIFICIAL BUTTER FLAVOR; GREEN FOOD COLOR; RED FOOD COLOR; YELLOW FOOD COLOR	**ARTIFICIAL COLORING**
CHOCOLATE EXTRACT	**ARTIFICIAL COLORS;** PROPYLENE GLYCOL ALGINATE

BRANDS	ADDITIVES-OF-CONCERN BOLD—CONCERN TO ALL LIGHT—CONCERN TO SOME
—SCOTCH BUY ARTIFICIAL VANILLA FLAVORING	NO ADDITIVES OF CONCERN
STOP & SHOP **—STOP & SHOP** PURE LEMON EXTRACT; PURE VANILLA EXTRACT	NO ADDITIVES OF CONCERN
ITEMS IN HEALTH FOOD STORES **WALNUT ACRES** PURE VANILLA EXTRACT	NO ADDITIVES OF CONCERN

FLOURS & LEAVENING AGENTS, STABILIZERS

Most of the items in this section did not contain any additives of concern, and these appear first, without an additives-of-concern column.

The items that do contain additives of concern follow, in the usual style.

FLOUR

GOLD MEDAL ALL PURPOSE BLEACHED; SELF-RISING BLEACHED; WHOLE WHEAT; WONDRA BLEACHED

KING ARTHUR UNBLEACHED

PILLSBURY'S BEST ALL PURPOSE BLEACHED; MEDIUM RYE; SELF RISING BLEACHED; UNBLEACHED; WHOLE WHEAT

SWANS DOWN BLEACHED CAKE

—SUPERMARKET PRIVATE BRANDS
A & P ANN PAGE ALL PURPOSE BLEACHED

FIRST NATIONAL FINAST ALL PURPOSE BLEACHED

LUCKY LADY LEE ALL PURPOSE BLEACHED

SAFEWAY MRS. WRIGHT'S BLEACHED; KITCHEN CRAFT BLEACHED

STOP & SHOP ALL PURPOSE BLEACHED

WINN DIXIE THRIFTY MAID ALL PURPOSE BLEACHED; BLEACHED (SELF-RISING)

—ITEMS IN HEALTH FOOD STORES
EREWHON BROWN RICE; BUCKWHEAT DARK; RYE; SOYBEAN

OLDE MILL WHOLE WHEAT PASTRY; WHOLE WHEAT STONE GROUND

SHILOH FARMS BUCKWHEAT; CORN; OAT

WALNUT ACRES GRAHAM; OAT

LEAVENING AGENTS
ARM & HAMMER BAKING SODA

CALUMET BAKING POWDER

DAVIS BAKING POWDER

RUMFORD BAKING POWDER

—SUPERMARKET PRIVATE BRANDS
FIRST NATIONAL EDWARDS BAKING SODA

FIRST NATIONAL FINAST BAKING SODA

STOP & SHOP BAKING SODA

—ITEMS IN HEALTH FOOD STORES
EL MOLINO MILLS ACTIVE DRY YEAST

EREWHON AGAR

RED STAR DRY YEAST

WALNUT ACRES BAKON YEAST

The following items contain additives of concern.

BRANDS	ADDITIVES-OF-CONCERN **BOLD—CONCERN TO ALL** LIGHT—CONCERN TO SOME
FLEISCHMANN'S ACTIVE DRY YEAST	**BHA**
ITEMS IN HEALTH FOOD STORES	
SHILOH FARMS NATURAL STABILIZER #2	GUAR [GUM]; LOCUST BEAN [GUM]

OILS & SHORTENINGS

CRISCO OIL; VEGETABLE SHORTENING	NO ADDITIVES OF CONCERN
FILIPPO BERIO OLIO D'OLIVA	NO ADDITIVES OF CONCERN
HOLLYWOOD SAFFLOWER OIL	NO ADDITIVES OF CONCERN
MAZOLA CORN OIL; NO STICK VEGETABLE SPRAY-ON	NO ADDITIVES OF CONCERN
PAM VEGETABLE COOKING SPRAY	NO ADDITIVES OF CONCERN
PLANTERS PEANUT OIL	NO ADDITIVES OF CONCERN

BRANDS	**ADDITIVES-OF-CONCERN** **BOLD—CONCERN TO ALL** LIGHT—CONCERN TO SOME
WESSON VEGETABLE OIL	NO ADDITIVES OF CONCERN
SUPERMARKET PRIVATE BRANDS	
A & P	
—A & P DEXOLA VEGETABLE OIL	**BHA; BHT**
—ANN PAGE OLIVE OIL; PURE CORN OIL	NO ADDITIVES OF CONCERN
FIRST NATIONAL	
—FINAST VEGETABLE OIL	NO ADDITIVES OF CONCERN
VEGETABLE SHORTENING	**BHA; BHT**
LUCKY	
—HARVEST DAY SHORTENING	**BHA; BHT; MEAT FATS & VEGETABLE OILS**
—LADY LEE ALL PURPOSE VEGETABLE OIL; CORN OIL; PAN COATING	NO ADDITIVES OF CONCERN
SAFEWAY	
—EMPRESS OLIVE OIL	NO ADDITIVES OF CONCERN
—NUMADE VEGETABLE OIL; VEGETABLE SHORTENING	NO ADDITIVES OF CONCERN

BRANDS	**ADDITIVES-OF-CONCERN** **BOLD—CONCERN TO ALL** LIGHT—CONCERN TO SOME
STOP & SHOP —STOP & SHOP	
CORN OIL; SALAD OIL	NO ADDITIVES OF CONCERN
WINN DIXIE —ASTOR	
SHORTENING	NO ADDITIVES OF CONCERN
ITEMS IN HEALTH FOOD STORES	
ARROWHEAD MILLS *OIL*	
BLENDED; PEANUT; SOYBEAN	NO ADDITIVES OF CONCERN
EREWHON *OIL*	
CORN; SAFFLOWER	NO ADDITIVES OF CONCERN
HAIN *OIL*	
ALL BLEND; CORN; OLIVE; SAFFLOWER; SUNFLOWER; WALNUT	NO ADDITIVES OF CONCERN
COCOANUT	**COCONUT OIL**

V. BAKING MIXES, DESSERT TOPPINGS, GELATIN, & PUDDING

BISCUIT, BREAD, & MUFFIN BAKING MIXES

BRANDS	**ADDITIVES-OF-CONCERN** **BOLD—CONCERN TO ALL** LIGHT—CONCERN TO SOME
AUNT JEMIMA CORN BREAD MIX	NO ADDITIVES OF CONCERN
BISQUICK BUTTERMILK BAKING MIX	**(BEEF FAT, LARD);* BHA; BHT**
BETTY CROCKER *MUFFIN MIX*	
CORN	**(BEEF TALLOW, LARD);* BHA; BHT**
WILD BLUEBERRY	**BHA; BHT; MODIFIED CORN STARCH**
CRUTCHFIELD'S BRAN MUFFIN MIX	**BHA; BHT**
POPOVER MIX	NO ADDITIVES OF CONCERN
DUNCAN HINES WILD BLUEBERRY MUFFIN MIX	NO ADDITIVES OF CONCERN

*Blend may contain one or more of these saturated fats.

BRANDS	**ADDITIVES-OF-CONCERN** **BOLD—CONCERN TO ALL** LIGHT—CONCERN TO SOME
FLAKO CORN MUFFIN MIX	NO ADDITIVES OF CONCERN
POPOVER MIX	**LARD**
JIFFY BISCUIT BAKING MIX	**BHA; BHT; CALCIUM CASEINATE; (LARD, TALLOW);*** WHEY
—MUFFIN MIX APPLE-CINNAMON; BLUEBERRY	**ARTIFICIAL COLOR; BHA; BHT; (LARD, TALLOW)***
BRAN W/ DATES	**BHA; BHT; (LARD, TALLOW)***
PILLSBURY HOT ROLL MIX	**(BEEF FAT, LARD; HYDROGENATED COTTONSEED, PALM, SOYBEAN OILS);* BHA; BHT;** WHEY
—QUICK BREAD MIX APRICOT NUT; BANANA	**(BEEF FAT, HYDROGENATED SOYBEAN OIL, LARD);* BHA; BHT**
BLUEBERRY NUT	**ARTIFICIAL COLOR; (BEEF FAT, HYDROGENATED SOYBEAN OIL, LARD);* BHA; BHT;** SODIUM ALGINATE

BRANDS	ADDITIVES-OF-CONCERN BOLD—CONCERN TO ALL LIGHT—CONCERN TO SOME
CHERRY NUT	**ARTIFICIAL COLOR; (BEEF FAT, HYDROGENATED SOYBEAN OIL, LARD);* BHA; BHT; MODIFIED FOOD STARCH**
DATE; NUT	**BHA; BHT**
SUPERMARKET PRIVATE BRANDS	
WINN DIXIE	
—DIXIE DARLING	
BRAN MUFFIN MIX	**BHA;** WHEY (DRIED SWEET DAIRY)
BUTTERMILK BISCUIT MIX	**(BEEF FAT);* BHA;** WHEY (DRIED SWEET DAIRY)
CORN BREAD MIX	**BHA**
ITEMS IN HEALTH FOOD STORES	
ARROWHEAD MILLS	
MIX	
MULTIGRAIN CORN BREAD; STONE GROUND WHOLE WHEAT BREAD; WHOLE GRAIN BISCUIT; WHOLE GRAIN BRAN MUFFIN	NO ADDITIVES OF CONCERN
JOLLY JOAN *MIX*	
BARLEY	NO ADDITIVES OF CONCERN

*Blend may contain one or more of these saturated fats and/or oils.

BRANDS	ADDITIVES-OF-CONCERN BOLD—CONCERN TO ALL LIGHT—CONCERN TO SOME
JOLLY JOAN *MIX (Continued)*	
LOW PROTEIN BREAD	**CARRAGEENAN; MODIFIED VEGETABLE GUM**
WALNUT ACRES	
CORNELL BREAD FLOUR MIX	NO ADDITIVES OF CONCERN

DESSERT BAKING MIXES (BROWNIES, CAKES, COOKIES, PIE CRUSTS, PIES, & SIMILAR ITEMS)

AUNT JEMIMA	
COFFEE CAKE EASY MIX	NO ADDITIVES OF CONCERN
BETTY CROCKER	
BOSTON CREAM PIE MIX	**ARTIFICIAL COLOR; (BEEF TALLOW, LARD);* BHA; BHT; CALCIUM CASEINATE;** GUAR GUM; SODIUM ALGINATE; WHEY
COCONUT MACAROON MIX	NO ADDITIVES OF CONCERN
GINGERBREAD MIX	**(BEEF TALLOW, LARD);* BHA; BHT**
PIE CRUST MIX; PIE CRUST STICKS	**(BEEF TALLOW, LARD);* BHA; BHT; MODIFIED CORN STARCH; SODIUM CASEINATE; TURMERIC EXTRACT**

BRANDS	**ADDITIVES-OF-CONCERN** **BOLD—CONCERN TO ALL** LIGHT—CONCERN TO SOME
—BIG BATCH COOKIE MIX	
DOUBLE CHOCOLATE; PEANUT BUTTER	NO ADDITIVES OF CONCERN
SUGAR	WHEY
—BROWNIE MIX	
CHOCOLATE CHIP BUTTERSCOTCH	**BHA; BHT**
FUDGE	**ARTIFICIAL COLOR; (BEEF TALLOW, LARD);* BHA; BHT**
WALNUT	**BHA; BHT;** EGG WHITES; **SODIUM CASEINATE**
—CAKE MIX	
ANGEL FOOD	EGG WHITES; **MODIFIED CORN STARCH**
CHOCOLATE PUDDING	**(BEEF TALLOW, LARD);* BHA; BHT**
GOLDEN POUND	**TURMERIC EXTRACT**
LEMON PUDDING	**ARTIFICIAL COLOR; (BEEF TALLOW, LARD);* BHA; BHT**
—SNACKIN' CAKE MIX	
APPLESAUCE RAISIN; GOLDEN CHOCOLATE CHIP	**BHA**

*Blend may contain one or more of these saturated fats.

BRANDS	ADDITIVES-OF-CONCERN BOLD—CONCERN TO ALL LIGHT—CONCERN TO SOME
BETTY CROCKER—*SNACKIN' CAKE MIX (Continued)*	
BANANA WALNUT; COCONUT PECAN	**ARTIFICIAL COLOR; BHA**
—STIR 'N FROST MIX	
CHOCOLATE CAKE/CHOCOLATE FROSTING	**BHA; BHT; SODIUM CASEINATE**
LEMON CAKE/LEMON FROSTING	**ARTIFICIAL COLOR; (BEEF TALLOW, LARD);* BHA; BHT;** EGG WHITES; **MODIFIED CORN STARCH**
YELLOW CAKE/CHOCOLATE FROSTING	**(BEEF TALLOW, LARD);* BHA; BHT; MODIFIED CORN STARCH; TURMERIC EXTRACT**
—SUPER MOIST CAKE MIX	
CHERRY CHIP	**ARTIFICIAL COLOR; (BEEF TALLOW, LARD);* BHA; BHT;** EGG WHITES; **MODIFIED CORN STARCH**
CHOCOLATE FUDGE; DEVIL'S FOOD	**(BEEF TALLOW; LARD);* BHA; BHT;** EGG WHITES; GUAR GUM; **MODIFIED CORN STARCH**
MARBLE; YELLOW	**(BEEF TALLOW, LARD);* BHA; BHT;** EGG WHITES; **MODIFIED CORN STARCH; TURMERIC EXTRACT**

BRANDS	ADDITIVES-OF-CONCERN **BOLD—CONCERN TO ALL** LIGHT—CONCERN TO SOME
WHITE	**(BEEF TALLOW, LARD);* BHA; BHT;** EGG WHITES; **MODIFIED CORN STARCH**
DROMEDARY GINGERBREAD MIX	**BHA**
POUND CAKE MIX	**(BEEF FAT, LARD);* BHA**
DUNCAN HINES ANGEL FOOD DELUXE CAKE MIX	EGG WHITES (DRIED); **MODIFIED FOOD STARCH**
DOUBLE FUDGE BROWNIE MIX	NO ADDITIVES OF CONCERN
MOIST & EASY SPICY APPLE RAISIN SNACK CAKE MIX	**ARTIFICIAL COLORING;** EGG WHITES (DRIED); **MODIFIED FOOD STARCH**
—DELUXE II CAKE MIX BANANA SUPREME; FUDGE MARBLE; LEMON SUPREME; YELLOW	**ARTIFICIAL COLORING; MODIFIED FOOD STARCH**
DEEP CHOCOLATE; DEVIL'S FOOD	**ARTIFICIAL COLORING; MODIFIED FOOD STARCH;** WHEY SOLIDS (DRIED)
STRAWBERRY SUPREME	**ARTIFICIAL COLORING; CONFECTIONER'S GLAZE;** WHEY SOLIDS (DRIED)
SWISS CHOCOLATE	**MODIFIED FOOD STARCH**

*Blend may contain one or more of these saturated fats.

BRANDS	ADDITIVES-OF-CONCERN BOLD—CONCERN TO ALL LIGHT—CONCERN TO SOME
DUNCAN HINES—*DELUXE II CAKE MIX (Continued)*	
WHITE	GUAR GUM; **MODIFIED FOOD STARCH;** WHEY SOLIDS (DRIED)
—PUDDING RECIPE CAKE MIX	
DEVIL'S FOOD	GUAR GUM; **MODIFIED FOOD STARCH;** WHEY SOLIDS (DRIED)
GERMAN CHOCOLATE	**MODIFIED FOOD STARCH**
LEMON; YELLOW	**ARTIFICIAL COLOR; MODIFIED FOOD STARCH**
JELL-O CHEESECAKE MIX	**ARTIFICIAL COLOR; BHA; MODIFIED TAPIOCA STARCH; SODIUM CASEINATE;** WHEY
JIFFY	
FUDGE BROWNIE MIX	**BHA; BHT; (TALLOW)***
PIE CRUST MIX	**BHA; BHT; (LARD, TALLOW)***
—CAKE MIX	
DEVIL'S FOOD; WHITE	**BHA; BHT; CALCIUM CASEINATE; (LARD, TALLOW);*** WHEY
GOLDEN YELLOW	**BHA; BHT; CALCIUM CASEINATE;** FD & C YELLOW NO. 5; **(LARD, TALLOW);*** WHEY

BRANDS	ADDITIVES-OF-CONCERN BOLD—CONCERN TO ALL LIGHT—CONCERN TO SOME
NESTLÉ *COOKIE MIX* CHOCOLATE CHIP; PEANUT BUTTER	NO ADDITIVES OF CONCERN
SUGAR	**(HYDROGENATED COTTONSEED, PALM, SOYBEAN OILS);*** WHEY
PILLSBURY DELUXE FUDGE BROWNIE MIX	**BEEF FAT; BHA**
PIE CRUST MIX	**ARTIFICIAL COLOR; (BEEF FAT, LARD);* BHA; BHT**
SOUR CREAM COFFEE CAKE MIX	**(BEEF FAT, HYDROGENATED SOYBEAN OIL, LARD);* BHA; BHT**
—BUNDT CAKE FILLING AND GLAZE MIX FUDGE NUT CROWN	**ARTIFICIAL COLOR; (BEEF FAT, LARD; HYDROGENATED COCONUT, COTTONSEED, PALM, SOYBEAN OILS);* BHA; BHT; MODIFIED TAPIOCA STARCH**
LEMON BLUEBERRY	**ARTIFICIAL COLOR; BHA; BHT; MODIFIED TAPIOCA STARCH**

*Blend may contain one or more of these saturated fats and/or oils.

BRANDS	ADDITIVES-OF-CONCERN BOLD—CONCERN TO ALL LIGHT—CONCERN TO SOME
PILLSBURY *(Continued)*	
—PLUS CAKE MIX	
DARK CHOCOLATE	**(BEEF FAT, LARD; HYDROGENATED COTTONSEED, PALM, SOYBEAN OILS);* BHA; BHT; MODIFIED TAPIOCA STARCH;** WHEY PROTEIN
FUDGE MARBLE; LEMON; YELLOW	**ARTIFICIAL COLOR; (BEEF FAT, LARD; HYDROGENATED COTTONSEED, PALM, SOYBEAN OILS);* BHA; BHT; MODIFIED TAPIOCA STARCH**
WHITE	**(BEEF FAT, LARD; HYDROGENATED COTTONSEED, PALM, SOYBEAN OILS);* BHA; BHT;** EGG WHITES; **MODIFIED TAPIOCA STARCH**
—STREUSEL SWIRL CAKE MIX	
CINNAMON; LEMON SUPREME	**ARTIFICIAL COLOR; (BEEF FAT, LARD; HYDROGENATED COCONUT, COTTONSEED, SOYBEAN OILS);* BHA; BHT; MODIFIED TAPIOCA STARCH;** WHEY PROTEIN

BRANDS	ADDITIVES-OF-CONCERN **BOLD—CONCERN TO ALL** LIGHT—CONCERN TO SOME
FUDGE MARBLE	**ARTIFICIAL COLOR; (BEEF FAT, LARD; HYDROGENATED COCONUT, COTTONSEED, PALM, SOYBEAN OILS);* BHA; BHT; MODIFIED TAPIOCA STARCH;** WHEY PROTEIN
ROYAL REAL CHEESE CAKE MIX	**ARTIFICIAL COLOR; MODIFIED FOOD STARCH; SODIUM CASEINATE;** WHEY
QUAKER *COOKIE MIX*	
CHOCOLATE CHIP	WHEY (DRIED)
OATMEAL	NO ADDITIVES OF CONCERN
SUPERMARKET PRIVATE BRANDS	
FIRST NATIONAL —EDWARDS-FINAST	
FUDGE BROWNIE MIX	**ARTIFICIAL COLOR; (BEEF FAT, LARD);* BHA; MODIFIED CORN STARCH**
—DELUXE CAKE MIX	
FUDGE MARBLE; LEMON; YELLOW	**ARTIFICIAL COLOR; (BEEF FAT, LARD);* BHA;** GUAR GUM; **MODIFIED CORN STARCH**
WHITE	**(BEEF FAT, LARD);* BHA;** GUAR GUM; **MODIFIED CORN STARCH;** WHEY
—FINAST FLAKY PIE CRUST MIX	**COCONUT OIL**

*Blend may contain one or more of these saturated fats and/or oils.

BRANDS	ADDITIVES-OF-CONCERN **BOLD—CONCERN TO ALL** LIGHT—CONCERN TO SOME
LUCKY —LADY LEE *CAKE MIX*	
DEVIL'S FOOD; WHITE	**CALCIUM CASEINATE;** GUAR GUM; **SODIUM CASEINATE;** WHEY (SWEET)
LEMON; YELLOW	**ARTIFICIAL COLORING; CALCIUM CASEINATE;** GUAR GUM; **MODIFIED WHEAT STARCH; SODIUM CASEINATE;** WHEY (SWEET)
SAFEWAY —MRS. WRIGHT'S *CAKE MIX*	
BUTTER RECIPE	**ARTIFICIAL COLOR; (BEEF FAT, LARD);* BHA; MODIFIED CORN STARCH**
PUDDING DELIGHT DEVIL'S FOOD	**(BEEF FAT, LARD);* BHA; MODIFIED FOOD STARCH**
—DELUXE CAKE MIX	
ANGEL FOOD	EGG WHITES
DEVIL'S FOOD	**BHA;** GUAR GUM; **MODIFIED CORN STARCH**
LEMON; YELLOW	**ARTIFICIAL COLOR; (BEEF FAT, LARD);* BHA;** GUAR GUM; **MODIFIED CORN STARCH**
WHITE	**(BEEF FAT, LARD);* BHA;** GUAR GUM; **MODIFIED CORN STARCH;** WHEY

BRANDS	**ADDITIVES-OF-CONCERN** BOLD—CONCERN TO ALL LIGHT—CONCERN TO SOME
STOP & SHOP —STOP & SHOP FUDGE BROWNIE MIX	**ARTIFICIAL COLOR; (BEEF FAT, LARD);* BHA; MODIFIED CORN STARCH**
—DELUXE CAKE MIX DEVIL'S FOOD	**BHA;** GUAR GUM; **MODIFIED CORN STARCH**
FUDGE MARBLE; LEMON	**ARTIFICIAL COLOR; (BEEF FAT, COCONUT OIL, LARD);* BHA; BHT;** GUAR GUM; **MODIFIED CORN STARCH**
WHITE	**(BEEF FAT, COCONUT OIL, LARD);* BHA; BHT;** GUAR GUM; **MODIFIED CORN STARCH;** WHEY
WINN DIXIE —DIXIE DARLING FUDGE BROWNIE MIX	**(BEEF FAT, COCONUT OIL, LARD);* BHA; BHT; COLOR; MODIFIED CORN STARCH**
—CAKE MIX DEVIL'S FOOD	GUAR GUM; **MODIFIED CORN STARCH**
LEMON; YELLOW	**ARTIFICIAL COLOR; (BEEF FAT, COCONUT OIL, LARD);* BHA; BHT;** GUAR GUM; **MODIFIED CORN STARCH**

*Blend may contain one or more of these saturated fats and/or oils.

BRANDS	ADDITIVES-OF-CONCERN **BOLD—CONCERN TO ALL** LIGHT—CONCERN TO SOME
ITEMS IN HEALTH FOOD STORES	
ARROWHEAD MILLS	
WHOLE GRAIN CAROB CAKE	NO ADDITIVES OF CONCERN
FEARN *CAKE MIX*	
CAROB	WHEY (SWEET DAIRY)
CARROT; SPICE	**NUTMEG;** WHEY (SWEET DAIRY)

DESSERT TOPPINGS (OTHER THAN WHIPPED CREAM OR WHIPPED-CREAM SUBSTITUTES)

BRANDS	ADDITIVES-OF-CONCERN
EVAN'S WALNUT DESSERT TOPPING	NO ADDITIVES OF CONCERN
JOHNSTON'S	
BUTTERSCOTCH-READY TOPPING	NO ADDITIVES OF CONCERN
KRAFT *TOPPING*	
BUTTERSCOTCH; STRAWBERRY; WALNUT	**ARTIFICIAL COLOR**
CHOCOLATE	WHEY
PINEAPPLE	NO ADDITIVES OF CONCERN
SMUCKER'S *TOPPING*	
BUTTERSCOTCH; PINEAPPLE	**ARTIFICIAL COLOR**
CHOCOLATE FUDGE	NO ADDITIVES OF CONCERN

BRANDS	ADDITIVES-OF-CONCERN **BOLD—CONCERN TO ALL** LIGHT—CONCERN TO SOME

FROSTINGS (CANNED & MIXES)

BETTY CROCKER *—FROSTING MIX* CHOCOLATE FUDGE	**BHA; BHT**
FLUFFY WHITE	EGG WHITES; **MODIFIED CORN STARCH**
—CREAMY DELUXE READY TO SPREAD FROSTING CHOCOLATE	**(BEEF TALLOW, LARD);* BHA; BHT**
SOUR CREAM WHITE; SUNKIST LEMON	**ARTIFICIAL COLOR; (BEEF TALLOW, LARD);* BHA; BHT**
JIFFY *FROSTING MIX* FUDGE	**BHA; BHT; (TALLOW);*** WHEY
WHITE	**BHA; BHT; CALCIUM CASEINATE; (TALLOW);*** WHEY
PILLSBURY *—READY TO SPREAD FROSTING SUPREME* CHOCOLATE FUDGE; MILK CHOCOLATE	**(BEEF FAT, HYDROGENATED COTTONSEED, PALM, SOYBEAN OILS);* BHA; BHT**; GUAR GUM

*Blend may contain one or more of these saturated fats and/or oils.

BRANDS	ADDITIVES-OF-CONCERN BOLD—CONCERN TO ALL LIGHT—CONCERN TO SOME
PILLSBURY *(Continued)*	
VANILLA	**ARTIFICIAL COLOR; (BEEF FAT, HYDROGENATED COTTONSEED, PALM, SOYBEAN OILS);* BHA; BHT;** GUAR GUM
SUPERMARKET PRIVATE BRANDS	
A & P	
—ANN PAGE	
FROSTING MIX	
CHOCOLATE FUDGE; MILK CHOCOLATE	NO ADDITIVES OF CONCERN
CREAMY WHITE	**ARTIFICIAL COLOR**
FIRST NATIONAL	
—EDWARDS	
FROSTING	
CREAMY FUDGE; CREAMY WHITE	**ARTIFICIAL COLOR; (BEEF FAT, COCONUT OIL, LARD);* BHA; MODIFIED FOOD STARCH**
LUCKY	
—LADY LEE	
CREAMY WHITE FROSTING MIX	**ARTIFICIAL COLORING**
SAFEWAY	
—MRS. WRIGHT'S	
FROSTING MIX	
CHOCOLATE FUDGE	**BHA**
WHITE	**ARTIFICIAL COLOR; (BEEF FAT, LARD);* BHA; MODIFIED CORN STARCH**

BRANDS	ADDITIVES-OF-CONCERN **BOLD—CONCERN TO ALL** LIGHT—CONCERN TO SOME
STOP & SHOP —STOP & SHOP *DELUXE FROSTING MIX*	
CHOCOLATE FUDGE	**BHA**
CREAMY WHITE	**ARTIFICIAL COLOR; (BEEF FAT, COCONUT OIL, LARD);* BHA; BHT**
—READY TO SPREAD FROSTING	
CHOCOLATE; VANILLA	**ARTIFICIAL COLOR; (BEEF FAT, COCONUT OIL, LARD);* BHA; MODIFIED FOOD STARCH**
WINN DIXIE —DIXIE DARLING *FROSTING MIX*	
FUDGE	**BHA; BHT**
WHITE	**ARTIFICIAL COLOR; BHA; BHT; MODIFIED CORN STARCH**

GELATIN & GELATIN DESSERTS

JELL-O *GELATIN DESSERT*	
APRICOT; CHERRY; PEACH; RASPBERRY; STRAWBERRY; STRAWBERRY/BANANA	**ARTIFICIAL COLOR**
LEMON; LIME; MIXED FRUIT; ORANGE	**ARTIFICIAL COLOR; BHA**

*Blend may contain one or more of these saturated fats and/or oils.

BRANDS	ADDITIVES-OF-CONCERN BOLD—CONCERN TO ALL LIGHT—CONCERN TO SOME
KNOX	
ORANGE DRINKING GELATINE	**ARTIFICIAL COLOR;** GUM ARABIC
UNFLAVORED GELATINE	NO ADDITIVES OF CONCERN
ROYAL *GELATIN DESSERT*	
CHERRY; LEMON; ORANGE; PEACH; RASPBERRY; STRAWBERRY	**ARTIFICIAL COLOR**
SUPERMARKET PRIVATE BRANDS	
A & P	
—ANN PAGE	
UNFLAVORED GELATIN	NO ADDITIVES OF CONCERN
—*GELATIN DESSERT*	
APRICOT; CHERRY; RED RASPBERRY	**ARTIFICIAL COLOR**
LEMON; LIME	**ARTIFICIAL COLOR; BHA**
LUCKY	
—LADY LEE *GELATIN DESSERT*	
LEMON; LIME; ORANGE; RASPBERRY; STRAWBERRY	**ARTIFICIAL COLOR** OR **COLORS**
SAFEWAY	
—JELL-WELL	
GELATIN DESSERT	
BLACK RASPBERRY;	**ARTIFICIAL COLOR**

(continues)

BRANDS	**ADDITIVES-OF-CONCERN** **BOLD—CONCERN TO ALL** LIGHT—CONCERN TO SOME
CHERRY; LEMON; MIXED-FRUIT; ORANGE; RASPBERRY; STRAWBERRY	**ARTIFICIAL COLOR**
STOP & SHOP —STOP & SHOP KITCHENS *GELATIN DESSERT*	
CHERRY APPLE; FRUIT SALAD; PINEAPPLE CARROT	**ARTIFICIAL COLOR**

PANCAKE & WAFFLE MIXES

AUNT JEMIMA *PANCAKE & WAFFLE MIX*	
BUCKWHEAT	NO ADDITIVES OF CONCERN
BUTTERMILK	**ARTIFICIAL COLORING;** BUTTERMILK SOLIDS (DRIED)
BUTTERMILK COMPLETE	**ARTIFICIAL COLORING;** BUTTERMILK SOLIDS (DRIED); **CARRAGEENAN;** WHEY (DRIED)
COMPLETE	**ARTIFICIAL COLORING;** WHEY (DRIED)
LOG CABIN	
COMPLETE PANCAKE & WAFFLE MIX	WHEY

BRANDS	ADDITIVES-OF-CONCERN BOLD—CONCERN TO ALL LIGHT—CONCERN TO SOME
PILLSBURY *HUNGRY JACK* COMPLETE PANCAKE MIX	EGG WHITES; **HYDROGENATED SOYBEAN OIL; SODIUM CASEINATE**
EXTRA LIGHTS PANCAKE & WAFFLE MIX	NO ADDITIVES OF CONCERN
SUPERMARKET PRIVATE BRANDS	
FIRST NATIONAL —EDWARDS-FINAST *PANCAKE & WAFFLE MIX*	
BUTTERMILK	**MODIFIED CORN STARCH**
COMPLETE	**BHA;** EGG WHITES; WHEY
OLD FASHIONED	**ARTIFICIAL COLOR**
STOP & SHOP —STOP & SHOP *PANCAKE & WAFFLE MIX*	
COMPLETE	**BHA; BHT;** EGG WHITES; WHEY
OLD FASHIONED	**ARTIFICIAL COLOR**
ITEMS IN HEALTH FOOD STORES	
ARROWHEAD MILLS	
MULTIGRAIN PANCAKE & WAFFLE MIX	NO ADDITIVES OF CONCERN

BRANDS	ADDITIVES-OF-CONCERN **BOLD—CONCERN TO ALL** LIGHT—CONCERN TO SOME
FEARN SOY/O *PANCAKE MIX* BUCKWHEAT; WHOLE WHEAT	NO ADDITIVES OF CONCERN
JOLLY JOAN WHEAT & SOY PANCAKE & WAFFLE MIX	BUTTERMILK SOLIDS (DRY); MILK SOLIDS (NONFAT DRY); WHEY (SWEET DAIRY)
KRUSTEAZ WHOLE WHEAT 'N HONEY COMPLETE PANCAKE MIX	**CALCIUM CASEINATE; CALCIUM STEARATE; CARRAGEENAN;** EGG WHITES; **SODIUM CASEINATE;** WHEY (SWEET)
SHILOH FARMS WHOLE WHEAT PANCAKE MIX	NO ADDITIVES OF CONCERN

PIE & PUDDING FILLINGS (CANNED & MIXES)

BETTY CROCKER *PUDDING* CHOCOLATE; TAPIOCA	**ARTIFICIAL COLOR; MODIFIED TAPIOCA STARCH**
RICE	**ARTIFICIAL COLOR; MODIFIED CORN STARCH**
BORDEN NONE SUCH MINCE MEAT	**MODIFIED FOOD STARCH**
COMSTOCK APPLE PIE FILLING	**MODIFIED FOOD STARCH; TURMERIC**

BRANDS	ADDITIVES-OF-CONCERN **BOLD—CONCERN TO ALL** LIGHT—CONCERN TO SOME
DEL MONTE ***PUDDING CUP***	
CHOCOLATE; VANILLA	**ARTIFICIAL COLOR;** FD & C YELLOW NO. 5; **MODIFIED FOOD STARCH**
HUNT'S ***SNACK PACK PUDDING***	
BUTTERSCOTCH; TAPIOCA; VANILLA	**ARTIFICIAL COLORS; MODIFIED FOOD STARCH**
CHOCOLATE; CHOCOLATE FUDGE	**MODIFIED FOOD STARCH**
JELL-O	
—AMERICANA	
CHOCOLATE TAPIOCA PUDDING	NO ADDITIVES OF CONCERN
GOLDEN EGG CUSTARD	**ARTIFICIAL COLOR; BHA; CALCIUM CARRAGEENAN;** WHEY
RICE PUDDING	**ARTIFICIAL COLOR; CALCIUM CARRAGEENAN**
—INSTANT PUDDING & PIE FILLING	
BUTTERSCOTCH; VANILLA	**ARTIFICIAL COLOR; BHA; MODIFIED CORN & TAPIOCA STARCHES**
CHOCOLATE; CHOCOLATE FUDGE	**BHA; MODIFIED CORN & TAPIOCA STARCHES**
—PUDDING & PIE FILLING	

(continues)

BRANDS	**ADDITIVES-OF-CONCERN** **BOLD—CONCERN TO ALL** LIGHT—CONCERN TO SOME
BANANA CREAM; BUTTERSCOTCH; COCONUT CREAM; VANILLA	**ARTIFICIAL COLOR; CALCIUM CARRAGEENAN; MODIFIED CORN STARCH**
CHOCOLATE; CHOCOLATE FUDGE	**CALCIUM CARRAGEENAN; MODIFIED CORN STARCH**
LEMON	**ARTIFICIAL COLOR; BHA; MODIFIED TAPIOCA STARCH**
JUNKET	
DANISH DESSERT PUDDING-PIE FILLING RASPBERRY-CURRANT; RENNET CUSTARD RASPBERRY	**ARTIFICIAL COLOR** OR **COLORS**
RENNET CUSTARD VANILLA	NO ADDITIVES OF CONCERN
MINUTE TAPIOCA	NO ADDITIVES OF CONCERN
MY-T-FINE *PUDDING & PIE FILLING*	
CHOCOLATE; VANILLA	**ARTIFICIAL COLOR; CALCIUM CARRAGEENAN; MODIFIED CORN STARCH**
LEMON	**ARTIFICIAL COLOR; MODIFIED TAPIOCA STARCH**
ROYAL	
TAPIOCA VANILLA PUDDING	**ARTIFICIAL COLOR**

BRANDS	ADDITIVES-OF-CONCERN **BOLD—CONCERN TO ALL** LIGHT—CONCERN TO SOME
ROYAL *(Continued)*	
—INSTANT PUDDING & PIE FILLING	
BUTTERSCOTCH	**ARTIFICIAL COLOR; MODIFIED FOOD STARCH;** SODIUM ALGINATE
CHOCOLATE; DARK 'N' SWEET CHOCOLATE	**ARTIFICIAL COLOR; MODIFIED FOOD STARCH**
—PIE FILLING	
KEY LIME; LEMON	**ARTIFICIAL COLOR; CALCIUM CARRAGEENAN; MODIFIED TAPIOCA STARCH**
—PUDDING & PIE FILLING	
BUTTERSCOTCH; CHOCOLATE; VANILLA	**ARTIFICIAL COLOR; CALCIUM CARRAGEENAN**
THANK YOU *PUDDING*	
BUTTERSCOTCH; TAPIOCA	**ARTIFICIAL COLORING** W/ FD & C YELLOW NO. 5; **MODIFIED FOOD STARCH**
CHOCOLATE FUDGE	**MODIFIED FOOD STARCH**
WYMAN'S *PIE FILLING*	
APPLE; WILD BLUEBERRY	NO ADDITIVES OF CONCERN
STRAWBERRY	**ARTIFICIAL COLORING**
SUPERMARKET PRIVATE BRANDS	
A & P	
—ANN PAGE	
EGG CUSTARD MIX	**ARTIFICIAL COLORS; CALCIUM CARRAGEENAN;** MILK SOLIDS (NONFAT DRY)

BRANDS	ADDITIVES-OF-CONCERN **BOLD—CONCERN TO ALL** LIGHT—CONCERN TO SOME
—INSTANT PUDDING BUTTERSCOTCH; CHOCOLATE; TOASTED COCONUT; VANILLA	**ARTIFICIAL COLOR; MODIFIED STARCH (PRECOOKED)**
—PUDDING & PIE FILLING BANANA CREAM; BUTTERSCOTCH; CHOCOLATE; LEMON; VANILLA	**ARTIFICIAL COLOR**
COCONUT	**ARTIFICIAL COLOR; CALCIUM CARRAGEENAN**
—TAPIOCA PUDDING CHOCOLATE; VANILLA	**ARTIFICIAL COLOR**
SAFEWAY **—JELL-WELL** *INSTANT PUDDING & PIE FILLING* CHOCOLATE	**BHA;** MILK SOLIDS (NONFAT); **MODIFIED CORN & TAPIOCA STARCHES**
LEMON; PISTACHIO; VANILLA	**ARTIFICIAL COLOR; BHA;** MILK SOLIDS (NONFAT); **MODIFIED CORN & TAPIOCA STARCHES**

BRANDS	ADDITIVES-OF-CONCERN **BOLD—CONCERN TO ALL** LIGHT—CONCERN TO SOME
—TOWN HOUSE *PUDDING* BUTTERSCOTCH; RICE; TAPIOCA; VANILLA	**ARTIFICIAL COLOR; MODIFIED FOOD STARCH**
CHOCOLATE	**MODIFIED FOOD STARCH**
WINN DIXIE **—THRIFTY MAID** ***PUDDING SNACK*** BUTTERSCOTCH; VANILLA	**ARTIFICIAL COLOR; MODIFIED FOOD STARCH**
CHOCOLATE; CHOCOLATE FUDGE	**MODIFIED FOOD STARCH**

VI. BEVERAGE MIXES & BEVERAGES

CARBONATED BEVERAGES

Sugars like sucrose (cane and beet sugar), corn sweeteners (corn syrup, dextrose, fructose, etc.), and honey have not been treated in this volume as additives of concern in the amounts they are used as additives in many foods. However, when they comprise a substantial percentage of a food, a caution is warranted and has been noted for candy and gum, ready-to-eat cereals, carbonated beverages, and jellies and other sweet spreads. A Department of Agriculture study (Nutrient Composition Laboratory, USDA, June 10, 1980, news release) revealed a sugar content of 1 to 1½ ounces per 12-ounce can of some sodas. In almost all sodas the most abundant ingredient next to water was sugar, except for plain club soda, which did not contain sugar.

BRANDS	**ADDITIVES-OF-CONCERN** BOLD—CONCERN TO ALL LIGHT—CONCERN TO SOME
BARRELHEAD ROOT BEER	ACACIA GUM
CANADA DRY CLUB SODA; COLLINS MIXER; GINGER ALE	NO ADDITIVES OF CONCERN
HALF & HALF	**ESTER GUM**
JAMAICA COLA	ACACIA GUM; CAFFEINE
ORANGE SODA	**ARTIFICIAL COLORING; ESTER GUM**
TONIC WATER	QUININE
COCA-COLA	CAFFEINE

BRANDS	ADDITIVES-OF-CONCERN BOLD—CONCERN TO ALL LIGHT—CONCERN TO SOME
COTT COLA	ACACIA GUM; CAFFEINE
CREAM SODA; DRY GINGER ALE; SPARKLING WATER	NO ADDITIVES OF CONCERN
ENERGADE; HALF & HALF	ACACIA GUM; **BROMINATED VEGETABLE OIL; GLYCERYL ABIETATE**
ORANGE SODA; FRUIT PUNCH SODA	ACACIA GUM; **ARTIFICIAL COLOR; BROMINATED VEGETABLE OIL; GLYCERYL ABIETATE**
TONIC WATER	EXTRACT OF QUININE
CRUSH ORANGE SODA	ACACIA GUM; **ARTIFICIAL COLOR; BROMINATED VEGETABLE OIL; GLYCEROL ESTER OF WOOD ROSIN**
DR PEPPER	CAFFEINE
HIRES ROOT BEER	NO ADDITIVES OF CONCERN
KEEFERS CLUB SODA BUBBLY WATER; GINGER ALE	NO ADDITIVES OF CONCERN
TONIC MIXER QUININE WATER	QUININE SALTS
MELLO YELLO CITRUS SODA	ACACIA GUM; **ARTIFICIAL COLOR; BROMINATED VEGETABLE OIL;** CAFFEINE; LOCUST BEAN GUM

BRANDS	ADDITIVES-OF-CONCERN **BOLD—CONCERN TO ALL** LIGHT—CONCERN TO SOME
MOUNTAIN DEW	**BROMINATED VEGETABLE OIL;** CAFFEINE; FD & C YELLOW NO. 5; GUM ARABIC
PEPSI-COLA	CAFFEINE
SCHWEPPES	
BITTER LEMON	QUININE
CLUB SODA; GINGER ALE; ROOT BEER	NO ADDITIVES OF CONCERN
ORANGE	**ARTIFICIAL COLOR**
TONIC WATER	QUININE
7-UP	NO ADDITIVES OF CONCERN
SUNKIST ORANGE SODA	ACACIA GUM; **ARTIFICIAL COLOR; BHA;** CAFFEINE; **GLYCERYL ABIETATE**
WELCH'S GRAPE SODA	**ARTIFICIAL COLORS**
SUPERMARKET PRIVATE BRANDS	
A & P	
—YUKON	
COLA	CAFFEINE; GUM ARABIC
GINGER ALE	NO ADDITIVES OF CONCERN
ORANGE SODA	**ARTIFICIAL COLOR; BROMINATED VEGETABLE OIL; GLYCERYL ABIETATE;** GUM ARABIC
LUCKY	
—LADY LEE	
BLACK CHERRY SODA	ACACIA GUM; **ARTIFICIAL COLOR**

BRANDS	ADDITIVES-OF-CONCERN BOLD—CONCERN TO ALL LIGHT—CONCERN TO SOME
—LADY LEE *(Continued)*	
COLA	CAFFEINE
CREME SODA; GINGER ALE	NO ADDITIVES OF CONCERN
ORANGE SODA	ACACIA GUM; **ARTIFICIAL COLOR; BROMINATED VEGETABLE OIL; GLYCERYL ABIETATE**
ROOT BEER	**MODIFIED FOOD STARCH**
PUBLIX	
—PIX	
BLACK CHERRY SODA; STRAWBERRY SODA	**ARTIFICIAL COLOR**
COLA	ACACIA [GUM]; CAFFEINE
CREAM SODA; GINGER ALE; LEMON-HI SODA	NO ADDITIVES OF CONCERN
GRAPE SODA	ACACIA [GUM]; **ARTIFICIAL COLOR**
ORANGE SODA	ACACIA [GUM]; **ARTIFICIAL COLOR; BROMINATED VEGETABLE OIL; GLYCEROL ESTER OF WOOD ROSIN**
QUININE TONIC WATER	QUININE

BRANDS	ADDITIVES-OF-CONCERN **BOLD—CONCERN TO ALL** LIGHT—CONCERN TO SOME
SAFEWAY —CRAGMONT	
BLACK CHERRY SODA; CHERRY COLA; GRAPE SODA; STRAWBERRY SODA	**ARTIFICIAL COLOR** OR **COLORING;** GUM ARABIC
COLLINS MIX	**BROMINATED VEGETABLE OIL; GLYCERYL ABIETATE;** GUM ARABIC
CREAM SODA; GINGER ALE; LEMON LIME SODA	NO ADDITIVES OF CONCERN
ORANGE SODA; SPARKLING PUNCH	**ARTIFICIAL COLOR; BROMINATED VEGETABLE OIL; GLYCERYL ABIETATE;** GUM ARABIC
TONIC MIX	QUININE HYDROCHLORIDE
STOP & SHOP —SUN GLORY	
COLA	ACACIA GUM; CAFFEINE
CREME SODA; GINGER ALE; SQUEEZ O' LEMON SODA	NO ADDITIVES OF CONCERN
LEMON & LIME SODA; ORANGE SODA	ACACIA GUM; **ARTIFICIAL COLOR; BROMINATED VEGETABLE OIL; GLYCERYL ABIETATE**
ROOT BEER	ACACIA GUM
WINN DIXIE —CHEK COLA	CAFFEINE

BRANDS	ADDITIVES-OF-CONCERN **BOLD—CONCERN TO ALL** LIGHT—CONCERN TO SOME
WINN DIXIE—CHEK *(Continued)*	
CREME SODA; GINGER ALE; LEMON LIME SODA; ROOT BEER	NO ADDITIVES OF CONCERN
GRAPE SODA	ACACIA GUM; **ARTIFICIAL COLOR**
ORANGE SODA	**ARTIFICIAL COLOR; BROMINATED VEGETABLE OIL; GLYCEROL ESTER OF WOOD ROSIN**
ITEMS IN HEALTH FOOD STORES	
CORR'S GINSENG RUSH	NO ADDITIVES OF CONCERN
DR. TIMA KOLA	NO ADDITIVES OF CONCERN
HANSEN'S *SODA*	
GRAPEFRUIT; LEMON LIME	NO ADDITIVES OF CONCERN
HEALTH VALLEY	
HONEY PINEAPPLE SODA; IMITATION CREME COLA	NO ADDITIVES OF CONCERN
HONEY PURE	
ENGLISH GINGER; MOUNTAIN ROOT BEER	NO ADDITIVES OF CONCERN
NECTAREL	
LIME ORANGE; ROOT BEER	NO ADDITIVES OF CONCERN
ORIGINAL	**TURMERIC**
SOHO BLACK CHERRY SODA	NO ADDITIVES OF CONCERN

BRANDS	ADDITIVES-OF-CONCERN **BOLD—CONCERN TO ALL** LIGHT—CONCERN TO SOME

CHOCOLATE DRINKS, INSTANT BREAKFAST MIXES, LIQUID MEALS, MILK MODIFIERS (COCOA MIXES, SYRUPS)

BRANDS	ADDITIVES-OF-CONCERN
BOSCO CHOCOLATE FLAVORED SYRUP	ARTIFICIAL COLOR
CARNATION	
—INSTANT BREAKFAST	
EGGNOG	AMMONIUM CARRAGEENAN; ARTIFICIAL COLOR; FERRIC ORTHOPHOSPHATE; SODIUM CASEINATE
CHOCOLATE; CHOCOLATE MALT	AMMONIUM CARRAGEENAN; FERRIC ORTHOPHOSPHATE; SODIUM CASEINATE
COFFEE; VANILLA	AMMONIUM CARRAGEENAN; FERRIC ORTHOPHOSPHATE
—INSTANT HOT COCOA MIX	
COCO SUPREME; RICH CHOCOLATE	SODIUM CASEINATE
W/ MINI-MARSHMALLOWS	ARTIFICIAL COLOR; SODIUM CASEINATE
HERSHEY'S	
HOT COCOA MIX	ARTIFICIAL COLORING; SODIUM CASEINATE; STABILIZERS

BRANDS	ADDITIVES-OF-CONCERN BOLD—CONCERN TO ALL LIGHT—CONCERN TO SOME
HERSHEY'S *(Continued)*	
INSTANT; SYRUP CHOCOLATE	NO ADDITIVES OF CONCERN
MILK MATE	
CHOCOLATE SYRUP	**ARTIFICIAL COLOR; MODIFIED FOOD STARCH**
NESTLÉ	
QUIK CHOCOLATE	NO ADDITIVES OF CONCERN
QUIK STRAWBERRY	**ARTIFICIAL COLOR**
HOT COCOA MIX	WHEY
HOT COCOA MIX W/ MINI-MARSHMALLOWS	**ARTIFICIAL COLOR;** WHEY
NUTRAMENT	
DUTCH CHOCOLATE	**ARTIFICIAL COLOR; CALCIUM CASEINATE; CARRAGEENAN; SODIUM CASEINATE**
VANILLA	**CALCIUM CASEINATE; CARRAGEENAN; SODIUM CASEINATE**
OVALTINE	
CHOCOLATE; MALT	**FERRIC SODIUM PYROPHOSPHATE;** WHEY
HOT COCOA MIX	**ARTIFICIAL COLOR; CARRAGEENAN; (COCONUT OIL);* FERRIC SODIUM PYROPHOSPHATE; SODIUM CASEINATE;** WHEY

*Blend may contain this saturated oil.

BRANDS	**ADDITIVES-OF-CONCERN** **BOLD—CONCERN TO ALL** LIGHT—CONCERN TO SOME
PARTY TIME CHOCOMITE	**ARTIFICIAL COLOR; CARRAGEENAN; COCONUT FAT [OIL];** GUAR GUM; **SODIUM CASEINATE**
PDQ CHOCOLATE MILK FLAVORING	**ARTIFICIAL COLOR**
QWIP CHOCOLATE SHAKE	**CARRAGEENAN; COCONUT OIL;** WHEY
SWISS MISS *HOT COCOA MIX* LITE; MILK CHOCOLATE; W/ MINI-MARSHMALLOWS	NO ADDITIVES OF CONCERN
SUPERMARKET PRIVATE BRANDS	
FIRST NATIONAL —FINAST INSTANT CHOCOLATE FLAVORED MIX	WHEY SOLIDS
PUBLIX —DAIRI-FRESH CHOCOLATE DRINK	**CARRAGEENAN;** GUAR GUM; **SODIUM CASEINATE;** WHEY
SAFEWAY —LUCERNE *INSTANT BREAKFAST* CHOCOLATE MALT; COFFEE; VANILLA	**CARRAGEENAN**
CHOCOLATE FUDGE; EGGNOG; STRAWBERRY	**ARTIFICIAL COLOR; CARRAGEENAN**

BRANDS	ADDITIVES-OF-CONCERN BOLD—CONCERN TO ALL LIGHT—CONCERN TO SOME
—LUCERNE *INSTANT HOT COCOA MIX* (REGULAR); W/ MINI-MARSHMALLOWS	**ARTIFICIAL COLOR; SODIUM CASEINATE**
STOP & SHOP	
—STOP & SHOP HOT COCOA MIX	**ARTIFICIAL COLOR; SODIUM CASEINATE**
INSTANT CHOCOLATE FLAVORED MIX	NO ADDITIVES OF CONCERN
—FLAVORED SYRUP	
COFFEE	NO ADDITIVES OF CONCERN
FRUIT PUNCH; GRAPE; ORANGE	**U.S. CERTIFIED FOOD COLORS**
WINN DIXIE	
—SUPERBRAND CHOCO-RIFFIC DRINK	**ARTIFICIAL COLORS; CARRAGEENAN;** GUAR GUM; **SODIUM CASEINATE;** WHEY
ITEMS IN HEALTH FOOD STORES	
CHRISTOPHER'S	
MALTED MILK	
CAROB	MILK PROTEIN; SODIUM ALGINATE
VANILLA	ALGIN
HORLICKS INSTANT MALTED MILK	NO ADDITIVES OF CONCERN

COFFEE BEVERAGES & COFFEE SUBSTITUTES, DECAFFEINATED COFFEE, GROUND & INSTANT COFFEE

Most of the items in this section did not contain any additives of concern, and these appear first, without an additives-of-concern column.

The items that do contain additives of concern follow, in the usual style.

Regular and instant coffees—and some coffee substitutes—contain a significant amount of caffeine as a natural component, not as an additive. Whether present naturally or added, caffeine is believed to be of concern to the health of some individuals.

BRIM DECAFFEINATED; FREEZE-DRIED DECAFFEINATED
CHOCK FULL O' NUTS ALL-METHOD GRIND; INSTANT
FOLGER'S FLAKED; REGULAR
FRENCH MARKET COFFEE & CHICORY
HILLS BROS. REGULAR
KAVA INSTANT
MAXIM FREEZE-DRIED
MAX-PAX REGULAR
MAXWELL HOUSE INSTANT; REGULAR
MEDAGLIA D'ORO CAFFÉ ESPRESSO
MELLOW ROAST INSTANT; REGULAR
NESCAFÉ DECAFFEINATED INSTANT; INSTANT
PARADE DELUXE BLEND REGULAR; INSTANT
POSTUM ARTIFICIAL COFFEE FLAVOR; REGULAR
SANKA FREEZE-DRIED INSTANT; INSTANT; REGULAR
SUNRISE INSTANT COFFEE MELLOWED W/ CHICORY
TASTER'S CHOICE FREEZE-DRIED; FREEZE-DRIED DECAFFEINATED
YUBAN INSTANT; REGULAR/DRIP

SUPERMARKET PRIVATE BRANDS

FIRST NATIONAL FINAST DRIP GRIND; ELECTRIC PERK; FREEZE DRIED; INSTANT; REGULAR GRIND
LUCKY LADY LEE DRIP-FILTER GRIND; ELECTRIC PERK; FREEZE-DRIED; REGULAR GRIND
PUBLIX BREAKFAST CLUB 100% COLOMBIAN ALL PURPOSE GRIND

SAFEWAY ALL-PURPOSE GRIND; INSTANT
SAFEWAY EDWARDS DRIP OR FINE GRIND; FOR ELECTRIC PERCOLATORS; REGULAR GRIND; FLAKE GRIND; WHOLE-BEAN
SAFEWAY SCOTCH BUY INSTANT
STOP & SHOP DECAFFEINATED ALL METHOD GRIND; DRIP GRIND; ELECTRIC PERCOLATOR GRIND; INSTANT PRESIDENT'S BLEND; NEW; REGULAR GRIND
STOP & SHOP SUN GLORY ALL METHOD GRIND
WINN DIXIE ASTOR INSTANT

ITEMS IN HEALTH FOOD STORES
BAMBU SWISS COFFEE-SUBSTITUTE
CAFIX NATURAL CEREAL BEVERAGE
PERO INSTANT CEREAL BEVERAGE
PIONIER INSTANT SWISS COFFEE SUBSTITUTE

The following items contain additives of concern.

BRANDS	**ADDITIVES-OF-CONCERN** BOLD—CONCERN TO ALL LIGHT—CONCERN TO SOME
GENERAL FOODS *—INTERNATIONAL COFFEES*	
CAFÉ FRANÇAIS; CAFÉ VIENNA; ORANGE CAPPUCCINO; SUISSE MOCHA	**COCONUT OIL; SODIUM CASEINATE SOLIDS**
IRISH MOCHA MINT	**CARRAGEENAN; COCONUT OIL; SODIUM CASEINATE SOLIDS**
HILLS BROS.	
BAVARIAN MINT; CAFE MOCHA	**CARRAGEENAN; SODIUM CASEINATE**
ITEMS IN HEALTH FOOD STORES	
SYMINGTONS	
HON-Y-CUP	WHEY

BRANDS	ADDITIVES-OF-CONCERN **BOLD—CONCERN TO ALL** LIGHT—CONCERN TO SOME

FRUIT DRINK MIXES & CONCENTRATES

COUNTRY TIME	
LEMONADE; PINK LEMONADE	**ARTIFICIAL COLOR; BHA; MODIFIED CORN & TAPIOCA STARCHES**
HAWAIIAN PUNCH	
CHERRY; GRAPE; RED; STRAWBERRY	**ARTIFICIAL COLOR**
LEMONADE	**ARTIFICIAL COLOR; BHA; BHT**
HI-C	
FRUIT PUNCH; LEMONADE; ORANGE	**ARTIFICIAL COLOR; MODIFIED CORN STARCH**
KOOL-AID	
UNSWEETENED TROPICAL PUNCH	**ARTIFICIAL COLOR; BHA; MODIFIED CORN & TAPIOCA STARCHES**
—SUGAR SWEETENED	
CHERRY; ORANGE; STRAWBERRY	**ARTIFICIAL COLOR; BHA**
GRAPE	**ARTIFICIAL COLOR**
LEMONADE	**ARTIFICIAL COLOR; BHA; MODIFIED CORN & TAPIOCA STARCHES**
MINUTE MAID	
LEMONADE CRYSTALS	NO ADDITIVES OF CONCERN
TANG ORANGE	**ARTIFICIAL COLOR; BHA**

BRANDS	ADDITIVES-OF-CONCERN BOLD—CONCERN TO ALL LIGHT—CONCERN TO SOME
WELCH'S GRAPE	**U.S. CERTIFIED ARTIFICIAL COLORS**
WYLER'S CHERRY; GRAPE; PINK LEMONADE	**ARTIFICIAL COLOR** OR **U.S. CERTIFIED ARTIFICIAL COLORS**
LEMONADE	**ARTIFICIAL COLOR; CARRAGEENAN**
SUPERMARKET PRIVATE BRANDS **A & P** **—ANN PAGE** LEMONADE SUPREME	**ARTIFICIAL COLOR**
TROPICAL FRUIT PUNCH W/ SUGAR	ACACIA GUM; **ARTIFICIAL COLOR; COCONUT OIL**
—CHEERI AID CHERRY; GRAPE	**ARTIFICIAL COLOR; DIOCYTL SODIUM SULFOSUCCINATE (DSS)**
IMITATION LEMONADE; ORANGE	**ARTIFICIAL COLOR; BHA; DIOCYTL SODIUM SULFOSUCCINATE (DSS)**
STRAWBERRY	**ARTIFICIAL COLOR**
—CHEERI AID COMPLETE W/ SUGAR IMITATION LEMONADE; RASPBERRY; STRAWBERRY	**ARTIFICIAL COLOR**

BRANDS	**ADDITIVES-OF-CONCERN** **BOLD—CONCERN TO ALL** LIGHT—CONCERN TO SOME
TROPICAL FRUIT PUNCH	ACACIA GUM; **ARTIFICIAL COLOR; COCONUT OIL**
FIRST NATIONAL	
—EDWARDS	
INSTANT BREAKFAST DRINK ORANGE	ARTIFICIAL COLOR FD & C YELLOW NO. 5 & NO. 6
LUCKY	
—LADY LEE	
CHERRY; GRAPE; ORANGE; PUNCH	**ARTIFICIAL FOOD COLOR**
LEMONADE	**ARTIFICIAL FOOD COLOR; VEGETABLE GUM**
PUBLIX	
—FLAVOR PERFECT	
CHERRY; GRAPE	**ARTIFICIAL COLOR; BHA;** GUM ARABIC
FRUIT PUNCH	**ARTIFICIAL COLOR; BHA**
IMITATION LEMONADE	**ARTIFICIAL COLOR; BHA; CARRAGEENAN;** GUM ARABIC
SAFEWAY	
—CRAGMONT	
CHERRY; GRAPE; STRAWBERRY	**ARTIFICIAL COLOR**
LEMONADE	**ARTIFICIAL COLOR; BHA**
ORANGE	**BHA**
—TOWN HOUSE	
INSTANT BREAKFAST DRINK ORANGE	**ARTIFICIAL COLORS** W/ FD & C YELLOW NO. 5; **MODIFIED FOOD STARCH**

BRANDS	**ADDITIVES-OF-CONCERN** **BOLD—CONCERN TO ALL** LIGHT—CONCERN TO SOME
STOP & SHOP **—STOP & SHOP**	
CHERRY; GRAPE; STRAWBERRY	**U.S. CERTIFIED ARTIFICIAL COLOR**
FRUIT PUNCH; IMITATION PINK LEMONADE	**MODIFIED FOOD STARCH; U.S. CERTIFIED ARTIFICIAL COLOR; VEGETABLE GUMS**
IMITATION LEMONADE; ORANGEADE	**MODIFIED FOOD STARCH; U.S. CERTIFIED ARTIFICIAL COLOR**
ITEMS IN HEALTH FOOD STORES	
HAIN *CONCENTRATE*	
APRICOT; BLACK CHERRY; RED RASPBERRY; STRAWBERRY	NO ADDITIVES OF CONCERN

INSTANT ICED TEA & TEA MIXES

FANTA ICED TEA LEMON FLAVORED*	NO ADDITIVES OF CONCERN
LIPTON	
ICED TEA LEMON FLAVORED*	**ARTIFICIAL COLOR**
ICED TEA MIX LEMON FLAVOR & SUGAR; INSTANT LEMON FLAVORED	NO ADDITIVES OF CONCERN
NESTEA	
ICED TEA MIX SUGAR & LEMON FLAVORED	NO ADDITIVES OF CONCERN

BRANDS	ADDITIVES-OF-CONCERN **BOLD—CONCERN TO ALL** LIGHT—CONCERN TO SOME
ICED TEA SUGAR & LEMON FLAVOR*	**MODIFIED FOOD STARCH**
INSTANT TEA MIX LEMON FLAVOR	GUM ARABIC
TETLEY ICED TEA MIX SUGAR & LEMON FLAVOR	**BHA; CERTIFIED FOOD COLORING**
SUPERMARKET PRIVATE BRANDS **A & P** **—OUR OWN** ICED TEA MIX LEMON FLAVORED W/ SUGAR	**ARTIFICIAL COLOR; BHA**
FIRST NATIONAL **—FINAST** ICED TEA MIX W/ SUGAR & LEMON FLAVOR	**ARTIFICIAL FOOD COLOR**
INSTANT TEA	NO ADDITIVES OF CONCERN
LUCKY **—LADY LEE** ICED TEA MIX SUGAR & LEMON FLAVOR	**ARTIFICIAL FOOD COLOR**
PUBLIX **—FLAVOR PERFECT** ICED TEA MIX SUGAR & LEMON FLAVOR	**BHA; CERTIFIED FOOD COLORING**
SAFEWAY **—CROWN COLONY** ICED TEA MIX LEMON FLAVORED W/ SUGAR	**ARTIFICIAL FOOD COLOR**

*Noncarbonated, in can.

BRANDS	ADDITIVES-OF-CONCERN BOLD—CONCERN TO ALL LIGHT—CONCERN TO SOME
STOP & SHOP **—STOP & SHOP** INSTANT ICED TEA MIX	**BHA**
WINN DIXIE **—ASTOR** ICED TEA MIX W/ SUGAR & LEMON FLAVOR	**ARTIFICIAL FOOD COLOR**

VII. BREAD & BREAD PRODUCTS; CRACKERS

R alongside a brand indicates regional (not national) distribution, based on shopping experience.

BAGELS, BUNS, MUFFINS, ROLLS

	BRANDS	ADDITIVES-OF-CONCERN **BOLD—CONCERN TO ALL** LIGHT—CONCERN TO SOME
	ARNOLD BRAN'NOLA MUFFINS W/ BRAN; DUTCH EGG SANDWICH BUNS; SOFT SANDWICH ROLLS	GLUTEN (VITAL WHEAT)
	DELI-TWIST ROLLS	GLUTEN (WHEAT); **VEGETABLE SHORTENING (LIQUID)**
	DINNER PARTY ROLLS; DINNER PARTY PARKERHOUSE ROLLS; EXTRA CRISP MUFFINS	NO ADDITIVES OF CONCERN
R	AUGUST BROS. ASSORTED DINNER ROLLS; EGG ROLLS; ONION ROLLS	**PAPRIKA; TURMERIC**
	JR. BAGELS	HIGH GLUTEN FLOUR (BLEACHED)
R	COLOMBO SOUR FRENCH ROLLS; SOUR *(continues)*	NO ADDITIVES OF CONCERN

	BRANDS	ADDITIVES-OF-CONCERN BOLD—CONCERN TO ALL LIGHT—CONCERN TO SOME
	COLUMBO *(Continued)*	
	SANDWICH ROLLS; SWEET SANDWICH ROLLS	NO ADDITIVES OF CONCERN
R	**COUNTRY FAIR**	
	BUNS	GLUTEN (WHEAT); **LARD**
	ENGLISH MUFFINS	GLUTEN (WHEAT)
R	**DANDEE** BROWN 'N SERVE ROLLS	NO ADDITIVES OF CONCERN
	FRANCISCO	
	FRENCH ROLLS; SANDWICH ROLLS; SOUR DOUGH FRENCH ROLLS	NO ADDITIVES OF CONCERN
R	**GREEN FREEDMAN**	
	ONION DELI ROLLS	**PAPRIKA; TURMERIC**
R	**HOLSUM**	
	BAKE & SERVE ONION HOT BREAD	NO ADDITIVES OF CONCERN
	BAKE & SERVE CINNAMON HOT BREAD	EGG WHITES (DRIED); GLUTEN; GUAR GUM; **MODIFIED FOOD STARCH**
	BRAN GRANOLA ENGLISH MUFFINS; BROWN 'N SERVE ROLLS	GLUTEN OR WHEAT GLUTEN
	SOF-BUNS; YOU'LL LIKE BAHAMA BUNS	**CALCIUM PEROXIDE;** GLUTEN (WHEAT)
R	**MERITA** GOLDEN HONEY BUNS	GLUTEN (WHEAT)

	BRANDS	ADDITIVES-OF-CONCERN BOLD—CONCERN TO ALL LIGHT—CONCERN TO SOME
R	**NISSEN**	
	BAKE 'N BROWN BUNS	ARTIFICIAL COLOR FD & C YELLOW NO. 5; EGG WHITES; GLUTEN (WHEAT)
	BLUEBERRY MUFFINS	EGG WHITES; GLUTEN (WHEAT); GUAR GUM; WHEY (MILK)
	BROWN 'N SERVE ROLLS (WHITE)	**(BEEF TALLOW, LARD);* CALCIUM PEROXIDE;** EGG WHITES; GLUTEN (WHEAT)
	CRANBERRY-ORANGE MUFFINS; HAMBURGER BUNS	NO ADDITIVES OF CONCERN
	NEW ENGLAND STYLE FRANKFURT ROLLS	**(BEEF TALLOW, LARD)***
R	**OLD COUNTRY**	
	DINNER ROLLS	**ARTIFICIAL COLOR; PAPRIKA; TURMERIC**
	EGG TWIST ROLLS; HARD ROLLS	NO ADDITIVES OF CONCERN
	PEPPERIDGE FARM	
	BUTTER CRESCENT ROLLS; GOLDEN TWIST ROLLS BROWN & SERVE	**BUTTER**
	CINNAMON RAISIN MUFFINS; HAMBURGER ROLLS; SOFT FAMILY ROLLS	GLUTEN (WHEAT)

*Blend may contain one or more of these saturated fats.

	BRANDS	ADDITIVES-OF-CONCERN **BOLD—CONCERN TO ALL** LIGHT—CONCERN TO SOME
	PEPPERIDGE FARM *(Continued)*	
	CLUB ROLLS BROWN & SERVE; FRENCH ROLLS; WHITE SANDWICH POCKETS	NO ADDITIVES OF CONCERN
	ENGLISH MUFFINS	WHEY
	OLD FASHIONED ROLLS; PARKER HOUSE ROLLS; PARTY ROLLS	EGG WHITES
R	**SUNBEAM**	
	ENGLISH MUFFIN LOAF	WHEY (DRIED)
	ENGLISH MUFFINS	NO ADDITIVES OF CONCERN
	FRANKFURTER ROLLS; SLICED ROLLS	**(LARD)***
	THOMAS'	
	BRAN TOAST-R-CAKES W/ RAISINS; ENGLISH MUFFINS	WHEY SOLIDS
	WHEAT ENGLISH MUFFINS	NO ADDITIVES OF CONCERN
R	**TOSCANA'S**	
	ALL PURPOSE SAN'WICH ROLLS; CLUSTER ROLLS; ONION ROLLS; SEEDED ROLLS	NO ADDITIVES OF CONCERN

*Blend may contain this saturated fat.

BRANDS	ADDITIVES-OF-CONCERN BOLD—CONCERN TO ALL LIGHT—CONCERN TO SOME
WONDER	
BROWN N SERVE ROLLS	NO ADDITIVES OF CONCERN
ENGLISH MUFFINS	GLUTEN (WHEAT); WHEY
HOT DOG BUNS	GLUTEN (WHEAT); **LARD**
SUPERMARKET PRIVATE BRANDS	
FIRST NATIONAL —FINAST	
BAKE & SERVE CINNAMON HOT BREAD	EGG WHITES (DRIED); GLUTEN; GUAR GUM; **MODIFIED FOOD STARCH**
LUCKY —HARVEST DAY	
ENGLISH MUFFINS; SOURDOUGH MUFFINS	**ENZYME FROM ASPERGILLUS ORYZAE;** GLUTEN (WHEAT)
GOURMET ROLLS	ARTIFICIAL COLOR FD & C YELLOW NO. 5; WHEY
RAISIN MUFFINS	NO ADDITIVES OF CONCERN
SESAME BUNS; SLICED BUNS	WHEY
—BROWN 'N SERVE	
BUTTERMILK TWINS ROLLS	NO ADDITIVES OF CONCERN
CLOVERLEAF ROLLS; FLAKY GEMS ROLLS	WHEY
PUBLIX —BREAKFAST CLUB	
HAMBURGER BUNS	NO ADDITIVES OF CONCERN

BRANDS	ADDITIVES-OF-CONCERN BOLD—CONCERN TO ALL LIGHT—CONCERN TO SOME
—THE DELI AT PUBLIX	
DINNER ROLLS	GLUTEN (VITAL WHEAT)
EGG BAGELS; GARLIC BAGELS; ONION BAGELS; PUMPERNICKEL BAGELS	NO ADDITIVES OF CONCERN
ONION ROLLS	GLUTEN (VITAL WHEAT); **MODIFIED STARCH;** WHEY
SAFEWAY	
—MRS. WRIGHT'S	
CRUSHED WHEAT HAMBURGER BUNS	**CALCIUM PEROXIDE;** GLUTEN (WHEAT)
ENGLISH MUFFINS; SOURDOUGH MUFFINS	NO ADDITIVES OF CONCERN
HONEY WHEAT BERRY ROLLS	GLUTEN (WHEAT)
HAMBURGER BUNS; HOT DOG BUNS; SESAME ENRICHED ROLLS	**CALCIUM PEROXIDE;** GLUTEN (WHEAT); WHEY
SWEET FARMSTYLE ROLLS	WHEY
STOP & SHOP	
—(NO BRAND NAME)	
DINNER ROLLS	**PAPRIKA; TURMERIC**
GRINDER ROLLS	NO ADDITIVES OF CONCERN

BRANDS	ADDITIVES-OF-CONCERN **BOLD—CONCERN TO ALL** LIGHT—CONCERN TO SOME
—OUR OWN	
BRAN TOASTIES; CORN TOASTIES; NEWFANGLED CORN MUFFINS	WHEY
NEWFANGLED BLUEBERRY MUFFINS	**MODIFIED FOOD STARCH;** WHEY
—STOP & SHOP	
BIG BURGERS SANDWICH ROLLS; FRANKFURTER ROLLS; SLICED ONION ROLLS; SLICED SESAME ROLLS; WHEATBERRY ENGLISH MUFFINS	GLUTEN (WHEAT)
BLUEBERRY FLAVORED ENGLISH MUFFINS	**ARTIFICIAL COLOR;** SODIUM ALGINATE; WHEY
BULKIE ROLLS	NO ADDITIVES OF CONCERN
BUTTER ENGLISH MUFFINS	**BUTTER; PAPRIKA; TURMERIC**
ENGLISH MUFFINS	**ARTIFICIAL COLOR;** WHEY
WINN DIXIE	
—DIXIE DARLING	
CLUSTER BROWN 'N SERVE ROLLS; HAMBURGER BUNS; TWIN BROWN 'N SERVE ROLLS	**CALCIUM PEROXIDE;** GLUTEN (VITAL WHEAT)

BRANDS	ADDITIVES-OF-CONCERN BOLD—CONCERN TO ALL LIGHT—CONCERN TO SOME
—DIXIE DARLING *(Continued)*	
SEEDED FRENCH BROWN 'N SERVE HARD ROLLS	NO ADDITIVES OF CONCERN
—PRESTIGE	
BISCUITS MADE W/ BUTTERMILK BROWN 'N SERVE	**ARTIFICIAL COLOR;** WHEY
BONANZA STEAK ROLLS; ONION ROLLS	NO ADDITIVES OF CONCERN
BROWN 'N SERVE ROLLS	GLUTEN (VITAL WHEAT)
ENGLISH MUFFINS	MILK SOLIDS (NONFAT DRY)
WEINER BUNS	**CALCIUM PEROXIDE; PAPRIKA; TURMERIC**
ITEMS IN HEALTH FOOD STORES	
GIUSTO'S VITA-GRAIN	
HONEY-PRO DINNER ROLLS	HI PROTEIN FLOUR (UNBLEACHED)
SOYA BUNS; WHOLE WHEAT DINNER ROLLS	NO ADDITIVES OF CONCERN
MATTHEW'S *WHOLE WHEAT ENGLISH MUFFINS*	
CINNAMON RAISIN; PLAIN	NO ADDITIVES OF CONCERN
STAFF OF LIFE	
SESAME WHOLE WHEAT ROLLS	NO ADDITIVES OF CONCERN

BRANDS	ADDITIVES-OF-CONCERN BOLD—CONCERN TO ALL LIGHT—CONCERN TO SOME
VITAL VITTLES REAL BREAD ROLLS	NO ADDITIVES OF CONCERN

BREAD

BRANDS	ADDITIVES-OF-CONCERN
ANADAMA	
CORNMEAL & MOLASSES	GLUTEN (VITAL WHEAT)
ARNOLD	
BRICK OVEN 100% WHOLE WHEAT; BRICK OVEN WHITE; MELBA THIN DIETSLICE 100% WHOLE WHEAT; MELBA THIN DIETSLICE WHITE	NO ADDITIVES OF CONCERN
JEWISH RYE W/ CARAWAY SEEDS; NATURÉL; PUMPERNICKLE; RAISIN TEA LOAF; STONE GROUND 100% WHOLE WHEAT	GLUTEN (VITAL WHEAT)
R **AUGUST BROS.**	
GENUINE JEWISH RYE; ITALIAN; ONION RYE; PITA; SLICED PUMPERNICKLE	NO ADDITIVES OF CONCERN
STONE GROUND 100% WHOLE WHEAT	GLUTEN FLOUR (WHEAT)
BRAN'NOLA (OROWEAT)	
COUNTRY OAT; HEARTY WHEAT	GLUTEN (VITAL WHEAT); WHEY

	BRANDS	ADDITIVES-OF-CONCERN **BOLD—CONCERN TO ALL** LIGHT—CONCERN TO SOME
	BRAN'NOLA *(Continued)*	
	GOLDEN SESAME	WHEY
	YOGURT	GLUTEN (VITAL WHEAT)
	BREADS INTERNATIONAL GERMAN PUMPERNICKEL; JEWISH RYE	NO ADDITIVES OF CONCERN
	SAN FRANCISCO SOURDOUGH	EGG WHITES; GLUTEN (WHEAT)
R	**COLOMBO** ENRICHED FRENCH SWEET	NO ADDITIVES OF CONCERN
R	**COTE'S** STONE GROUND WHEAT	**CALCIUM PEROXIDE;** GLUTEN (WHEAT)
	COUNTY FAIR KING SIZE SANDWICH	WHEY
	COUNTRY HEARTH BRAN 'N HONEY; 7 WHOLE GRAIN	GLUTEN (WHEAT)
	BUTTERSPLIT; MOUNTAIN MEAL; OLD FASHIONED WHITE	GLUTEN (WHEAT); WHEY (SWEET DAIRY)
R	**FLOWERS** BUTTERMAID; DANDEE OLD FASHIONED WHITE; NEW YORK JEWISH RYE	NO ADDITIVES OF CONCERN

	BRANDS	ADDITIVES-OF-CONCERN **BOLD—CONCERN TO ALL** LIGHT—CONCERN TO SOME
	NEW ORLEANS STYLE VIENNA	EGG WHITES; GLUTEN (WHEAT)
	—*NATURE'S OWN*	
	BUTTERBREAD; OLD FASHIONED WHITE; REAL BRAN 'N HONEY	NO ADDITIVES OF CONCERN
	HONEY WHEAT; REAL GRAIN	GLUTEN (WHEAT); WHEY
	FRANCISCO	
	FRENCH STYLE; VIENNA FRENCH SLICED	NO ADDITIVES OF CONCERN
R	**GREEN FREEDMAN**	
	CARAWAY RYE; RUSSIAN RYE	NO ADDITIVES OF CONCERN
R	**GROSSINGER'S**	
	PUMPERNICKEL; RYE NO SEEDS	HIGH PROTEIN WHEAT FLOUR
	HOLLYWOOD	
	DARK	GLUTEN (WHEAT); KELP (DEHYDRATED)
	LIGHT	GLUTEN (WHEAT); KELP (DEHYDRATED); WHEY
R	**HOLSUM**	
	SANDWICH THIN	WHEY
	WHITE NO SALT	**CALCIUM PEROXIDE**
	BAHAMA	**CALCIUM PEROXIDE;** GLUTEN (WHEAT); WHEY

BRANDS		ADDITIVES-OF-CONCERN **BOLD—CONCERN TO ALL** LIGHT—CONCERN TO SOME
	HOME PRIDE BUTTER TOP; WHEAT BUTTER TOP	**CALCIUM CASEINATE;** GLUTEN (WHEAT); WHEY
R	**JOSEPH'S** MIDDLE EAST STYLE SYRIAN	NO ADDITIVES OF CONCERN
R	**KASANOF'S** GENUINE LIGHT RYE; GENUINE PUMPERNICKEL RYE; JEWISH CARAWAY RYE; RUSSIAN BLACK RYE	NO ADDITIVES OF CONCERN
R	**MERITA** OLD FASHIONED	WHEY
R	**MONK'S** HI-FIBRE; RAISIN; WHITE	WHEY (DRIED SWEET DAIRY)
R	**MUNZENMAIER'S** KÜMMELBROT SEEDED RYE	HIGH GLUTEN WHEAT FLOUR
	RUSSIAN PUMPERNICKEL	NO ADDITIVES OF CONCERN
R	**NISSEN** BRAN; BUTTER TOP OATMEAL; CANADIAN BROWN; HONEY TOP CRACKED WHEAT	GLUTEN (WHEAT)
	BUTTER TOP; OLD SETTLER WHITE; RYE; TENDER CRUST GIANT WHITE	NO ADDITIVES OF CONCERN

	BRANDS	ADDITIVES-OF-CONCERN **BOLD—CONCERN TO ALL** LIGHT—CONCERN TO SOME
	CINNAMON W/ RAISINS	GUM ARABIC; WHEY
	ITALIAN; TENDER CRUST GIANT SANDWICH WHITE; VIENNA	**(BEEF TALLOW, LARD)***
R	**OLD COUNTRY**	
	FRENCH; ITALIAN; ONION RYE; PUMPERNICKEL	NO ADDITIVES OF CONCERN
	OROWEAT	
	RUSSIAN RYE; SCHWARZWÄLDER DARK RYE; SPROUTED WHEAT	GLUTEN (VITAL WHEAT)
	PEPPERIDGE FARM	
	CORN & MOLASSES; HONEY BRAN; ITALIAN BROWN & SERVE; 100% WHOLE WHEAT; RAISIN W/ CINNAMON; SANDWICH RYE; WHITE	NO ADDITIVES OF CONCERN
	CRACKED WHEAT; FAMILY RYE W/ SEEDS; HONEY WHEAT BERRY; OATMEAL; PUMPERNICKEL; SEEDLESS RYE; WHEAT	GLUTEN (WHEAT)

*Blend may contain one or more of these saturated fats.

	BRANDS	ADDITIVES-OF-CONCERN BOLD—CONCERN TO ALL LIGHT—CONCERN TO SOME
	ROMAN MEAL BREAD	GLUTEN (VITAL WHEAT); WHEY SOLIDS
R	**SUNBEAM** BATTER WHIPPED	**(LARD);* SODIUM CASEINATE;** WHEY
	BUTTERTOP OATMEAL	GLUTEN (WHEAT)
	BUTTERTOP WHEAT	**CALCIUM PEROXIDE;** GLUTEN (WHEAT)
	HONEY CRUSHED WHEAT	NO ADDITIVES OF CONCERN
	VIENNA	**(LARD)***
R	**THOMAS'** PROTEIN	GLUTEN FLOUR
R	**TOSCANA'S** ENRICHED FRENCH SWEET	NO ADDITIVES OF CONCERN
R	**WEIGHT WATCHERS** THIN SLICED	GLUTEN (WHEAT); WHEY
R	**WONDER** ENRICHED	WHEY

*Blend may contain this saturated fat.

BRANDS	**ADDITIVES-OF-CONCERN** BOLD—CONCERN TO ALL LIGHT—CONCERN TO SOME
GOLDEN WHEAT; BEEFSTEAK SOFT RYE	GLUTEN (WHEAT)
100% WHOLE WHEAT	GLUTEN (WHEAT); WHEY
SUPERMARKET PRIVATE BRANDS	
LUCKY	
—LADY LEE	
PREMIUM SANDWICH WHEAT	NO ADDITIVES OF CONCERN
PREMIUM SANDWICH WHITE	MILK SOLIDS (NONFAT DRY)
—HARVEST DAY	
CHUCK WAGON; LARGE DELUXE WHITE; VIENNA	MILK SOLIDS (NONFAT DRY); WHEY
CRUSHED WHEAT SANDWICH; THIN SLICED SANDWICH	WHEY
DARK THINLY SLICED SLIMLINE; LIGHT THINLY SLICED SLIMLINE	**CALCIUM CASEINATE;** GLUTEN (WHEAT); WHEY
LA TORTILLA CORN TORTILLAS; MADE W/ BUTTERMILK; 100% WHOLE WHEAT; POTATO	NO ADDITIVES OF CONCERN
LA TORTILLA FLOUR TORTILLAS	**LARD;** LOCUST BEAN GUM
RAISIN; RYE	GLUTEN (WHEAT)

BRANDS	ADDITIVES-OF-CONCERN BOLD—CONCERN TO ALL LIGHT—CONCERN TO SOME
LUCKY—HARVEST DAY *(Continued)*	
SPLIT TOP WHITE	**ARTIFICIAL COLOR; BUTTER**
PUBLIX	
—BREAKFAST CLUB	
WHITE	NO ADDITIVES OF CONCERN
—PUBLIX *SPECIAL RECIPE*	
BUTTER CRUST WHITE; HONEY WHEAT; THIN SLICED STONE GROUND WHEAT; 100% WHOLE WHEAT STONE GROUND	NO ADDITIVES OF CONCERN
—THE DELI AT PUBLIX	
CHALLAH EGG TWIST; ITALIAN; PUMPERNICKLE	GLUTEN (VITAL WHEAT)
FRENCH; NATURAL WHEAT ITALIAN	**MODIFIED STARCH**
JEWISH STYLE ONION RYE	NO ADDITIVES OF CONCERN
SAFEWAY	
—MRS. WRIGHT'S	
BUTTER & EGG; HOMESTYLE BUTTER TOP; 100% WHOLE WHEAT; SPLIT TOP WHITE	WHEY
CRUSHED WHEAT; ENRICHED	NO ADDITIVES OF CONCERN

(continues)

BRANDS	ADDITIVES-OF-CONCERN BOLD—CONCERN TO ALL LIGHT—CONCERN TO SOME
BUTTERMILK; HOMESTYLE WHITE; MALT-O-WHEAT; OATMEAL; OLD FASHIONED ITALIAN; POTATO; RAISIN NUT; SANDWICH RYE W/ SEEDS; SOURDOUGH WHITE ENRICHED	NO ADDITIVES OF CONCERN
ENGLISH MUFFIN STYLE; GRAIN BELT; GRANOLA BRAN; RAISIN; SPECIAL FORMULA DARK STYLE	GLUTEN (WHEAT)
SANDWICH WHEAT; SUPER SOFT WHEAT	**CALCIUM PEROXIDE;** GLUTEN (WHEAT)
SUPER SOFT SANDWICH WHITE; SUPER SOFT WHITE	**CALCIUM PEROXIDE**
—(NO BRAND NAME)	
7 GRAIN; SOYA; STONE GROUND WHOLE WHEAT; WHEAT GERM	NO ADDITIVES OF CONCERN
STONEHEDGE FARM WHOLE WHEAT	**BUTTER;** WHEY
—SAFEWAY	
PREMIUM WHEAT	WHEY
PREMIUM WHITE THIN SLICED SANDWICH	**CALCIUM PEROXIDE;** WHEY

BRANDS	ADDITIVES-OF-CONCERN **BOLD—CONCERN TO ALL** LIGHT—CONCERN TO SOME
STOP & SHOP —OUR OWN	
BANANA TEA	**ARTIFICIAL COLOR**
CRANBERRY NUT	NO ADDITIVES OF CONCERN
DATE NUT	EGG WHITES
—STOP & SHOP	
HOT CINNAMON LOAF BAKE AT HOME	EGG WHITES (DRIED); GLUTEN; GUAR GUM; **MODIFIED FOOD STARCH**
ITALIAN BROWN 'N' SERVE; MINI LOAVES BROWN 'N' SERVE WHITE	NO ADDITIVES OF CONCERN
—STOP & SHOP BAKERY	
BUTTER TOP; DAISY; YAH YAH BUTTERCREST WHITE	WHEY
BUTTER TOP HONEY WHEAT; CHEESE;* ENGLISH MUFFIN LOAF; HIGH FIBER WHEAT W/ BRAN; HOME KITCHEN; SANDWICH; ITALIAN; PLAIN RYE	GLUTEN (WHEAT)

*By law, at this writing, it is not required of a manufacturer to state whether cheeses contain artificial colors, unless the color is FD & C Yellow No. 5 (Tartrazine). Their presence in bread containing cheese therefore remains an uncertainty, unless it is voluntarily declared on the label that artificial colors have been used or have not. For a more detailed explanation of artificial color in cheese, refer to the beginning of the cheese section in "Dairy Products & Substitutes."

BRANDS	ADDITIVES-OF-CONCERN BOLD—CONCERN TO ALL LIGHT—CONCERN TO SOME
CANADIAN STYLE OATMEAL; FRENCH STICKS; KOSHER PUMPERNICKEL; KOSHER RYE; MADE W/ BUTTERMILK; OATMEAL N' MOLASSES; 100% WHOLE WHEAT; SWEDISH	NO ADDITIVES OF CONCERN
CRACKED WHEAT; OATMEAL; VIENNA	GLUTEN (WHEAT); WHEY
LOTS 'O RAISIN	GLUTEN (WHEAT); **PAPRIKA; TURMERIC**
WINN DIXIE —DIXIE DARLING	
BERMUDA	GLUTEN (VITAL WHEAT)
LARGE WHITE; WHITE; WHITE MADE W/ BUTTERMILK	**CALCIUM PEROXIDE**
—PRESTIGE	
COUNTRY WHITE; VERY THIN WHITE	WHEY SOLIDS
DELUXE	**CALCIUM PEROXIDE; PAPRIKA; TURMERIC**
NATURAL FIBER; 100% WHOLE WHEAT	GLUTEN (VITAL WHEAT OR WHEAT)
RAISIN	GLUTEN (VITAL WHEAT); **MACE**
V-10 PROTEIN	GLUTEN (WHEAT); **SODIUM CASEINATE;** WHEY

BRANDS	ADDITIVES-OF-CONCERN **BOLD—CONCERN TO ALL** LIGHT—CONCERN TO SOME
ITEMS IN HEALTH FOOD STORES	
BALDWIN HILL	
NATURAL SOURDOUGH	
RYE; WHOLE WHEAT	NO ADDITIVES OF CONCERN
EREWHON MIDDLE EASTERN WHOLE WHEAT; 100% STONE GROUND WHOLE WHEAT	NO ADDITIVES OF CONCERN
FOOD FOR LIFE BAKING	
BRAN FOR LIFE	GLUTEN (VITAL WHEAT); MILK SOLIDS
7-GRAIN SPROUTED WHEAT	GLUTEN (VITAL WHEAT)
GARDEN OF EATIN'	
BIBLE; CORNTILLAS	NO ADDITIVES OF CONCERN
GIUSTO'S VITA-GRAIN	
HI PROTEIN GLUTEN; HI PROTEIN GLUTEN & SOYA	GLUTEN FLOUR
HI PROTEIN WHEAT; SOURDOUGH WHOLE WHEAT & CORN	HIGH PROTEIN FLOUR (UNBLEACHED OR WHOLE WHEAT)
—*UNSALTED LOW SODIUM*	
WHEAT; WHITE	NO ADDITIVES OF CONCERN
LIFESTREAM	
ESSENE; ESSENE RAISIN	NO ADDITIVES OF CONCERN

BRANDS	ADDITIVES-OF-CONCERN **BOLD—CONCERN TO ALL** LIGHT—CONCERN TO SOME
QUEBRADA OATMEAL; RAISIN WHOLE WHEAT; VITA-LOAF	NO ADDITIVES OF CONCERN
SPRUCE TREE ONION RYE; WHOLE WHEAT	NO ADDITIVES OF CONCERN
STAFF OF LIFE APPLE; SOURDOUGH RYE; VEGETABLE HERB; WHEATLESS PUMPERNICKEL RYE; WHOLE WHEAT	NO ADDITIVES OF CONCERN
THE GRAIN-BIN CASHEW DATE; WHOLE WHEAT	NO ADDITIVES OF CONCERN
VITAL VITTLES BUTTERMILK CORNBREAD; REAL LITTLE; REAL SESAME-MILLET; SOURDOUGH	NO ADDITIVES OF CONCERN

BREAD & CORN-FLAKE CRUMBS; CRACKER MEAL

ARNOLD ALL PURPOSE BREAD CRUMBS	GLUTEN (VITAL WHEAT)
ITALIAN BREAD CRUMBS	GLUTEN (VITAL WHEAT); MONOSODIUM GLUTAMATE

	BRANDS	ADDITIVES-OF-CONCERN **BOLD—CONCERN TO ALL** LIGHT—CONCERN TO SOME
R	**COLOMBO** BREAD CRUMBS; BREAD CRUMBS W/ ITALIAN SEASONING	**VEGETABLE SHORTENING**
R	**COLONNA** FLAVORED BREAD CRUMBS	MILK SOLIDS; **SHORTENING**
	CONTADINA SEASONED BREAD CRUMBS	NO ADDITIVES OF CONCERN
	DEVONSHEER PLAIN BREAD CRUMBS	NO ADDITIVES OF CONCERN
	SEASONED BREAD CRUMBS	MONOSODIUM GLUTAMATE
R	**4C** BREAD CRUMBS PLAIN	NO ADDITIVES OF CONCERN
	BREAD CRUMBS REDI-FLAVORED	MONOSODIUM GLUTAMATE
R	**INTROVIGNE'S** "5 in 1" MIX FLAVORED BREAD CRUMBS	NO ADDITIVES OF CONCERN
R	**JASON** READY FLAVORED BREAD CRUMBS	MILK SOLIDS
	KELLOGG'S CORN FLAKE CRUMBS	**BHA**

	BRANDS	ADDITIVES-OF-CONCERN **BOLD—CONCERN TO ALL** LIGHT—CONCERN TO SOME
	NABISCO CRACKER MEAL	NO ADDITIVES OF CONCERN
R	**NISSEN**	
	BREAD CRUMBS	WHEY
	FLAVORED BREAD CRUMBS	MONOSODIUM GLUTAMATE; **PAPRIKA;** WHEY
R	**OLD LONDON**	
	BREAD CRUMBS ITALIAN	MONOSODIUM GLUTAMATE; WHEY
	BREAD CRUMBS REGULAR	WHEY
R	**PASTENE** ITALIAN BREAD CRUMBS	MONOSODIUM GLUTAMATE
R	**PRINCE** ITALIAN BREAD CRUMB MIX	MONOSODIUM GLUTAMATE
	PROGRESSO	
	ITALIAN BREAD CRUMBS	MONOSODIUM GLUTAMATE; **PAPRIKA;** WHEY
	PLAIN BREAD CRUMBS	WHEY
R	**VIGO**	
	ITALIAN BREAD CRUMBS	MONOSODIUM GLUTAMATE

BRANDS	ADDITIVES-OF-CONCERN **BOLD—CONCERN TO ALL** LIGHT—CONCERN TO SOME
SUPERMARKET PRIVATE BRANDS	
FIRST NATIONAL —FINAST	
BREAD CRUMBS	**SHORTENING**
BREAD CRUMBS FLAVORED	MONOSODIUM GLUTAMATE
STOP & SHOP —STOP & SHOP	
FLAVORED BREAD CRUMBS	MONOSODIUM GLUTAMATE; WHEY
TOASTED BREAD CRUMBS	WHEY
ITEMS IN HEALTH FOOD STORES	
BALDWIN HILL	
NATURAL SOURDOUGH WHOLE WHEAT BREAD CRUMBS	NO ADDITIVES OF CONCERN

BREADSTICKS

	BRANDS	ADDITIVES-OF-CONCERN
	ANGONOA'S	
	CHEESE	**ARTIFICIAL COLOR;** WHEY
	GARLIC; ITALIAN; SESAME ROYALE	NO ADDITIVES OF CONCERN
R	**BEST QUALITY**	
	SESAME	NO ADDITIVES OF CONCERN
R	**COLOMBO** ITALIAN STYLE	GLUTEN (WHEAT)

BRANDS		ADDITIVES-OF-CONCERN BOLD—CONCERN TO ALL LIGHT—CONCERN TO SOME
R	**GARDETTO'S** CHEESE;* GARLIC; ONION; SALTED	NO ADDITIVES OF CONCERN
R	**PAN D'OR** SLIM ITALIAN	NO ADDITIVES OF CONCERN
	PEPPERIDGE FARM LIGHTLY SALTED SNACK STICKS; PUMPERNICKEL SNACK STICKS; SESAME SNACK STICKS	**COCONUT OIL;** GLUTEN (WHEAT)
	STELLA D'ORO BREADSTICKS; DIETETIC FOR SODIUM RESTRICTED DIETS SESAME; SESAME	NO ADDITIVES OF CONCERN
R	**TOSCANA'S** BREAD STICKS	NO ADDITIVES OF CONCERN
	ITEMS IN HEALTH FOOD STORES	
	DE BOLES ARTI-SESAME-STIX	HIGH GLUTEN FLOUR; **100% PURE VEGETABLE OIL**

*By law, at this writing, it is not required of a manufacturer to state whether cheeses contain artificial colors, unless the color is FD & C Yellow No. 5 (Tartrazine). Their presence in breadsticks containing cheese therefore remains an uncertainty, unless it is voluntarily declared on the label that artificial colors have been used or have not. For a more detailed explanation of artificial color in cheese, refer to the beginning of the cheese section of "Dairy Products & Substitutes."

BRANDS	ADDITIVES-OF-CONCERN BOLD—CONCERN TO ALL LIGHT—CONCERN TO SOME

CRACKERS

By law, at this writing, it is not required of a manufacturer to state whether cheeses contain artificial colors, unless the color is FD & C Yellow No. 5 (Tartrazine). Their presence in crackers containing cheese therefore remains an uncertainty, unless it is voluntarily declared on the label that artificial colors have been used or have not. For a more detailed explanation of artificial color in cheese, refer to the beginning of the cheese section of "Dairy Products & Substitutes."

	Brands	Additives-of-Concern
	BREMNER	
	POPPYSEED; WAFERS	**(BEEF FAT, LARD)***
	BURRY	
	EUPHRATES	**(COCONUT, HYDROGENATED PALM OILS);*** WHEY (DRIED DAIRY)
	MARINER SEA	**(HYDROGENATED PALM OIL)***
	CARR'S TABLE WATER	**(COCONUT, PALM KERNEL OILS)***
R	**CELLU** LOW SODIUM UNSALTED	**(LARD)***
	CHE-CRI CHEESE CRISPIES	**(COCONUT OIL; HYDROGENATED PALM, SOYBEAN OILS)***
	CRAQUELINS	
	SESAME & ONION	**COCONUT OIL**
	STONED WHEAT THINS	WHEY

BRANDS	ADDITIVES-OF-CONCERN **BOLD—CONCERN TO ALL** LIGHT—CONCERN TO SOME
DEVONSHEER	
GARLIC ROUNDS; ONION ROUNDS; PLAIN MELBA TOAST; PLAIN MELBA TOAST NO SALT; RYE MELBA TOAST; SESAME MELBA ROUNDS; UNSALTED PLAIN MELBA TOAST; WHEAT MELBA TOAST	NO ADDITIVES OF CONCERN
FFV	
APPETIZER THINS	**(BEEF FAT, COCONUT OIL)***
BLEU CHEESE	**(BEEF FAT, COCONUT OIL);*** MONOSODIUM GLUTAMATE; WHEY
OCEAN CRISP UNSALTED TOPS	NO ADDITIVES OF CONCERN
ONION STIX	**ARTIFICIAL COLORS; (BEEF FAT, COCONUT OIL);* PAPRIKA**
STONED WHEAT	**(BEEF FAT);*** WHEY
WHEAT	**(BEEF FAT, COCONUT OIL);*** MONOSODIUM GLUTAMATE
FINN CRISP	
LIGHT; DARK W/ CARAWAY	NO ADDITIVES OF CONCERN

*Blend may contain one or more of these saturated fats and/or oils.

	BRANDS	ADDITIVES-OF-CONCERN **BOLD—CONCERN TO ALL** LIGHT—CONCERN TO SOME
	FIRESIDE ANIMAL	**(BEEF FAT, LARD);*** WHEY
	SALTINE	**(COCONUT OIL, LARD)***
	SUGAR HONEY GRAHAM	NO ADDITIVES OF CONCERN
	FLAVOR TREE CHEDDAR CHIPS; SESAME CHIPS	NO ADDITIVES OF CONCERN
	FRENCH ONION CRISPS; PARTY MIX	**TURMERIC**
	GOODMAN'S CARAWAY & RYE MATZOS	NO ADDITIVES OF CONCERN
R	**GRIELLE** CRISP TOAST W/ GLUTEN	GLUTEN FLOUR; **RAPESEED OIL**
	CRISP TOAST TOASTED WHEATBREAD	**RAPESEED OIL**
	HOROWITZ MARGARETEN UNSALTED MATZOHS	NO ADDITIVES OF CONCERN
R	**HOVIS** EXTRA WHEAT GERM	**BUTYLATED HYDROXYANISOLE; (COCONUT, PALM KERNEL OILS; HYDROGENATED PALM OIL, LARD);*** WHEY (DRIED SALT)

BRANDS	ADDITIVES-OF-CONCERN BOLD—CONCERN TO ALL LIGHT—CONCERN TO SOME
IDEAL CRISPBREAD WAFERS ULTRA-THIN; CRISPBREAD WAFERS WHOLE GRAIN	NO ADDITIVES OF CONCERN
JACOB'S CREAM	**BHA; (COCONUT OIL, HYDROGENATED MARINE OIL)***
KAVLI NORWEGIAN FLATBREAD THIN	NO ADDITIVES OF CONCERN
KEEBLER ANIMAL; ZESTA SALTINE	**(BEEF FAT, LARD)***
CHEESE SHINDIGS	**ARTIFICIAL COLORING; (COCONUT OIL, LARD);*** MONOSODIUM GLUTAMATE; WHEY
CINNAMON CRISP	WHEY
CLUB; ONION TOAST; SESAME STICKS; WHEAT TOAST	**(COCONUT OIL, LARD);*** WHEY
HONEY GRAHAMS	**LARD**
RYE TOAST; SESAME TOAST; TOWN HOUSE	**(COCONUT OIL, LARD)***

*Blend may contain one or more of these saturated fats and/or oils.

	BRANDS	ADDITIVES-OF-CONCERN BOLD—CONCERN TO ALL LIGHT—CONCERN TO SOME
	KEEBLER *(Continued)*	
	SOUR CREAM & ONION SHINDIGS	**(COCONUT OIL, LARD);*** MILK SOLIDS (NONFAT); MONOSODIUM GLUTAMATE; **PAPRIKA;** SOUR CREAM SOLIDS; WHEY
	WHEAT CRISPS	**(COCONUT OIL, LARD);*** GLUTEN (WHEAT); MONOSODIUM GLUTAMATE
R	**KINGS BREAD** CRISP BREAD BROWN	NO ADDITIVES OF CONCERN
	KITANIHON	
	SEAWEED CRUNCH	NORI
	SHRIMP RICE CRUNCH	**ARTIFICIAL COLOR FD & C RED NO. 3**
	MANISCHEWITZ	
	MATZO-THINS; UNSALTED MATZOS; WHOLE WHEAT MATZOS	NO ADDITIVES OF CONCERN
	GARLIC TAMS; ONION TAMS; TAM TAM	**(COCONUT, HYDROGENATED COTTONSEED, SOYBEAN OILS)***
R	**MARUKAI**	
	SESAME RICE	**SEAWEED**
	CHEESE RICE	**U.S. CERTIFIED COLOR**
	MASTER	
	HOL-RY	**COCONUT OIL**

BRANDS	**ADDITIVES-OF-CONCERN** **BOLD—CONCERN TO ALL** LIGHT—CONCERN TO SOME
LOW SODIUM RYE; LOW SODIUM TOAST; OLD COUNTRY HARDTACK RYE	NO ADDITIVES OF CONCERN
NABISCO	
BACON FLAVORED THINS	**(COCONUT OIL, LARD);* NATURAL SMOKE FLAVOR;** WHEY
BARNUM'S ANIMALS	WHEY
BUTTERY SESAME; TRISCUIT WAFERS	**(COCONUT OIL)***
CHEESE NIPS; NABS CHEESE SANDWICH	**ARTIFICIAL COLOR; (COCONUT OIL, LARD);*** MONOSODIUM GLUTAMATE; **PAPRIKA;** WHEY
CHICKEN IN A BISKIT; DIP IN A CHIP	**(COCONUT OIL, LARD);*** MONOSODIUM GLUTAMATE; WHEY
CINNAMON TREATS GRAHAM; PREMIUM SALTINE; RITZ; SOCIABLES; WAVERLY WAFERS	**(COCONUT OIL, LARD)***
CROWN PILOT CHOWDER; DANDY SOUP & OYSTER; HONEY MAID GRAHAM; *(continues)*	**(LARD)***

*Blend may contain one or more of these saturated fats and/or oils.

BRANDS	ADDITIVES-OF-CONCERN **BOLD—CONCERN TO ALL** LIGHT—CONCERN TO SOME
NABISCO *(Continued)*	
OYSTERETTES; ROYAL LUNCH MILK	**(LARD)***
DIXIES	**(COCONUT OIL, LARD);*** EGG WHITES; MONOSODIUM GLUTAMATE; **NATURAL SMOKE FLAVOR; TURMERIC OLEORESIN;** WHEY
NABS CHEESE PEANUT BUTTER SANDWICH	**ARTIFICIAL COLOR; (COCONUT, HYDROGENATED PEANUT OILS, LARD);* PAPRIKA**
NABS MALTED MILK & PEANUT BUTTER SANDWICH	**(COCONUT, HYDROGENATED PEANUT OILS, LARD)***
GRAHAM; SEA ROUNDS	NO ADDITIVES OF CONCERN
SESAME MEAL MATES	**(COCONUT OIL);*** KARAYA GUM
SESAME WHEATS!	**(COCONUT OIL, LARD);* TURMERIC OLEORESIN;** WHEY
SWISS CHEESE	**(COCONUT OIL, LARD);* MODIFIED CORN STARCH;** MONOSODIUM GLUTAMATE; WHEY
WHEATSWORTH	GLUTEN (WHEAT)

BRANDS	ADDITIVES-OF-CONCERN BOLD—CONCERN TO ALL LIGHT—CONCERN TO SOME
WHEAT THINS	**ARTIFICIAL COLOR; (COCONUT OIL, LARD)***
ZWIEBACK TOAST	EGG WHITES; LACTALBUMIN; **MACE; NUTMEG**
OLD LONDON	
RYE MELBA TOAST; SESAME MELBA TOAST; WHITE MELBA TOAST	**PROTEOLYTIC ENZYMES DERIVED FROM ASPERGILLUS ORYZAE**
WHITE MELBA TOAST UNSALTED	**(HYDROGENATED COTTONSEED, PALM, SOYBEAN OILS)***
PEEK FREANS	
CRACKERS FOR SNACKERS	**BHA; (COCONUT, RAPESEED OILS)***
PEPPERIDGE FARM	
BUTTER THINS	**BUTTER; COCONUT OIL**
CHEDDAR CHEESE GOLDFISH; LIGHTLY SALTED GOLDFISH	**COCONUT OIL**
GOLDFISH THINS RYE	**MODIFIED FOOD STARCH**
GOLDFISH THINS LIGHTLY SALTED	**(COCONUT OIL);* MODIFIED FOOD STARCH**
GOLDFISH THINS CHEESE; MIXED SUITS SNACK HERB; MIXED SUITS SNACK SESAME	**COCONUT OIL; MODIFIED FOOD STARCH**

*Blend may contain one or more of these saturated fats and/or oils.

BRANDS	ADDITIVES-OF-CONCERN BOLD—CONCERN TO ALL LIGHT—CONCERN TO SOME
PEPPERIDGE FARM *(Continued)*	
PRETZEL GOLDFISH	**(COCONUT OIL)***
TACO GOLDFISH	**COCONUT OIL; PAPRIKA**
RY KRISP	
NATURAL	NO ADDITIVES OF CONCERN
SEASONED	**BHA; COCONUT OIL**
SOME OF EACH	
CRACKED WHEAT/RYE/GOLDEN	**ARTIFICIAL COLOR; (BEEF FAT, COCONUT OIL, LARD);*** MONOSODIUM GLUTAMATE
SUNSHINE	
ANIMAL; KRISPY SALTINE; OYSTER & SOUP	WHEY
CHEEZ-IT	**(COCONUT OIL);* PAPRIKA**
WHEAT WAFERS	**(COCONUT OIL)***
HI-HO	**(COCONUT OIL);*** WHEY (DAIRY)
CINNAMON GRAHAM; HONEY GRAHAM; KRISPY UNSALTED TOPS	NO ADDITIVES OF CONCERN
VENUS	
CRACKED WHEAT WAFERS; CRACKED *(continues)*	NO ADDITIVES OF CONCERN

*Blend may contain one or more of these saturated fats and/or oils.

BRANDS	**ADDITIVES-OF-CONCERN** **BOLD—CONCERN TO ALL** LIGHT—CONCERN TO SOME
WHEAT WAFERS NO SALT; NEW ENGLAND SODA; WHEAT WAFERS	
WASA	
CRISP RYE BREAD HEARTY RYE; SOURDOUGH TOAST; SWEDISH CRISPBREAD SEASONED; SWEDISH CRISPBREAD SESAME	NO ADDITIVES OF CONCERN
R WESTON	
ONION STIX	**ANIMAL & VEGETABLE SHORTENING; ARTIFICIAL FOOD COLOR;** MONOSODIUM GLUTAMATE
WHEAT O'S	**(BEEF TALLOW, LARD; COCONUT OIL, VEGETABLE OIL SHORTENING);*** WHEY (DRIED)
SUPERMARKET PRIVATE BRANDS	
A & P	
—A & P	
ANIMAL; SOUP & OYSTER	NO ADDITIVES OF CONCERN
CHEESE	**ARTIFICIAL COLOR;** FD & C YELLOW NO. 5; **(COCONUT OIL)***
GRAHAM	**ARTIFICIAL COLOR;** FD & C YELLOW NO. 5

*Blend may contain one or more of these saturated fats and/or oils.

BRANDS	ADDITIVES-OF-CONCERN BOLD—CONCERN TO ALL LIGHT—CONCERN TO SOME
—A & P *(Continued)*	
SALTINES; SNACK	**(COCONUT OIL)***
FIRST NATIONAL	
—EDWARDS OYSTER	NO ADDITIVES OF CONCERN
—FINAST ANIMAL	NO ADDITIVES OF CONCERN
SAFEWAY	
—BUSY BAKER	
BACON	**(COCONUT OIL, HYDROGENATED SOYBEAN OIL);* NATURAL SMOKE FLAVORED YEAST;** WHEY
GARDEN VEGETABLE	**(COCONUT, HYDROGENATED SOYBEAN OILS);* TURMERIC**
ONION; RYE; SESAME; WHEAT	**(COCONUT, HYDROGENATED SOYBEAN OILS)***
SESAME CHEDDAR	**(COCONUT, HYDROGENATED SOYBEAN OILS);* PAPRIKA**
STOP & SHOP	
—STOP & SHOP	
ANIMAL; SALTINES; SOUP & OYSTER	NO ADDITIVES OF CONCERN
CHEESE	**ARTIFICIAL COLOR;** FD & C YELLOW NO. 5; **(COCONUT OIL)***
SUGAR HONEY GRAHAM	**ARTIFICIAL COLOR;** FD & C YELLOW NO. 5

BRANDS	ADDITIVES-OF-CONCERN BOLD—CONCERN TO ALL LIGHT—CONCERN TO SOME
WINN DIXIE —CRACKIN GOOD	
BLEU CHEESE	**ARTIFICIAL COLOR** W/ FD & C YELLOW NO. 5; **(BEEF FAT, LARD);* BHA; COCONUT OIL;** MONOSODIUM GLUTAMATE; WHEY (DAIRY)
CHEEZ BITS	**ARTIFICIAL COLOR** W/ FD & C YELLOW NO. 5; **(BEEF FAT, LARD);* BHA; COCONUT OIL; PAPRIKA;** WHEY (DAIRY)
SUGAR HONEY GRAHAM	**BHA**
ORLEANS WAFERS; SALTINES	**(BEEF FAT, LARD);* BHA;** WHEY (DAIRY)
WHEAT BITS	**ARTIFICIAL COLOR; (BEEF FAT, LARD);* BHA;** WHEY (DAIRY)
ITEMS IN HEALTH FOOD STORES	
CHICO-SAN *RICE CAKES*	
SALTED W/ MILLET ADDED; UNSALTED	NO ADDITIVES OF CONCERN
EL MOLINO MILLS	
HONEY BRAN	GLUTEN FLOUR; WHEY POWDER
EREWHON	
BROWN RICE TAMARI FLAVORED; KOISHI RICE	NO ADDITIVES OF CONCERN

*Blend may contain one or more of these saturated fats and oils.

BRANDS	ADDITIVES-OF-CONCERN BOLD—CONCERN TO ALL LIGHT—CONCERN TO SOME
EREWHON *(Continued)*	
NORI MAKI RICE	NORI SEAWEED
HEALTH VALLEY	
CHEESE	**PAPRIKA**
HERB; SESAME; WHOLE WHEAT; YOGURT	NO ADDITIVES OF CONCERN
HONEY GRAHAM	WHEY POWDER
HUG DAR-VIDA SWISS WHOLE-WHEAT BREAD	**VEGETABLE SHORTENING**
MI-DEL HONEY GRAHAMS	NO ADDITIVES OF CONCERN
SOKEN-SHA	
BROWN RICE	NO ADDITIVES OF CONCERN
SEAWEED CRUNCH	GREEN NORI; HIJIKI SEAWEED; KOMBU SEAWEED; WAKAME SEAWEED
SOVEX *PROTHINS SNACK CHIPS*	
BARBEQUE	**HICKORY SMOKED TORULA YEAST;*** KELP; **PAPRIKA**

*Although "smoked" does not refer to a specific additive, it represents a process utilizing wood smoke, as "smoke flavoring" does, and therefore the foods may contain some cancer-causing benzopyrene chemicals. For this reason, when a food or any of its constituents have been smoked, this has been noted in the additives-of-concern column.

BRANDS	**ADDITIVES-OF-CONCERN** **BOLD—CONCERN TO ALL** LIGHT—CONCERN TO SOME
CELERY; ONION	KELP
SPIRAL BRAND SALTED RICE CAKES	NO ADDITIVES OF CONCERN

CROUTONS

By law, at this writing, it is not required of a manufacturer to state whether cheeses contain artificial colors, unless the color is FD & C Yellow No. 5 (Tartrazine). Their presence in croutons containing cheese therefore remains an uncertainty, unless it is voluntarily declared on the label that artificial colors have been used or have not. For a more detailed explanation of artificial color in cheese, refer to the beginning of the cheese section of "Dairy Products & Substitutes."

BRANDS	ADDITIVES-OF-CONCERN
ARNOLD CANADIAN CHEESE GARLIC; SPICY ITALIAN SALAD FIXIN'S	**ARTIFICIAL COLOR; BHA;** MONOSODIUM GLUTAMATE; WHEY
FRENCH GARLIC	**BHA**
BROWNBERRY BLEU CHEESE; TOASTED	**BHA;** WHEY (DRIED)
CAESAR SALAD	WHEY (DRIED)
ONION & GARLIC; SEASONED	MONOSODIUM GLUTAMATE; WHEY (DRIED)
DEVONSHEER CHEESE & GARLIC; ONION & GARLIC; SEASONED	MONOSODIUM GLUTAMATE; WHEY

BRANDS	ADDITIVES-OF-CONCERN BOLD—CONCERN TO ALL LIGHT—CONCERN TO SOME
DEVONSHEER *(Continued)*	
PLAIN	NO ADDITIVES OF CONCERN
FRANCISCO	
CHEESE & GARLIC	**VEGETABLE OIL**
SEASONED	**(COCONUT OIL);*** MONOSODIUM GLUTAMATE; WHEY (SWEET DAIRY)
FRENCH'S BEL-AIR CHEESE & GARLIC	**ARTIFICIAL COLOR; COCONUT OIL**
KELLOGG'S	
CROUTETTES STUFFING HERB SEASONED	**BHA;** MONOSODIUM GLUTAMATE
PEPPERIDGE FARM	
CHEDDAR & ROMANO CHEESE; CHEESE & GARLIC; ONION & GARLIC; SEASONED	**COCONUT OIL**
SALAD CRISPINS	
FRENCH STYLE	**ARTIFICIAL COLOR;** BUTTERMILK SOLIDS; **PAPRIKA;** WHEY
ITALIAN STYLE	**ARTIFICIAL COLOR; (HYDROGENATED COTTONSEED, PALM, SOYBEAN OILS);* PAPRIKA**
SUPERMARKET PRIVATE BRANDS	
LUCKY	
—HARVEST DAY CROUTON STUFFING MIX	NO ADDITIVES OF CONCERN

	BRANDS	ADDITIVES-OF-CONCERN **BOLD—CONCERN TO ALL** LIGHT—CONCERN TO SOME
	SAFEWAY —MRS. WRIGHT'S HERB SEASONED	**ARTIFICIAL COLOR;** BUTTERMILK SOLIDS; **LARD;** MONOSODIUM GLUTAMATE; WHEY
	UNSEASONED BREAD CUBES	NO ADDITIVES OF CONCERN
	STOP & SHOP —STOP & SHOP CROUTONS	WHEY

CRUMB COATING MIXES

	BRANDS	ADDITIVES-OF-CONCERN
	GOLDEN DIPT *MIX* BATTER	**CALCIUM CASEINATE;** SODIUM ALGINATE; WHEY
	BREADING SEASONED COATING	**OLEORESIN PAPRIKA**
	SPOON 'N FRY DONUT BATTER	EGG WHITES; **SODIUM CASEINATE;** WHEY
	SEAFOOD SEASONED COATING	NO ADDITIVES OF CONCERN
R	KRISPPE BATTER MIX	WHEY
	FRYING BATTER MIX	**CERTIFIED COLOR;** WHEY
	FRYING BREADER	MONOSODIUM GLUTAMATE; **PAPRIKA;** WHEY

BRANDS	ADDITIVES-OF-CONCERN **BOLD—CONCERN TO ALL** LIGHT—CONCERN TO SOME
OVEN FRY *COATING FOR CHICKEN*	
CRISPY CRUMB RECIPE	MONOSODIUM GLUTAMATE; **PAPRIKA**
HOME STYLE FLOUR RECIPE	**ARTIFICIAL COLOR; MODIFIED CORN STARCH; PAPRIKA**
SHAKE 'N BAKE *—SEASONED COATING MIX*	
BARBEQUE CHICKEN	**ARTIFICIAL COLOR;** GUM TRAGACANTH; **MODIFIED CORN STARCH;** MONOSODIUM GLUTAMATE; **PAPRIKA (& EXTRACTIVES);** WHEY
BARBEQUE PORK & RIBS	**MODIFIED CORN STARCH**
CHICKEN ORIGINAL	**ARTIFICIAL COLOR;** MONOSODIUM GLUTAMATE; **NATURAL HICKORY SMOKE FLAVOR; PAPRIKA**
FISH	**ARTIFICIAL COLOR; BHA;** MONOSODIUM GLUTAMATE
PORK	**ARTIFICIAL COLOR; MODIFIED CORN STARCH;** MONOSODIUM GLUTAMATE
SUPERMARKET PRIVATE BRANDS	
SAFEWAY —MRS. WRIGHT'S	
POULTRY, MEAT & FISH DRESSING	NO ADDITIVES OF CONCERN

BRANDS	ADDITIVES-OF-CONCERN **BOLD—CONCERN TO ALL** LIGHT—CONCERN TO SOME
ITEMS IN HEALTH FOOD STORES	
DR. TIMA SEASONED COATING MIX	**PAPRIKA**

QUICK STUFFING & STUFFING MIXES

	ARNOLD	
	GREAT STUFF CHICKEN STUFFING MIX	**BHA; MODIFIED CORN STARCH;** MONOSODIUM GLUTAMATE; **TURMERIC**
	SEASONED STUFFIN'	**BHA;** GLUTEN (VITAL WHEAT); MONOSODIUM GLUTAMATE
R	**BELL'S**	
	PAN TO TABLE QUICK STUFFING MIX CHICKEN	MONOSODIUM GLUTAMATE; **TURMERIC**
	STUFFING	MONOSODIUM GLUTAMATE
R	**GOLDEN DIPT**	
	SEASONED STUFFIN' DRESSING	**(COCONUT, HYDROGENATED SOYBEAN OILS);*** MSG
	PEPPERIDGE FARM	
	CHICKEN & HERB PAN STYLE STUFFING MIX	MONOSODIUM GLUTAMATE
	CORN BREAD STUFFING; HERB SEASONED STUFFING	NO ADDITIVES OF CONCERN
	SEASONED PAN STYLE STUFFING MIX	MONOSODIUM GLUTAMATE; **NATURAL SMOKE FLAVOR**

*Blend may contain one or more of these saturated oils.

BRANDS	ADDITIVES-OF-CONCERN **BOLD—CONCERN TO ALL** LIGHT—CONCERN TO SOME
STOVE TOP *STUFFING MIX* CHICKEN FLAVOR; FOR PORK; W/ RICE	**BHA; MODIFIED WHEAT STARCH;** MONOSODIUM GLUTAMATE; **TURMERIC;** WHEY
CORNBREAD	**ARTIFICIAL COLOR; BHA;** GLUTEN (WHEAT); **MODIFIED WHEAT STARCH;** MONOSODIUM GLUTAMATE; **TURMERIC**
UNCLE BEN'S STUFF 'N SUCH STUFFING MIX TRADITIONAL SAGE	**FERRIC ORTHOPHOSPHATE;** MONOSODIUM GLUTAMATE; **TURMERIC**
SUPERMARKET PRIVATE BRANDS SAFEWAY —MRS. WRIGHT'S CORNBREAD STUFFING MIX	**(HYDROGENATED SOYBEAN OIL);*** MSG

*Blend may contain this saturated oil.

VIII. CANDY & GUM

Although sugars (both sucrose and corn sweeteners) and honey are not treated as additives of concern in this volume when present in moderate amounts as additives, excessive amounts of these or other carbohydrates in the diet can be of health concern. Confections, including chewing gum (other than sugarless gum), are likely to consist of from 20% to 75% sugar.

CANDY BARS

BRANDS	ADDITIVES-OF-CONCERN **BOLD—CONCERN TO ALL** LIGHT—CONCERN TO SOME
BABY RUTH	**ARTIFICIAL COLOR;** WHEY (DAIRY)
CADBURY'S CARAMELLO	WHEY
PEPPERMINT	**ARTIFICIAL COLOR**
—MILK CHOCOLATE BRAZIL NUT; FRUIT & NUT; HAZEL NUT; REGULAR	NO ADDITIVES OF CONCERN
CHARLESTON CHEW! BITE SIZE; VANILLA	**(COCONUT, PALM KERNEL OILS);*** EGG ALBUMIN; MILK SOLIDS (NONFAT)
CHUNKY	NO ADDITIVES OF CONCERN
FOREVER YOURS	EGG WHITES
HEATH ENGLISH TOFFEE	NO ADDITIVES OF CONCERN

*Blend may contain one or more of these saturated oils.

BRANDS	ADDITIVES-OF-CONCERN **BOLD—CONCERN TO ALL** LIGHT—CONCERN TO SOME
HERSHEY'S KRACKEL; MILK CHOCOLATE; MILK CHOCOLATE W/ ALMONDS; MR. GOODBAR	NO ADDITIVES OF CONCERN
KIT KAT	NO ADDITIVES OF CONCERN
MILKY WAY	EGG WHITES
NECCO SKY BAR	EGG WHITES
NESTLÉ CHOCO'LITE; CRUNCH; MILK CHOCOLATE; $100,000	NO ADDITIVES OF CONCERN
PETER PAUL ALMOND JOY	EGG WHITES; WHEY
MOUNDS	EGG WHITES
SNICKERS	EGG WHITES
3 MUSKETEERS	EGG WHITES
ITEMS IN HEALTH FOOD STORES	
APLETS	NO ADDITIVES OF CONCERN
BURRY HEALTH FOODS CAROB CRUNCH; CAROB MINT	**FRACTIONATED PALM KERNEL OIL**
CAROB MILK	NO ADDITIVES OF CONCERN

BRANDS	ADDITIVES-OF-CONCERN BOLD—CONCERN TO ALL LIGHT—CONCERN TO SOME
CAROB CRUNCH	NO ADDITIVES OF CONCERN
GRAPELETS	NO ADDITIVES OF CONCERN
HALVAH CAROB-SESAME; SESAME-HONEY	NO ADDITIVES OF CONCERN
HOFFMAN'S CAROB CLUSTER	**VEGETABLE OIL**
HI-PROTEEN ENERGY	EGG WHITES (DRIED)
SUPER HI-PROTEEN DIET	EGG WHITES (DRIED); **SODIUM CASEINATE**
LIFESTREAM CASHEW HALVAH; HIKERS; RASPBERRY DELITE; SESAME DREAM; SUNSHINE HONEY	NO ADDITIVES OF CONCERN
NATURE'S FANTASY	NO ADDITIVES OF CONCERN
NATURE NOUGATS	WHEY
TANIA'S *CHEWS* MAPLE NUT; SESAME COCONUT	NO ADDITIVES OF CONCERN
WHA GURU CHEW CASHEW-ALMOND; ORIGINAL; SESAME-ALMOND	NO ADDITIVES OF CONCERN

BRANDS	ADDITIVES-OF-CONCERN **BOLD—CONCERN TO ALL** LIGHT—CONCERN TO SOME

CANDY: OTHER (CHOCOLATE, HARD CANDIES, JELLIED & OTHER SOFT CANDIES, MARSHMALLOWS, MINTS, NUT CANDIES)

BRANDS	ADDITIVES-OF-CONCERN
ANDES CREME DE MENTHE'S	**CERTIFIED COLORS;** MILK SOLIDS (NONFAT & WHOLE); **PALM KERNEL OIL**
BORDEN CAMPFIRE MINIATURE MARSHMALLOWS	**ARTIFICIAL COLOR**
BRACH'S	
BURGUNDY; ROYALS	**ARTIFICIAL COLORS;** FD & C YELLOW NO. 5; **BHA;** EGG WHITES; **MODIFIED STARCH**
LICORICE TWISTS	**ARTIFICIAL COLOR**
CLARK PEANUT BUTTER LOGS	**ARTIFICIAL COLOR**
HERSHEY'S KISSES	NO ADDITIVES OF CONCERN
KRAFT	
BUTTER MINTS	**ARTIFICIAL COLOR; BHA; BUTTER**
FUDGIES	WHEY
JETS MARSHMALLOWS	**MODIFIED FOOD STARCH**

BRANDS	ADDITIVES-OF-CONCERN **BOLD—CONCERN TO ALL** LIGHT—CONCERN TO SOME
MINIATURE MARSHMALLOWS (LEMON, LIME, ORANGE, STRAWBERRY)	**ARTIFICIAL COLOR; MODIFIED FOOD STARCH**
TOFFEE	NO ADDITIVES OF CONCERN
LIFE SAVERS	
BREATH SAVERS WINTERGREEN	ACACIA [GUM]; **ARTIFICIAL COLOR; CALCIUM STEARATE**
BUTTER RUM; FIVE FLAVOR; STIKOPEP; TROPICAL FRUITS	**ARTIFICIAL COLORS**
CRYSTOMINT; PEPOMINT; WINTOGREEN	NO ADDITIVES OF CONCERN
M & M'S *CHOCOLATE CANDIES*	
PEANUT; PLAIN	**ARTIFICIAL COLORS**
NECCO	
ASSORTED WAFERS	**ARTIFICIAL COLORS;** GUM ARABIC; GUM TRAGACANTH
CANADA MINTS	GUM TRAGACANTH
CANADA SPEARMINT; CANADA WINTERGREEN	**ARTIFICIAL COLOR** OR **COLORS;** GUM TRAGACANTH
PEARSON	
CARAMEL NIP; COFFEE NIP	WHEY (DAIRY)

BRANDS	ADDITIVES-OF-CONCERN BOLD—CONCERN TO ALL LIGHT—CONCERN TO SOME
PEARSON *(Continued)*	
CHOCOLATE PARFAIT	**(COCONUT, PALM KERNEL, HYDROGENATED SOYBEAN OILS);*** WHEY (DAIRY)
PETER PAUL YORK	
PEPPERMINT PATTIE	EGG WHITES
REESE'S PEANUT BUTTER CUP	NO ADDITIVES OF CONCERN
RICHARDSON	
CLUB MINTS; PARTY JELLIES; PASTEL MINTS	**ARTIFICIAL COLORS**
ROLO	NO ADDITIVES OF CONCERN
SOPHIE MAE PEANUT BRITTLE	NO ADDITIVES OF CONCERN
SPANGLER	
DUM-DUM POPS	**ARTIFICIAL COLOR**
STARBURST FRUIT CHEWS	**ARTIFICIAL COLORS; MODIFIED FOOD STARCH; PALM KERNEL OIL**
TOOTSIE ROLL	
POP DROPS	**ARTIFICIAL COLORS; (COCONUT OIL);*** WHEY
ROLL; ROLL MIDGEES	WHEY
ROLL POPS	**ARTIFICIAL COLOR;** WHEY

*Blend may contain one or more of these saturated oils.

BRANDS	ADDITIVES-OF-CONCERN **BOLD—CONCERN TO ALL** LIGHT—CONCERN TO SOME
SUPERMARKET PRIVATE BRANDS	
A & P —ANN PAGE	
BUTTERSCOTCH; CLEARMINTS; 50 POPS; FILLED BERRIES; JELLY BEANS; LICORICE GUMS; MINT PUFFS; PEPPERMINT LOZENGES; SALT WATER TAFFY; SOUR BALLS; SPEARMINT LEAVES; SPICE DROPS; STRAWBERRY LACES	**ARTIFICIAL COLOR (COLORS)** OR **CERTIFIED COLOR** OR **U.S. CERTIFIED COLOR (COLORS)**
ASSORTED TOFFEES	**ARTIFICIAL COLORS;** EGG ALBUMIN; WHEY SOLIDS
BLACK LACES; CARAMELS; MARY JANES; MILK CHOCOLATE STARS; MINT WAFERS; PEANUT CLUSTERS	NO ADDITIVES OF CONCERN
BUTTERMINTS	**ARTIFICIAL COLOR; BHA; BUTTER**
CANDY CORN	**CONFECTIONER'S GLAZE;** EGG ALBUMIN; **U.S. CERTIFIED COLORS**
CHOCOLATE BRIDGE MIX	**ARTIFICIAL COLOR;** EGG WHITES (DRIED); GUM ARABIC; **RESINOUS GLAZE**
CHOCOLATE PEANUTS	**BHA; HYDROGENATED VEGETABLE OIL;** WHEY POWDER

BRANDS	ADDITIVES-OF-CONCERN **BOLD—CONCERN TO ALL** LIGHT—CONCERN TO SOME
—ANN PAGE *(Continued)*	
CREAM DROPS; THIN MINTS	EGG WHITES (DRIED)
LICORICE DROPS	**ARTIFICIAL COLORS; BUTTER**
MALTED MILK BALLS	**CONFECTIONER'S GLAZE; HYDROGENATED VEGETABLE OIL**
PEANUT BUTTER KISSES	**ARTIFICIAL COLOR; BHA**
STOP & SHOP	
—STOP & SHOP	
CARAMEL RUM	**ARTIFICIAL COLOR; BUTTER**
DUTCHESS MIX; MINT MOLASSES; PEPPERMINT STARLITES; ROOT BEER; SOUR BALLS	**ARTIFICIAL COLOR**
FRUIT & SPICE; ORANGE SLICES; SPEARMINT LEAVES	**ARTIFICIAL COLORS** W/ FD & C YELLOW NO. 5
ITEMS IN HEALTH FOOD STORES	
BARBARA'S	
DELICIOUS FUDGE; DELICIOUS SESAME CRUNCH; PEANUT BRITTLE	NO ADDITIVES OF CONCERN

BRANDS	ADDITIVES-OF-CONCERN **BOLD—CONCERN TO ALL** LIGHT—CONCERN TO SOME
NIK'S SWEETS *TREATS* BARLEY; GINGER; PEANUT	NO ADDITIVES OF CONCERN
PANDA LICORICE ALL NATURAL	NO ADDITIVES OF CONCERN
YINNIES CARAMEL CANDY; TAFFY	NO ADDITIVES OF CONCERN

CHEWING GUM

Chewing-gum base: Chewing gum contains chewing-gum base; 18 specific bases are used, and a few (for example, glycerol esters of wood rosin) are regarded as additives of concern when identified on labels of other foods. Since there is no way of determining which base has been used in a particular gum, and the likelihood of much being ingested is minimal, chewing-gum bases are not noted in the additives-of-concern column.

BRANDS	ADDITIVES-OF-CONCERN
BEECH-NUT SPEARMINT	**ARTIFICIAL COLORS; BHA**
BUBBLE YUM *BUBBLE GUM* REGULAR; SPEARMINT	**ARTIFICIAL COLORS; BHA; SOFTENERS**
GRAPE	**ARTIFICIAL COLORS; BHT; SOFTENERS**
BUBBLICIOUS *SOFT BUBBLE GUM*	**ARTIFICIAL COLOR; SOFTENERS**

BRANDS	ADDITIVES-OF-CONCERN **BOLD—CONCERN TO ALL** LIGHT—CONCERN TO SOME
CARE-FREE *SUGARLESS GUM* BUBBLE	**ARTIFICIAL COLORS; BHA; SOFTENER**
CINNAMON	**ARTIFICIAL COLORS; SOFTENER**
FRUIT; PEPPERMINT; SPEARMINT	**SOFTENER**
CHICLETS FRUIT FLAVORS; SPEARMINT; TINY SIZE FLAVOR COATED	**ARTIFICIAL COLORS**
PEPPERMINT	NO ADDITIVES OF CONCERN
CLORETS	**ARTIFICIAL COLORS**
DENTYNE REGULAR	**ARTIFICIAL COLORS; SOFTENERS**
SPEARMINT	**ARTIFICIAL COLORS**
FRESHEN-UP CINNAMON; SPEARMINT	**ARTIFICIAL COLORS; SOFTENERS**
ORBIT *SUGAR FREE GUM* CINNAMON	**ARTIFICIAL COLORS; BHA; SOFTENERS**
PEPPERMINT; SPEARMINT	**SOFTENERS**

BRANDS	ADDITIVES-OF-CONCERN **BOLD—CONCERN TO ALL** LIGHT—CONCERN TO SOME
TRIDENT *SUGARLESS GUM*	
BUBBLE; SPEARMINT	ACACIA [GUM]; **ARTIFICIAL COLOR** OR **COLORS**
CINNAMON; FRUIT FLAVOR	ACACIA [GUM]; **ARTIFICIAL COLORS; XYLITOL**
ORIGINAL FLAVOR	ACACIA [GUM]
WRIGLEY'S	
BIG RED	**ARTIFICIAL COLORS; SOFTENERS**
DOUBLEMINT; JUICY FRUIT; SPEARMINT	**SOFTENERS**
FREEDENT PEPPERMINT; FREEDENT SPEARMINT	**ARTIFICIAL COLORS; BHA; SOFTENERS**
ITEMS IN HEALTH FOOD STORES	
XYLITOL	
CINNAMON	GUM ARABIC; **XYLITOL**
PEPPERMINT; SPEARMINT	**XYLITOL**

IX. CEREAL, CEREAL BARS (BREAKFAST & SNACK)

CEREAL: HOT

BRANDS	**ADDITIVES-OF-CONCERN** BOLD—CONCERN TO ALL LIGHT—CONCERN TO SOME
MAYPO	
30-SECOND OATMEAL W/ MAPLE FLAVOR; VERMONT STYLE HOT OAT CEREAL	**FERRIC ORTHOPHOSPHATE**
NABISCO	
INSTANT CREAM OF WHEAT	NO ADDITIVES OF CONCERN
QUICK CREAM OF WHEAT	**IRON PHOSPHATE**
—MIX 'N EAT CREAM OF WHEAT	
PLAIN	**BHA;** GUAR GUM; **IRON PHOSPHATE**
BANANA & SPICE	**BHA;** GUAR GUM; **IRON PHOSPHATE; NUTMEG**
PILLSBURY FARINA	NO ADDITIVES OF CONCERN
QUAKER	
ENRICHED WHITE HOMINY GRITS; OLD FASHIONED OATS; QUICK OATS	NO ADDITIVES OF CONCERN
INSTANT GRITS	**BHA**

BRANDS	ADDITIVES-OF-CONCERN **BOLD—CONCERN TO ALL** LIGHT—CONCERN TO SOME
—INSTANT OATMEAL APPLES & CINNAMON; MAPLE & BROWN SUGAR; REGULAR FLAVOR	GUAR GUM
RALSTON INSTANT	**BHA**
WHEATENA	NO ADDITIVES OF CONCERN
SUPERMARKET PRIVATE BRANDS	
LUCKY **—LADY LEE** QUICK OATS	NO ADDITIVES OF CONCERN
SAFEWAY **—MRS. WRIGHT'S** HOMINY GRITS	NO ADDITIVES OF CONCERN
STOP & SHOP **—STOP & SHOP** QUICK OATS	NO ADDITIVES OF CONCERN
ITEMS IN HEALTH FOOD STORES	
CREAM OF RYE	NO ADDITIVES OF CONCERN
EREWHON BROWN RICE CREAM; SCOTCH OATS; SWEET CREAM CEREAL	NO ADDITIVES OF CONCERN
ROMAN MEAL 5 MINUTE	NO ADDITIVES OF CONCERN

BRANDS	ADDITIVES-OF-CONCERN **BOLD—CONCERN TO ALL** LIGHT—CONCERN TO SOME

CEREAL: READY-TO-EAT; WHEAT GERM

When used in reasonable amounts, sugars have not been treated here as additives of concern. However, when they are present as a sizable proportion of the total content of a cereal, a caution may be warranted. Percentages of sugar content have been provided in parentheses for the cereals below for which information was available.

BRANDS	ADDITIVES-OF-CONCERN
FAMILIA	NO ADDITIVES OF CONCERN
GENERAL MILLS BOO BERRY (46);* CHEERIOS (3); LUCKY CHARMS (39); TRIX (31)	**ARTIFICIAL COLOR** OR **COLORS**
BUC WHEATS (11); COCOA PUFFS (34); COUNT CHOCULA (37); GOLDEN GRAHAMS (29); KIX (3)	NO ADDITIVES OF CONCERN
CRAZY COW (CHOCOLATE) (44)	**ARTIFICIAL COLOR; CARRAGEENAN;** GUAR GUM
NATURE VALLEY GRANOLA CINNAMON & RAISINS; NATURE VALLEY GRANOLA TOASTED OAT MIXTURE	**COCONUT OIL**

BRANDS	ADDITIVES-OF-CONCERN BOLD—CONCERN TO ALL LIGHT—CONCERN TO SOME
TOTAL (7); WHEATIES (8)	**BHT**
KELLOGG'S	
ALL-BRAN (19); BRAN BUDS (30);* 40% BRAN FLAKES (16);* RAISIN BRAN (29); SUGAR SMACKS (54)	NO ADDITIVES OF CONCERN
APPLE JACKS (52); FROOT LOOPS (48)	**ARTIFICIAL COLORING; BHA**
COCOA KRISPIES (41)	**BHA**; WHEY
CORN FLAKES (5); CRACKLIN' BRAN (28); FROSTED MINI-WHEATS BROWN SUGAR CINNAMON (16);* FROSTED MINI-WHEATS SUGAR FROSTED (34);* FROSTED RICE (34); PRODUCT 19 (10); RICE KRISPIES (7); SUGAR CORN POPS *(continues)*	**BHA**

*Source of sugar content (sucrose only): Ira L. Shannon, *Brand Name Guide to Sugar* (Nelson-Hall Publishers, 1977). Source for unstarred entries (sucrose and corn sweeteners): B. W. Li and P. J. Shuhmann, *Gas-Liquid Chromatographic Analysis of Sugars in Ready-to-Eat Breakfast Cereals* (Institute of Food Technologists; reprinted in *J. Food Science,* 45:138-141, 1980).

BRANDS	ADDITIVES-OF-CONCERN BOLD—CONCERN TO ALL LIGHT—CONCERN TO SOME
KELLOGG'S *(Continued)*	
(44); SUGAR FROSTED FLAKES (38); TOASTED MINI-WHEATS PLAIN	**BHA**
SPECIAL K (5)	**BHA; CALCIUM CASEINATE;** GLUTEN (WHEAT); WHEY
—NUTRI-GRAIN	
BARLEY; CORN; RYE; WHEAT	**BHA**
KRETSCHMER	
WHEAT GERM REGULAR; WHEAT GERM W/ SUGAR & HONEY	NO ADDITIVES OF CONCERN
NABISCO	
100% BRAN (19)	NO ADDITIVES OF CONCERN
SHREDDED WHEAT (1)	**BHT**
TEAM (13)	**BHT; IRON PHOSPHATE**
POST	
ALPHA-BITS (35)	**ARTIFICIAL COLOR; BHA**
C. W. POST (27); C. W. POST W/ RAISINS (27)	WHEY
COCOA PEBBLES (41)	**(COCONUT, PALM KERNEL OILS)***
FORTIFIED OAT FLAKES (18)	**BHA;** MILK PROTEIN

*Blend may contain one or more of these saturated oils.

BRANDS	**ADDITIVES-OF-CONCERN** **BOLD—CONCERN TO ALL** LIGHT—CONCERN TO SOME
40% BRAN FLAKES (12); GRAPE-NUTS FLAKES (12); HONEY-COMB (36)	**BHA**
FRUITY PEBBLES (41)	**ARTIFICIAL COLOR; (COCONUT, PALM KERNEL OILS)***
GRAPE-NUTS (7); RAISIN BRAN (28); SUPER SUGAR CRISP (45)	NO ADDITIVES OF CONCERN
QUAKER	
LIFE (15)	**ARTIFICIAL COLORING; BHA;** CASEIN
LIFE CINNAMON FLAVOR (21)	**ARTIFICIAL COLOR; BHA;** LACTALBUMIN
PUFFED RICE (0.1); PUFFED WHEAT (0.5)	NO ADDITIVES OF CONCERN
—CAP'N CRUNCH	
CRUNCH BERRIES (43)	**ARTIFICIAL COLORING; BHA;** WHEY (DRIED)
PEANUT BUTTER (31)	**BHA**
REGULAR (39)	**ARTIFICIAL COLOR; BHA**
CORN BRAN	**ARTIFICIAL COLOR** FD & C BLUE NO. 1, **RED NO. 3,** YELLOW NO. 5 & NO. 6
—100% NATURAL CEREAL	
REGULAR; W/ APPLES & CINNAMON (17);* W/ RAISINS & DATES	NO ADDITIVES OF CONCERN

BRANDS	ADDITIVES-OF-CONCERN BOLD—CONCERN TO ALL LIGHT—CONCERN TO SOME
RALSTON PURINA	
—CHEX	
BRAN; CORN (3); RICE (4); WHEAT (4)	**BHA**
—COOKIE CRISP	
CHOCOLATE CHIP (41)	**ARTIFICIAL COLOR;** FD & C YELLOW NO. 5; **BHT**
OATMEAL COOKIE (40)	**ARTIFICIAL COLORS; BHT**
SUPERMARKET PRIVATE BRANDS	
LUCKY	
—LADY LEE	
CRISPY RICE; RAISIN BRAN;	**BHT**
SUGAR FROSTED FLAKES	**BHT; IRON PHOSPHATE**
TASTEEOS	**ARTIFICIAL COLOR;** FD & C YELLOW NO. 5; **BHA; BHT**
—SAFEWAY	
CORN FLAKES; CRISPY RICE	**BHT; IRON PHOSPHATE**
TASTEEOS	**ARTIFICIAL COLOR;** FD & C YELLOW NO. 5; **BHA; BHT**

*Source of sugar content (sucrose only): Ira L. Shannon, *Brand Name Guide to Sugar* (Nelson-Hall Publishers, 1977). Source for unstarred entries (sucrose and corn sweeteners): B. W. Li and P. J. Shuhmann, *Gas-Liquid Chromatographic Analysis of Sugars in Ready-to-Eat Breakfast Cereals* (Institute of Food Technologists; reprinted in *J. Food Science,* 45:138-141, 1980).

BRANDS	**ADDITIVES-OF-CONCERN** **BOLD—CONCERN TO ALL** LIGHT—CONCERN TO SOME
STOP & SHOP **—STOP & SHOP** CORN FLAKES; CRISPY RICE; SUGAR FROSTED FLAKES	**BHT; IRON PHOSPHATE**
—SUN GLORY PUFFED RICE	ACACIA GUM; **FERRIC ORTHOPHOSPHATE**
ITEMS IN HEALTH FOOD STORES	
EL MOLINO PUFFED CORN; PUFFED WHEAT	NO ADDITIVES OF CONCERN
EREWHON SUNFLOWER CRUNCH	**UNREFINED VEGETABLE OIL**
—GRANOLA CAROB COCONUT; DATE NUT; MAPLE; NUMBER 9 SALT-FREE; SPICED APPLE	**UNREFINED VEGETABLE OIL**
HONEY ALMOND	NO ADDITIVES OF CONCERN
FAMILIA SWISS GRANOLA	NO ADDITIVES OF CONCERN
FEARN SOY/O WHEAT CEREAL & SOYA	NO ADDITIVES OF CONCERN
GOOD SHEPHERD ALMONDS 'N' MOLASSES	**COCONUT OIL**
TRADITIONAL W/ ADDED BRAN	NO ADDITIVES OF CONCERN

BRANDS	ADDITIVES-OF-CONCERN BOLD—CONCERN TO ALL LIGHT—CONCERN TO SOME
GOOD SHEPARD *(Continued)*	
UNSWEETENED GRANOLA	**COCONUT OIL;** WHEY
—8 GRAIN CEREAL 'N' SNACK	
APPLE CINNAMON; RAISIN CINNAMON	WHEY
HEALTH VALLEY	
BRAN W/ APPLES & CINNAMON; ORANGEOLA; SPROUTS 7	NO ADDITIVES OF CONCERN
ROSE ENRIGHT	
GREAT GRAINS 'N HONEY	NO ADDITIVES OF CONCERN
SOVEX	
HONEY ALMOND GRANOLA	NO ADDITIVES OF CONCERN
MAPLE WALNUT TRITICALE	WHEY

CEREAL BARS (BREAKFAST & SNACK)

CARNATION	
BREAKFAST BAR	
CHOCOLATE CHIP	**CALCIUM CASEINATE;** MONOSODIUM GLUTAMATE
PEANUT BUTTER CRUNCH	**CALCIUM CASEINATE**

BRANDS	ADDITIVES-OF-CONCERN **BOLD—CONCERN TO ALL** LIGHT—CONCERN TO SOME
CRUNCHOLA PEANUT BUTTER & GRANOLA BARS CHOCOLATE CHIP	**HYDROGENATED PALM OIL; PALM KERNEL OIL; U.S. CERTIFIED COLOR;** WHEY
LITTLE DEBBIE GRANOLA BARS	NO ADDITIVES OF CONCERN
NATURE VALLEY *GRANOLA BARS* CINNAMON; COCONUT; OATS 'N HONEY	**COCONUT OIL**

X. DAIRY PRODUCTS & SUBSTITUTES

R alongside a brand indicates regional (not national) distribution, based on shopping experience.

Note on Cheese; Cheese: Cottage & Ricotta; Cheese: Cream (& Imitations). Declaration of the presence of artificial color in these dairy products by the manufacturer is not required unless the FDA has sufficient evidence that there is a question of safety, in which case the specific color must be identified. FD & C Yellow No. 5 (Tartrazine) is the only artificial color now subjected to this requirement. There is disagreement about the possibility of harmful effects caused by other artificial colors. When artificial color appears on a label, this has been noted in the additives-of-concern column. But when it does not, its absence remains uncertain unless it is voluntarily declared on the label that artificial color has or has not been used.

It may be helpful to compare the color of cheese with milk, its basic component, taking into account that an animal's diet, if rich in grass and other greens, can cause a light yellow hue. Also, a color derived from annatto seed, which produces a yellow to orange hue, is frequently added to cheese: it is not regarded here as an additive of concern. For a review and assessment of food colors, see page 534.

CHEESE

BRANDS	**ADDITIVES-OF-CONCERN** BOLD—CONCERN TO ALL LIGHT—CONCERN TO SOME
ALOUETTE W/ GARLIC & SPICES	LOCUST BEAN GUM
AUSTRIAN ALPS SWISS SLICED	NO ADDITIVES OF CONCERN
BORDEN GRATED PARMESAN; LITE-LINE SINGLE WRAP SLICES; *(continues)*	NO ADDITIVES OF CONCERN

BRANDS	ADDITIVES-OF-CONCERN BOLD—CONCERN TO ALL LIGHT—CONCERN TO SOME
MOZZARELLA SLICED; ROQUEFORT;	
—PASTEURIZED PROCESS AMERICAN	
EASY-PEEL SLICES (ORANGE)	**ARTIFICIALLY COLORED**
EASY-PEEL SLICES (WHITE)	NO ADDITIVES OF CONCERN
CHEESE FOOD SINGLE WRAP SLICES (ORANGE)	**ARTIFICIALLY COLORED;** WHEY & MODIFIED WHEY SOLIDS
—SKIM AMERICAN PASTEURIZED PROCESS CHEESE PRODUCT	
SINGLE WRAP SLICES (ORANGE)	**ARTIFICALLY COLORED;** CAROB BEAN GUM; GUAR GUM
SINGLE WRAP SLICES (WHITE)	CAROB BEAN GUM; GUAR GUM
CASINO	
BRICK; MONTEREY JACK; MOZZARELLA; MUENSTER; ROMANO	NO ADDITIVES OF CONCERN
CHURNY	
CHEESE BALL COLD PACK CHEESE FOOD W/ ALMONDS	GUAR GUM; WHEY
RUGGED SHARP CHEDDAR	NO ADDITIVES OF CONCERN
SMOKSTIK	**HICKORY SMOKED***

BRANDS		ADDITIVES-OF-CONCERN **BOLD—CONCERN TO ALL** LIGHT—CONCERN TO SOME
R	**COON BRAND** EXTRA SHARP CHEDDAR	NO ADDITIVES OF CONCERN
	CRACKER BARREL	
	EXTRA SHARP CHEDDAR	**ARTIFICIAL COLOR**
	EXTRA SHARP-WHITE CHEDDAR; MELLO-WHITE CHEDDAR; MILD-WHITE CHEDDAR; SHARP-WHITE CHEDDAR	NO ADDITIVES OF CONCERN
	SHARP CHEDDAR COLD PACK CHEESE FOOD	**ARTIFICIAL COLOR;** GUAR GUM; WHEY
	KAUKAUNA KLUB	
	CHEDDAR; CHEDDAR (WINE)	**ARTIFICIAL COLOR;** GUAR GUM; WHEY
	KRAFT	
	AGED SWISS SLICED; MOZZARELLA; NEW YORK EXTRA SHARP CHEDDAR IN WAX; 100% GRATED PARMESAN; SWITZERLAND SWISS	NO ADDITIVES OF CONCERN

*Although "smoked" does not refer to a specific additive, it represents a process utilizing wood smoke, as "smoke flavoring" does, and therefore may contain some cancer-causing benzopyrene chemicals. For this reason, when a food or any of its constituents have been smoked, this has been noted in the additives-of-concern column.

BRANDS	ADDITIVES-OF-CONCERN **BOLD—CONCERN TO ALL** LIGHT—CONCERN TO SOME
PROVOLONE SLICED	**SMOKE FLAVOR**
—CHEESE SPREAD PASTEURIZED NEUFCHÂTEL	
W/ OLIVE & PIMIENTO; W/ PINEAPPLE	CAROB BEAN GUM; WHEY
—CHEESE SPREAD PASTEURIZED PROCESS	
CHEEZ WHIZ; VELVEETA	**ARTIFICIAL COLOR;** WHEY
SHARP OLD ENGLISH; W/ BACON	**ARTIFICIAL COLOR**
—PASTEURIZED PROCESS AMERICAN	
CHEESE FOOD SINGLES (ORANGE)	**ARTIFICIAL COLOR;** WHEY
CHEESE FOOD SINGLES (WHITE)	WHEY
DELUXE CHOICE (ORANGE); DELUXE CHOICE SHARP OLD ENGLISH; DELUXE CHOICE (WHITE)	NO ADDITIVES OF CONCERN
LIGHT N' LIVELY AMERICAN FLAVORED CHEESE PRODUCT	**ARTIFICIAL COLOR;** SODIUM ALGINATE; WHEY

BRANDS	ADDITIVES-OF-CONCERN **BOLD—CONCERN TO ALL** LIGHT—CONCERN TO SOME
LAUGHING COW CHEEZBITS	NO ADDITIVES OF CONCERN
MAY-BUD EDAM; GOUDA	NO ADDITIVES OF CONCERN
PARADE AMERICAN SLICES SINGLE WRAPPED (ORANGE); AMERICAN SLICES SINGLE WRAPPED (WHITE)	**ARTIFICIAL COLOR;** GUAR GUM; WHEY & MODIFIED WHEY
EXTRA SHARP CHEDDAR; MILD CHEDDAR	NO ADDITIVES OF CONCERN
RONDELÉ SPICED W/ GARLIC & HERBS; SPICED W/ PEPPER	NO ADDITIVES OF CONCERN
SARGENTO PROVOLONE SLICES; SHREDDED CHEDDAR; SHREDDED MOZZARELLA FOR PIZZA	NO ADDITIVES OF CONCERN
STELLA CRUMBLED BLUE; MOZZARELLA; PARMESAN; ROMANO	NO ADDITIVES OF CONCERN
SWISS KNIGHT (ASSORTED CHEESES)	NO ADDITIVES OF CONCERN

BRANDS	ADDITIVES-OF-CONCERN **BOLD—CONCERN TO ALL** LIGHT—CONCERN TO SOME
WEIGHT WATCHERS PASTEURIZED PROCESS CHEESE PRODUCT	**ARTIFICIAL COLOR**
WISPRIDE *COLD PACK CHEESE FOOD* HICKORY SMOKED	**ARTIFICIAL COLOR;** GUAR GUM; **SMOKE FLAVOR;** WHEY SOLIDS (CHEESE)
SHARP CHEDDAR; W/ PORT WINE	**ARTIFICIAL COLOR** OR **COLORING;** GUAR GUM; WHEY SOLIDS (CHEESE)
WOODY'S *COLD PACK CHEESE FOOD* SHARP CHEDDAR; W/ PORT WINE	**ARTIFICIAL COLOR;** GUAR GUM; WHEY
SUPERMARKET PRIVATE BRANDS **A & P** **—ANN PAGE** GRATED PARMESAN; NATURAL MUENSTER; NATURAL SHARP CHEDDAR (ORANGE); NATURAL SHARP CHEDDAR (WHITE)	NO ADDITIVES OF CONCERN
—AMERICAN PASTEURIZED PROCESS CHED-O-BIT CHEESE FOOD	**ARTIFICIALLY COLORED;** WHEY & MODIFIED WHEY SOLIDS

BRANDS	ADDITIVES-OF-CONCERN **BOLD—CONCERN TO ALL** LIGHT—CONCERN TO SOME
ANN PAGE—*PASTEURIZED PROCESS (Continued)*	
CHED-O-BIT WHITE CHEESE FOOD	WHEY & MODIFIED WHEY SOLIDS
MEL-O-BIT; MEL-O-BIT PIMENTO	NO ADDITIVES OF CONCERN
FIRST NATIONAL	
—EDWARDS	
MOZZARELLA	NO ADDITIVES OF CONCERN
—FINAST	
DELUXE AMERICAN PASTEURIZED PROCESS (ORANGE)	**ARTIFICIALLY COLORED**
DELUXE WHITE AMERICAN PASTEURIZED PROCESS	NO ADDITIVES OF CONCERN
LUCKY	
—LADY LEE	
COLBY; LONGHORN CHEDDAR; MEDIUM TILAMOOK CHEDDAR; MILD CHEDDAR; MONTEREY JACK; MOZZARELLA (SHREDDED); MUENSTER; 100% GRATED PARMESAN; PEPPER JACK; SWISS	NO ADDITIVES OF CONCERN
LONGHORN STYLE CHEDDAR; MILD LONGHORN STYLE CHEDDAR	**ARTIFICIAL COLOR**

BRANDS	ADDITIVES-OF-CONCERN **BOLD—CONCERN TO ALL** LIGHT—CONCERN TO SOME
—PASTEURIZED PROCESS	
SLICES AMERICAN	**ARTIFICIAL COLOR**
SLICES AMERICAN CHEESE SPREAD	**ARTIFICIAL COLOR;** GUAR GUM; WHEY
SLICES SWISS	NO ADDITIVES OF CONCERN
SAFEWAY	
—LUCERNE	
MELLOW SHREDDED CHEDDAR; SHARP SHREDDED CHEDDAR	**COLORING**
SHREDDED MOZZARELLA; SWISS	NO ADDITIVES OF CONCERN
—MIDGET LONGHORN	
NATURAL CHEDDAR; NATURAL COLBY	**COLORING**
—PASTEURIZED PROCESS CHEESE FOOD	
SINGLE SLICES PIMENTO	**ARTIFICIAL COLOR;** WHEY
—(NO BRAND NAME)	
EXTRA SHARP NATURAL CHEDDAR; LONGHORN STYLE NATURAL CHEDDAR; MEDIUM NATURAL CHEDDAR; MILD NATURAL CHEDDAR; MUENSTER; SHARP NATURAL CHEDDAR	**COLORING**

BRANDS	**ADDITIVES-OF-CONCERN** BOLD—CONCERN TO ALL LIGHT—CONCERN TO SOME
SAFEWAY—(NO BRAND NAME) *(Continued)*	
KUMINOST SPICED; MILD CHEDDAR CLUB	NO ADDITIVES OF CONCERN
—(NO BRAND NAME) *PASTEURIZED PROCESS* SMOKED CHEDDAR; SMOKED SWISS	**SMOKED***
—SAFEWAY SWISS (UNSLICED)	NO ADDITIVES OF CONCERN
—*PASTEURIZED PROCESS* AMERICAN; SLICES SHARP AMERICAN	**ARTIFICIAL COLOR**
SWISS	NO ADDITIVES OF CONCERN
STOP & SHOP	
—STOP & SHOP *PASTEURIZED PROCESS* AMERICAN CHEESE FOOD (ORANGE)	**ARTIFICIAL COLOR;** WHEY

*Although "smoked" does not refer to a specific additive, it represents a process utilizing wood smoke, as "smoke flavoring" does, and therefore the food may contain some cancer-causing benzopyrene chemicals. For this reason, when a food or any of its constituents have been smoked, this has been noted in the additives-of-concern column.

BRANDS	ADDITIVES-OF-CONCERN **BOLD—CONCERN TO ALL** LIGHT—CONCERN TO SOME
AMERICAN CHEESE FOOD (WHITE)	WHEY
ITEMS IN HEALTH FOOD STORES	
ALTA-DENA	
MONTEREY JACK; NATURAL SWISS; RAW MILK MILD CHEDDAR; RENNETLESS MILD CHEDDAR	NO ADDITIVES OF CONCERN
CABOT	
NATURAL CHEDDAR	NO ADDITIVES OF CONCERN
VERMONT SMOKED NATURAL CHEDDAR	**NATURAL SMOKE FLAVORING**
GREENBANK FARMS	
—RAW MILK RENNETLESS NATURAL	
JALAPEÑO JACK; LOW SALT JACK; MILD CHEDDAR; MUENSTER; SWISS; TACO CHEDDAR	NO ADDITIVES OF CONCERN
REDWOOD	
—NATURAL RAW WHOLE MILK CHEESE	
LONGHORN CHEDDAR; MEDIUM CHEDDAR; MONTEREY JACK; MUENSTER; SWISS	NO ADDITIVES OF CONCERN

BRANDS	ADDITIVES-OF-CONCERN **BOLD—CONCERN TO ALL** LIGHT—CONCERN TO SOME
TILLAMOOK MEDIUM AGED CHEDDAR	**VEGETABLE COLORING**

CHEESE: COTTAGE & RICOTTA

BELLA COTTA RICOTTA	WHEY (DRY)
COLOMBO RICOTTA	NO ADDITIVES OF CONCERN
HOOD *COTTAGE CHEESE* W/ PINEAPPLE	CAROB BEAN GUM; WHEY (COTTAGE CHEESE)
NUFORM LOWFAT	CAROB BEAN GUM; **CARRAGEENAN;** GUAR GUM; WHEY (COTTAGE CHEESE)
—ALL NATURAL COTTAGE CHEESE CHIVE; GARDEN SALAD; PLAIN LARGE CURD; PLAIN SMALL CURD	CAROB BEAN GUM; WHEY (COTTAGE CHEESE)
THE SEWARD FAMILY VERMONT HILL COUNTRY COTTAGE CHEESE	**CARRAGEENAN;** GUAR GUM

BRANDS	ADDITIVES-OF-CONCERN **BOLD—CONCERN TO ALL** LIGHT—CONCERN TO SOME
SUPERMARKET PRIVATE BRANDS A & P —A & P *COTTAGE CHEESE*	
SMALL CURD	CAROB [BEAN] GUM; WHEY (COTTAGE CHEESE)
FIRST NATIONAL —FINAST *COTTAGE CHEESE*	
LARGE CURD; SMALL CURD	**CARRAGEENAN;** LOCUST BEAN GUM
LUCKY —LADY LEE *COTTAGE CHEESE*	
LARGE CURD; LOWFAT; LOWFAT W/ CHIVES; SMALL CURD	NO ADDITIVES OF CONCERN
PUBLIX —DAIRI-FRESH *COTTAGE CHEESE*	
CHIVES; LARGE CURD	**CARRAGEENAN;** GUAR GUM; LOCUST BEAN GUM
LOWFAT	NO ADDITIVES OF CONCERN
PINEAPPLE	**CARRAGEENAN;** GUAR GUM; LOCUST BEAN GUM; **MODIFIED STARCH**
SAFEWAY —LUCERNE *COTTAGE CHEESE*	
FRUIT SALAD	**MODIFIED STARCH**
PINEAPPLE	GUAR GUM

BRANDS	ADDITIVES-OF-CONCERN BOLD—CONCERN TO ALL LIGHT—CONCERN TO SOME
STOP & SHOP **—STOP & SHOP** ALL NATURAL RICOTTA	NO ADDITIVES OF CONCERN
—COTTAGE CHEESE LARGE CURD; LOWFAT; SMALL CURD	NO ADDITIVES OF CONCERN
WINN DIXIE **—SUPERBRAND** COUNTRY STYLE COTTAGE CHEESE	**CARRAGEENAN;** GUAR GUM
ITEMS IN HEALTH FOOD STORES	
CABOT COTTAGE CHEESE	NO ADDITIVES OF CONCERN
NANCY'S CULTURED RENNETLESS COTTAGE CHEESE	NO ADDITIVES OF CONCERN
ROSEDALE BRAND COTTAGE CHEESE DRY CURD; RICOTTA	NO ADDITIVES OF CONCERN
VALLEY GOLD *COTTAGE CHEESE* LARGE CURD; SMALL CURD	NO ADDITIVES OF CONCERN
LOWFAT	**CARRAGEENAN;** GUAR GUM; LOCUST BEAN GUM
ZAUSNER'S COTTAGE CHEESE	**CARRAGEENAN**

BRANDS	ADDITIVES-OF-CONCERN **BOLD—CONCERN TO ALL** LIGHT—CONCERN TO SOME

CHEESE: CREAM (& IMITATIONS)

BORDEN LITE-LINE NEUFCHÂTEL	CAROB-BEAN GUM
KING SMOOTHIE IMITATION	**ARTIFICIAL COLOR; CARRAGEENAN; (COCONUT OIL);*** GUAR GUM; **SODIUM CASEINATE**
PHILADELPHIA IMITATION; PLAIN; W/ CHIVES	CAROB BEAN GUM
—WHIPPED CREAM CHEESE & BLUE CHEESE; PLAIN; W/ BACON & HORSERADISH; W/ CHIVES; W/ PIMENTOS	CAROB BEAN GUM; GUAR GUM
W/ ONION	CAROB BEAN GUM; GUAR GUM; **TURMERIC;** WHEY
W/ SMOKED SALMON	CAROB BEAN GUM; GUAR GUM; MONOSODIUM GLUTAMATE; **PAPRIKA; SMOKED SALMON†**

*Blend may contain this saturated oil.

†Although "smoked" does not refer to a specific additive, it represents a process utilizing wood smoke, as "smoke flavoring" does, and therefore the food may contain some cancer-causing benzopyrene chemicals. For this reason, when a food or any of its constituents have been smoked, this has been noted in the additives-of-concern column.

BRANDS	ADDITIVES-OF-CONCERN BOLD—CONCERN TO ALL LIGHT—CONCERN TO SOME
SUPERMARKET PRIVATE BRANDS	
LUCKY	
—LADY LEE CREAM CHEESE	CAROB BEAN GUM
WINN DIXIE	
—SUPERBRAND CREAM CHEESE	CAROB BEAN GUM
ITEMS IN HEALTH FOOD STORES	
ZAUSNER'S WHIPPED CREAM CHEESE & CHIVES	**VEGETABLE GUM**

CREAM, HALF & HALF, WHIPPED CREAM

	BRANDS	ADDITIVES-OF-CONCERN
R	**BORDEN** LIGHT	NO ADDITIVES OF CONCERN
	WHIPPING	SODIUM ALGINATE
R	**DAIRI PRIDE**	
	FOUNT-WIP WHIPPED CREAM	**CARRAGEENAN;** MILK SOLIDS (NONFAT)
	HALF & HALF	NO ADDITIVES OF CONCERN
	WHIPPING	ALGIN
R	**DAIRYLEA**	
	HALF & HALF; HEAVY; LIGHT	**CALCIUM CARRAGEENAN**
R	**HOME** HEAVY WHIPPING	SODIUM ALGINATE

	BRANDS	ADDITIVES-OF-CONCERN BOLD—CONCERN TO ALL LIGHT—CONCERN TO SOME
R	**HOOD**	
	HALF & HALF; LIGHT	NO ADDITIVES OF CONCERN
	HEAVY; WHIPPING	SODIUM ALGINATE
	HUNT'S	
	REDDI WIP INSTANT WHIPPED CREAM	**CARRAGEENAN;** MILK SOLIDS (NONFAT)
R	**McARTHUR DAIRY**	
	HEAVY WHIPPING	SODIUM ALGINATE
	LIGHT	MILK SOLIDS (NONFAT)
R	**SEALTEST**	
	HALF & HALF	NO ADDITIVES OF CONCERN
	LIGHT; WHIPPING	SODIUM ALGINATE
	SUPERMARKET PRIVATE BRANDS	
	FIRST NATIONAL —FINAST	
	HALF & HALF; HEAVY; LIGHT	**CALCIUM CARRAGEENAN**
	LUCKY —LADY LEE	
	HALF & HALF	NO ADDITIVES OF CONCERN
	CREAM TOPPING	**CARRAGEENAN;** WHEY
	HEAVY WHIPPING; WHIPPING	**CARRAGEENAN;** GUAR [GUM]
	PUBLIX —DAIRI-FRESH	
	COFFEE; WHIPPING	ALGIN

BRANDS	ADDITIVES-OF-CONCERN **BOLD—CONCERN TO ALL** LIGHT—CONCERN TO SOME
PUBLIX—DAIRY-FRESH *(Continued)*	
HALF & HALF	NO ADDITIVES OF CONCERN
SAFEWAY —LUCERNE	
CREAM TOPPING	**ARTIFICIAL COLOR; CARRAGEENAN;** MILK SOLIDS (NONFAT)
HALF & HALF	NO ADDITIVES OF CONCERN
WHIPPING	**CARRAGEENAN**
STOP & SHOP —STOP & SHOP	
HALF & HALF; LIGHT	NO ADDITIVES OF CONCERN
HEAVY; WHIPPING	ALGIN
ITEMS IN HEALTH FOOD STORES VALLEY GOLD	
HALF & HALF; WHIPPING CREAM	NO ADDITIVES OF CONCERN

CREAM: NONDAIRY; WHIPPED-CREAM SUBSTITUTES

BORDEN CREMORA	**(COCONUT, PALM KERNEL OILS);* SODIUM CASEINATE**
CARNATION	
COFFEE-MATE	**ARTIFICIAL COLORS; (COCONUT, PALM KERNEL OILS);* SODIUM CASEINATE**

*This blend may contain one or more of these saturated oils.

BRANDS	ADDITIVES-OF-CONCERN BOLD—CONCERN TO ALL LIGHT—CONCERN TO SOME
COOL WHIP NON-DAIRY WHIPPED TOPPING	**ARTIFICIAL COLOR; COCONUT & PALM KERNEL OILS;** GUAR GUM; **SODIUM CASEINATE**
DREAM WHIP WHIPPED TOPPING MIX	**ARTIFICIAL COLOR; BHA; PALM KERNEL OIL; SODIUM CASEINATE;** WHEY
KRAFT WHIPPED TOPPING	**ARTIFICIAL COLOR; CALCIUM & SODIUM CASEINATE;** GUAR GUM; **PALM KERNEL OIL;** SODIUM ALGINATE
LUCKY WHIP DESSERT TOPPING	**COCONUT OIL; SODIUM CASEINATE**
MITCHELL'S PERX	**COCONUT OIL; SODIUM CASEINATE**
PARADE NON-DAIRY CREAMER	**ARTIFICIAL COLORS; COCONUT OIL; SODIUM CASEINATE**
RICH'S COFFEE RICH	**ARTIFICIAL COLOR; COCONUT OIL**
WHIPPED TOPPING (CAN)	**ARTIFICIAL COLOR; COCONUT & PALM KERNEL OILS; METHYL ETHYL CELLULOSE**

BRANDS	ADDITIVES-OF-CONCERN BOLD—CONCERN TO ALL LIGHT—CONCERN TO SOME
RICH'S *(Continued)*	
WHIPPED TOPPING SPOON 'N SERVE	**ARTIFICIAL COLOR; COCONUT & PALM KERNEL OILS;** GUAR GUM; **SODIUM CASEINATE**
SUPERMARKET PRIVATE BRANDS	
A & P—A & P	
NON-DAIRY CREAMER	**ARTIFICIAL COLOR; COCONUT OIL**
EIGHT O'CLOCK NON-DAIRY CREAMER	**ARTIFICIAL COLOR; (COCONUT, PALM KERNEL OILS);* SODIUM CASEINATE**
—ANN PAGE	
HANDI WHIP NON-DAIRY WHIPPED TOPPING	**COCONUT & PALM KERNEL OILS;** GUAR GUM; **SODIUM CASEINATE**
FIRST NATIONAL	
—EDWARDS	
NON-DAIRY COFFEE CREAMER	**ARTIFICIAL COLORS; (COCONUT, PALM KERNEL OILS);* SODIUM CASEINATE**
—FINAST CREEM RITE	**ARTIFICIAL COLORS; CASEINATE; COCONUT OIL**
LUCKY	
—LADY LEE	
NON-DAIRY CREAMER	**ARTIFICIAL COLOR; (COCONUT, PALM KERNEL OILS);* SODIUM CASEINATE**
NON-DAIRY PRODUCT FOR CEREAL OR COFFEE	NO ADDITIVES OF CONCERN

BRANDS	ADDITIVES-OF-CONCERN **BOLD—CONCERN TO ALL** LIGHT—CONCERN TO SOME
NON-DAIRY WHIPPED TOPPING	**(COCONUT, PALM KERNEL OILS);*** GUAR GUM; **SODIUM CASEINATE**
PUBLIX	
—FLAVOR PERFECT	
NON-DAIRY CREAMER	**ARTIFICIAL COLOR; (COCONUT, PALM KERNEL OILS);* SODIUM CASEINATE**
SAFEWAY	
—BLOSSOM TIME	
DESSERT TOPPING	**CALCIUM & SODIUM CASEINATE;** GUAR GUM; **PALM KERNEL OIL;** SODIUM ALGINATE
—LUCERNE	
CEREAL BLEND	NO ADDITIVES OF CONCERN
COFFEE TONE	**ARTIFICIAL COLORS; (COCONUT, PALM KERNEL OILS);* SODIUM CASEINATE**
WHIPPING BLEND	**ARTIFICIAL COLOR; HYDROXYPROPYL CELLULOSE; PALM KERNEL OIL; SODIUM CASEINATE**
STOP & SHOP	
—STOP & SHOP	
COFFEE CREAMER	**ARTIFICIAL COLORS; (COCONUT, PALM KERNEL OILS);* SODIUM CASEINATE**

*This blend may contain one or more of these saturated oils.

BRANDS	ADDITIVES-OF-CONCERN **BOLD—CONCERN TO ALL** LIGHT—CONCERN TO SOME
WINN DIXIE —ASTOR NON-DAIRY CREAMER	**ARTIFICIAL COLORS; (COCONUT, PALM KERNEL OILS);* SODIUM CASEINATE**

ICE CREAM & NONDAIRY ICE-CREAM SUBSTITUTES

Note on Ice Cream and Nondairy Ice-Cream Substitutes; Ice-Cream Bars, Cake Rolls, Pies, Popsicles, Sandwiches; Ice Milk, Ices, Yogurt (Frozen); Juice Bars; Sherbet; Yogurt Bars: Declaration of the presence of artificial color in these dairy products by the manufacturer is not required unless the FDA has sufficient evidence that there is a question of safety, in which case the specific color must be identified. FD & C Yellow No. 5 (Tartrazine) is the only artificial color now subjected to this requirement. There is disagreement about the possibility of harmful effects caused by other artificial colors. When artificial color appears on a label, this has been noted in the additives-of-concern column. But when it does not, its presence remains uncertain unless it is voluntarily declared on the label that artificial color has or has not been used.

	BRANDS	ADDITIVES-OF-CONCERN
R	BORDEN COFFEE; DUTCH CHOCOLATE CHIP; VANILLA	**ARTIFICIAL COLOR; CARRAGEENAN;** GUAR GUM; LOCUST BEAN GUM; WHEY
	—ALL NATURAL BUTTERED ALMOND; DUTCH CHOCOLATE; STRAWBERRIES 'N CREAM	**CARRAGEENAN;** GUAR GUM; LOCUST BEAN GUM

*Blend may contain one or more of these saturated oils.

BRANDS	ADDITIVES-OF-CONCERN **BOLD—CONCERN TO ALL** LIGHT—CONCERN TO SOME
CHERRY VANILLA	**CARRAGEENAN;** GUAR GUM; LOCUST BEAN GUM; **MODIFIED FOOD STARCH**
—LADY BORDEN BUTTER FUDGE SWIRL; VANILLA	**ARTIFICIAL COLOR; CARRAGEENAN;** GUAR GUM; LOCUST BEAN GUM
—OLD FASHIONED CHOCOLATE FUDGE SWIRL	**ARTIFICIAL COLOR; CARRAGEENAN;** GUAR GUM; LOCUST BEAN GUM; **MODIFIED FOOD STARCH;** WHEY
DUTCH CHOCOLATE	**CARRAGEENAN;** GUAR GUM; LOCUST BEAN GUM; WHEY
STRAWBERRIES 'N CREAM SWIRL	**ARTIFICIAL COLOR; CARRAGEENAN;** GUAR GUM; LOCUST BEAN GUM; WHEY
BREYERS BUTTER ALMOND; CHERRY VANILLA; CHOCOLATE; COFFEE; NATURAL MINT CHOCOLATE CHIP; NATURAL PEACH; NATURAL STRAWBERRY; NATURAL VANILLA; VANILLA FUDGE TWIRL	NO ADDITIVES OF CONCERN

	BRANDS	ADDITIVES-OF-CONCERN BOLD—CONCERN TO ALL LIGHT—CONCERN TO SOME
	ESKIMO	
	HALF N' HALF CHOCOLATE-VANILLA	**ARTIFICIAL COLOR; CARRAGEENAN;** GUAR GUM; LOCUST BEAN GUM; WHEY
	HÄAGEN-DAZS	
	CAROB; CHOCOLATE; COFFEE; HONEY; RUM RAISIN; STRAWBERRY; VANILLA	NO ADDITIVES OF CONCERN
R	**HOOD**	
	CHOCOLATE; COFFEE; HEAVENLY HASH	**CALCIUM CARRAGEENAN;** CAROB BEAN GUM; GUAR GUM; WHEY
	BUTTERSCOTCH WHIRL	**ARTIFICIAL COLOR; CALCIUM CARRAGEENAN;** CAROB BEAN GUM; GUAR GUM; **MODIFIED FOOD STARCH;** PROPYLENE GLYCOL ALGINATE; WHEY
	CHIPPEDY CHOCOLATY; PEANUT BUTTER CUP; VANILLA	**CALCIUM CARRAGEENAN;** CAROB BEAN GUM; FD & C YELLOW NO. 5 & NO. 6; GUAR GUM; WHEY
	HOODSIE CUPS CHOCOLATE & VANILLA; STRAWBERRY	**ARTIFICIAL COLOR; CALCIUM CARRAGEENAN;** CAROB BEAN GUM; GUAR GUM; WHEY

	BRANDS	ADDITIVES-OF-CONCERN BOLD—CONCERN TO ALL LIGHT—CONCERN TO SOME
	—CORONET	
	BAVARIAN CHOCOLATE; BUTTER ALMOND	CAROB BEAN GUM
	HOWARD JOHNSON'S	
	BURGUNDY CHERRY	**ARTIFICIAL COLOR; CARRAGEENAN;** GUAR GUM; LOCUST BEAN GUM; **MODIFIED FOOD STARCH**
	COFFEE; MOCHA CHIP; STRAWBERRY; VANILLA	**CARRAGEENAN;** GUAR GUM; LOCUST BEAN GUM
R	JUBILEE	
	CHOCOLATE	**CARRAGEENAN;** GUAR GUM; LOCUST BEAN GUM; WHEY
	STRAWBERRY	**CARRAGEENAN; CERTIFIED COLOR;** GUAR GUM; LOCUST BEAN GUM; WHEY
	VANILLA	**CARRAGEENAN;** FD & C YELLOW NO. 5; GUAR GUM; LOCUST BEAN GUM; WHEY
R	McARTHUR DAIRY	
	BUTTER PECAN; CHOCOLATE CHIP	**CARRAGEENAN;** FD & C YELLOW NO. 5; GUAR GUM; LOCUST BEAN GUM; WHEY
	STRAWBERRY	**CARRAGEENAN; CERTIFIED FOOD COLOR;** GUAR GUM; LOCUST BEAN GUM; WHEY

BRANDS		**ADDITIVES-OF-CONCERN** **BOLD—CONCERN TO ALL** LIGHT—CONCERN TO SOME
R	McCARTHY'S *OLD FASHIONED* BUTTERCRUNCH; CHOCOLATE CHIP; COFFEE; STRAWBERRY; VANILLA	**CARRAGEENAN;** GUAR GUM; LOCUST BEAN GUM
	DOUBLE CHOCOLATE	**CARRAGEENAN;** GUAR GUM; LOCUST BEAN GUM; MILK SOLIDS (NONFAT)
R	RICH'S HALF & HALF CHOCOLATE-VANILLA	**ARTIFICIAL COLOR; CARRAGEENAN;** GUAR GUM; WHEY
	DELUXE STRAWBERRY	**ARTIFICIAL COLOR; CARRAGEENAN;** GUAR GUM; WHEY
	SEALTEST BUTTER BRICKLE; MAPLE WALNUT; VANILLA	**ARTIFICIAL COLOR;** CAROB BEAN GUM; **CARRAGEENAN;** GUAR GUM; WHEY
	CHOCOLATE CHIP	CAROB BEAN GUM; **CARRAGEENAN;** GUAR GUM; WHEY
	CHOCOLATE ECLAIR	**ARTIFICIAL COLOR; CALCIUM CASEINATE;** CAROB BEAN GUM; **CARRAGEENAN;** GUAR GUM; PROPYLENE GLYCOL ALGINATE; WHEY

BRANDS	ADDITIVES-OF-CONCERN **BOLD—CONCERN TO ALL** LIGHT—CONCERN TO SOME
FUDGE ROYALE	**ARTIFICIAL COLOR;** CAROB BEAN GUM; **CARRAGEENAN;** GUAR GUM; SODIUM ALGINATE; WHEY & MODIFIED WHEY
STRAWBERRY	**ARTIFICIAL COLOR;** CAROB BEAN GUM; **CARRAGEENAN;** GUAR GUM; GUM KARAYA; WHEY
SUPERMARKET PRIVATE BRANDS	
A & P	
—A & P	
BUTTER PECAN; CHOCOLATE	**CARRAGEENAN;** GUAR GUM; WHEY
VANILLA-CHOCOLATE-STRAWBERRY	**ARTIFICIAL COLOR; CARRAGEENAN;** GUAR GUM; WHEY
—ANN PAGE	
COFFEE; VANILLA	**CARRAGEENAN;** GUAR GUM; WHEY
FIRST NATIONAL	
—EDWARDS-FINAST	
FUDGE MARBLE	**CARRAGEENAN;** GUAR GUM; LOCUST BEAN GUM; **MODIFIED FOOD STARCH;** WHEY
NEAPOLITAN VANILLA/CHOCOLATE/STRAWBERRY	**CARRAGEENAN;** GUAR GUM; LOCUST BEAN GUM; WHEY

BRANDS	**ADDITIVES-OF-CONCERN** **BOLD—CONCERN TO ALL** LIGHT—CONCERN TO SOME
LUCKY	
—HARVEST DAY CHOCOLATE; NEAPOLITAN; VANILLA	**ARTIFICIAL COLOR** OR **COLORING; CARRAGEENAN;** GUAR GUM; WHEY
—LADY LEE	
CHOCOLATE MARSHMALLOW	CAROB [BEAN] GUM; GUAR GUM; WHEY (DRIED)
DUTCH CHOCOLATE; MAPLE NUT; STRAWBERRY	**CARRAGEENAN;** GUAR GUM; WHEY (DRIED)
ENGLISH TOFFEE	**BHA; CARRAGEENAN;** GUAR GUM; WHEY (DRIED)
MARBLE FUDGE	GUAR GUM; **MODIFIED FOOD STARCH;** WHEY (DRIED)
MINT CHOCOLATE CHIP	ARTIFICIAL COLORING FD & C BLUE NO. 1 & YELLOW NO. 5; **CARRAGEENAN;** GUAR GUM; WHEY (DRIED)
VANILLA	CAROB [BEAN] GUM; GUAR GUM
SAFEWAY	
—JOYETT *IMITATION* CHOCOLATE	**CALCIUM CARRAGEENAN;** GUAR GUM; **(COCONUT, HYDROGENATED COTTONSEED, SOYBEAN OILS);*** LOCUST BEAN GUM; WHEY

BRANDS	ADDITIVES-OF-CONCERN **BOLD—CONCERN TO ALL** LIGHT—CONCERN TO SOME
CHOCOLATE CHIP; VANILLA	**ARTIFICIAL COLOR; CALCIUM CARRAGEENAN;** GUAR GUM; **(COCONUT, HYDROGENATED COTTONSEED, SOYBEAN OILS);*** LOCUST BEAN GUM; WHEY
—LUCERNE	
BUTTER BRICKLE; CHOCOLATE CHIP; STRAWBERRY; VANILLA	**ARTIFICIAL COLOR; CALCIUM CARRAGEENAN;** GUAR GUM; LOCUST BEAN GUM
CHOCOLATE MARBLE	**ARTIFICIAL COLOR; CALCIUM CARRAGEENAN;** GUAR GUM; LOCUST BEAN GUM; **MODIFIED FOOD STARCH**
COFFEE	**CALCIUM CARRAGEENAN;** GUAR GUM; LOCUST BEAN GUM; **NATURAL COLOR**
NEAPOLITAN	**ARTIFICIAL COLOR; CALCIUM CARRAGEENAN;** GUAR GUM; LOCUST BEAN GUM; WHEY
—GOURMET	
BURGUNDY CHERRY	**ARTIFICIAL COLOR; CALCIUM CARRAGEENAN;** GUAR GUM; LOCUST BEAN GUM; **MODIFIED STARCH**

*This blend may contain one or more of these saturated oils.

BRANDS	**ADDITIVES-OF-CONCERN** BOLD—CONCERN TO ALL LIGHT—CONCERN TO SOME
—LUCERNE *GOURMET (Continued)*	
ROCKY ROAD	**CALCIUM CARRAGEENAN;** GUAR GUM; LOCUST BEAN GUM; **MODIFIED STARCH;** WHEY
—PARTY PRIDE	
CHOCOLATE FUDGE SUNDAE	**ARTIFICIAL COLOR; CALCIUM CARRAGEENAN;** GUAR GUM; LOCUST BEAN GUM; **MODIFIED FOOD STARCH;** WHEY
—SNOW STAR	
CHOCOLATE MARBLE	**ARTIFICIAL COLOR; CALCIUM CARRAGEENAN;** GUAR GUM; LOCUST BEAN GUM; **MODIFIED FOOD STARCH;** WHEY
STOP & SHOP	
—STOP & SHOP	
COFFEE; COUNTRY CLUB; HARLEQUIN; STRAWBERRY; VANILLA	**ARTIFICIAL COLOR;** CAROB BEAN GUM; **CARRAGEENAN;** GUAR GUM; WHEY
—100% NATURAL	
CHOCOLATE; CHUNKY CHOCOLATE CHIP; VANILLA BEAN	CAROB BEAN GUM; GUAR GUM
WINN DIXIE	
—KOUNTRY FRESH	
PRESTIGE	
BUTTER PECAN; CHOCOLATE CHIP; VANILLA	GUAR GUM; LOCUST BEAN GUM

BRANDS	ADDITIVES-OF-CONCERN BOLD—CONCERN TO ALL LIGHT—CONCERN TO SOME
—SUPERBRAND FUDGE ROYALE	**CARRAGEENAN;** GUAR GUM; LOCUST BEAN GUM; **MODIFIED FOOD STARCH;** WHEY
STRAWBERRY ROYALE	**ARTIFICIAL COLOR; CARRAGEENAN;** GUAR GUM; GUM KARAYA; LOCUST BEAN GUM; **MODIFIED FOOD STARCH;** WHEY
ITEMS IN HEALTH FOOD STORES	
DAMIAN'S CHICORY COFFEE; COUNTRY VANILLA; DANISH TOASTED ALMOND; NATURAL CHOCOLATE; PURE STRAWBERRY; VERMONT MAPLE WALNUT	**IRISH SEA MOSS**
FARM FOODS ICE BEAN HONEY VANILLA	**VEGETABLE GUM**
NATURAL NECTAR BRAZILIAN MOCHA; MAPLE ALMOND CRUNCH; ROYAL DUTCH COCOA; VANILLA	GUAR [GUM]; MILK SOLIDS

	BRANDS	ADDITIVES-OF-CONCERN **BOLD—CONCERN TO ALL** LIGHT—CONCERN TO SOME

ICE-CREAM BARS, CAKE ROLLS, PIES, POPSICLES, SANDWICHES; JUICE BARS; YOGURT BARS

	BRANDS	ADDITIVES-OF-CONCERN
R	CHARLIE FROZEN YOGURT ON-A-STICK RASPBERRY	NO ADDITIVES OF CONCERN
	DANNY SQUARES STRAWBERRY	NO ADDITIVES OF CONCERN
	—FROZEN LOWFAT YOGURT ON-A-STICK CAROB COATED BOYSENBERRY; PIÑA COLADA	NO ADDITIVES OF CONCERN
	CHOCOLATE COATED STRAWBERRY	MILK SOLIDS (NONFAT)
	ESKIMO THIN MINTS	**ARTIFICIAL COLOR; CARRAGEENAN;** GUAR GUM; LOCUST BEAN GUM; SKIM MILK-DERIVED SOLIDS; WHEY
	TOASTED ALMOND CRUNCH!	NO ADDITIVES OF CONCERN
	HIRES ROOT BEER FLOAT BARS	GUM TRAGACANTH; LOCUST BEAN GUM
R	HOOD FUDGSICLE	CAROB BEAN GUM; **CARRAGEENAN;** GUAR GUM; SODIUM ALGINATE; WHEY

	BRANDS	ADDITIVES-OF-CONCERN **BOLD—CONCERN TO ALL** LIGHT—CONCERN TO SOME
	POPSICLE	**ARTIFICIAL COLORS;** GUAR GUM; GUM TRAGACANTH
	VANILLA ICE CREAM SANDWICHES	**ARTIFICIAL COLOR; CALCIUM CARRAGEENAN;** CAROB BEAN GUM; GUAR GUM; WHEY
	—FIRM 'N FROSTY FROZEN YOGURT PUSHUPS	
	ORANGE	BUTTERMILK SOLIDS; CAROB BEAN GUM
	RASPBERRY; STRAWBERRY	CAROB BEAN GUM
R	**KLONDIKES** KRISPY ICE CREAM BARS	**CARRAGEENAN;** GUAR GUM; LOCUST BEAN GUM
R	**LIFESTYLE** VANILLA YOGURT BARS	**CARRAGEENAN; CERTIFIED COLOR;** GUAR GUM; WHEY
R	**LOTTA POPS** SIX FLAVORS	**ARTIFICIAL COLOR;** GUAR GUM
R	**NESTLÉ** CRUNCH VANILLA ICE CREAM BARS	**ARTIFICIAL COLOR; CARRAGEENAN;** GUAR GUM; LOCUST BEAN GUM; WHEY
R	**RICH'S**	
	ICE CREAM CAKE ROLL	NO ADDITIVES OF CONCERN
	LIME JUICE BARS	ALGIN STABILIZER; **U.S. CERTIFIED COLOR**

BRANDS	ADDITIVES-OF-CONCERN **BOLD—CONCERN TO ALL** LIGHT—CONCERN TO SOME
RICH'S *(Continued)*	
ORANGE JUICE BARS	**CARRAGEENAN;** LOCUST BEAN GUM
SEALTEST	
FROZEN YOGURT STRAWBERRY BARS	CAROB BEAN GUM; GUAR GUM
—SMACKERS	
ASSORTED TWIN POPS	**ARTIFICIAL COLOR;** GUAR & KARAYA GUMS
BAVARIAN FUDGE KRUNCH BARS; ICE CREAM SANDWICHES; ORANGE TREAT BARS	**ARTIFICIAL COLOR;** CAROB BEAN GUM; **CARRAGEENAN;** GUAR GUM; WHEY
FUDGE BARS	CAROB BEAN GUM; **CARRAGEENAN;** GUAR GUM; WHEY & MODIFIED WHEY
R **TUSCAN FARMS**	
TUSCAN POPS FROZEN YOGURT ON A STICK CHOCOLATE	CAROB [BEAN] GUM; **CARRAGEENAN;** MILK SOLIDS (NONFAT)
SUPERMARKET PRIVATE BRANDS	
A & P	
—ANN PAGE	
POPSICLE TWIN TREATS	**ARTIFICIAL COLOR;** GUAR GUM
FUDGSICLE TREATS	ALGIN; **ARTIFICIAL COLOR; CARRAGEENAN;** GUAR GUM; LOCUST BEAN GUM; WHEY
VANILLA ICE CREAM BARS	**CARRAGEENAN;** GUAR GUM; WHEY

BRANDS	ADDITIVES-OF-CONCERN **BOLD—CONCERN TO ALL** LIGHT—CONCERN TO SOME
SAFEWAY	
—BEL-AIR FRUIT BARS	GUAR GUM; LOCUST BEAN GUM
—PARTY PRIDE	
ICE CREAM SANDWICHES VANILLA	**ARTIFICIAL COLOR; CALCIUM CARRAGEENAN;** GUAR GUM; LOCUST BEAN GUM; WHEY
MINI POPS ASSORTED FLAVORS	**ARTIFICIAL COLORS;** GUAR GUM
TOFFEE BRITTLE ICE CREAM BARS	**ARTIFICIAL COLOR; CALCIUM CARRAGEENAN;** GUAR GUM; LOCUST BEAN GUM; MILK SOLIDS (NONFAT & WHOLE); **MODIFIED FOOD STARCH;** WHEY
STOP & SHOP	
—STOP & SHOP	
CERT'NLY CITRUS JUICE STICKS	**ARTIFICIAL COLOR;** GUAR GUM
FUDGE & POP	ALGIN; **ARTIFICIAL COLORING; CARRAGEENAN;** GUAR GUM; LOCUST BEAN GUM; **SODIUM CASEINATE;** WHEY
WINN DIXIE	
—SUPERBRAND	
ASSORTED TWIN POPS	**ARTIFICIAL COLOR;** GUAR GUM

BRANDS	ADDITIVES-OF-CONCERN **BOLD—CONCERN TO ALL** LIGHT—CONCERN TO SOME
WINN DIXIE —SUPERBRAND *(Continued)*	
ICE CREAM BARS	**ARTIFICIAL COLOR; CARRAGEENAN;** GUAR GUM; LOCUST BEAN GUM; WHEY
ITEMS IN HEALTH FOOD STORES	
NATURAL NECTAR	
BANANA NUGGET; COCOA CHIP CRUNCH; YULOVIT	GUAR GUM; **IRISH MOSS;** LOCUST BEAN [GUM]; MILK SOLIDS
CAROB COATED VANILLA HONEY ICE CREAM BAR	GUAR [GUM]; MILK SOLIDS (& NONFAT)
INCREDIBLE EDIBLE; NECTAR PIE	GUAR [GUM]; **IRISH MOSS;** LOCUST BEAN [GUM]; MILK SOLIDS (FAT FREE AND/OR NONFAT)
MOCHA PIE	GUAR GUM; **IRISH MOSS;** LOCUST BEAN [GUM]; MILK SOLIDS (& FAT FREE)
—CREAM PIE	
BLUEBERRY CUSTARD	**CARRAGEENAN;** GUAR [GUM]; LOCUST BEAN GUM; MILK SOLIDS (FAT FREE); WHEY
CAPPUCCINO ESPRESSO; FRENCH PECAN	GUAR [GUM]; LOCUST BEAN [GUM]; MILK SOLIDS (FAT FREE); WHEY

	BRANDS	**ADDITIVES-OF-CONCERN** **BOLD—CONCERN TO ALL** LIGHT—CONCERN TO SOME

ICE MILK, ICES, YOGURT (FROZEN)

	BRANDS	ADDITIVES-OF-CONCERN
R	BORDEN NEAPOLITAN VANILLA-CHOCOLATE-STRAWBERRY ICE MILK	**ARTIFICIAL COLOR; CARRAGEENAN;** GUAR GUM; LOCUST BEAN GUM; WHEY
	DANNY *—IN-A-CUP FROZEN LOWFAT YOGURT* BANANA; CHERRY; LEMON; PINEAPPLE-ORANGE; RED RASPBERRY; VANILLA	NO ADDITIVES OF CONCERN
R	GUIDO'S *REAL ITALIAN ICES* CHERRY; LEMON	**ARTIFICIAL COLOR; VEGETABLE GUM**
R	HOOD VANILLA ICE MILK BARS	**ARTIFICIAL COLOR; CALCIUM CARRAGEENAN;** CAROB BEAN GUM; GUAR GUM; WHEY
	—FIRM 'N FROSTY FROZEN YOGURT BLUEBERRY; RASPBERRY	CAROB BEAN GUM
	—NUFORM *ICE MILK* CHIPPEDY CHOCOLATY; VANILLA CHOCOLATE STRAWBERRY	**ARTIFICIAL COLOR; CARRAGEENAN;** GUAR GUM

	BRANDS	ADDITIVES-OF-CONCERN **BOLD—CONCERN TO ALL** LIGHT—CONCERN TO SOME
R	LIFESTYLE *ICE MILK*	
	COFFEE	**CARRAGEENAN;** GUAR GUM; LOCUST BEAN GUM; WHEY
	VANILLA	**CARRAGEENAN;** FD & C YELLOW NO. 5; GUAR GUM; LOCUST BEAN GUM; WHEY
R	RICH'S *ICE MILK* *CAKE BARS*	
	CHOCOLATE SHORTCAKE	**ARTIFICIAL COLOR; CARRAGEENAN;** GUAR GUM; LOCUST BEAN GUM; WHEY
	STRAWBERRY SHORTCAKE	**ARTIFICIAL COLOR; VEGETABLE GUM STABILIZER**
	SEALTEST	
	ICE MILK FUDGE SUNDAE	**ARTIFICIAL COLOR;** CAROB BEAN GUM; **CARRAGEENAN;** GUAR GUM; SODIUM ALGINATE; WHEY & MODIFIED WHEY
	—FROZEN YOGURT	
	BLACK CHERRY; VANILLA BEAN	CAROB BEAN GUM; GUAR GUM
	—LIGHT N' LIVELY ICE MILK	
	CARAMEL NUT; VANILLA, CHOCOLATE, STRAWBERRY	CAROB BEAN GUM; **CARRAGEENAN;** GUAR GUM; WHEY
	CHOCOLATE MARSHMALLOW	CAROB BEAN GUM; **CARRAGEENAN;** EGG ALBUMIN; GUAR GUM; WHEY

BRANDS	**ADDITIVES-OF-CONCERN** **BOLD—CONCERN TO ALL** LIGHT—CONCERN TO SOME
SUPERMARKET PRIVATE BRANDS	
LUCKY	
—LADY LEE *ICE MILK*	
MARBLE FUDGE	**CARRAGEENAN;** GUAR GUM; **MODIFIED FOOD STARCH;** WHEY (DRIED)
NEAPOLITAN	**CARRAGEENAN;** GUAR GUM; WHEY (DRIED)
VANILLA FLAVORED ICE MILK	**CARRAGEENAN;** GUAR GUM; WHEY (DRIED)
SAFEWAY	
—LUCERNE *ICE MILK*	
ROCKY ROAD	**CALCIUM CARRAGEENAN;** GUAR GUM; LOCUST BEAN GUM; **MODIFIED STARCH;** WHEY
TRIPLE TREAT	**ARTIFICIAL COLOR** OR **COLORS; CALCIUM CARRAGEENAN;** GUAR GUM; LOCUST BEAN GUM; WHEY
—LOWFAT FROZEN YOGURT DESSERT	
LEMON; RED RASPBERRY	**ARTIFICIAL COLOR; CALCIUM CARRAGEENAN;** GUAR GUM; LOCUST BEAN GUM
STOP & SHOP	
—STOP & SHOP	
CHOC-LIT COVERS	**CARRAGEENAN;** GUAR GUM; LOCUST BEAN GUM; SKIM MILK SOLIDS

BRANDS	**ADDITIVES-OF-CONCERN** BOLD—CONCERN TO ALL LIGHT—CONCERN TO SOME
WINN DIXIE —THRIFTY MAID *ICE MILK*	
FUDGE ROYALE	**CARRAGEENAN;** GUAR GUM; LOCUST BEAN GUM; **MODIFIED FOOD STARCH;** WHEY
NEAPOLITAN	**ARTIFICIAL COLOR; CARRAGEENAN;** GUAR GUM; LOCUST BEAN GUM; WHEY
ITEMS IN HEALTH FOOD STORES SHILOH FARMS *FROZEN YOGURT*	
BLUEBERRY; HONEY; PEACH; RASPBERRY; STRAWBERRY	**CARRAGEENAN;** GUAR [GUM]; LOCUST BEAN [GUM]; MILK SOLIDS (NONFAT); WHEY POWDER

MARGARINE

Artificial color in margarine has not been noted because the coloring agent most often used, natural or synthetic beta carotene, is not regarded as an additive of concern. Beta carotene adds a yellowish-orange hue and a source of vitamin A to margarine.

AUTUMN NATURAL	NO ADDITIVES OF CONCERN
BLUE BONNET	
REGULAR; SOFT	**POTASSIUM CASEINATE;** WHEY
CHIFFON	
SOFT; WHIPPED	NO ADDITIVES OF CONCERN

	BRANDS	ADDITIVES-OF-CONCERN **BOLD—CONCERN TO ALL** LIGHT—CONCERN TO SOME
	FLEISCHMANN'S DIET IMITATION; SWEET UNSALTED	NO ADDITIVES OF CONCERN
	CORN OIL; LIGHT CORN OIL SPREAD; SOFT	**POTASSIUM CASEINATE;** WHEY
	IMPERIAL REGULAR; SOFT	**POTASSIUM CASEINATE;** WHEY
	KRAFT PARKAY; SOFT PARKAY; SQUEEZE PARKAY	**POTASSIUM CASEINATE;** WHEY
	LAND O LAKES CORN OIL; REGULAR; SOFT	NO ADDITIVES OF CONCERN
	MAZOLA REGULAR; SWEET-UNSALTED	NO ADDITIVES OF CONCERN
	MRS FILBERTS CORN OIL; SOFT GOLDEN; SPREAD 25	NO ADDITIVES OF CONCERN
	GOLDEN QUARTERS; SOFT	**POTASSIUM CASEINATE;** WHEY
R	**PROMISE** REGULAR	NO ADDITIVES OF CONCERN
R	**SHEDD'S** SPREAD	NO ADDITIVES OF CONCERN
	WEIGHT WATCHERS IMITATION	NO ADDITIVES OF CONCERN

BRANDS	ADDITIVES-OF-CONCERN **BOLD—CONCERN TO ALL** LIGHT—CONCERN TO SOME
SUPERMARKET PRIVATE BRANDS	
A & P	
—A & P	
CORN OIL; SOFT	NO ADDITIVES OF CONCERN
—ANN PAGE	
REGULAR	NO ADDITIVES OF CONCERN
FIRST NATIONAL	
—**EDWARDS** CORN OIL	MILK SOLIDS (NONFAT DRY)
—**FINAST** REGULAR	NO ADDITIVES OF CONCERN
LUCKY	
—LADY LEE	
CORN OIL; REGULAR; SOFT	**POTASSIUM CASEINATE;** WHEY
PUBLIX	
—BREAKFAST CLUB	
GOLDEN CORN OIL; REGULAR	NO ADDITIVES OF CONCERN
60% SPREAD	WHEY
SAFEWAY	
—**EMPRESS** CORN OIL	**POTASSIUM CASEINATE;** WHEY
—SCOTCH BUY	
REGULAR; SOFT	**POTASSIUM CASEINATE;** WHEY
SOFT SPREAD	NO ADDITIVES OF CONCERN
STOP & SHOP	
—STOP & SHOP	
REGULAR; SOFT	MILK SOLIDS (NONFAT DRY)

BRANDS	ADDITIVES-OF-CONCERN **BOLD—CONCERN TO ALL** LIGHT—CONCERN TO SOME
WINN DIXIE —SUPERBRAND LIGHT SPREAD; SOFT	NO ADDITIVES OF CONCERN
REGULAR	**POTASSIUM CASEINATE;** WHEY
ITEMS IN HEALTH FOOD STORES	
FRAZIER FARMS SOYBEAN OIL	NO ADDITIVES OF CONCERN
HAIN SAFFLOWER OIL	NO ADDITIVES OF CONCERN
SHEDD'S WILLOW RUN SOY BEAN	NO ADDITIVES OF CONCERN

BRANDS	ADDITIVES-OF-CONCERN BOLD—CONCERN TO ALL LIGHT—CONCERN TO SOME

MILK OTHER THAN WHOLE MILK* (ACIDOPHILUS & KEFIR, BUTTERMILK, CHOCOLATE, CONDENSED, EVAPORATED, LOW-FAT, NONFAT, NONFAT DRY, SKIM)

Note on Milk Other Than Whole Milk; Yogurt (Refrigerated): Declaration of the presence of artificial color in these dairy products by the manufacturer is not required unless the FDA has sufficient evidence that there is a question of safety, in which case the specific color must be identified. FD & C Yellow No. 5 (Tartrazine) is the only artificial color now subjected to this requirement. There is disagreement about the possibility of harmful effects caused by other artificial colors. When artificial color appears on a label, this has been noted in the additives-of-concern column. But when it does not, its presence remains uncertain unless it is voluntarily declared on the label that artificial color has or has not been used.

	Brands	Additives-of-concern
	ALBA NONFAT DRY	NO ADDITIVES OF CONCERN
R	BORDEN	
	BUTTERMILK	MILK SOLIDS (NONFAT)
	DUTCH CHOCOLATE	**CARRAGEENAN;** GUAR GUM
	—EAGLE BRAND CONDENSED	NO ADDITIVES OF CONCERN
	CARNATION	
	NONFAT DRY; EVAPORATED SKIMMED	NO ADDITIVES OF CONCERN

*Whole milk has not been included. Inspection of labels in various parts of the country found none that contained any additives of concern.

	BRANDS	ADDITIVES-OF-CONCERN **BOLD—CONCERN TO ALL** LIGHT—CONCERN TO SOME
	EVAPORATED	**CARRAGEENAN**
R	**FRIENDSHIP** BUTTERMILK	NO ADDITIVES OF CONCERN
R	**HOME** STA-TRIM LOWFAT	NO ADDITIVES OF CONCERN
	HOOD LOWFAT; SIL-OU-ET	NO ADDITIVES OF CONCERN
	CHOCOLATE LOWFAT	**CARRAGEENAN**
	NUFORM	MILK SOLIDS (NONFAT)
R	**McARTHUR DAIRY** BUTTERMILK	**CARRAGEENAN;** GUAR GUM; MILK SOLIDS (NONFAT)
	CHOCOLATE LOWFAT	**CARRAGEENAN**
	SWEET ACIDOPHILUS LOWFAT	NO ADDITIVES OF CONCERN
R	**McINTIRE DAIRY FARM** TRU FORM LOWFAT	NO ADDITIVES OF CONCERN
	MIX 'N DRINK NONFAT DRY	NO ADDITIVES OF CONCERN
	PARADE EVAPORATED	**CARRAGEENAN**
	PET EVAPORATED	**CARRAGEENAN**
	EVAPORATED SKIMMED	NO ADDITIVES OF CONCERN

	BRANDS	ADDITIVES-OF-CONCERN **BOLD—CONCERN TO ALL** LIGHT—CONCERN TO SOME
	SANALAC NONFAT DRY	NO ADDITIVES OF CONCERN
R	**SEALTEST**	
	CHOCOLATE LOWFAT	**CALCIUM CARRAGEENAN**
	LIGHT 'N' LIVELY LOWFAT	NO ADDITIVES OF CONCERN
R	T. G. LEE	
	COUNTRY STYLE BUTTERMILK	NO ADDITIVES OF CONCERN
	HI-LO LOWFAT	MILK SOLIDS (NONFAT)
	SUPERMARKET PRIVATE BRANDS	
	A & P	
	—A & P	
	LOOK FIT LOW FAT; NONFAT DRY	NO ADDITIVES OF CONCERN
	NONFAT	MILK SOLIDS (NONFAT)
	FIRST NATIONAL	
	—FINAST	
	EVAPORATED	**CARRAGEENAN**
	LUCKY	
	—LADY LEE	
	BUTTERMILK	NO ADDITIVES OF CONCERN
	LOWFAT W/ LACTOBACILLUS ACIDOPHILUS CULTURE ADDED; NONFAT	MILK SOLIDS (NONFAT)
	EVAPORATED	**CARRAGEENAN**

BRANDS	ADDITIVES-OF-CONCERN **BOLD—CONCERN TO ALL** LIGHT—CONCERN TO SOME
SAFEWAY —LUCERNE	
BUTTERMILK; LOWFAT W/ LACTOBACILLUS ACIDOPHILUS CULTURE ADDED; NONFAT	MILK SOLIDS (NONFAT)
CHOCOLATE LOWFAT; EVAPORATED	**CARRAGEENAN**
STOP & SHOP —STOP & SHOP	
EVAPORATED	**CARRAGEENAN**
GREAT SHAPE LOWFAT; NONFAT	MILK SOLIDS (NONFAT)
ITEMS IN HEALTH FOOD STORES	
ALTA-DENA KEFIR	
BLACK CHERRY; PEACH; RED RASPBERRY	NO ADDITIVES OF CONCERN
PINEAPPLE	**TURMERIC EXTRACT**
EAST COAST KEFIR	
BLUEBERRY; RASPBERRY; STRAWBERRY	**VEGETABLE COLOR**
MAPLE SYRUP	NO ADDITIVES OF CONCERN

	BRANDS	ADDITIVES-OF-CONCERN **BOLD—CONCERN TO ALL** LIGHT—CONCERN TO SOME
	SHERBET*	
R	BORDEN LIME; RAINBOW	**ARTIFICIAL COLOR; CARRAGEENAN;** GUAR GUM; WHEY
R	HÄAGEN-DAZS BOYSENBERRY	NO ADDITIVES OF CONCERN
R	HOOD LEMON-ORANGE-LIME	**ARTIFICIAL COLOR; CALCIUM CARRAGEENAN;** CAROB BEAN GUM; GUAR GUM; GUM KARAYA
R	McARTHUR DAIRY LIME	**ARTIFICIAL COLOR;** FD & C YELLOW NO. 5; GUAR GUM; LOCUST BEAN GUM; WHEY
	SEALTEST LEMON-LIME; ORANGE; RAINBOW	**ARTIFICIAL COLOR;** CAROB BEAN GUM; **CARRAGEENAN;** GUAR GUM; WHEY
	SUPERMARKET PRIVATE BRANDS A & P —A & P LEMON; LIME	**ARTIFICIAL COLOR;** GUAR GUM; LOCUST BEAN GUM; **TURMERIC** OR **TURMERIC COLOR**

*See note on page 222.

BRANDS	**ADDITIVES-OF-CONCERN** **BOLD—CONCERN TO ALL** LIGHT—CONCERN TO SOME
LUCKY **—LADY LEE** PINEAPPLE	CAROB [BEAN] GUM; GUAR GUM
RAINBOW	ARTIFICIAL COLORING FD & C BLUE NO. 1, YELLOW NO. 5 & NO. 6; CAROB [BEAN] GUM; GUAR GUM
RASPBERRY SHERBET/VANILLA ICE MILK	**CARRAGEENAN;** GUAR GUM; LOCUST BEAN GUM; WHEY (DRIED)
SAFEWAY **—LUCERNE** ORANGE; RASPBERRY; TRIPLE TREAT	**ARTIFICIAL COLOR** OR **COLORS; CALCIUM CARRAGEENAN;** GUAR GUM; LOCUST BEAN GUM
STOP & SHOP **—STOP & SHOP** ORANGE/LEMON/LIME; RASPBERRY	**ARTIFICIAL COLOR;** GUAR GUM; WHEY
WINN DIXIE **—SUPERBRAND** LIME; ORANGE	**ARTIFICIAL COLOR; CARRAGEENAN;** GUAR GUM; LOCUST BEAN GUM; WHEY

BRANDS	ADDITIVES-OF-CONCERN BOLD—CONCERN TO ALL LIGHT—CONCERN TO SOME
WINN DIXIE—SUPERBRAND *(Continued)*	
PINEAPPLE	**CARRAGEENAN;** GUAR GUM; LOCUST BEAN GUM; WHEY

SOUR CREAM & IMITATIONS

HOOD SOUR CREAM	LOCUST BEAN GUM
THE SEWARD FAMILY SOUR CREAM	**VEGETABLE GUM**
SUPERMARKET PRIVATE BRANDS	
A & P **—A & P** SOUR CREAM	**CARRAGEENAN;** LOCUST BEAN GUM
LUCKY **—LADY LEE** SOUR CREAM	**CARRAGEENAN;** LOCUST BEAN GUM; MILK SOLIDS (NONFAT); **MODIFIED STARCH**
SOUR DRESSING	**COCONUT OIL;** LOCUST BEAN GUM; **SODIUM CASEINATE;** WHEY SOLIDS
PUBLIX **—DAIRI-FRESH** SOUR CREAM; SOUR HALF & HALF	**CARRAGEENAN;** LOCUST BEAN GUM
SAFEWAY **—LUCERNE** IMITATION SOUR CREAM	**CARRAGEENAN; COCONUT OIL;** PROPYLENE GLYCOL ALGINATE

BRANDS	ADDITIVES-OF-CONCERN **BOLD—CONCERN TO ALL** LIGHT—CONCERN TO SOME
SOUR HALF & HALF	CAROB [BEAN] GUM; **CARRAGEENAN;** MILK SOLIDS (NONFAT); **MODIFIED STARCH**
ITEMS IN HEALTH FOOD STORES	
VALLEY GOLD SOUR CREAM	**CARRAGEENAN;** GUAR GUM; MILK SOLIDS (NONFAT DRY)

YOGURT (REFRIGERATED)*

BREYER'S PLAIN	NO ADDITIVES OF CONCERN
COLOMBO	
BLACK CHERRY; BLUEBERRY; PEACH MELBA; STRAWBERRY	MILK SOLIDS (NONFAT DRY); **MODIFIED FOOD STARCH**
NATURAL; VANILLA HONEY	MILK SOLIDS (NONFAT DRY)
DANNON	
BANANA; BLUEBERRY; COFFEE; DUTCH APPLE; PEACH; PLAIN; RED RASPBERRY; STRAWBERRY; VANILLA	MILK SOLIDS (NONFAT)
HOOD NUFORM	NO ADDITIVES OF CONCERN
KNUDSEN	
APRICOT PINEAPPLE	**ARTIFICIAL COLOR (VEGETABLE);** MILK SOLIDS (NONFAT); **MODIFIED FOOD STARCH**

*See note on page 222.

BRANDS	ADDITIVES-OF-CONCERN **BOLD—CONCERN TO ALL** LIGHT—CONCERN TO SOME
KNUDSEN *(Continued)*	
PLAIN	MILK SOLIDS (NONFAT)
RED RASPBERRY; STRAWBERRY	**ARTIFICIALLY COLORED; MODIFIED FOOD STARCH**
SPICED APPLE	MILK SOLIDS (NONFAT); **MODIFIED CORN STARCH**
NEW COUNTRY	
BLUEBERRY RIPPLE; PEACHES 'N CREAM	**MODIFIED FOOD STARCH**
HAWAIIAN SALAD	**MODIFIED FOOD STARCH; TURMERIC**
SIPPITY	
BLUEBERRY; PEACH; RASPBERRY	GUAR GUM; LOCUST BEAN GUM
SUPERMARKET PRIVATE BRANDS	
A & P	
—A & P	
APRICOT; BLUEBERRY; STRAWBERRY	MILK SOLIDS (NONFAT)
LUCKY	
—LADY LEE *BLENDED*	
CHERRY; LEMON; PEACH; RASPBERRY	**ARTIFICIAL COLOR;** CAROB [BEAN GUM]; MILK SOLIDS (NONFAT)
—FRUIT ON BOTTOM	
BLACKBERRY	**ARTIFICIALLY COLORED W/ VEGETABLE COLORS;** LOCUST BEAN GUM; MILK SOLIDS (NONFAT)

BRANDS	ADDITIVES-OF-CONCERN **BOLD—CONCERN TO ALL** LIGHT—CONCERN TO SOME
CHERRY; PEACH; STRAWBERRY	**ARTIFICIAL COLOR;** LOCUST BEAN GUM; MILK SOLIDS (NONFAT)
PUBLIX	
—DAIRI-FRESH	
PLAIN	LOCUST BEAN GUM; **MODIFIED STARCH**
—DAIRI-FRESH *SWISS STYLE*	
BLACK CHERRY; BOYSENBERRY; LEMON; MANDARIN ORANGE; PEACH; RASPBERRY	**CERTIFIED COLOR;** LOCUST BEAN GUM; **MODIFIED STARCH**
PRUNE	LOCUST BEAN GUM; **MODIFIED STARCH**
SAFEWAY	
—LUCERNE	
APRICOT-PINEAPPLE; LEMON	**TURMERIC COLOR**
BANANA; BLACKBERRY; BOYSENBERRY; ORANGE; PEACH; PLAIN; SPICED APPLE	NO ADDITIVES OF CONCERN
CHERRY; RED RASPBERRY; STRAWBERRY	**CARMINE COLOR**
CHERRY VANILLA	**CARMINE COLOR;** MILK SOLIDS (SKIM); **MODIFIED FOOD STARCH**

BRANDS	ADDITIVES-OF-CONCERN **BOLD—CONCERN TO ALL** LIGHT—CONCERN TO SOME
STOP & SHOP —STOP & SHOP	
APRICOT; BLUEBERRY; PEACH; PINEAPPLE; RASPBERRY; STRAWBERRY	NO ADDITIVES OF CONCERN
—SWISS STYLE	
BLUEBERRY; PEACH; PINEAPPLE; RASPBERRY; STRAWBERRY	**ARTIFICIAL COLOR;** MILK SOLIDS (NONFAT); **MODIFIED FOOD STARCH**
WINN DIXIE —SUPERBRAND *ALL NATURAL*	
MANDARIN ORANGE	**CARRAGEENAN**
VANILLA	NO ADDITIVES OF CONCERN
—SWISS STYLE	
CHERRY VANILLA; LEMON; RASPBERRY	**ARTIFICIAL COLOR; MODIFIED FOOD STARCH**
ITEMS IN HEALTH FOOD STORES	
ALTA-DENA *—MAYA*	
BLACK CHERRY; BLUEBERRY; BOYSENBERRY; RED RASPBERRY; STRAWBERRY	CAROB BEAN [GUM]; **CARRAGEENAN;** GUAR [GUM]; [GUM] TRAGACANTH
LEMON; PEACH	CAROB BEAN [GUM]; **CARRAGEENAN;** GUAR [GUM]; [GUM] TRAGACANTH; **TURMERIC EXTRACT**

BRANDS	**ADDITIVES-OF-CONCERN** **BOLD—CONCERN TO ALL** LIGHT—CONCERN TO SOME
PLAIN	**CARRAGEENAN**
—*NAJA LOWFAT* BLACK CHERRY; BLUEBERRY; PEACH; STRAWBERRY	CAROB BEAN [GUM]; **CARRAGEENAN;** GUAR [GUM]; [GUM] TRAGACANTH
LEMON	CAROB BEAN [GUM]; **CARRAGEENAN;** GUAR [GUM]; [GUM] TRAGACANTH; **TURMERIC EXTRACT**
PLAIN; VANILLA	**CARRAGEENAN**
BROWN COW FARM BLUEBERRY; PLAIN; STRAWBERRY; TUPELO HONEY; VANILLA	NO ADDITIVES OF CONCERN
MOONDANCE	WHEY SOLIDS

XI. FISH & SHELLFISH

CANNED FISH & SHELLFISH

BRANDS	**ADDITIVES-OF-CONCERN** **BOLD—CONCERN TO ALL** LIGHT—CONCERN TO SOME
BUMBLE BEE BRAND —*SALMON* ALASKAN SOCKEYE RED; PINK	NO ADDITIVES OF CONCERN
—*TUNA* CHUNK LIGHT; CHUNK WHITE; SOLID WHITE; SOLID WHITE IN WATER	NO ADDITIVES OF CONCERN
CHICKEN OF THE SEA *TUNA* CHUNK LIGHT	**VEGETABLE OIL**
CHUNK LIGHT IN WATER; SOLID WHITE IN WATER	NO ADDITIVES OF CONCERN
CONNORS KIPPERED SNACKS	**LIQUID SMOKE FLAVORED**
DOXSEE MINCED CLAMS	MONOSODIUM GLUTAMATE
DURKEE *SARDINES* GRANADAISA; KING DAVID	NO ADDITIVES OF CONCERN
EAST POINT SHRIMP	NO ADDITIVES OF CONCERN

BRANDS	ADDITIVES-OF-CONCERN BOLD—CONCERN TO ALL LIGHT—CONCERN TO SOME
EMPRESS WHOLE OYSTERS	NO ADDITIVES OF CONCERN
GEISHA KING CRABMEAT; NORWAY SARDINES; TINY SHRIMP; WHOLE OYSTERS	NO ADDITIVES OF CONCERN
—*TUNA* LIGHT IN WATER; SOLID WHITE IN WATER	NO ADDITIVES OF CONCERN
GORTON'S CODFISH CAKES	**MODIFIED FOOD STARCH**
MINCED CLAMS	MONOSODIUM GLUTAMATE
KING OSCAR KIPPER SNACKS; LIGHTLY SMOKED BRISLING SARDINES	**LIGHTLY SMOKED***
NORWAY SARDINES	**SMOKED***
NORSE PRINCE KIPPER SNACKS	**SMOKED***
NORWAY MAID *NORWEGIAN SARDINES* IN MUSTARD SAUCE; IN OIL; IN TOMATO SAUCE	NO ADDITIVES OF CONCERN

*Although "smoked" does not refer to a specific additive, it represents a process utilizing wood smoke, as "smoke flavoring" does, and therefore the food may contain some cancer-causing benzopyrene chemicals. For this reason, when a food or any of its constituents has been smoked, this has been noted in the additives-of-concern-column.

BRANDS	ADDITIVES-OF-CONCERN BOLD—CONCERN TO ALL LIGHT—CONCERN TO SOME
PORTLOCK SMOKED SALMON	**SMOKED***
ROMANOFF *ICELANDIC CAVIAR* BLACK LUMPFISH; RED LUMPFISH	**ARTIFICIAL COLOR**
SALTSEA MINCED CLAMS	NO ADDITIVES OF CONCERN
SNOW'S MINCED CLAMS	NO ADDITIVES OF CONCERN
STAR-KIST *TUNA* CHUNK LIGHT; CHUNK LIGHT IN SPRING WATER; CHUNK WHITE IN WATER	NO ADDITIVES OF CONCERN
SOLID WHITE TUNA	**VEGETABLE OIL**
UNDERWOOD SARDINES IN OIL	**SMOKED***
SUPERMARKET PRIVATE BRANDS	
A & P	
—A & P SOLID WHITE TUNA	NO ADDITIVES OF CONCERN

BRANDS	ADDITIVES-OF-CONCERN **BOLD—CONCERN TO ALL** LIGHT—CONCERN TO SOME
FIRST NATIONAL **—EDWARDS** CHUNK LIGHT TUNA	NO ADDITIVES OF CONCERN
—EDWARDS-FINAST CHUNK LIGHT TUNA	NO ADDITIVES OF CONCERN
SAFEWAY **—SEA TRADER** CHUNK LIGHT TUNA IN WATER	NO ADDITIVES OF CONCERN
STOP & SHOP **—STOP & SHOP** BRISLING SARDINES	NO ADDITIVES OF CONCERN
—*TUNA* CHUNK LIGHT; WHITE IN WATER	NO ADDITIVES OF CONCERN
ITEMS IN HEALTH FOOD STORES	
FEATHERWEIGHT PINK SALMON; UNSALTED LIGHT TUNA CHUNKS	NO ADDITIVES OF CONCERN
UNSALTED NORWAY SMOKED BRISLING SARDINES IN WATER	**SMOKED***

*Although "smoked" does not refer to a specific additive, it represents a process utilizing wood smoke, as "smoke flavoring" does, and therefore the food may contain some cancer-causing benzopyrene chemicals. For this reason, when a food or any of its constituents has been smoked, this has been noted in the additives-of-concern column.

BRANDS	**ADDITIVES-OF-CONCERN** BOLD—CONCERN TO ALL LIGHT—CONCERN TO SOME
HEALTH VALLEY BEST OF SEAFOOD ALASKAN PINK SALMON; RED SALMON; SOLID WHITE TUNA IN WATER	NO ADDITIVES OF CONCERN
SPRUCE ANCHOVIES	NO ADDITIVES OF CONCERN

FROZEN FISH & SHELLFISH (DINNERS & SINGLE ITEMS)

BRANDS	ADDITIVES-OF-CONCERN
GORTON'S	
BAKED STUFFED SCROD	**BUTTER**
CRUNCHY FISH FILLETS	**MODIFIED CORN STARCH;** MONOSODIUM GLUTAMATE; SODIUM ALGINATE; WHEY
HADDOCK IN LEMON BUTTER; SOLE IN LEMON BUTTER	**BUTTER;** WHEY
TINY FISH CAKES	GUAR GUM; **MODIFIED CORN STARCH; PAPRIKA EXTRACT;** WHEY
—BATTER FRIED	
FISH & CHIPS; FISH KABOBS; SCALLOPS	**MODIFIED CORN STARCH;** MONOSODIUM GLUTAMATE; SODIUM ALGINATE; WHEY
—FISH STICKS	
REGULAR	**MODIFIED CORN STARCH;** MONOSODIUM GLUTAMATE; **PAPRIKA EXTRACT**

BRANDS	**ADDITIVES-OF-CONCERN** **BOLD—CONCERN TO ALL** LIGHT—CONCERN TO SOME
W/ SHRIMP STUFFING	**MODIFIED CORN STARCH;** MONOSODIUM GLUTAMATE; SODIUM ALGINATE; WHEY
HOWARD JOHNSON'S	
FRIED CLAMS	NO ADDITIVES OF CONCERN
HADDOCK AU GRATIN*	**PAPRIKA**
MATLAW'S STUFFED CLAMS	MONOSODIUM GLUTAMATE
MRS. PAUL'S	
BUTTERED FISH FILLETS	**BUTTER**
FISH CAKES	**SMOKE FLAVOR;** WHEY
FISH STICKS; FRIED FISH FILLETS; FRIED SCALLOPS	WHEY
FRIED CLAMS	NO ADDITIVES OF CONCERN
—LIGHT BATTER	
CLAM FRITTERS; FISH FILLETS; SCALLOPS	**MODIFIED FOOD STARCH**
STOUFFER'S	
ALASKA KING CRAB NEWBURG; LOBSTER NEWBURG	MONOSODIUM GLUTAMATE

*By law, at this writing, it is not required of a manufacturer to state whether cheeses contain artificial colors, unless the color is FD & C Yellow No. 5 (Tartrazine). Their presence in frozen fish dinners containing cheese therefore remains an uncertainty, unless it is voluntarily declared on the label that artificial colors have been used or have not. For a more detailed explanation of artificial color in cheese, refer to the beginning of the cheese section of "Dairy Products & Substitutes."

BRANDS	**ADDITIVES-OF-CONCERN** BOLD—CONCERN TO ALL LIGHT—CONCERN TO SOME
STOUFFER'S *(Continued)*	
SCALLOPS & SHRIMP MARINER W/ RICE	**MODIFIED CORN STARCH**
SWANSON	
FISH 'N' CHIPS DINNER	GLUTEN (WHEAT); **MODIFIED FOOD STARCH**
HUNGRY-MAN DINNER FISH 'N' CHIPS	GLUTEN (WHEAT); **MODIFIED FOOD STARCH;** WHEY POWDER
TASTE O' SEA	
FISH CAKES	**ARTIFICIAL COLOR; BHA; BHT; MODIFIED FOOD STARCH; OLEORESIN PAPRIKA;** WHEY
FRIED CLAMS	**OLEORESIN PAPRIKA;** WHEY
"KRUNCHEE" FISH PORTIONS; SEA SCALLOPS; SHRIMP	**ARTIFICIAL COLOR; OLEORESIN PAPRIKA;** WHEY
—*BATTER DIPT*	
FISH PORTIONS; FISH STICKS; SHRIMP	EGG WHITES; **MODIFIED FOOD STARCH;** MONOSODIUM GLUTAMATE; SODIUM ALGINATE; WHEY
—*DINNER*	
BATTER DIPT SCROD	EGG WHITES; **MODIFIED FOOD STARCH;** MONOSODIUM GLUTAMATE; SODIUM ALGINATE; WHEY

BRANDS	ADDITIVES-OF-CONCERN BOLD—CONCERN TO ALL LIGHT—CONCERN TO SOME
FISH; FLOUNDER; HADDOCK; SOLE	**MODIFIED FOOD STARCH; OLEORESIN PAPRIKA**
SCALLOP; SHRIMP	**ARTIFICIAL COLOR; OLEORESIN PAPRIKA;** WHEY
—NATURAL FILLETS	
COD; FILLET OF SOLE; HADDOCK	NO ADDITIVES OF CONCERN
—PLATTER	
SEAFOOD	**ARTIFICIAL COLOR; BHA; BHT; MODIFIED FOOD STARCH; OLEORESIN PAPRIKA;** WHEY
CLAM	**OLEORESIN PAPRIKA;** WHEY
VAN DE KAMP'S	
FISH FILLETS	ALGINATE; **MODIFIED STARCH;** MONOSODIUM GLUTAMATE; WHEY
FISH KABOBS	**MODIFIED STARCH;** MONOSODIUM GLUTAMATE; WHEY
FISH STICKS	EGG WHITES (PASTEURIZED); **MODIFIED FOOD STARCH;** MONOSODIUM GLUTAMATE; WHEY
WAKEFIELD *ALASKA CRABMEAT*	
KING	NO ADDITIVES OF CONCERN

BRANDS	ADDITIVES-OF-CONCERN **BOLD—CONCERN TO ALL** LIGHT—CONCERN TO SOME
WEIGHT WATCHERS	
FLOUNDER	CAROB BEAN GUM; GUAR GUM; GUM TRAGACANTH; **PAPRIKA**
HADDOCK	CAROB BEAN GUM; GUAR GUM; GUM TRAGACANTH
SUPERMARKET PRIVATE BRANDS	
A & P	
—A & P	
FLOUNDER FILLETS; HADDOCK FILLETS	NO ADDITIVES OF CONCERN
CRISPY SCALLOPS; CRISPY SHRIMP	GUAR GUM; **OLEORESIN PAPRIKA;** WHEY
FISH STICKS	**MODIFIED FOOD STARCH;** MONOSODIUM GLUTAMATE; **OLEORESIN PAPRIKA;** WHEY
SOLE IN LEMON BUTTER	**BUTTER;** MONOSODIUM GLUTAMATE; WHEY
—BATTER DIPPED	
COD PORTIONS; FISH & CHIPS; FISH PORTIONS; FISH STICKS	GUAR GUM; **MODIFIED FOOD STARCH;** MONOSODIUM GLUTAMATE; WHEY
FIRST NATIONAL	
—FINAST	
FISH STICKS	**MODIFIED FOOD STARCH;** MONOSODIUM GLUTAMATE; **OLEORESIN PAPRIKA;** WHEY

BRANDS	**ADDITIVES-OF-CONCERN** BOLD—CONCERN TO ALL LIGHT—CONCERN TO SOME
—BATTER DIPPED FISH & CHIPS; FISH FINGERS; FISH PORTIONS	GUAR GUM; **MODIFIED FOOD STARCH;** MONOSODIUM GLUTAMATE; WHEY
SAFEWAY —CAPTAIN'S CHOICE FISH & CHIPS	GUAR GUM; **MODIFIED FOOD STARCH;** MONOSODIUM GLUTAMATE; WHEY
GOURMET BREADED FANTAIL SHRIMP	MONOSODIUM GLUTAMATE; SODIUM ALGINATE; WHEY
GOURMET FANCY FRENCH FRIED SHRIMP	**ARTIFICIAL COLORING;** GUAR GUM; **OLEORESIN PAPRIKA**
SHRIMP SCAMPI	**BUTTER; MODIFIED FOOD STARCH;** MONOSODIUM GLUTAMATE; **OLEORESIN PAPRIKA;** WHEY
SOLE IN LEMON BUTTER SAUCE	**BUTTER;** MONOSODIUM GLUTAMATE; WHEY
—BATTER FRIED FISH STICKS	**MODIFIED FOOD STARCH;** MONOSODIUM GLUTAMATE; WHEY
SCALLOPS	GUAR GUM; **OLEORESIN PAPRIKA;** WHEY

BRANDS	**ADDITIVES-OF-CONCERN** BOLD—CONCERN TO ALL LIGHT—CONCERN TO SOME
SAFEWAY (Continued)	
—SCOTCH BUY	
BREADED FANTAIL SHRIMP	GUAR GUM
GOLDEN FRIED FISH STICKS	**MODIFIED FOOD STARCH;** MONOSODIUM GLUTAMATE; **OLEORESIN PAPRIKA;** WHEY
SEAFOOD ASSORTMENT	GUAR GUM; **MODIFIED FOOD STARCH;** MONOSODIUM GLUTAMATE; **OLEORESIN PAPRIKA;** SODIUM ALGINATE; WHEY
STOP & SHOP	
—STOP & SHOP	
BREADED SHRIMP; SALAD SHRIMP	NO ADDITIVES OF CONCERN
FISH-NICS	**MODIFIED FOOD STARCH; OLEORESIN PAPRIKA**
FRIED CLAMS	**OLEORESIN PAPRIKA;** WHEY
SCALLOPS	**ARTIFICIAL COLOR; OLEORESIN PAPRIKA;** WHEY
—BATTER DIPPED	
FISH NUGGETS; SCALLOPS	EGG WHITES; **MODIFIED FOOD STARCH;** MONOSODIUM GLUTAMATE; SODIUM ALGINATE; WHEY

BRANDS	**ADDITIVES-OF-CONCERN** **BOLD—CONCERN TO ALL** LIGHT—CONCERN TO SOME
—GOLDEN FRIED FISH CAKES; MINI FISH CAKES	**ARTIFICIAL COLOR; BHA; BHT; MODIFIED FOOD STARCH; OLEORESIN PAPRIKA;** WHEY
FISH "MIDGETS"	**ARTIFICIAL COLOR; OLEORESIN PAPRIKA;** WHEY
FISH STICKS	**MODIFIED FOOD STARCH; OLEORESIN PAPRIKA**
ITEMS IN HEALTH FOOD STORES	
BEST OF SEAFOOD (HEALTH VALLEY) ALASKAN FISH FILLETS; ALASKAN FISH STICKS; COOKED SHRIMP	**PAPRIKA**
ALASKAN SALMON STEAK	NO ADDITIVES OF CONCERN

REFRIGERATED FISH

SAU-SEA SHRIMP COCKTAIL	**MODIFIED FOOD STARCH**
VITA HERRING IN CREAM SAUCE	MILK SOLIDS (NONFAT)
HERRING PARTY SNACKS; TASTEE BITS	NO ADDITIVES OF CONCERN
HERRING SALAD	MONOSODIUM GLUTAMATE

XII. FROZEN DINNERS, PIZZA, POT PIES

FROZEN DINNERS

By law, at this writing, it is not required of a manufacturer to state whether cheeses contain artificial colors, unless the color is FD & C Yellow No. 5 (Tartrazine). Their presence in frozen dinners containing cheese therefore remains an uncertainty, unless it is voluntarily declared on the label that artificial colors have been used or have not. For a more detailed explanation of artificial color in cheese, refer to the beginning of the cheese section of "Dairy Products & Substitutes."

BRANDS	**ADDITIVES-OF-CONCERN** BOLD—CONCERN TO ALL LIGHT—CONCERN TO SOME
BANQUET	
—DINNER	
CHOPPED BEEF	**COLORING;** WHEY
FRIED CHICKEN	MONOSODIUM GLUTAMATE; WHEY
MEAT LOAF	**CALCIUM CASEINATE; PAPRIKA;** WHEY (DRIED)
—MAN-PLEASER DINNER	
CHOPPED BEEF	**ARTIFICIAL COLOR; COLORING; MODIFIED FOOD STARCH**
FRIED CHICKEN	**MODIFIED FOOD STARCH;** MONOSODIUM GLUTAMATE
TURKEY	**ARTIFICIAL COLOR; MODIFIED FOOD STARCH;** MONOSODIUM GLUTAMATE; **OLEORESIN OF TURMERIC; PAPRIKA**

BRANDS	**ADDITIVES-OF-CONCERN** **BOLD—CONCERN TO ALL** LIGHT—CONCERN TO SOME
VEAL PARMIGIAN	**ARTIFICIAL COLORS; BHA; CALCIUM CASEINATE; COLORING;** MONOSODIUM GLUTAMATE; **PAPRIKA; SODIUM CASEINATE;** WHEY & DRY WHEY
CHUN KING *DINNER*	
BEEF PEPPER ORIENTAL	**CALCIUM CASEINATE; MODIFIED FOOD STARCH;** MONOSODIUM GLUTAMATE; SODIUM ALGINATE; WHEY (DRIED)
CHICKEN CHOW MEIN	**CALCIUM CASEINATE; MODIFIED FOOD STARCH;** MONOSODIUM GLUTAMATE; **PAPRIKA;** SODIUM ALGINATE; WHEY (DRIED)
FREEZER QUEEN *ENTRÉE*	
BREADED VEAL PARMIGIAN	**MODIFIED FOOD STARCH;** MONOSODIUM GLUTAMATE; **OLEORESIN PAPRIKA & PAPRIKA**
GRAVY & SLICED BEEF	**HYDROGENATED SOYBEAN OIL; MODIFIED FOOD STARCH;** MONOSODIUM GLUTAMATE; **OLEORESIN PAPRIKA**

BRANDS	ADDITIVES-OF-CONCERN **BOLD—CONCERN TO ALL** LIGHT—CONCERN TO SOME
MORTON *—DINNER*	
BEEF; MEAT LOAF; SALISBURY STEAK; VEAL PARMIGIANA	**ARTIFICIAL COLOR; MODIFIED FOOD STARCH;** MONOSODIUM GLUTAMATE; WHEY SOLIDS
BONELESS CHICKEN	**ARTIFICIAL COLOR; BHA; MODIFIED FOOD STARCH;** MONOSODIUM GLUTAMATE; **PAPRIKA OLEORESIN; TURMERIC;** WHEY SOLIDS
MACARONI & CHEESE	**ARTIFICIAL COLOR; MODIFIED FOOD STARCH;** WHEY SOLIDS
TURKEY	**ARTIFICIAL COLOR; MODIFIED FOOD STARCH;** MONOSODIUM GLUTAMATE; **PAPRIKA OLEORESIN; TURMERIC;** WHEY SOLIDS
—COUNTRY TABLE ENTRÉE	
FRIED CHICKEN	**ARTIFICIAL COLOR; MODIFIED FOOD STARCH; PAPRIKA;** WHEY SOLIDS
SLICED TURKEY	**ARTIFICIAL COLOR; MODIFIED FOOD STARCH;** MONOSODIUM GLUTAMATE; **PAPRIKA OLEORESIN; TURMERIC;** WHEY SOLIDS
—STEAK HOUSE	
SIRLOIN STRIP STEAK; TENDERLOIN STEAK	**MODIFIED FOOD STARCH;** SODIUM ALGINATE

BRANDS	ADDITIVES-OF-CONCERN **BOLD—CONCERN TO ALL** LIGHT—CONCERN TO SOME
SWANSON	
—DINNER	
BEEF; TURKEY	**MODIFIED FOOD STARCH;** MONOSODIUM GLUTAMATE; WHEY POWDER
LOIN OF PORK; VEAL PARMIGIANA	**MODIFIED FOOD STARCH**
CRISPY FRIED CHICKEN; FRIED CHICKEN	**MODIFIED FOOD STARCH; VEGETABLE OIL;** WHEY POWDER
HAM	**MODIFIED FOOD STARCH; SODIUM NITRITE;** WHEY POWDER
MACARONI & CHEESE	**MODIFIED FOOD STARCH; OLEORESIN PAPRIKA**
POLYNESIAN STYLE	**MODIFIED FOOD STARCH;** MONOSODIUM GLUTAMATE
SALISBURY STEAK	**CALCIUM CASEINATE; MODIFIED FOOD STARCH;** WHEY & WHEY POWDER
—ENTRÉE	
CHICKEN NIBBLES; FRIED CHICKEN	**VEGETABLE OIL**
GRAVY & SLICED BEEF	**MODIFIED FOOD STARCH;** MONOSODIUM GLUTAMATE
MEATBALLS	**MODIFIED FOOD STARCH**
SALISBURY STEAK	**CALCIUM CASEINATE; SODIUM CASEINATE;** WHEY

BRANDS	ADDITIVES-OF-CONCERN **BOLD—CONCERN TO ALL** LIGHT—CONCERN TO SOME
SWANSON *(Continued)*	
—*HUNGRY-MAN DINNER*	
BONELESS CHICKEN; SLICED BEEF; TURKEY	**MODIFIED FOOD STARCH;** MONOSODIUM GLUTAMATE; WHEY POWDER
SALISBURY STEAK	**CALCIUM CASEINATE; MODIFIED FOOD STARCH; SODIUM CASEINATE;** WHEY & WHEY POWDER
VEAL PARMIGIANA	**MODIFIED FOOD STARCH; OLEORESIN PAPRIKA;** WHEY POWDER
—*HUNGRY-MAN ENTRÉE*	
FRIED CHICKEN	**VEGETABLE OIL**
LASAGNA W/ MEAT	**MODIFIED FOOD STARCH**
WEIGHT WATCHERS	
CHICKEN LIVERS & ONIONS	CAROB BEAN GUM; GUAR GUM; GUM TRAGACANTH; **MODIFIED FOOD STARCH; NUTMEG**
VEAL PARMIGIANA & ZUCCHINI IN SAUCE	CAROB BEAN GUM; GUAR GUM; GUM TRAGACANTH
SUPERMARKET PRIVATE BRANDS	
A & P	
—ANN PAGE *DINNER*	
BEEF	**MODIFIED FOOD STARCH**
FRIED CHICKEN	**PAPRIKA;** WHEY

BRANDS	**ADDITIVES-OF-CONCERN** BOLD—CONCERN TO ALL LIGHT—CONCERN TO SOME
MEAT LOAF	**CALCIUM CASEINATE;** MONOSODIUM GLUTAMATE; **PAPRIKA;** WHEY
TURKEY	**MODIFIED FOOD STARCH;** MONOSODIUM GLUTAMATE; **OLEORESIN OF PAPRIKA; TURMERIC**
SAFEWAY —BEL-AIR *DINNER*	
FRIED CHICKEN	BUTTERMILK SOLIDS; MONOSODIUM GLUTAMATE; **OLEORESIN PAPRIKA;** SODIUM ALGINATE; WHEY SOLIDS
MEAT LOAF	BUTTERMILK SOLIDS; WHEY SOLIDS
SALISBURY STEAK	MONOSODIUM GLUTAMATE; **PAPRIKA;** WHEY (DRIED)
SPAGHETTI & MEATBALLS	EGG WHITES; MILK SOLIDS (NONFAT DRY)
TURKEY	WHEY (DRIED)

BRANDS	ADDITIVES-OF-CONCERN BOLD—CONCERN TO ALL LIGHT—CONCERN TO SOME

FROZEN PIZZA

By law, at this writing, it is not required of a manufacturer to state whether cheeses contain artificial colors, unless the color is FD & C Yellow No. 5 (Tartrazine). Their presence in frozen pizza therefore remains an uncertainty, unless it is voluntarily declared on the label that artificial colors have been used or have not. For a more detailed explanation of artificial color in cheese, refer to the beginning of the cheese section of "Dairy Products & Substitutes."

BRANDS	ADDITIVES-OF-CONCERN
CELESTE	
CHEESE	**MODIFIED FOOD STARCH**
CHEESE SICILIAN STYLE	**BHA; BHT;** GLUTEN; **MODIFIED FOOD STARCH**
PIZZA-FOR-ONE PEPPERONI	**BHA; BHT; MODIFIED FOOD STARCH; OLEORESIN OF PAPRIKA; SODIUM NITRITE**
ELLIO'S CHEESE	**MODIFIED FOOD STARCH**
JENO'S	
ITALIAN BREAD CHEESE	GLUTEN (WHEAT); **MODIFIED FOOD STARCH;** WHEY
CHEESE	**ARTIFICIAL COLOR;** LOCUST BEAN GUM; **MODIFIED FOOD STARCH;** WHEY POWDER
PIZZA SNACKS ASSORTED (SAUSAGE-PEPPERONI-CHEESE)	**ARTIFICIAL COLORING; BHA; BHT; CALCIUM & SODIUM CASEINATE;** LOCUST BEAN GUM;

(continues)

BRANDS	**ADDITIVES-OF-CONCERN** BOLD—CONCERN TO ALL LIGHT—CONCERN TO SOME
	MODIFIED FOOD STARCH; MONOSODIUM GLUTAMATE; **PAPRIKA (& OLEORESIN OF); SODIUM NITRITE**
JOHN'S	
HOMESTYLE THICK CRUST CHEESE; ORIGINAL CHEESE	**MODIFIED FOOD STARCH; PAPRIKA**
LA PIZZERIA	
CHEESE	NO ADDITIVES OF CONCERN
COMBINATION	**BHA; BHT; PAPRIKA; SODIUM NITRITE**
SALUTO *FRENCH BREAD PIZZA*	
CHEESE	NO ADDITIVES OF CONCERN
DELUXE; PEPPERONI	**BHA; BHT; PAPRIKA; SODIUM NITRITE**
STOUFFER'S *FRENCH BREAD PIZZA*	
CHEESE	**ARTIFICIAL COLOR; MODIFIED CORN STARCH;** MONOSODIUM GLUTAMATE
DELUXE SAUSAGE & PEPPERONI	**BHT; MODIFIED CORN STARCH;** MONOSODIUM GLUTAMATE; **OLEORESIN OF PAPRIKA; SODIUM NITRITE**
PEPPERONI	**MODIFIED CORN STARCH;** MONOSODIUM GLUTAMATE; **OLEORESIN OF PAPRIKA; SODIUM NITRITE**

BRANDS	ADDITIVES-OF-CONCERN **BOLD—CONCERN TO ALL** LIGHT—CONCERN TO SOME
SUPERMARKET PRIVATE BRANDS	
A & P—ANN PAGE	
CHEESE	GUAR GUM; GUM TRAGACANTH; **MODIFIED FOOD STARCH**
MINI	**ARTIFICIAL COLORING;** GUAR GUM; GUM TRAGACANTH; **MODIFIED FOOD STARCH;** RENNET CASEIN
SAFEWAY—BEL-AIR	
CHEESE	NO ADDITIVES OF CONCERN
CHEESE & PEPPERONI	**BHA; BHT; SODIUM NITRITE**
CHEESE & SAUSAGE	**SODIUM NITRATE; SODIUM NITRITE**
STOP & SHOP —STOP & SHOP	
CHEESE; SAUSAGE & CHEESE	**MODIFIED FOOD STARCH;** MONOSODIUM GLUTAMATE
CHEESE PIZZA 10 PAK	**MODIFIED FOOD STARCH; VEGETABLE OIL**
ITEMS IN HEALTH FOOD STORES	
PIZZA NATURALLY	
DELUXE VEGETABLE CHEESE	HIGH GLUTEN FLOUR
SHILOH FARMS	
WHOLE WHEAT PIZZA	
PLAIN; W/ SWEET PEPPERS; W/ SAUSAGE	GLUTEN (WHEAT OR VITAL WHEAT)

BRANDS	ADDITIVES-OF-CONCERN **BOLD—CONCERN TO ALL** LIGHT—CONCERN TO SOME
WORTHINGTON PIZZA ITALIANA	**ARTIFICIAL COLOR; CALCIUM CASEINATE;** EGG WHITES; GLUTEN (WHEAT); GUAR GUM; MONOSODIUM GLUTAMATE; **SODIUM CASEINATE;** WHEY

FROZEN POT PIES

BRANDS	ADDITIVES-OF-CONCERN
BANQUET	
BEEF	NO ADDITIVES OF CONCERN
CHICKEN; TURKEY	MONOSODIUM GLUTAMATE; **PAPRIKA; OLEORESIN OF TURMERIC**
MORTON	
BEEF	**MODIFIED FOOD STARCH;** MONOSODIUM GLUTAMATE; **SHORTENING**
TURKEY	**BHA; MODIFIED FOOD STARCH;** MONOSODIUM GLUTAMATE; **OLEORESINS OF PAPRIKA & TURMERIC;** WHEY SOLIDS
STOUFFER'S	
BEEF	**MODIFIED CORN STARCH;** MONOSODIUM GLUTAMATE
CHICKEN; TURKEY	**MODIFIED CORN STARCH;** MONOSODIUM GLUTAMATE; **TURMERIC**

BRANDS	ADDITIVES-OF-CONCERN BOLD—CONCERN TO ALL LIGHT—CONCERN TO SOME
SWANSON	
BEEF	**MODIFIED FOOD STARCH**
CHICKEN; TURKEY	**MODIFIED FOOD STARCH;** MONOSODIUM GLUTAMATE
SUPERMARKET PRIVATE BRANDS	
A & P	
—ANN PAGE	
BEEF	**MODIFIED FOOD STARCH**
CHICKEN; TURKEY	**MODIFIED FOOD STARCH;** MONOSODIUM GLUTAMATE; **OLEORESIN OF PAPRIKA; TURMERIC**
FIRST NATIONAL	
—FINAST	
BEEF	**MODIFIED FOOD STARCH**
CHICKEN; TURKEY	**MODIFIED FOOD STARCH;** MONOSODIUM GLUTAMATE; **OLEORESIN OF PAPRIKA; TURMERIC**
SAFEWAY	
—MANOR HOUSE	
BEEF	**(BEEF FAT, LARD)***
CHICKEN	**(BEEF FAT, LARD);*** MONOSODIUM GLUTAMATE; **PAPRIKA; OLEORESIN OF TURMERIC**

*Blend may contain one or more of these saturated fats.

BRANDS	**ADDITIVES-OF-CONCERN** **BOLD—CONCERN TO ALL** LIGHT—CONCERN TO SOME
ITEMS IN HEALTH FOOD STORES	
WORTHINGTON *PIE*	
TUNO	**MODIFIED CORN STARCH; MODIFIED TAPIOCA STARCH;** MONOSODIUM GLUTAMATE; **TURMERIC**
VEGETARIAN	**ARTIFICIAL COLORS;** EGG WHITE SOLIDS; **MODIFIED CORN STARCH;** MONOSODIUM GLUTAMATE

XIII. FRUIT, FRUIT DRINKS, FRUIT JUICES

FRUIT: CANNED

Most of the items in this section do not contain any additives of concern, and these appear first, without an additives-of-concern column.

The items that do contain additives of concern follow, in the usual style.

DEL MONTE APRICOT HALVES; BARTLETT PEAR HALVES; ROYAL ANNE CHERRIES; SLICED PINEAPPLE; WHOLE FIGS; YELLOW CLING SLICED PEACHES

DOLE CHUNK PINEAPPLE IN HEAVY SYRUP; CHUNK PINEAPPLE IN UNSWEETENED PINEAPPLE JUICE

GEISHA MANDARIN ORANGE SEGMENTS

LIBBY'S GRAPEFRUIT SECTIONS; PEAR HALVES; YELLOW CLING SLICED PEACHES

MOTT'S APPLE SAUCE; APPLE SAUCE NATURAL STYLE

OCEAN SPRAY CRAN RASPBERRY JELLIED SAUCE; JELLIED CRANBERRY SAUCE; WHOLE BERRY CRANBERRY SAUCE

SENECA CINNAMON APPLESAUCE

STOKELY VAN CAMP'S APPLESAUCE; YELLOW CLING SLICED PEACHES

SUNSWEET COOKED PRUNES

3 DIAMONDS MANDARIN ORANGE SEGMENTS

VERYFINE APPLE SAUCE NATURAL STYLE; APPLE SAUCE 100% McINTOSH

SUPERMARKET PRIVATE BRANDS
A & P ANN PAGE APPLE SAUCE; PEACHES YELLOW CLING SLICED

FIRST NATIONAL FINAST APPLE SAUCE; APRICOT HALVES; BARTLETT PEARS; FANCY APPLE SAUCE; GRAPEFRUIT SECTIONS; YELLOW CLING PEACHES

LUCKY HARVEST DAY APPLE SAUCE; APRICOTS; FRUIT MIX; PEACHES; PEARS

LUCKY LADY LEE APPLE SAUCE; APRICOTS; BARTLETT PEAR HALVES; CRUSHED PINEAPPLE; FANCY GRAVENSTEIN APPLE SAUCE; FREESTONE PEACHES; MANDARIN ORANGES; PEACH HALVES; PEACH SLICES; PINEAPPLE CHUNKS; UNSWEETENED APPLE SAUCE; WHOLE CRANBERRY SAUCE

SAFEWAY SCOTCH BUY APPLE SAUCE; FRUIT MIX; HALVES BARTLETT PEARS; HALVES YELLOW CLING PEACHES; MANDARIN ORANGE SEGMENTS; UNPEELED HALVES APRICOTS; WHOLE & SPLIT KADOTA FIGS; YELLOW ELBERTA FREESTONE PEACHES

SAFEWAY TOWN HOUSE APPLE SAUCE; CRANBERRY SAUCE; CRUSHED PINEAPPLE; GRAPEFRUIT SECTIONS; GRAVENSTEIN APPLE SAUCE; HALVES BARTLETT PEARS; MANDARIN ORANGE SEGMENTS; PINEAPPLE CHUNKS IN HEAVY SYRUP; PINEAPPLE CHUNKS IN UNSWEETENED PINEAPPLE JUICE; RED SOUR PITTED CHERRIES; SLICED CLING PEACHES; WHOLE APRICOTS; WHOLE PURPLE PLUMS

STOP & SHOP APPLE SAUCE; APRICOT HALVES; BARTLETT PEARS; CHUNKY APPLE SAUCE; CLING PEACHES; CRANBERRY SAUCE; CRUSHED

(continues)

STOP & SHOP *(Continued)*

PINEAPPLE; ELBERTA PEACHES HALVES; GRAPEFRUIT & ORANGE SECTIONS IN LIGHT SYRUP; GRAPEFRUIT IN UNSWEETENED GRAPEFRUIT JUICE; PINEAPPLE CHUNKS; SLICED PEARS; SLICED PINEAPPLE IN HEAVY SYRUP

STOP & SHOP SUN GLORY HALVES PEACHES; SLICED PEACHES; SLICED PEARS

WINN DIXIE ASTOR APRICOTS; BARTLETT PEARS; CRUSHED PINEAPPLE; SLICED PINEAPPLE; YELLOW CLING PEACHES

WINN DIXIE THRIFTY MAID APPLE SAUCE; BARTLETT PEARS; GRAPEFRUIT SECTIONS; SLICED PINEAPPLE; YELLOW CLING SLICED PEACHES; WHOLE APRICOTS

ITEMS IN HEALTH FOOD STORES

CELLU BARTLETT PEARS

FEATHERWEIGHT KADOTA FIGS; SLICED PINEAPPLE; STEWED PRUNES

HEALTH VALLEY BARTLETT PEARS; CHUNK PINEAPPLE; FRUIT MIX; HALVES APRICOTS; HALVES PEACHES; SLICED PINEAPPLE

WALNUT ACRES CRANBERRY-HONEY SAUCE; WHOLE CRANBERRY SAUCE

The following items contain additives of concern.

BRANDS	**ADDITIVES-OF-CONCERN** **BOLD—CONCERN TO ALL** LIGHT—CONCERN TO SOME
DEL MONTE FRUIT COCKTAIL; FRUITS FOR SALAD; TROPICAL FRUIT SALAD	CHERRIES **ARTIFICIALLY COLORED RED**

BRANDS	**ADDITIVES-OF-CONCERN** **BOLD—CONCERN TO ALL** LIGHT—CONCERN TO SOME
LIBBY'S FRUIT COCKTAIL	CHERRIES **ARTIFICIALLY COLORED RED**
OCEAN SPRAY CRAN ORANGE RELISH	**MODIFIED STARCH**
ROYAL WILLAMETTE MARASCHINO CHERRIES	**ARTIFICIAL COLOR**
STOKELY VAN CAMP'S FRUIT COCKTAIL	CHERRIES **ARTIFICIALLY COLORED RED**
SUPERMARKET PRIVATE BRANDS	
A & P—ANN PAGE FRUIT COCKTAIL	CHERRIES **ARTIFICIALLY COLORED RED**
FIRST NATIONAL —FINAST FRUIT COCKTAIL	CHERRIES **ARTIFICIALLY COLORED RED**
LUCKY—LADY LEE FRUIT COCKTAIL	CHERRIES **ARTIFICIALLY COLORED RED**
SAFEWAY —TOWN HOUSE FRUIT COCKTAIL	CHERRIES **ARTIFICIALLY COLORED RED**
STOP & SHOP —STOP & SHOP FRUIT COCKTAIL	CHERRIES **ARTIFICIALLY COLORED RED**
MARASCHINO CHERRIES	**CERTIFIED COLOR**

BRANDS	**ADDITIVES-OF-CONCERN** **BOLD—CONCERN TO ALL** LIGHT—CONCERN TO SOME
WINN DIXIE **—ASTOR** FRUIT COCKTAIL	CHERRIES **ARTIFICIALLY COLORED RED**

FRUIT: DRIED*

DEL MONTE EVAPORATED APPLES; APRICOTS; LARGE PRUNES; SEEDLESS RAISINS

DROMEDARY CHOPPED DATES

SUN MAID GOLDEN RAISINS; RAISINS; ZANTE CURRANTS

SUNSWEET APRICOTS; LARGE PRUNES

SUPERMARKET PRIVATE BRANDS
LUCKY LADY LEE SEEDLESS RAISINS

SAFEWAY SCOTCH BUY DRIED FIGS

SAFEWAY TOWN HOUSE APPLES; LARGE PEACHES; LARGE PRUNES; MEDIUM APRICOTS; MEDIUM PRUNES; SEEDLESS GOLDEN RAISINS; SEEDLESS RAISINS

STOP & SHOP LARGE PRUNES; PITTED PRUNES; SEEDLESS RAISINS

ITEMS IN HEALTH FOOD STORES
BARBARA'S APPLES

*See note on page 280.

EREWHON BREAD DATES; APRICOT; BLACK MISSION FIGS

SONOMA PEARS; PINEAPPLE

The following item contains an additive of concern.

BRANDS	ADDITIVES-OF-CONCERN BOLD—CONCERN TO ALL LIGHT—CONCERN TO SOME
SUN MAID MUSCAT RAISINS	**TREATED W/ VEGETABLE OIL**

FRUIT: FROZEN

None of the items in this section contain any additives of concern, and all the items appear without an additives-of-concern column.

BIRDS EYE MIXED FRUIT; PEACHES; RED RASPBERRIES; WHOLE STRAWBERRIES

NATURIPE SLICED STRAWBERRIES

NEWTON ACRES MELON BALLS

STEWART'S MAINE WILD BLUEBERRIES

SUPERMARKET PRIVATE BRANDS
A & P SLICED STRAWBERRIES

LUCKY LADY LEE SLICED STRAWBERRIES

SAFEWAY BEL-AIR BLUEBERRIES; BOYSENBERRIES; MIXED MELON BALLS; RASPBERRIES; RHUBARB; SLICED PEACHES; STRAWBERRIES

BRANDS	ADDITIVES-OF-CONCERN BOLD—CONCERN TO ALL LIGHT—CONCERN TO SOME

FRUIT DRINKS: CANNED

DEL MONTE *JUICE DRINK*	
PINEAPPLE GRAPEFRUIT	NO ADDITIVES OF CONCERN
PINEAPPLE ORANGE	**ARTIFICIAL COLOR**
DOLE	
PINEAPPLE PINK GRAPEFRUIT JUICE-DRINK	**ARTIFICIAL COLORING;** GUM ARABIC
HAWAIIAN PUNCH *FRUIT PUNCH*	
GRAPE; ORANGE; RED; VERY BERRY	**ARTIFICIAL COLOR** OR **COLORS**
HI-C	
FRUIT PUNCH	**ARTIFICIAL COLORS**
—DRINK	
APPLE; CHERRY; CITRUS COOLER; GRAPE; ORANGE; WILD BERRY	**ARTIFICIAL COLOR** OR **COLORS**
LEMON TREE	
LEMONADE FLAVOR DRINK	ACACIA GUM; **ARTIFICIAL COLOR; BHA; GLYCERYL ABIETATE**
MOTT'S *FRUIT DRINK*	
A.M.	**ARTIFICIAL COLOR;** GUM ARABIC; GUM TRAGACANTH
P.M. APPLE-GRAPE	**ARTIFICIAL COLOR;** GUM ARABIC

BRANDS	ADDITIVES-OF-CONCERN **BOLD—CONCERN TO ALL** LIGHT—CONCERN TO SOME
MUSSELMAN'S BREAKFAST COCKTAIL ORANGE APRICOT FRUIT DRINK	**ARTIFICIAL COLORS; BHT;** GUM ARABIC
OCEAN SPRAY CRANAPPLE; CRANBERRY JUICE COCKTAIL; CRAN-GRAPE; CRANICOT	NO ADDITIVES OF CONCERN
PARADE CRANBERRY COCKTAIL	NO ADDITIVES OF CONCERN
STOKELY VAN CAMP'S GATORADE (LEMON-LIME); GATORADE (ORANGE)	**ARTIFICIAL COLOR** W/ FD & C YELLOW NO. 5; **ESTER GUM**
TROPI-CAL-LO ORANGE DRINK	**ARTIFICIAL COLOR; MODIFIED STARCH; SODIUM SACCHARIN; VEGETABLE GUMS**
TROPICANA GRAPE DRINK	**ARTIFICIAL COLOR; VEGETABLE GUMS**
WAGNER BREAKFAST GRAPEFRUIT DRINK	**MODIFIED FOOD STARCH; VEGETABLE GUM**
THIRST QUENCHER	**ARTIFICIAL COLOR;** GUAR GUM; **MODIFIED FOOD STARCH**

BRANDS	**ADDITIVES-OF-CONCERN** **BOLD—CONCERN TO ALL** LIGHT—CONCERN TO SOME
WELCHADE	
FRUIT PUNCH	**ESTER GUM;** GUM ARABIC; **U.S. CERTIFIED ARTIFICIAL COLOR**
GRAPE DRINK	**ARTIFICIAL COLOR**
SUPERMARKET PRIVATE BRANDS	
A & P	
—A & P CRANBERRY APPLE DRINK	NO ADDITIVES OF CONCERN
—ANN PAGE	
TROPICAL FRUIT PUNCH	**ARTIFICIAL COLOR**
DRINK	
APPLE	NO ADDITIVES OF CONCERN
CHERRY; GRAPE	**ARTIFICIAL COLOR**
CITRUS COOLER; ORANGE	**ARTIFICIAL COLOR;** GUM ARABIC; GUM TRAGACANTH
FIRST NATIONAL	
—EDWARDS *DRINK*	
CHERRY; GRAPE; ORANGE	**ARTIFICIAL COLOR; VEGETABLE GUM**
—FINAST *DRINK*	
CRANBERRY APPLE	NO ADDITIVES OF CONCERN
WILD BERRY	**U.S. CERTIFIED COLOR**
LUCKY	
—LADY LEE FRUIT PUNCH CONCENTRATE	**ARTIFICIAL COLOR**

BRANDS	**ADDITIVES-OF-CONCERN** BOLD—CONCERN TO ALL LIGHT—CONCERN TO SOME
SAFEWAY —CRAGMONT LEMONADE FLAVORED DRINK	**ARTIFICIAL COLOR; BROMINATED VEGETABLE OIL; GLYCEROL ESTER OF WOOD ROSIN;** GUM ARABIC
—SCOTCH BUY GRAPE DRINK	**ARTIFICIAL COLOR;** GUM ARABIC
—TOWN HOUSE CRAN-APPLE DRINK	NO ADDITIVES OF CONCERN
STOP & SHOP —STOP & SHOP TROPICAL PUNCH	**U.S. CERTIFIED COLOR**
DRINK APPLE	NO ADDITIVES OF CONCERN
GRAPE; PINEAPPLE GRAPEFRUIT JUICE	**ARTIFICIAL COLOR**
WINN DIXIE —THRIFTY MAID *DRINK* APPLE; PINEAPPLE GRAPEFRUIT	NO ADDITIVES OF CONCERN
FRUIT PUNCH; GRAPE; ORANGE	**ARTIFICIAL COLOR** OR **COLORS**

FRUIT DRINKS: FROZEN

BIRDS EYE AWAKE; ORANGE PLUS	**ARTIFICIAL COLOR; BHA; MODIFIED CORN OR FOOD STARCH**

BRANDS	ADDITIVES-OF-CONCERN **BOLD—CONCERN TO ALL** LIGHT—CONCERN TO SOME
COUNTRY TIME	
LEMONADE	**ARTIFICIAL COLOR; BHA; MODIFIED STARCHES**
HAWAIIAN PUNCH	
RED	**ARTIFICIAL COLOR**
MINUTE MAID	
LEMONADE; LIMEADE; PINK LEMONADE	NO ADDITIVES OF CONCERN
OCEAN SPRAY	
CRANBERRY JUICE COCKTAIL; CRANORANGE	NO ADDITIVES OF CONCERN
WELCHADE GRAPE DRINK	NO ADDITIVES OF CONCERN
SUPERMARKET PRIVATE BRANDS	
A & P	
—ANN PAGE	
LEMONADE	NO ADDITIVES OF CONCERN
PINK LEMONADE	**ARTIFICIALLY COLORED**
LUCKY	
—LADY LEE	
FRUIT PUNCH	ACACIA GUM; **ARTIFICIAL COLOR**
PINK LEMONADE	**U.S. CERTIFIED ARTIFICIAL COLOR**

BRANDS	**ADDITIVES-OF-CONCERN** **BOLD—CONCERN TO ALL** LIGHT—CONCERN TO SOME
SAFEWAY —BEL-AIR CONCENTRATE FOR LEMONADE; CONCENTRATE FOR LIMEADE	NO ADDITIVES OF CONCERN
FRUIT PUNCH	ACACIA GUM; **ARTIFICIAL COLOR; GLYCERYL ABIETATE**
PINK LEMONADE	**ARTIFICIAL COLOR**

FRUIT DRINKS: REFRIGERATED

HOOD ALL NATURAL LEMONADE	NO ADDITIVES OF CONCERN
FRUIT PUNCH	**ARTIFICIAL COLOR**
SUPERMARKET PRIVATE BRANDS LUCKY —LADY LEE FRUIT PUNCH; ORANGE DRINK	**ARTIFICIAL COLOR; GLYCERYL ABIETATE; MODIFIED FOOD STARCH**
SAFEWAY —LUCERNE *DRINK* LEMON; ORANGE	ACACIA ESTER GUM; **ARTIFICIAL COLOR**
GRAPE	**ARTIFICIAL COLOR; MODIFIED FOOD STARCH**

BRANDS	ADDITIVES-OF-CONCERN **BOLD—CONCERN TO ALL** LIGHT—CONCERN TO SOME
WINN DIXIE —SUPERBRAND SWANEE *DRINK* GRAPE; LEMON; ORANGE	**ARTIFICIAL COLOR**

FRUIT JUICES: CANNED

Most of the items in this section do not contain any additives of concern, and these appear first, without an additives-of-concern column.

The items that do contain additives of concern follow, in the usual style.

APPLE & EVE APPLE

BIG TEX GRAPEFRUIT RUBY RED

DEL MONTE PINEAPPLE UNSWEETENED

DOLE PINEAPPLE UNSWEETENED

GLORIETTA APRICOT NECTAR; PEAR NECTAR

HEART'S DELIGHT APRICOT NECTAR; PEACH NECTAR; PEAR NECTAR

JUICY JUICE *100% JUICES* GOLDEN; PURPLE

LIBBY'S APRICOT NECTAR; ORANGE UNSWEETENED

MOTT'S APPLE McINTOSH; APPLE NATURAL STYLE; PRUNE SUPER

OCEAN SPRAY GRAPEFRUIT UNSWEETENED

PARADE HAWAIIAN UNSWEETENED PINEAPPLE; UNSWEETENED APPLE; UNSWEETENED GRAPE;

(continues)

100% UNSWEETENED GRAPEFRUIT; 100% UNSWEETENED ORANGE; UNSWEETENED PRUNE

REALEMON RECONSTITUTED LEMON

REALIME RECONSTITUTED LIME

SENECA APPLE

SUNSWEET PRUNE UNSWEETENED

TREESWEET GRAPEFRUIT UNSWEETENED

WELCH'S GRAPE; GRAPE RED; WHITE GRAPE

SUPERMARKET PRIVATE BRANDS
A & P APPLE CIDER; CRANBERRY COCKTAIL

A & P ANN PAGE APPLE; 100% GRAPEFRUIT UNSWEETENED

FIRST NATIONAL EDWARDS GRAPE; RECONSTITUTED LEMON; UNSWEETENED GRAPEFRUIT

FIRST NATIONAL FINAST APPLE; CRANBERRY COCKTAIL; 100% GRAPEFRUIT UNSWEETENED; 100% ORANGE UNSWEETENED; UNSWEETENED PINK GRAPEFRUIT

LUCKY LADY LEE APRICOT NECTAR; CRANBERRY COCKTAIL; GRAPE; HAWAIIAN PINEAPPLE UNSWEETENED; 100% GRAPEFRUIT UNSWEETENED; PINK GRAPEFRUIT; PRUNE UNSWEETENED; PURE APPLE CIDER; PURE APPLE; RECONSTITUTED LEMON

SAFEWAY TOWN HOUSE APPLE; APPLE CIDER; APRICOT NECTAR; CRANBERRY COCKTAIL; GRAPEFRUIT 100% UNSWEETENED; ORANGE; ORANGE 100% UNSWEETENED; PINEAPPLE; PINK GRAPEFRUIT 100% UNSWEETENED; PRUNE; RECONSTITUTED LEMON

STOP & SHOP APRICOT NECTAR

WINN DIXIE ASTOR UNSWEETENED PRUNE

WINN DIXIE THRIFTY MAID CRANBERRY; 100% FLORIDA GRAPEFRUIT UNSWEETENED; 100% FLORIDA ORANGE & GRAPEFRUIT UNSWEETENED; 100% FLORIDA ORANGE UNSWEETENED; RECONSTITUTED LEMON; SWEET APPLE CIDER; UNSWEETENED APPLE; UNSWEETENED FANCY HAWAIIAN PINEAPPLE; UNSWEETENED GRAPE

ITEMS IN HEALTH FOOD STORES
AFTER THE FALL APPLE; APPLE-APRICOT; APPLE BLACKBERRY; APPLE-CHERRY; APPLE-GRAPE; APPLE-PINEAPPLE; APPLE RASPBERRY; APPLE STRAWBERRY; PEAR; PURE GRAPE

CHARISMA GRAPE; PRUNE

EREWHON APPLE; APPLE APRICOT; APPLE BANANA; APPLE CRANBERRY; APPLE LIME; APPLE RASPBERRY; APPLE STRAWBERRY

HAIKU GUAVA; GUAVA-BERRY; PAPAYA; PASSION

HAIN CRANBERRY COCKTAIL; FIG

HEALTH VALLEY APRICOT ZAPPER

HEINKE'S APPLE; APPLE-APRICOT BLEND; APRICOT; BLACK CHERRY; PEACH; PEAR; PLUM; POMEGRANATE

KEDEM GRAPE

KNUDSEN'S APPLE; APPLE-BOYSENBERRY; APPLE-STRAWBERRY; CONCORD GRAPE

LAKEWOOD APPLE LOGANBERRY; CARROT ORANGE; LEMON-LIME; "PINEAPPLE N' PAPAYA"; "SECOND WIND"; "STRAWBERRIES N' CREME"

LEHR'S PURE RED GRAPE

TAP 'N APPLE APPLE

WALNUT ACRES CRANBERRY NECTAR

WESTBRAE MACHU-PICCHU PUNCH

The following items contain additives of concern.

BRANDS	**ADDITIVES-OF-CONCERN** **BOLD—CONCERN TO ALL** LIGHT—CONCERN TO SOME
ITEMS IN HEALTH FOOD STORES	
LAKEWOOD BANANA COLADA; COCONUT MILK; PINA COLADA; PINEAPPLE AMBROSIA	**VEGETABLE GUM**
CRANBERRY; LOGANBERRY; MANGO; PASSION FRUIT; RED PAPAYA RICA; SÃO PAULO PUNCH	PACIFIC SEA KELP (ALGINATE)

FRUIT JUICES: FROZEN

None of the items in this section contain any additives of concern, and all the items appear without an additives-of-concern column.

DONALD DUCK ORANGE

MINUTE MAID GRAPE; 100% LEMON; ORANGE; PINEAPPLE; UNSWEETENED GRAPEFRUIT

SENECA APPLE; GRAPE

TREESWEET ORANGE

TROPICANA ORANGE

WELCH'S GRAPE

SUPERMARKET PRIVATE BRANDS
A & P ORANGE

LUCKY LADY LEE FROZEN CONCENTRATE FOR APPLE; 100% FROZEN CONCENTRATED ORANGE; SWEETENED CONCORD FROZEN CONCENTRATED GRAPE

SAFEWAY BEL-AIR FROZEN CONCENTRATED SWEETENED CONCORD GRAPE; FROZEN CONCENTRATE FOR APPLE; 100% FROZEN CONCENTRATED GRAPEFRUIT; 100% FROZEN CONCENTRATED ORANGE

SAFEWAY SCOTCH BUY 100% FROZEN CONCENTRATED ORANGE

FRUIT JUICES: REFRIGERATED

None of the items in this section contain any additives of concern, and all the items appear without an additives-of-concern column.

HOOD 100% PURE ORANGE

KRAFT ORANGE; UNSWEETENED GRAPEFRUIT

MINUTE MAID GRAPEFRUIT; ORANGE

TROPICANA GRAPEFRUIT; ORANGE

SUPERMARKET PRIVATE BRANDS
LUCKY LADY LEE REAL ORANGE

SAFEWAY LUCERNE GRAPEFRUIT; ORANGE

XIV. GRAVIES, SAUCES, & SEASONINGS

GLAZE & MARINADE MIXES

BRANDS	ADDITIVES-OF-CONCERN BOLD—CONCERN TO ALL LIGHT—CONCERN TO SOME
ADOLPH'S CHICKEN MARINADE	**MODIFIED FOOD STARCH; PAPRIKA; TURMERIC**
—MARINADE IN MINUTES BARBECUE FLAVOR	**EXTRACTIVES OF PAPRIKA; MODIFIED FOOD STARCH**
GARLIC FLAVOR	**MODIFIED FOOD STARCH**
STEAK SAUCE FLAVOR	**MODIFIED FOOD STARCH; NATURAL SMOKE FLAVOR**
TEMPO HAM GLAZE	NO ADDITIVES OF CONCERN
—CHICKEN GLAZE HERB HONEY; SAVORY ORANGE	NO ADDITIVES OF CONCERN

GRAVIES & GRAVY MIXES

DURKEE *—GRAVY MIX* BROWN; MUSHROOM	**MODIFIED FOOD STARCH;** MONOSODIUM GLUTAMATE

BRANDS	ADDITIVES-OF-CONCERN **BOLD—CONCERN TO ALL** LIGHT—CONCERN TO SOME
DURKEE *GRAVY MIX (Continued)*	
CHICKEN	BUTTERMILK SOLIDS; **MODIFIED FOOD STARCH;** MONOSODIUM GLUTAMATE; **TURMERIC**
FOR TURKEY	**ARTIFICIAL COLOR; MODIFIED FOOD STARCH;** MONOSODIUM GLUTAMATE
—ROASTIN' BAG GRAVY MIX	
FOR CHICKEN; POT ROAST	**MODIFIED FOOD STARCH;** MONOSODIUM GLUTAMATE
FRANCO-AMERICAN	
—GRAVY	
AU JUS; BEEF; BROWN W/ ONIONS; CHICKEN GIBLET	**MODIFIED FOOD STARCH;** MONOSODIUM GLUTAMATE
CHICKEN	MONOSODIUM GLUTAMATE; **OLEORESIN TURMERIC**
FRENCH'S	
GRAVY MAKINS MIX FOR TURKEY	**BHA; BHT; MODIFIED CORN STARCH;** MONOSODIUM GLUTAMATE; **PAPRIKA; SODIUM CASEINATE**
—GRAVY MIX	
BROWN	**LARD; MODIFIED CORN STARCH;** WHEY
MUSHROOM	**MODIFIED CORN STARCH;** MONOSODIUM GLUTAMATE; **SODIUM CASEINATE**

BRANDS	**ADDITIVES-OF-CONCERN** BOLD—CONCERN TO ALL LIGHT—CONCERN TO SOME
GRAVY MASTER SEASONING & BROWNING SAUCE	NO ADDITIVES OF CONCERN
HEINZ *—HOME STYLE GRAVY*	
CHICKEN	**MODIFIED FOOD STARCH;** MONOSODIUM GLUTAMATE; **PAPRIKA OLEORESIN; TURMERIC; SODIUM CASEINATE**
ONION	GUM ARABIC; **MODIFIED FOOD STARCH;** MONOSODIUM GLUTAMATE
McCORMICK GRAVY MIX FOR PORK	MILK SOLIDS (BLENDED); MONOSODIUM GLUTAMATE
SUPERMARKET PRIVATE BRANDS **A & P** **—ANN PAGE** *GRAVY MIX*	
BROWN	**BUTYLATED HYDROXYANISOLE; MODIFIED STARCH;** MONOSODIUM GLUTAMATE; WHEY SOLIDS
CHICKEN	**MODIFIED STARCH;** MONOSODIUM GLUTAMATE; **TURMERIC**
MUSHROOM	**HYDROGENATED SHORTENING; MODIFIED STARCH**

BRANDS	ADDITIVES-OF-CONCERN BOLD—CONCERN TO ALL LIGHT—CONCERN TO SOME
SAFEWAY —CROWN COLONY *GRAVY MIX* AU JUS	NO ADDITIVES OF CONCERN
CHICKEN	**ARTIFICIAL COLOR; BHA; MODIFIED WHEAT STARCH;** MONOSODIUM GLUTAMATE; **TURMERIC**
MUSHROOM	**MODIFIED WHEAT STARCH**
ONION	MILK SOLIDS (NONFAT); **MODIFIED WHEAT STARCH**

SAUCE MIXES & SAUCES

By law, at this writing, it is not required of a manufacturer to state whether cheeses contain artificial colors, unless the color is FD & C Yellow No. 5 (Tartrazine). Their presence in sauce mixes and sauces containing cheese therefore remains an uncertainty, unless it is voluntarily declared on the label that artificial colors have been used or have not. For a more detailed explanation of artificial color in cheese, refer to the beginning of the cheese section of "Dairy Products & Substitutes."

A.1. STEAK SAUCE	NO ADDITIVES OF CONCERN
CHEF BOYARDEE SPAGHETTI SAUCE W/ GROUND BEEF	**MODIFIED FOOD STARCH**
CHUN KING SOY SAUCE	NO ADDITIVES OF CONCERN

BRANDS	ADDITIVES-OF-CONCERN BOLD—CONCERN TO ALL LIGHT—CONCERN TO SOME
CROSSE & BLACKWELL SEAFOOD COCKTAIL SAUCE	**MODIFIED FOOD STARCHES; PAPRIKA EXTRACT**
DAWN FRESH MUSHROOM STEAK SAUCE	**ARTIFICIAL COLOR; CARRAGEENAN; MODIFIED CORN STARCH**
DURKEE REDHOT! SAUCE	NO ADDITIVES OF CONCERN
—SAUCE MIX CHEESE	**MODIFIED FOOD STARCH**
HOLLANDAISE	**ARTIFICIAL COLORS** W/ FD & C YELLOW NO. 5; **MODIFIED FOOD STARCH**
SPAGHETTI	ALGIN; **MODIFIED FOOD STARCH;** MONOSODIUM GLUTAMATE; **PAPRIKA**
WHITE	**ARTIFICIAL COLOR; BHA; BHT; MODIFIED FOOD STARCH**
FRENCH'S WORCESTERSHIRE SAUCE; SLOPPY JOE SEASONING MIX	NO ADDITIVES OF CONCERN
HEINZ CHILI SAUCE; WORCESTERSHIRE SAUCE	NO ADDITIVES OF CONCERN

BRANDS	ADDITIVES-OF-CONCERN BOLD—CONCERN TO ALL LIGHT—CONCERN TO SOME
HEINZ *(Continued)*	
57 SAUCE	GUAR GUM; **TURMERIC**
HELLMAN'S BIG H	
BURGER SAUCE	**ARTIFICIAL COLOR;** PROPYLENE GLYCOL ALGINATE
HUNT'S *PRIMA SALSA* *SPAGHETTI SAUCE EXTRA THICK & ZESTY!*	
MEAT FLAVORED; (PLAIN); W/ MUSHROOMS	**MODIFIED FOOD STARCH;** MONOSODIUM GLUTAMATE
KIKKOMAN	
SOY SAUCE; TERIYAKI	NO ADDITIVES OF CONCERN
KRAFT *BARBECUE SAUCE*	
GARLIC FLAVORED; ONION BITS	CAROB BEAN GUM; **MODIFIED FOOD STARCH; PAPRIKA**
HICKORY SMOKE FLAVORED; (PLAIN)	CAROB BEAN GUM; **HICKORY SMOKE FLAVOR; MODIFIED FOOD STARCH; PAPRIKA**
LEA & PERRINS	
WORCESTERSHIRE SAUCE	NO ADDITIVES OF CONCERN
McCORMICK *SAUCE MIX*	
SPAGHETTI	**ARTIFICIAL COLORS;** MILK SOLIDS (BLENDED); **PAPRIKA**

BRANDS	ADDITIVES-OF-CONCERN **BOLD—CONCERN TO ALL** LIGHT—CONCERN TO SOME
WHITE	MONOSODIUM GLUTAMATE; **TURMERIC**
McILHENNY CO. TABASCO PEPPER SAUCE	NO ADDITIVES OF CONCERN
OPEN PIT ORIGINAL FLAVOR BARBECUE SAUCE	**ARTIFICIAL COLOR; MODIFIED TAPIOCA STARCH**
PRINCE ITALIAN COOKING SAUCE	NO ADDITIVES OF CONCERN
RAGU ITALIAN COOKING SAUCE; TABLE SAUCE	NO ADDITIVES OF CONCERN
—EXTRA THICK & ZESTY SPAGHETTI SAUCE (PLAIN); W/ MEAT; W/ MUSHROOMS	**MODIFIED FOOD STARCH**
—SPAGHETTI SAUCE (PLAIN); W/ MEAT; W/ MUSHROOMS & ONIONS	NO ADDITIVES OF CONCERN
SUPERMARKET PRIVATE BRANDS **A & P—ANN PAGE** SPAGHETTI SAUCE MIX	**CALCIUM STEARATE; MODIFIED STARCH;** MONOSODIUM GLUTAMATE
TARTAR SAUCE	NO ADDITIVES OF CONCERN

BRANDS	ADDITIVES-OF-CONCERN **BOLD—CONCERN TO ALL** LIGHT—CONCERN TO SOME
—ANN PAGE *(Continued)*	
—BARBEQUE SAUCE	
PLAIN; W/ MINCED ONIONS	**MODIFIED STARCH**
—SPAGHETTI SAUCE	
FLAVORED W/ MEAT; MARINARA; MEATLESS W/ MUSHROOMS	**MODIFIED STARCH**
LUCKY	
—LADY LEE	
SPAGHETTI SAUCE	
FLAVORED W/ MEAT; W/ MUSHROOMS	NO ADDITIVES OF CONCERN
SAFEWAY	
—CROWN COLONY	
SAUCE MIX	
SPAGHETTI	**PAPRIKA;** WHEY
SPAGHETTI ITALIAN STYLE W/ MUSHROOMS	**PAPRIKA**
—TOWN HOUSE	
CHILI SAUCE	NO ADDITIVES OF CONCERN
STOP & SHOP	
—STOP & SHOP	
CHILI SAUCE	NO ADDITIVES OF CONCERN
TARTAR SAUCE	**TURMERIC**
WINN DIXIE	
—DEEP SOUTH	
REGULAR BARBECUE SAUCE	**SMOKE FLAVOR**

BRANDS	ADDITIVES-OF-CONCERN **BOLD—CONCERN TO ALL** LIGHT—CONCERN TO SOME
—THRIFTY MAID *SPAGHETTI SAUCE* (PLAIN); W/ MEAT; W/ MUSHROOMS	**MODIFIED FOOD STARCH; PAPRIKA OLEORESIN**
ITEMS IN HEALTH FOOD STORES	
DE BOLES *SPAGHETTI SAUCE* IMITATION MEAT FLAVORED; MARINARA; MEATLESS; MUSHROOM	NO ADDITIVES OF CONCERN
ENRICO'S *SPAGHETTI SAUCE* NO SALT—ALL PURPOSE; (PLAIN); W/ MUSHROOMS	NO ADDITIVES OF CONCERN
FEATHERWEIGHT CHILI SAUCE	NO ADDITIVES OF CONCERN
HAIN NATURAL BAR-B-QUE SAUCE	ALGIN
SPAGHETTI SAUCE MIX	NO ADDITIVES OF CONCERN
JOHNSON'S SPAGHETTI SAUCE	NO ADDITIVES OF CONCERN
MARK'S NATURAL *SPAGHETTI SAUCE* (PLAIN); W/ ONION & MUSHROOMS	NO ADDITIVES OF CONCERN
SOKEN *SAUCE* BAR-B-QUE; PEANUT	NO ADDITIVES OF CONCERN

BRANDS	ADDITIVES-OF-CONCERN BOLD—CONCERN TO ALL LIGHT—CONCERN TO SOME

SEASONING MIXES & SEASONINGS

ACCENT	MONOSODIUM GLUTAMATE
ADOLPH'S	
MEAT TENDERIZER SEASONED	**PAPRIKA**
SALT SUBSTITUTE	GLUTAMIC ACID; MONOPOTASSIUM GLUTAMATE
BELL'S SEASONING	NO ADDITIVES OF CONCERN
DURKEE	
BUTTER FLAVORED SALT	**ARTIFICIAL COLOR** W/ FD & C YELLOW NO. 5
IMITATION BACON BITS	**ARTIFICIAL COLORS**
INSTANT MEAT TENDERIZER; SLOPPY JOE SEASONING MIX	MONOSODIUM GLUTAMATE
KNORR SWISS	
AROMAT ALL PURPOSE SEASONING	MONOSODIUM GLUTAMATE; **TURMERIC**
LAWRY'S	
GARLIC SALT	**MODIFIED FOOD STARCH;** MONOSODIUM GLUTAMATE
GARLIC SPREAD	NO ADDITIVES OF CONCERN
SEASONED SALT	MONOSODIUM GLUTAMATE; **PAPRIKA; TURMERIC**

BRANDS	ADDITIVES-OF-CONCERN BOLD—CONCERN TO ALL LIGHT—CONCERN TO SOME
McCORMICK CINNAMON SUGAR	**MODIFIED CORN STARCH; VEGETABLE OIL**
FANCY PAPRIKA	**PAPRIKA**
GARLIC SALT; ONION SALT	**CALCIUM STEARATE; VEGETABLE GUM**
IMITATION BACON BITS	**BHA; BHT; FD & C RED NO. 3**
IMITATION BUTTER FLAVORED SALT	**COCONUT OIL;** FD & C YELLOW NO. 5; **MODIFIED CORN STARCH; SODIUM CASEINATE**
POULTRY SEASONING	**NUTMEG**
SLOPPY JOE'S SEASONING MIX	**ARTIFICIAL COLORS;** MONOSODIUM GLUTAMATE
MORTON SALT SUBSTITUTE	NO ADDITIVES OF CONCERN
SPICE ISLANDS SEASONING SALT	MONOSODIUM GLUTAMATE
SWEET 'N LOW **—NU SALT** SALT SUBSTITUTE	NO ADDITIVES OF CONCERN
SUPERMARKET PRIVATE BRANDS	
A & P **—ANN PAGE** CELERY SALT	NO ADDITIVES OF CONCERN
GARLIC SALT; ONION SALT	**CALCIUM STEARATE**

BRANDS	ADDITIVES-OF-CONCERN **BOLD—CONCERN TO ALL** LIGHT—CONCERN TO SOME
—ANN PAGE *(Continued)*	
HAMBURGER & MEAT LOAF SEASONING	MONOSODIUM GLUTAMATE
IMITATION BACON BITS	**U.S. CERTIFIED COLOR**
SALAD SEASONING	MONOSODIUM GLUTAMATE; **OLEORESIN PAPRIKA**
—*MEAT TENDERIZER*	
SEASONED; UNSEASONED	**CALCIUM STEARATE**
—*SEASONING MIX*	
CHILI; GROUND BEEF W/ ONIONS	**MODIFIED STARCH**
SLOPPY JOE	ALGIN DERIVATIVE
SAFEWAY	
—CROWN COLONY	
IMITATION BACON BITS	**ARTIFICIAL COLOR**
—*SEASONING MIX*	
BEEF STEW; SLOPPY JOE	MONOSODIUM GLUTAMATE; **PAPRIKA**
CHILI; ENCHILADA	NO ADDITIVES OF CONCERN
TACO	MONOSODIUM GLUTAMATE; **PAPRIKA**; WHEY
STOP & SHOP	
—STOP & SHOP	
GARLIC SALT; ONION SALT	NO ADDITIVES OF CONCERN

BRANDS	**ADDITIVES-OF-CONCERN** **BOLD—CONCERN TO ALL** LIGHT—CONCERN TO SOME
WINN DIXIE —ASTOR	
BARBEQUE SEASONING; LEMON & PEPPER SEASONING	MONOSODIUM GLUTAMATE
FLAVOR SALT; MEAT TENDERIZER	MONOSODIUM GLUTAMATE; **PAPRIKA**
GARLIC SALT; ONION SALT	NO ADDITIVES OF CONCERN
IMITATION BACON BITS	**CERTIFIED COLORS; VEGETABLE OIL**
ITEMS IN HEALTH FOOD STORES	
ATLANTIC MARICULTURE	
ATLANTIC KELP FLAKES; ATLANTIC KELP POWDER	KELP
BIOFORCE *HERB SEASONING SALT*	
HERBA-MARE; TROCO-MARE	KELP
CHICO SAN LIMA SOY SAUCE	NO ADDITIVES OF CONCERN
EREWHON	
NATURAL SHOYU; TAMARI	NO ADDITIVES OF CONCERN
HAIN VEGETABLE SEASONED SALT	DULSE; KELP

BRANDS	ADDITIVES-OF-CONCERN BOLD—CONCERN TO ALL LIGHT—CONCERN TO SOME
INDO FLAVORIZES-SEASONS-TENDERIZES	NO ADDITIVES OF CONCERN
MARUSAN TAMARI	NO ADDITIVES OF CONCERN
PRIDE OF SZEGED HUNGARIAN PAPRIKA	**PAPRIKA**
VEGE-SAL VEGETIZED SEASONER	**PACIFIC SEA GREENS (SEAWEED)**
WESTBRAE TAMARI	NO ADDITIVES OF CONCERN

XV. JELLIES & OTHER SWEET SPREADS; NUT & SEED BUTTERS

FRUIT BUTTERS

None of the items in this section contain any additives of concern, and all the items appear without an additives-of-concern column.

MUSSELMAN'S APPLE

SUPERMARKET PRIVATE BRANDS
LUCKY LADY LEE APPLE

SAFEWAY EMPRESS APPLE

SAFEWAY SCOTCH BUY APPLE

ITEMS IN HEALTH FOOD STORES
ARROWHEAD MILLS APPLE; APRICOT; PEACH; PLUM; RASPBERRY; STRAWBERRY

HAIN APPLE

KIMES APPLE

SORRELL RIDGE DAMSON PLUM

TAP 'N APPLE APPLE

WESTBRAE *BUTTER & HONEY* APPLE APRICOT; CHERRY; MACHU PICCHU FRUIT; PEACH; PLUM; STRAWBERRY

JAMS, JELLIES, MARMALADE, PRESERVES

Although sugars (both sucrose and corn sweeteners) and honey are not treated as additives of concern in this volume when present in moderate amounts as additives, excessive amounts of these or other carbohydrates in the diet can be of health concern. Sugars usually are listed in second and third positions in order of quantity among the ingredients in the jams, jellies, marmalade, and preserves represented here—an indication that sugars are present in considerable amounts.

Most of the items in this section do not contain any additives of concern, and these appear first, without an additives-of-concern column.

The items that do contain additives of concern follow, in the usual style.

CROSSE & BLACKWELL PURE ORANGE MARMALADE

KRAFT
ORANGE MARMALADE
—*JELLY* APPLE; GRAPE; RED CURRANT; STRAWBERRY
—*PRESERVES* APRICOT; RED RASPBERRY; STRAWBERRY

SMUCKERS
SWEET ORANGE MARMALADE
—*JAM* GRAPE; SEEDLESS RED RASPBERRY; STRAWBERRY
—*JELLY* APPLE; GRAPE; STRAWBERRY
—*PRESERVES* APRICOT; BLUEBERRY; CHERRY; PEACH; RED RASPBERRY; STRAWBERRY

WELCH'S GRAPE JAM; GRAPE JELLY; STRAWBERRY PRESERVES

SUPERMARKET PRIVATE BRANDS
A & P ANN PAGE
ORANGE MARMALADE
—*JAM* BLACK RASPBERRY; GRAPE
—*JELLY* APPLE; BLACK RASPBERRY; CRAB APPLE;

(continues)

CURRANT; GRAPE; RED RASPBERRY; STRAWBERRY; WILD ELDERBERRY
—*PRESERVES* BLACKBERRY; BLUEBERRY; CHERRY; DAMSON PLUM; PEACH; PINEAPPLE; RED RASP-BERRY; STRAWBERRY

FIRST NATIONAL EDWARDS BLACKBERRY PRE-SERVES; STRAWBERRY JELLY

FIRST NATIONAL FINAST
ORANGE MARMALADE
—*JELLY* APPLE; CURRANT; GRAPE; RASPBERRY
—*PRESERVES* APRICOT; PEACH; PINEAPPLE; PLUM

LUCKY LADY LEE
ORANGE MARMALADE
—*JAM* BLACKBERRY; CONCORD GRAPE; PLUM
—*JELLY* APPLE; CONCORD GRAPE; CURRANT; MIXED FRUIT; STRAWBERRY
—*PRESERVES* APRICOT; APRICOT-PINEAPPLE; BOYSENBERRY; PEACH; RED RASPBERRY; STRAWBERRY

PUBLIX
ORANGE MARMALADE
—*JAM* GRAPE; RED RASPBERRY
—*JELLY* APPLE; CURRANT; GRAPE; RED RASP-BERRY; STRAWBERRY
—*PRESERVES* APRICOT; BLACK RASPBERRY; CHERRY; DAMSON PLUM; PINEAPPLE; RED RASP-BERRY; STRAWBERRY

SAFEWAY EMPRESS
—*JAM* CURRANT; PURE BLACKBERRY; PURE BLUE-BERRY; PURE STRAWBERRY
—*JELLY* APPLE; BLACKBERRY; BLACK RASPBERRY; BOYSENBERRY; CONCORD GRAPE GUAVA; MIXED-FRUIT; PLUM; RED CURRANT; RED RASPBERRY; STRAWBERRY
—*MARMALADE* CALIFORNIA STYLE SWEET ORANGE; PURE SEVILLE ORANGE
—*PRESERVES* APRICOT; APRICOT-PINEAPPLE;
(continues)

SAFEWAY EMPRESS PRESERVES *(Continued)*
BLACKBERRY; BLACK CHERRY; BLACK RASPBERRY; BLUEBERRY; BOYSENBERRY; PEACH; PEACH-PINEAPPLE; PLUM; RED CHERRY; RED RASPBERRY; STRAWBERRY

SAFEWAY SCOTCH BUY GRAPE JAM; GRAPE JELLY; STRAWBERRY PRESERVES

STOP & SHOP
GRAPE JAM; ORANGE MARMALADE
—*JELLY* APPLE; CRABAPPLE; CURRANT; GRAPE; STRAWBERRY
—*PRESERVES* APRICOT; BLACKBERRY; BLACK RASPBERRY; PEACH; PINEAPPLE; RED RASPBERRY; STRAWBERRY

STOP & SHOP SUN GLORY
ORANGE MARMALADE
—*PRESERVES* RED RASPBERRY; STRAWBERRY

WINN DIXIE DEEP SOUTH
—*JAM* BLACKBERRY; DAMSON PLUM; GRAPE
—*JELLY* APPLE; BLACKBERRY; CURRANT; GRAPE; GUAVA; STRAWBERRY
—*PRESERVES* PEACH; PINEAPPLE; STRAWBERRY

ITEMS IN HEALTH FOOD STORES
ARROWHEAD MILLS
ORANGE MARMALADE
—*JAM* CHERRY; GRAPE

CHARISMA
ORANGE MARMALADE
—*JAM* APRICOT; CHERRY; GRAPE; PEACH; RASPBERRY; STRAWBERRY

HAIN
ORANGE MARMALADE
—*PRESERVES* APRICOT; BLACKBERRY; GRAPE; RASPBERRY; STRAWBERRY

SORRELL RIDGE
—*PURE PRESERVES* RASPBERRY; STRAWBERRY; WILD BLUEBERRY; WILD PARTRIDGEBERRY

TREE OF LIFE
—*PRESERVES* APRICOT; BLACKBERRY; BLUEBERRY; CHERRY; RASPBERRY; STRAWBERRY

WESTBRAE NATURAL
—*UNSWEETENED SPREAD* APRICOT; BOYSENBERRY; RASPBERRY

WM ESCOTT'S
ORANGE MARMALADE
—*PRESERVES* APRICOT-PINEAPPLE; BOYSENBERRY; FIG & DATE; RED RASPBERRY; STRAWBERRY

The following items contain additives of concern.

BRANDS	**ADDITIVES-OF-CONCERN** **BOLD—CONCERN TO ALL** LIGHT—CONCERN TO SOME
KRAFT MINT FLAVORED APPLE JELLY	**ARTIFICIAL COLOR**
RAFFETTO MINT W/ LEAVES	**ARTIFICIAL COLOR**
SUPERMARKET PRIVATE BRANDS	
A & P	
—ANN PAGE MINT FLAVORED IMITATION JELLY	**ARTIFICIAL COLOR; VEGETABLE GUM STABILIZER**
FIRST NATIONAL	
—FINAST MINT FLAVORED IMITATION JELLY	**U.S. CERTIFIED FOOD COLORING**

BRANDS	ADDITIVES-OF-CONCERN **BOLD—CONCERN TO ALL** LIGHT—CONCERN TO SOME
LUCKY	
—LADY LEE MINT FLAVORED APPLE JELLY	**ARTIFICIAL COLOR**
PUBLIX	
—PUBLIX MINT JELLY	**ARTIFICIAL COLOR**
SAFEWAY	
—EMPRESS MINT FLAVORED APPLE JELLY	**ARTIFICIAL COLORING**
STOP & SHOP	
—STOP & SHOP MINT JELLY	**ARTIFICIAL COLORING**
WINN DIXIE	
—DEEP SOUTH MINT FLAVORED IMITATION JELLY	**U.S. CERTIFIED COLOR**

MARSHMALLOW CREAM, MOLASSES, SYRUPS

Although sugars (both sucrose and corn sweeteners) and honey are not treated as additives of concern in this volume when present in moderate amounts as additives, excessive amounts of these or other carbohydrates can be of health concern. Some of the brands of syrups listed below report their sugar content, ranging from 84% to 99% of total content.

Most of the items in this section do not contain any additives of concern, and these appear first, without an additives-of-concern column.

The items that do contain additives of concern follow, in the usual style.

AUNT JEMIMA SYRUP

BRER RABBIT MOLASSES

GOLDEN GRIDDLE PANCAKE SYRUP

GRANDMA'S *MOLASSES* DARK, RICH ROBUST STYLE; THE FAMOUS "UNSULPHURED"

KARO *SYRUP* DARK CORN; LIGHT CORN

LOG CABIN *SYRUP* COUNTRY KITCHEN; REGULAR

VERMONT MAID SYRUP

SUPERMARKET PRIVATE BRANDS
A & P ANN PAGE *SYRUP* PANCAKE & WAFFLE; REGULAR

FIRST NATIONAL EDWARDS PANCAKE & WAFFLE SYRUP

LUCKY LADY LEE PANCAKE & WAFFLE SYRUP

SAFEWAY SCOTCH BUY WAFFLE & PANCAKE SYRUP

STOP & SHOP PANCAKE & WAFFLE SYRUP

ITEMS IN HEALTH FOOD STORES
CROSBY'S MOLASSES

McCLURE'S MAPLE SYRUP

NIBLACK PURE MALT SYRUP

OLD COLONY PURE MAPLE SYRUP

PLANTATION *MOLASSES* BARBADOS; BLACKSTRAP

WESTBRAE NATURAL MALTED GRAIN SYRUP

YINNIES RICE SYRUP

BRANDS	ADDITIVES-OF-CONCERN BOLD—CONCERN TO ALL LIGHT—CONCERN TO SOME

The following items contain additives of concern.

KRAFT MARSHMALLOW CREME	**ARTIFICIAL COLOR;** EGG WHITES
MRS BUTTERWORTH'S THICK 'N RICH SYRUP	ALGIN DERIVATIVE
SAFEWAY EMPRESS *SYRUP* BOYSENBERRY	**CALCIUM CARRAGEENAN**
RASPBERRY; STRAWBERRY	**ARTIFICIAL COLORING; CALCIUM CARRAGEENAN**

PEANUT BUTTER & OTHER NUT & SEED BUTTERS

None of the items in this section contain any additives of concern, and all the items appear without an additives-of-concern column.

JIF *PEANUT BUTTER* CREAMY; EXTRA CRUNCHY

PETER PAN *PEANUT BUTTER* CREAMY; CRUNCHY

SKIPPY *PEANUT BUTTER* CREAMY; SUPER CHUNK

SMUCKER'S NATURAL PEANUT BUTTER

SUPERMARKET PRIVATE BRANDS
A & P ANN PAGE *PEANUT BUTTER* CREAMY SMOOTH; KRUNCHY

FIRST NATIONAL EDWARDS OLD FASHIONED PEANUT BUTTER

FIRST NATIONAL EDWARDS-FINAST CRUNCHY PEANUT BUTTER

LUCKY LADY LEE *PEANUT BUTTER* CHUNK STYLE; CHUNKY; CREAMY

PUBLIX CRUNCHY PEANUT BUTTER

STOP & SHOP *PEANUT BUTTER* CHUNK; CREAMY

WINN DIXIE DEEP SOUTH SMOOTH PEANUT BUTTER

ITEMS IN HEALTH FOOD STORES
ARROWHEAD MILLS *DEAF SMITH PEANUT BUTTER* CRUNCHY; OLD FASHIONED

EREWHON *BUTTER* ALMOND; CASHEW; CREAMY & SALTED PEANUT; SESAME; SUNFLOWER

XVI. LOW-CALORIE BEVERAGES & FOODS

LOW-CALORIE BAKED GOODS (COOKIES & PASTRIES)

BRANDS	ADDITIVES-OF-CONCERN BOLD—CONCERN TO ALL LIGHT—CONCERN TO SOME
AMUROL *DIETETIC/LOW SODIUM* FILLED WAFERS	**ARTIFICIAL COLORS; (COCONUT OIL)***
LEMON COOKIES; VANILLA COOKIES	**CERTIFIED COLOR**
OATMEAL RAISIN COOKIES	NO ADDITIVES OF CONCERN
ESTEE *DIETETIC* ASSORTED FILLED WAFERS	**ARTIFICIAL COLORS; (COCONUT, PALM KERNEL OILS);*** WHEY
CHOCOLATE CHIP COOKIES; VANILLA THINS	**ARTIFICIAL COLOR;** WHEY
COCONUT COOKIES	**BHA**
FUDGE COOKIES	**ARTIFICIAL COLOR; BHA**
LEMON SANDWICH COOKIES	**ARTIFICIAL COLOR; (COCONUT OIL);*** GLUTEN FLOUR

*Blend may contain one or more of these saturated oils.

BRANDS	**ADDITIVES-OF-CONCERN** **BOLD—CONCERN TO ALL** LIGHT—CONCERN TO SOME
OATMEAL RAISIN COOKIES	NO ADDITIVES OF CONCERN
STELLA D'ORO *DIETETIC* APPLE PASTRY; PEACH-APRICOT PASTRY	**ARTIFICIAL COLOR;** GUAR & LOCUST BEAN GUMS; **MACE**
COCONUT COOKIES; EGG BISCUITS; LOVE COOKIES	NO ADDITIVES OF CONCERN
KICHEL	**ARTIFICIAL COLOR**

LOW-CALORIE BEVERAGE MIXES & BEVERAGES

ALBA *—'66 HOT COCOA MIX* CHOCOLATE & MARSHMALLOW; MILK CHOCOLATE	**SODIUM SACCHARIN;** WHEY SOLIDS (DAIRY)
—'77 FIT 'N FROSTY CHOCOLATE FLAVOR	**SODIUM SACCHARIN;** WHEY SOLIDS (DAIRY)
STRAWBERRY	**ARTIFICIAL COLOR; SODIUM SACCHARIN;** WHEY SOLIDS (DAIRY)

BRANDS	ADDITIVES-OF-CONCERN BOLD—CONCERN TO ALL LIGHT—CONCERN TO SOME
BARRELHEAD SUGAR FREE ROOT BEER	ACACIA GUM; **SODIUM SACCHARIN**
CANADA DRY *DIET* GINGER ALE	**SODIUM SACCHARIN**
ORANGE	ACACIA GUM; **ARTIFICIAL COLORING; ESTER GUM; SODIUM SACCHARIN**
TONIC WATER	QUININE; **SODIUM SACCHARIN**
COTT *SUGAR FREE* COLA	CAFFEINE; **SACCHARIN**
ORANGE; PINK GRAPEFRUIT	ACACIA GUM; **ARTIFICIAL COLOR; BROMINATED VEGETABLE OIL; GLYCERYL ABIETATE; SACCHARIN**
ROOT BEER	ACACIA GUM; **SACCHARIN**
DR PEPPER SUGAR FREE	CAFFEINE; **SODIUM SACCHARIN**
FRESCA SUGAR FREE	**ARTIFICIAL COLOR; BROMINATED VEGETABLE OIL; GLYCEROL ESTER OF WOOD ROSIN;** GUM ARABIC; **SODIUM SACCHARIN**
HAWAIIAN PUNCH LOW SUGAR FRUIT PUNCH	**ARTIFICIAL COLOR; SODIUM SACCHARIN**

BRANDS	**ADDITIVES-OF-CONCERN** **BOLD—CONCERN TO ALL** LIGHT—CONCERN TO SOME
LIPTON ICED TEA LEMON FLAVORED SUGAR FREE; ICED TEA MIX LOW CALORIE LEMON FLAVORED	**SODIUM SACCHARIN**
MOXIE SUGAR FREE	CAFFEINE; **SODIUM SACCHARIN**
NESTEA LIGHT ICED TEA MIX	NO ADDITIVES OF CONCERN
LOW CALORIE ICED TEA MIX	GUM ARABIC; **SODIUM SACCHARIN**
OCEAN SPRAY *LOW CALORIE* CRANAPPLE; CRANBERRY JUICE COCKTAIL	**CALCIUM SACCHARIN**
OVALTINE REDUCED CALORIE HOT COCOA MIX	**ARTIFICIAL COLOR; CARRAGEENAN; (COCONUT OIL);* FERRIC SODIUM PYROPHOSPHATE; SODIUM CASEINATE;** WHEY
PEPSI-COLA DIET	CAFFEINE; **SODIUM SACCHARIN**
PEPSI LIGHT	CAFFEINE; **SODIUM SACCHARIN**

*Blend may contain this saturated oil.

BRANDS	ADDITIVES-OF-CONCERN **BOLD—CONCERN TO ALL** LIGHT—CONCERN TO SOME
SALADA LIGHT ICED TEA MIX LEMON FLAVOR	**BHA**
SCHWEPPES DIET GINGER ALE	**SODIUM SACCHARIN**
7 UP SUGAR FREE	**SODIUM SACCHARIN**
SHASTA *DIET* BLACK CHERRY; GRAPE	**ARTIFICIAL COLOR;** GUM ARABIC; **SODIUM SACCHARIN**
COLA	**BROMINATED VEGETABLE OIL;** CAFFEINE; GUM ARABIC; **SODIUM SACCHARIN**
GINGER ALE; LEMON LIME	**SODIUM SACCHARIN**
ORANGE	**ARTIFICIAL COLOR; BROMINATED VEGETABLE OIL; GLYCEROL ESTER OF WOOD ROSIN;** GUM ARABIC; **SODIUM SACCHARIN**
SWEET 'N LOW *LO-CALORIE SOFT DRINK MIX* CHERRY; GRAPE	**CERTIFIED COLOR; SODIUM SACCHARIN**
TAB SUGAR FREE GINGER ALE	**SODIUM SACCHARIN**

BRANDS	**ADDITIVES-OF-CONCERN** **BOLD—CONCERN TO ALL** LIGHT—CONCERN TO SOME
REGULAR	CAFFEINE; **SODIUM SACCHARIN**
ROOT BEER	GUM ARABIC; **SODIUM SACCHARIN**
TROPI-CAL-LO ORANGE DRINK	**ARTIFICIAL COLOR; MODIFIED STARCH; SODIUM SACCHARIN; VEGETABLE GUMS**
SUPERMARKET PRIVATE BRANDS A & P—YUKON *SUGAR FREE DIET* COLA	CAFFEINE; **SACCHARIN**
GINGER ALE; ROOT BEER	**SODIUM SACCHARIN**
ORANGE SODA	**ARTIFICIAL COLOR; BROMINATED VEGETABLE OIL; GLYCERYL ABIETATE;** GUM ARABIC; **SODIUM SACCHARIN**
FIRST NATIONAL —FINAST *SUGAR FREE* COLA	CAFFEINE; **SODIUM SACCHARIN**
GINGER ALE; OLD FASHIONED ROOT BEER	**SODIUM SACCHARIN**
ORANGE SODA	**ESTER GUM;** GUM ARABIC; **SODIUM SACCHARIN; U.S. CERTIFIED COLOR**

BRANDS	ADDITIVES-OF-CONCERN BOLD—CONCERN TO ALL LIGHT—CONCERN TO SOME
LUCKY —LADY LEE *SUGAR FREE DIET* CREME SODA	**SODIUM SACCHARIN**
GRAPE SODA	**ARTIFICIAL COLOR; SODIUM SACCHARIN; VEGETABLE GUM**
ORANGE SODA	ACACIA GUM; **ARTIFICIAL COLOR; GLYCERYL ABIETATE; SODIUM SACCHARIN**
PUBLIX —PIX *DIET* BLACK CHERRY SODA; ORANGE SODA; ROOT BEER	**ARTIFICIAL COLOR; SODIUM SACCHARIN**
COLA; CREAM SODA	**SODIUM SACCHARIN**
SAFEWAY —CRAGMONT *SUGAR FREE DIET* BLACK CHERRY	**ARTIFICIAL COLOR;** GUM ARABIC; **SODIUM SACCHARIN**
COLA; ROOT BEER	GUM ARABIC; **SODIUM SACCHARIN**
CREAM; GINGER ALE; LEMON LIME	**SODIUM SACCHARIN**

BRANDS	ADDITIVES-OF-CONCERN BOLD—CONCERN TO ALL LIGHT—CONCERN TO SOME
ORANGE	**ARTIFICIAL COLOR; BROMINATED VEGETABLE OIL; GLYCERYL ABIETATE;** GUM ARABIC; **SODIUM SACCHARIN**
TONIC MIX	QUININE HYDROCHLORIDE; **SODIUM SACCHARIN**
STOP & SHOP —SUN GLORY *SUGAR FREE*	
CITRUS	ACACIA GUM; **ARTIFICIAL COLOR; BROMINATED VEGETABLE OIL; GLYCERYL ABIETATE; SACCHARIN**
COLA	CAFFEINE; **SACCHARIN**
GINGER	**SODIUM SACCHARIN**
RASPBERRY	**ARTIFICIAL COLOR; SACCHARIN**
ROOT BEER	ACACIA GUM; **SACCHARIN**
WINN DIXIE —CHEK *SUGAR FREE*	
COLA	CAFFEINE; **SODIUM SACCHARIN**
FRESHY	**ARTIFICIAL COLOR; BROMINATED VEGETABLE OIL; GLYCERYL ABIETATE; MODIFIED FOOD STARCH; SODIUM SACCHARIN**

BRANDS	ADDITIVES-OF-CONCERN **BOLD—CONCERN TO ALL** LIGHT—CONCERN TO SOME
WINN DIXIE—CHEK *SUGAR FREE (Continued)*	
GINGER ALE; ROOT BEER	**SODIUM SACCHARIN**

LOW-CALORIE CANNED FRUIT

DIET DELIGHT	
APPLE SAUCE; APRICOTS; BARTLETT PEARS PACKED IN JUICE; BARTLETT PEARS PACKED IN WATER; CHERRIES; CLING PEACHES PACKED IN JUICE; CLING PEACHES PACKED IN WATER; ELBERTA PEACHES; GRAPEFRUIT SECTIONS; MANDARIN ORANGE SECTIONS; PINEAPPLE TIDBITS; PURPLE PLUMS	NO ADDITIVES OF CONCERN
FRUIT COCKTAIL; FRUITS FOR SALAD	CHERRIES **ARTIFICIALLY COLORED RED**

BRANDS	ADDITIVES-OF-CONCERN **BOLD—CONCERN TO ALL** LIGHT—CONCERN TO SOME

LOW-CALORIE DESSERT TOPPINGS, FROZEN DIETARY DAIRY DESSERTS,* GELATIN DESSERTS, PUDDING

BRANDS	ADDITIVES-OF-CONCERN
BORDEN *FROZEN DIETARY DAIRY DESSERT*	
CHOCOLATE	**CARRAGEENAN**
VANILLA	**ARTIFICIAL COLOR; CARRAGEENAN**
DIA-MEL	
CHOCOLATE FLAVOR INSTANT PUDDING	**CARRAGEENAN; COLOR; SODIUM SACCHARIN**
RASPBERRY GELATIN DESSERT	**CALCIUM SACCHARIN; U.S. CERTIFIED COLOR**
DIET DELIGHT	
CHOCOLATE TOPPING	**MODIFIED CORN STARCH**
D-ZERTA	
LOW CALORIE WHIPPED TOPPING MIX	**ARTIFICIAL COLOR; BHA; COCONUT & HYDROGENATED SOYBEAN OILS; SODIUM CASEINATE; SODIUM SACCHARIN;** WHEY SOLIDS

*Declaration by the manufacturer of the presence of artificial color in some dairy products, including frozen desserts, is not required by law except for FD & C Yellow No. 5 (Tartrazine). The presence or absence of other artificial colors in low-calorie frozen dairy desserts, therefore, remains uncertain unless it is declared on the label.

BRANDS	ADDITIVES-OF-CONCERN **BOLD—CONCERN TO ALL** LIGHT—CONCERN TO SOME
D-ZERTA *(Continued)*	
—LOW CALORIE GELATIN DESSERT	
CHERRY; STRAWBERRY	**ARTIFICIAL COLOR; SODIUM SACCHARIN**
LEMON; ORANGE	**ARTIFICIAL COLOR; BHA; SODIUM SACCHARIN**
—LOW CALORIE PUDDING	
BUTTERSCOTCH; VANILLA	**ARTIFICIAL COLOR; BHA; CALCIUM CARRAGEENAN; SODIUM SACCHARIN;** WHEY SOLIDS
CHOCOLATE	**CALCIUM CARRAGEENAN; SODIUM SACCHARIN**
ESKIMO DIETETIC BAR	GUAR GUM; MILK SOLIDS (NONFAT)
ESTEE	
—GEL	
CHERRY; STRAWBERRY	**ARTIFICIAL COLOR;** LOCUST BEAN GUM; **POTASSIUM CARRAGEENAN**
—PUDDING	
CHOCOLATE	**CALCIUM CARRAGEENAN; MODIFIED CORN STARCH**
VANILLA	**ARTIFICIAL COLOR; CALCIUM CARRAGEENAN;** EGG ALBUMIN; **MODIFIED CORN STARCH**
FEATHERWEIGHT WHIPPED TOPPING	NO ADDITIVES OF CONCERN

BRANDS	ADDITIVES-OF-CONCERN **BOLD—CONCERN TO ALL** LIGHT—CONCERN TO SOME
—*GELATIN DESSERT* CHERRY; LEMON; STRAWBERRY	**ARTIFICIAL COLOR; CALCIUM SACCHARIN**
—*PUDDING* BUTTERSCOTCH; CHOCOLATE; VANILLA	**ARTIFICIAL COLOR; CALCIUM SACCHARIN; CARRAGEENAN; MODIFIED CORN STARCH**
HOWARD JOHNSON'S COFFEE FROZEN DIETARY DAIRY DESSERT	GUAR GUM; GUM ARABIC; LOCUST BEAN GUM
SWEET 'N LOW WHITE FROSTING MIX	**MODIFIED FOOD STARCH**
THIN N' CREAMIE *DIETARY FROZEN DESSERT* BUTTER ALMOND; CHOCOLATE MINT; CREAMY ORANGE; VANILLA FUDGE	**CARRAGEENAN;** LOCUST BEAN GUM
WEIGHT WATCHERS CHOCOLATE FROZEN DIETARY DAIRY DESSERT	**ARTIFICIAL COLOR; CARRAGEENAN**
SUPERMARKET PRIVATE BRANDS **SAFEWAY** **—LUCERNE** *DIETETIC ICE CREAM* CHOCOLATE	**CALCIUM CARRAGEENAN;** GUAR GUM; LOCUST BEAN GUM

BRANDS	ADDITIVES-OF-CONCERN BOLD—CONCERN TO ALL LIGHT—CONCERN TO SOME
SAFEWAY—LUCERNE *DIETETIC (Continued)*	
VANILLA	**ARTIFICIAL COLOR; CALCIUM CARRAGEENAN;** GUAR GUM; LOCUST BEAN GUM

LOW-CALORIE DIET MEALS (FROZEN DINNERS, FROZEN ONE-COURSE DISHES, LIQUID MEALS, MEAL BARS, POWDERED MIXES)

By law, at this writing, it is not required of a manufacturer to state whether cheeses contain artificial colors, unless the color is FD & C Yellow No. 5 (Tartrazine). Their presence in low-calorie diet meals containing cheese therefore remains an uncertainty, unless it is voluntarily declared on the label that artificial colors have been used or have not. For a more detailed explanation of artificial color in cheese, refer to the beginning of the cheese section of "Dairy Products & Substitutes."

BRANDS	ADDITIVES-OF-CONCERN
CARNATION SLENDER	
—DIET MEAL BARS	
CHOCOLATE; CHOCOLATE PEANUT BUTTER; VANILLA	**ARTIFICIAL COLOR; CALCIUM CASEINATE;** EGG WHITES; WHEY PROTEIN CONCENTRATE
—LIQUID	
CHOCOLATE FLAVOR; CHOCOLATE FUDGE FLAVOR; COFFEE FLAVOR	**ARTIFICIAL COLOR** OR **COLORS; CALCIUM CARRAGEENAN; FERRIC ORTHOPHOSPHATE; SODIUM CASEINATE**
—POWDER MIX	
CHOCOLATE; DUTCH CHOCOLATE	**AMMONIUM CARRAGEENAN; ARTIFICIAL COLOR**

BRANDS	**ADDITIVES-OF-CONCERN** BOLD—CONCERN TO ALL LIGHT—CONCERN TO SOME
COFFEE; FRENCH VANILLA	**AMMONIUM CARRAGEENAN**
WILD STRAWBERRY	**AMMONIUM CARRAGEENAN; ARTIFICIAL COLOR; BHA**
PILLSBURY FIGURINES	
CHOCOLATE	**COCONUT & PALM KERNEL OILS;** MILK PROTEIN; GLUTEN (VITAL WHEAT)
CHOCOLATE CARAMEL	**BHA; COCONUT & PALM KERNEL OILS;** GLUTEN (VITAL WHEAT); MILK PROTEIN
VANILLA	**ARTIFICIAL COLOR; BHA; COCONUT & PALM KERNEL OILS;** GLUTEN (VITAL WHEAT); MILK PROTEIN
WEIGHT WATCHERS *—FROZEN DINNERS*	
CHICKEN LIVERS & ONIONS	CAROB BEAN GUM; GUAR GUM; GUM TRAGACANTH; **MODIFIED FOOD STARCH; NUTMEG**
FLOUNDER	CAROB BEAN GUM; GUAR GUM; GUM TRAGACANTH; **PAPRIKA**
HADDOCK; VEAL PARMIGIANA & ZUCCHINI IN SAUCE	CAROB BEAN GUM; GUAR GUM; GUM TRAGACANTH

BRANDS	ADDITIVES-OF-CONCERN **BOLD—CONCERN TO ALL** LIGHT—CONCERN TO SOME
WEIGHT WATCHERS *(Continued)*	
—FROZEN ONE-COURSE DISHES	
CHEESE & TOMATO PIES	NO ADDITIVES OF CONCERN
CHICKEN CREOLE	**VEGETABLE STABILIZERS**
EGGPLANT PARMIGIANA; ZITI MACARONI W/ VEAL, CHEESE & SAUCE	CAROB BEAN GUM; GUAR GUM; GUM TRAGACANTH
VEAL STUFFED PEPPER	CAROB BEAN GUM; GUAR GUM; GUM TRAGACANTH; WHEY

LOW-CALORIE JAMS, JELLIES, PRESERVES, SYRUPS

BRANDS	ADDITIVES-OF-CONCERN
CARY'S LOW CALORIE SYRUP	MONOSODIUM GLUTAMATE; **SODIUM SACCHARIN**
DIET DELIGHT *LOW CALORIE* PANCAKE SYRUP	NO ADDITIVES OF CONCERN
—JAM	
BLACKBERRY; STRAWBERRY	**ARTIFICIAL COLOR; CARRAGEENAN**
FEATHERWEIGHT *LOW CALORIE IMITATION*	
CHERRY JELLY; GRAPE JELLY	**CARRAGEENAN; SEAWEED EXTRACT**
STRAWBERRY JELLY; STRAWBERRY PRESERVES	**ARTIFICIAL COLOR; CARRAGEENAN; SEAWEED EXTRACT**

BRANDS	**ADDITIVES-OF-CONCERN** **BOLD—CONCERN TO ALL** LIGHT—CONCERN TO SOME
SMUCKER'S SLENDERELLA *LOW CALORIE IMITATION* ORANGE MARMALADE; STRAWBERRY JAM	**ARTIFICIAL COLOR; CARRAGEENAN**
—JELLY BLACKBERRY; CHERRY; GRAPE	**CARRAGEENAN**

LOW-CALORIE MAYONNAISE & SALAD DRESSINGS

By law, at this writing, it is not required of a manufacturer to state whether cheeses contain artificial colors, unless the color is FD & C Yellow No. 5 (Tartrazine). Their presence in low-calorie mayonnaise and salad dressings containing cheese therefore remains an uncertainty, unless it is voluntarily declared on the label that artificial colors have been used or have not. For a more detailed explanation of artificial color in cheese, refer to the beginning of the cheese section of "Dairy Products & Substitutes."

DIA-MEL MAYONNAISE	**VEGETABLE OIL**
DIET DELIGHT MAY-O-LITE	ALGIN DERIVATIVES; **MODIFIED CORN STARCH; PAPRIKA**
FEATHERWEIGHT *LOW CALORIE DRESSING* CREAMY ITALIAN	**BHA; BHT**
THOUSAND ISLAND	ALGIN DERIVATIVE; **BHA; BHT; MODIFIED CORN STARCH; SODIUM SACCHARIN; VEGETABLE GUM**

BRANDS	ADDITIVES-OF-CONCERN BOLD—CONCERN TO ALL LIGHT—CONCERN TO SOME
FRENCHETTE *LOW CALORIE DRESSING*	
ITALIAN	**ARTIFICIAL COLORING; BHA; BHT;** GUM TRAGACANTH
THOUSAND ISLAND	**BHA; BHT;** GUM TRAGACANTH; **MODIFIED FOOD STARCH; OLEORESIN OF PAPRIKA**
KRAFT *LOW CALORIE DRESSING*	
BLUE CHEESE	**MODIFIED FOOD STARCH**
CREAMY CUCUMBER	PROPYLENE GLYCOL ALGINATE
RUSSIAN	**ARTIFICIAL COLOR; OLEORESIN PAPRIKA;** PROPYLENE GLYCOL ALGINATE
ZESTY ITALIAN	**ARTIFICIAL COLOR**
WEIGHT WATCHERS	
CREAMY ITALIAN DRESSING	**MODIFIED FOOD STARCH**
IMITATION MAYONNAISE	**MODIFIED FOOD STARCH; NATURAL COLOR**
THOUSAND ISLAND DRESSING	**MODIFIED FOOD STARCH; NATURAL COLOR; PAPRIKA**

BRANDS	ADDITIVES-OF-CONCERN **BOLD—CONCERN TO ALL** LIGHT—CONCERN TO SOME
WISH-BONE *LOW CALORIE DRESSING*	
CHUNKY BLUE CHEESE	ALGIN DERIVATIVE; **VEGETABLE GUM**
CREAMY ITALIAN; THOUSAND ISLAND	ALGIN DERIVATIVE
FRENCH STYLE	**OLEORESIN PAPRIKA**
RUSSIAN	ALGIN DERIVATIVE; **ARTIFICIAL COLOR**
SUPERMARKET PRIVATE BRANDS	
A & P	
—ANN PAGE *LOW CALORIE DRESSING*	
BLUE CHEESE; IMITATION FRENCH STYLE	ALGIN DERIVATIVE
THOUSAND ISLAND	ALGIN DERIVATIVE; **ARTIFICIAL COLOR**
SAFEWAY	
—NUMADE *REDUCED CALORIE DRESSING*	
ITALIAN	**ARTIFICIAL COLOR**
1000 ISLAND DRESSING	**ARTIFICIAL COLOR;** PROPYLENE GLYCOL ALGINATE

LOW-CALORIE SUGAR SUBSTITUTES

DIA-MEL SUGAR-LIKE	**SACCHARIN**
SUCARYL	**SODIUM SACCHARIN**

BRANDS	ADDITIVES-OF-CONCERN BOLD—CONCERN TO ALL LIGHT—CONCERN TO SOME
SWAN	
SWEETEST EFFERVESCENT SACCHARIN	**SACCHARIN**
SWEET 'N LOW	
REGULAR	**SACCHARIN**
ZERO-CAL	**CALCIUM SACCHARIN**
WEIGHT WATCHERS	
SWEET'NER	**SODIUM SACCHARIN**

XVII. MEAT & POULTRY & SUBSTITUTES

CANNED MEAT & POULTRY & SUBSTITUTES

BRANDS	**ADDITIVES-OF-CONCERN** BOLD—CONCERN TO ALL LIGHT—CONCERN TO SOME
ARMOUR	
CORNED BEEF HASH	GUM ARABIC; **SODIUM NITRITE**
POTTED MEAT FOOD PRODUCT; VIENNA SAUSAGE	**SODIUM NITRITE**
TREET	**COLORING; SMOKE FLAVORING; SODIUM NITRITE**
CUDAHY BAR S HAM	**SODIUM NITRITE**
HORMEL	
CHOPPED HAM	**SODIUM NITRATE; SODIUM NITRITE**
HAM PATTIES	**HICKORY SMOKE FLAVORING; SODIUM NITRITE**
—*DINTY MOORE*	
BEEF STEW	**MODIFIED FOOD STARCH**
MEATBALL STEW	**MODIFIED FOOD STARCH;** MONOSODIUM GLUTAMATE

BRANDS	ADDITIVES-OF-CONCERN BOLD—CONCERN TO ALL LIGHT—CONCERN TO SOME
HORMEL *(Continued)*	
—*TENDER CHUNK* CHICKEN; TURKEY	MONOSODIUM GLUTAMATE
HAM	**SMOKE FLAVORING; SODIUM NITRITE**
KRAKUS POLISH HAM	**SODIUM NITRITE**
LIBBY'S	
CORNED BEEF; CORNED BEEF HASH	**SODIUM NITRITE**
SLOPPY JOE	**MODIFIED CORNSTARCH; PAPRIKA; TURMERIC**
MARY KITCHEN	
CORNED BEEF HASH	**SODIUM NITRITE**
ROAST BEEF HASH	NO ADDITIVES OF CONCERN
PLUMROSE DANISH HAM	**SODIUM NITRITE**
SELL'S LIVER PATÉ	NO ADDITIVES OF CONCERN
SPAM	
DEVILED LUNCHEON MEAT; REGULAR	**SODIUM NITRITE**
SWANSON	
BEEF STEW MAIN DISH; BONED CHICKEN; BONED TURKEY	NO ADDITIVES OF CONCERN
CHICKEN STEW MAIN DISH; CHUNK CHICKEN	MONOSODIUM GLUTAMATE

BRANDS	ADDITIVES-OF-CONCERN **BOLD—CONCERN TO ALL** LIGHT—CONCERN TO SOME
SWIFT PREMIUM CORNED BEEF	**SODIUM NITRITE**
UNDERWOOD CHUNKY CHICKEN SPREAD	**MODIFIED FOOD STARCH;** MONOSODIUM GLUTAMATE; **TURMERIC (& EXTRACTIVES OF)**
CORNED BEEF SPREAD; DEVILED HAM	**SODIUM NITRITE**
LIVERWURST SPREAD	**PORK FAT (COOKED)**
SUPERMARKET PRIVATE BRANDS	
SAFEWAY	
—SAFEWAY HAM PATTIES	**SMOKE FLAVORING; SODIUM NITRITE**
—TOWN HOUSE VIENNA SAUSAGE	**SODIUM NITRITE**
WINN DIXIE	
—THRIFTY MAID VIENNA SAUSAGE	**SODIUM NITRITE**
ITEMS IN HEALTH FOOD STORES	
FEATHERWEIGHT UNSALTED BONED CHICKEN	NO ADDITIVES OF CONCERN
SOVEX VEGE-PAT	NO ADDITIVES OF CONCERN
WORTHINGTON CHOPLETS	MONOSODIUM GLUTAMATE
SANDWICH SPREAD	CAROB BEAN & GUAR GUMS; **MODIFIED CORN STARCH**

BRANDS	ADDITIVES-OF-CONCERN **BOLD—CONCERN TO ALL** LIGHT—CONCERN TO SOME
WORTHINGTON *(Continued)*	
SOYAMEAT BEEF FLAVOR	**ARTIFICIAL COLOR;** EGG WHITES; GLUTEN (WHEAT); MONOSODIUM GLUTAMATE
SOYAMEAT CHICKEN FLAVOR	**ARTIFICIAL COLOR; CARRAGEENAN;** EGG WHITE SOLIDS; **MODIFIED CORN STARCH;** MONOSODIUM GLUTAMATE
SUPER-LINKS	**ARTIFICIAL COLORS;** EGG WHITES; GLUTEN (WHEAT); **MODIFIED CORN STARCH; NATURAL SMOKE FLAVOR; PAPRIKA**
209 SMOKED TURKEY FLAVOR	EGG WHITES; GLUTEN (WHEAT); MONOSODIUM GLUTAMATE; **NATURAL SMOKE FLAVOR**
VEJA-LINKS	**ARTIFICIAL COLORS; CARRAGEENAN;** EGG WHITES; GLUTEN (WHEAT); MONOSODIUM GLUTAMATE; **NATURAL SMOKE FLAVOR; PAPRIKA;** SODIUM ALGINATE

FROZEN MEAT & POULTRY & SUBSTITUTES

JONES	
COUNTRY PORK SAUSAGE; LITTLE PORK SAUSAGES; MINUTE BREAKFAST LINKS	NO ADDITIVES OF CONCERN

BRANDS	ADDITIVES-OF-CONCERN **BOLD—CONCERN TO ALL** LIGHT—CONCERN TO SOME
LOVITT'S SHAVED STEAK	NO ADDITIVES OF CONCERN
MAID-RITE	
BEEF PEPPER STEAKS	MONOSODIUM GLUTAMATE
BREADED VEAL STEAKS	MONOSODIUM GLUTAMATE; **PAPRIKA;** WHEY (DRIED)
VEAL STEAKS	NO ADDITIVES OF CONCERN
MORNINGSTAR FARMS	
BREAKFAST LINKS	EGG WHITES; GLUTEN (WHEAT); GUAR GUM; **MODIFIED CORN STARCH;** MONOSODIUM GLUTAMATE; **SODIUM CASEINATE**
BREAKFAST STRIPS	**ARTIFICIAL COLOR;** CAROB BEAN & GUAR GUMS; **CARRAGEENAN;** EGG WHITES; **MODIFIED CORN STARCH;** MONOSODIUM GLUTAMATE
SWIFT PREMIUM *BROWN 'N SERVE SAUSAGE*	
BACON 'N SAUSAGE	**BHA; BHT; SMOKE FLAVORING; SODIUM NITRITE**
BEEF; THE ORIGINAL	**BHA; BHT;** MONOSODIUM GLUTAMATE
HICKORY SMOKE FLAVORED	**BHA; BHT; HICKORY SMOKE FLAVORING;** MONOSODIUM GLUTAMATE

BRANDS	ADDITIVES-OF-CONCERN BOLD—CONCERN TO ALL LIGHT—CONCERN TO SOME
SWIFT PREMIUM *(Continued)*	
MAPLE FLAVORED	**BHA; BHT**
TABLE TREAT	
STEAK-UMM ALL BEEF SANDWICH STEAKS	NO ADDITIVES OF CONCERN
SUPERMARKET PRIVATE BRANDS	
SAFEWAY	
—MANOR HOUSE	
FRIED CHICKEN	
BREAST PORTIONS; FULLY COOKED ASSORTED PIECES; WING PORTIONS	MONOSODIUM GLUTAMATE
ITEMS IN HEALTH FOOD STORES	
HEALTH IS WEALTH	
BREADED TURKEY PATTIE; CHICK PUPS; TURKEY BREAST ROLL	NO ADDITIVES OF CONCERN
HEALTH VALLEY	
SLICED BREAKFAST BEEF; SLICED BREAKFAST PORK; WHOLE FRYING CHICKEN	NO ADDITIVES OF CONCERN
SMOKED DRIED BEEF	**SMOKED***

*Although "smoked" does not refer to a specific additive, it represents a process utilizing wood smoke, as "smoke flavoring" does, and therefore may contain some cancer-causing benzopyrene chemicals. For this reason, when a food or any of its constituents has been smoked, this has been noted in the additives-of-concern column.

BRANDS	ADDITIVES-OF-CONCERN **BOLD—CONCERN TO ALL** LIGHT—CONCERN TO SOME
—NATURAL SPICE UNCURED COOKED SAUSAGE BEEF WIENER FLAVORING; CHICKEN BOLOGNA FLAVORING; OUR SUPREME BRAND BEEF WIENER FLAVORING; TURKEY WIENER FLAVORING	**PAPRIKA**
SHILOH FARMS *SAUSAGE* BREAKFAST; UNCURED COOKED	NO ADDITIVES OF CONCERN
WORTHINGTON DINNER ROAST	**CALCIUM CASEINATE; CARRAGEENAN;** EGG WHITES; GLUTEN (WHEAT); **MODIFIED CORN STARCH; SODIUM CASEINATE; TURMERIC**
FILLETS	EGG WHITES; **MODIFIED CORN STARCH; MODIFIED TAPIOCA STARCH;** MONOSODIUM GLUTAMATE
FRI PATS	EGG WHITES; **MODIFIED TAPIOCA STARCH; SODIUM CASEINATE**
LUNCHEON SLICES SMOKED TURKEY-LIKE FLAVOR	EGG WHITES; GLUTEN (WHEAT); MONOSODIUM GLUTAMATE; **NATURAL SMOKE FLAVOR**

BRANDS	ADDITIVES-OF-CONCERN BOLD—CONCERN TO ALL LIGHT—CONCERN TO SOME
WORTHINGTON *(Continued)*	
MEATLESS CHICKEN	**ARTIFICIAL COLOR; CARRAGEENAN;** EGG WHITES; MONOSODIUM GLUTAMATE; **TURMERIC**
STAKELETS	EGG WHITE SOLIDS; **SODIUM CASEINATE**
STRIPPLES	**ARTIFICIAL COLOR; CARRAGEENAN;** EGG ALBUMIN (RECONSTITUTED); GLUTEN (WHEAT); **MODIFIED TAPIOCA STARCH; SODIUM CASEINATE**

REFRIGERATED MEAT & POULTRY

BRANDS	ADDITIVES-OF-CONCERN
ARMOUR	
BACON; 1877 CANADIAN STYLE BACON	**SODIUM NITRITE**
BEEF HOT DOGS; HOT DOGS	**OLEORESIN OF PAPRIKA; SODIUM NITRITE**
CASERTA BRAND PEPERONI	**BHA; BHT; OLEORESIN OF PAPRIKA; SODIUM NITRATE; SODIUM NITRITE**
GENOA SALAMI; HARD SALAMI	**SODIUM NITRATE; SODIUM NITRITE**
CARL BUDDIG	
CORNED BEEF	MONOSODIUM GLUTAMATE; **SODIUM NITRITE**

BRANDS	**ADDITIVES-OF-CONCERN** **BOLD—CONCERN TO ALL** LIGHT—CONCERN TO SOME
SMOKED BEEF; SMOKED HAM; SMOKED PASTRAMI	MONOSODIUM GLUTAMATE; **SODIUM NITRITE; WOOD SMOKED***
SMOKED CHICKEN	**MODIFIED FOOD STARCH;** MONOSODIUM GLUTAMATE; **TURMERIC; WOOD SMOKED***
HEBREW NATIONAL	
BEEF FRANKFURTERS; BEEF SALAMI; PASTRAMI	**PAPRIKA; SODIUM NITRITE**
JONES SLICED BACON	**SODIUM NITRITE**
OSCAR MAYER	
BEEF BOLOGNA; BOLOGNA	**PAPRIKA; SODIUM NITRITE**
BEEF COTTO SALAMI; BRAUNSCHWEIGER; CHOPPED HAM; HARD SALAMI; WIENERS	**SODIUM NITRITE**
HAM STEAKS	**SMOKE FLAVORING; SODIUM NITRITE**
HONEY LOAF; OLD FASHIONED LOAF	**CALCIUM CASEINATE;** MONOSODIUM GLUTAMATE; **SODIUM NITRITE;** WHEY

*Although "smoked" does not refer to a specific additive, it represents a process utilizing wood smoke, as "smoke flavoring" does, and therefore the food may contain some cancer-causing benzopyrene chemicals. For this reason, when a food or any of its constituents has been smoked, this has been noted in the additives-of-concern column.

BRANDS	ADDITIVES-OF-CONCERN BOLD—CONCERN TO ALL LIGHT—CONCERN TO SOME
OSCAR MAYER *(Continued)*	
MORTADELLA	MONOSODIUM GLUTAMATE; **PORK FAT; SODIUM NITRITE**
OLIVE LOAF; PICKLE & PIMENTO LOAF	**CALCIUM CASEINATE; SODIUM NITRITE;** WHEY
PLUMROSE *PREMIUM*	
AMERICAN COOKED HAM; SLICED BACON	**SODIUM NITRITE**
SWIFT SIZZLEAN PORK BREAKFAST STRIPS	MONOSODIUM GLUTAMATE; **SMOKE FLAVORING; SODIUM NITRITE**
WEAVER	
CHICKEN FRANKS	**PAPRIKA; SODIUM NITRITE**
WHITE MEAT CHICKEN ROLL	MONOSODIUM GLUTAMATE
SUPERMARKET PRIVATE BRANDS	
A & P—A & P	
BOLOGNA; COOKED SALAMI; DANISH COOKED HAM; NEW ENGLAND BRAND SAUSAGE; SKINLESS BEEF FRANKS	**SODIUM NITRITE**
BRAUNSCHWEIGER	**PORK FAT; SODIUM NITRITE**
CHICKEN ROLL	EGG ALBUMIN
HARD SALAMI	**BHA; BHT; NATURAL SMOKE FLAVOR; SODIUM NITRITE**

BRANDS	ADDITIVES-OF-CONCERN **BOLD—CONCERN TO ALL** LIGHT—CONCERN TO SOME
PEPPERONI	**BHA; BHT; PAPRIKA; NATURAL SMOKE FLAVOR; SODIUM NITRITE**
TURKEY BREAST ROLL	NO ADDITIVES OF CONCERN
—*SMOKED SLICED*	
BEEF; HAM; TURKEY	MONOSODIUM GLUTAMATE; **SMOKED;* SODIUM NITRITE**
PASTRAMI	**PAPRIKA; SMOKED;* SODIUM NITRITE**
—**ANN PAGE** SLICED BACON	**SODIUM NITRITE**
FIRST NATIONAL	
—EDWARDS-FINAST	
SLICED BACON	**SODIUM NITRITE**
—FINAST	
BEEF FRANKS; COOKED SALAMI	**SODIUM NITRITE**
SPICED LUNCHEON LOAF	MONOSODIUM GLUTAMATE; **SODIUM NITRITE**
LUCKY—LADY LEE	
BOLOGNA; CHICKEN BOLOGNA	**PAPRIKA; SMOKE FLAVORING; SODIUM NITRITE**

*Although "smoked" does not refer to a specific additive, it represents a process utilizing wood smoke, as "smoke flavoring" does, and therefore the food may contain some cancer-causing benzopyrene chemicals. For this reason, when a food or any of its constituents has been smoked, this has been noted in the additives-of-concern column.

BRANDS	ADDITIVES-OF-CONCERN BOLD—CONCERN TO ALL LIGHT—CONCERN TO SOME
LUCKY—LADY LEE *(Continued)*	
BEEF SALAMI; COOKED SALAMI	**SMOKE FLAVORING; SODIUM NITRITE**
CHOPPED HAM; CHOPPED PORK; SLICED BACON	**SODIUM NITRITE**
GARLIC SAUSAGE; HOT SAUSAGE LINKS; KNOCKWURST	**PAPRIKA; HICKORY SMOKE FLAVORING; SODIUM NITRITE**
HOT PORK SAUSAGE	**BHA**
POLISH SAUSAGE	**HICKORY SMOKE FLAVORING; SODIUM NITRITE**
—FRANKS	
CHICKEN; REGULAR; TURKEY	**PAPRIKA; HICKORY SMOKE FLAVORING; SODIUM NITRITE**
SAFEWAY	
—SAFEWAY	
ALL VEAL STEAKS	NO ADDITIVES OF CONCERN
BEEF BACON; BRAUNSCHWEIGER	**NATURAL SMOKE FLAVOR** OR **SMOKE FLAVORING; SODIUM NITRITE**
BEEF BOLOGNA; BOLOGNA; COOKED HAM; COTTO SALAMI; GERMAN BRAND SAUSAGE; KNOCKWURST; THURINGER	**SODIUM NITRITE**

BRANDS	ADDITIVES-OF-CONCERN BOLD—CONCERN TO ALL LIGHT—CONCERN TO SOME
BEEF BREAKFAST STRIPS	**HICKORY SMOKE FLAVORING; SODIUM NITRITE**
CHOPPED HAM	MONOSODIUM GLUTAMATE; **SODIUM NITRITE**
COMBINATION LOAF; OLIVE LOAF	MONOSODIUM GLUTAMATE; **SODIUM CASEINATE; SODIUM NITRITE;** WHEY (SWEET DAIRY)
HOT PORK SAUSAGE WHOLE HOG; MEDIUM PORK SAUSAGE WHOLE HOG	MONOSODIUM GLUTAMATE
—MANOR HOUSE	
CHICKEN BOLOGNA	**SODIUM NITRITE**
—(NO BRAND NAME) *SMOKED-SLICED-CHOPPED-PRESSED-COOKED*	
BEEF; HAM; PASTRAMI	MONOSODIUM GLUTAMATE; **SMOKED;* SODIUM NITRITE**
CHICKEN; TURKEY	MONOSODIUM GLUTAMATE; **SMOKED;* SODIUM CASEINATE; SODIUM NITRITE**

*Although "smoked" does not refer to a specific additive, it represents a process utilizing wood smoke, as "smoke flavoring" does, and therefore the food may contain some cancer-causing benzopyrene chemicals. For this reason, when a food or any of its constituents has been smoked, this has been noted in the additives-of-concern column.

BRANDS	ADDITIVES-OF-CONCERN BOLD—CONCERN TO ALL LIGHT—CONCERN TO SOME
SAFEWAY *(Continued)*	
—TROPHY ITALIAN BRAND BREADED VEAL PATTIES	**ARTIFICIALLY COLORED;** MONOSODIUM GLUTAMATE
STOP & SHOP **—STOP & SHOP** BEEF BOLOGNA	**OLEORESIN PAPRIKA; SODIUM NITRITE**
CHOPPED HAM; LUNCHEON LOAF	MONOSODIUM GLUTAMATE; **SODIUM NITRITE**
POLISH BRAND LOAF	MONOSODIUM GLUTAMATE; **PORK FAT; SODIUM CASEINATE; SODIUM NITRITE;** WHEY (DRIED)
—BACON MAPLE SUGAR CURED; THICK SLICED SUGAR CURED	**SODIUM NITRITE**
—FRANKS BEEF SKINLESS; EXTRA MILD SKINLESS	**OLEORESIN OF PAPRIKA; SODIUM NITRITE**
GET-A-LONG-DOGGIE	**OLEORESIN OF PAPRIKA; PORK FAT; SODIUM NITRITE**
—SUN GLORY SUGAR CURED SLICED BACON	**SODIUM NITRITE**

BRANDS	ADDITIVES-OF-CONCERN BOLD—CONCERN TO ALL LIGHT—CONCERN TO SOME
WINN DIXIE	
—W-D	
BEEF BOLOGNA; CHOPPED HAM; COOKED HAM; COOKED SALAMI; PRESTIGE SLICED BACON; SOUSE	**SODIUM NITRITE**
LEBANON BOLOGNA	**POTASSIUM NITRATE**
SMOKED SAUSAGE; SMOKED PORK SHOULDER PICNIC	**SMOKED;* SODIUM NITRITE**
SPICED LUNCHEON LOAF	MONOSODIUM GLUTAMATE; **SODIUM CASEINATE; SODIUM NITRITE;** WHEY SOLIDS
—WINN DIXIE	
BREAKFAST SAUSAGE MADE FROM BEEF	MONOSODIUM GLUTAMATE

*Although "smoked" does not refer to a specific additive, it represents a process utilizing wood smoke, as "smoke flavoring" does, and therefore the food may contain some cancer-causing benzopyrene chemicals. For this reason, when a food or any of its constituents has been smoked, this has been noted in the additives-of-concern column.

XVIII. ONE-COURSE DISHES

CANNED ONE-COURSE DISHES (BEANS, ETHNIC FOODS)

By law, at this writing, it is not required of a manufacturer to state whether cheeses contain artificial colors, unless the color is FD & C Yellow No. 5 (Tartrazine). Their presence in one-course dishes containing cheese therefore remains an uncertainty, unless it is voluntarily declared on the label that artificial colors have been used or have not. For a more detailed explanation of artificial color in cheese, refer to the beginning of the cheese section of "Dairy Products & Substitutes."

BRANDS	**ADDITIVES-OF-CONCERN** BOLD—CONCERN TO ALL LIGHT—CONCERN TO SOME
B & M *BAKED BEANS* REGULAR; W/ RED KIDNEY BEANS; W/ YELLOW EYE BEANS	NO ADDITIVES OF CONCERN
BUITONI SPAGHETTI TWISTS	**MODIFIED FOOD STARCH**
CAMPBELL'S *BEANS* BARBEQUE	**OLEORESIN TURMERIC; SMOKE FLAVORING**
W/ FRANKS	**OLEORESIN PAPRIKA; MODIFIED FOOD STARCH; SODIUM NITRITE**
OLD FASHIONED	NO ADDITIVES OF CONCERN
W/ PORK	**MODIFIED FOOD STARCH; OLEORESIN PAPRIKA**

BRANDS	ADDITIVES-OF-CONCERN **BOLD—CONCERN TO ALL** LIGHT—CONCERN TO SOME
CHEF BOY-AR-DEE	
BEEFARONI; LASAGNA; MACARONI SHELLS	**MODIFIED FOOD STARCH**
BEEF RAVIOLI; CHEESE RAVIOLI IN SAUCE; ROLLER COASTERS; SPAGHETTI & MEAT BALLS	**MODIFIED FOOD STARCH;** MONOSODIUM GLUTAMATE
CHUN KING	
BEAN SPROUTS; CHOW MEIN NOODLES; CHOW MEIN VEGETABLES	NO ADDITIVES OF CONCERN
FRANCO-AMERICAN	
BEEF RAVIOLIOS	**MODIFIED FOOD STARCH; OLEORESIN PAPRIKA**
BEEFY MAC	MONOSODIUM GLUTAMATE; **OLEORESIN PAPRIKA**
ELBOW MACARONI & CHEESE	MONOSODIUM GLUTAMATE; **OLEORESIN PAPRIKA;** WHEY
SPAGHETTI; SPAGHETTIOS; SPAGHETTIOS W/ LITTLE MEATBALLS	NO ADDITIVES OF CONCERN
SPAGHETTI W/ MEATBALLS	MONOSODIUM GLUTAMATE
SPAGHETTIOS W/ SLICED FRANKS	**OLEORESIN PAPRIKA; SODIUM NITRITE**

BRANDS	ADDITIVES-OF-CONCERN BOLD—CONCERN TO ALL LIGHT—CONCERN TO SOME
HEINZ VEGETARIAN BEANS	**MODIFIED FOOD STARCH**
HORMEL BEEF TAMALES IN CHILI SAUCE	NO ADDITIVES OF CONCERN
CHILI NO BEANS	**MODIFIED FOOD STARCH;** MONOSODIUM GLUTAMATE
CHILI W/ BEANS; HOT CHILI W/ BEANS	**MODIFIED FOOD STARCH**
LIBBY'S CHILI W/ BEANS	NO ADDITIVES OF CONCERN
DEEP BROWN PORK & BEANS	**MODIFIED CORN STARCH**
STEWART'S SATURDAY SUPPER PEA BEANS & PORK	NO ADDITIVES OF CONCERN
SUPERMARKET PRIVATE BRANDS **A & P** **—ANN PAGE** BOSTON STYLE BEANS	**MODIFIED CORN STARCH**
PORK & BEANS IN TOMATO SAUCE	NO ADDITIVES OF CONCERN
FIRST NATIONAL **—EDWARDS** BROWN BEANS IN CHILI GRAVY	NO ADDITIVES OF CONCERN

BRANDS	ADDITIVES-OF-CONCERN **BOLD—CONCERN TO ALL** LIGHT—CONCERN TO SOME
LUCKY—LADY LEE	
CHILI W/ BEANS HOT	**MODIFIED FOOD STARCH; NATURAL COLORING; PAPRIKA**
CORNED BEEF HASH	**SODIUM NITRITE**
REFRIED BEANS	**LARD**
SAFEWAY —TOWN HOUSE	
BEEF STEW	**MODIFIED FOOD STARCH**
CHILI CON CARNE W/ BEANS	MONOSODIUM GLUTAMATE; **PAPRIKA**
CHILI CON CARNE W/ OUT BEANS	**MODIFIED FOOD STARCH;** MONOSODIUM GLUTAMATE; **PAPRIKA**
CORNED BEEF HASH	**SODIUM NITRITE**
PORK & BEANS	**EXTRACTIVES OF PAPRIKA; MODIFIED FOOD STARCH**
—STOP & SHOP	
SALAD	
MACARONI; OIL & VINEGAR POTATO	NO ADDITIVES OF CONCERN
POTATO	PROPYLENE GLYCOL ALGINATE
WINN DIXIE —THRIFTY MAID	
BEEF STEW; CHILI W/ BEANS; MEXICAN STYLE CHILI BEANS	NO ADDITIVES OF CONCERN

BRANDS	ADDITIVES-OF-CONCERN BOLD—CONCERN TO ALL LIGHT—CONCERN TO SOME
WINN DIXIE—THRIFTY MAID *(Continued)*	
HOT DOG CHILI SAUCE	**BEEF FAT; MODIFIED FOOD STARCH;** MONOSODIUM GLUTAMATE
MEAT RAVIOLI	**MODIFIED FOOD STARCH**
SPAGHETTI; SPAGHETTI RINGS	**MODIFIED FOOD STARCH; PAPRIKA**
ITEMS IN HEALTH FOOD STORES	
FEATHERWEIGHT	
UNSALTED	
BEEF RAVIOLI	EGG WHITES; **MODIFIED FOOD STARCH**
BEEF STEW; LAMB STEW	NO ADDITIVES OF CONCERN
DUMPLINGS W/ CHICKEN	**BHA; BHT; MODIFIED FOOD STARCH**
SPAGHETTI W/ MEATBALLS	**MODIFIED FOOD STARCH**
HEALTH VALLEY	
HONEY BAKED BEANS	NO ADDITIVES OF CONCERN
—VEGETARIAN CHILI	
MILD; SPICY	**PAPRIKA**
WALNUT ACRES CHILI CON CARNE	**PAPRIKA**

BRANDS	ADDITIVES-OF-CONCERN **BOLD—CONCERN TO ALL** LIGHT—CONCERN TO SOME

DRY MIXES FOR ONE-COURSE DISHES

By law, at this writing, it is not required of a manufacturer to state whether cheeses contain artificial colors, unless the color is FD & C Yellow No. 5 (Tartrazine). Their presence in dry mixes for one-course dishes containing cheese therefore remains an uncertainty, unless it is voluntarily declared on the label that artificial colors have been used or have not. For a more detailed explanation of artificial color in cheese, refer to the beginning of the cheese section of "Dairy Products & Substitutes."

BRANDS	ADDITIVES-OF-CONCERN
BETTY CROCKER NOODLES ROMANOFF	**ARTIFICIAL COLOR; BHA**
—HAMBURGER HELPER FOR BEEF NOODLE	MONOSODIUM GLUTAMATE
FOR BEEF ROMANOFF	**BHA; MODIFIED CORN STARCH;** MONOSODIUM GLUTAMATE; WHEY
FOR CHEESEBURGER MACARONI	**ARTIFICIAL COLOR; BHA; MODIFIED CORN STARCH; SODIUM CASEINATE**
FOR HAMBURGER POTATOES AU GRATIN	**ARTIFICIAL COLOR; BHA;** MONOSODIUM GLUTAMATE; WHEY
FOR RICE ORIENTAL	**BHT**
—MUG-O-LUNCH BEEF NOODLES & GRAVY	**MODIFIED CORN & TAPIOCA STARCHES;** MONOSODIUM GLUTAMATE
MACARONI & CHEESE	**ARTIFICIAL COLOR; MODIFIED CORN STARCH;** WHEY

BRANDS	ADDITIVES-OF-CONCERN BOLD—CONCERN TO ALL LIGHT—CONCERN TO SOME
BETTY CROCKER *(Continued)*	
—*TUNA HELPER* FOR CREAMY NOODLES 'N TUNA	**ARTIFICIAL COLOR; BHA; SODIUM CASEINATE**
FOR NOODLES, CHEESE SAUCE 'N TUNA	**ARTIFICIAL COLOR; BHA; MODIFIED CORN STARCH;** MONOSODIUM GLUTAMATE; **SODIUM CASEINATE;** WHEY
BELL'S MEATLOAF	**MODIFIED FOOD STARCH;** MONOSODIUM GLUTAMATE; **PAPRIKA**
CHEF BOY-AR-DEE	
COMPLETE CHEESE PIZZA; COMPLETE CHEESE PIZZA IN A SKILLET	GUAR GUM; **MODIFIED FOOD STARCH;** WHEY (DRIED)
SPAGHETTI DINNER W/ MEAT SAUCE	**MODIFIED FOOD STARCH**
CHUN KING	
ORIENTAL VEGETABLES & SAUCE MIX FOR STIR-FRY PEPPER STEAK	**MODIFIED FOOD STARCH;** MONOSODIUM GLUTAMATE
GOLDEN GRAIN	
MACARONI & CHEDDAR	**ARTIFICIAL COLOR;** WHEY
NOODLE RONI STROGANOFF	MONOSODIUM GLUTAMATE; **PAPRIKA;** WHEY
KRAFT *DINNERS*	
MACARONI & CHEESE DELUXE	**ARTIFICIAL COLOR;** SODIUM ALGINATE; WHEY

BRANDS	**ADDITIVES-OF-CONCERN** **BOLD—CONCERN TO ALL** LIGHT—CONCERN TO SOME
TANGY ITALIAN STYLE SPAGHETTI	**ARTIFICIAL COLOR; MODIFIED FOOD STARCH;** MONOSODIUM GLUTAMATE
LIPTON *LITE-LUNCH*	
BEEF; STOCKPOT VEGETABLE	**(HYDROGENATED COTTONSEED, PALM, SOYBEAN OILS);*** MONOSODIUM GLUTAMATE
CHICKEN	**(HYDROGENATED COTTONSEED, PALM, SOYBEAN OILS);*** MONOSODIUM GLUTAMATE; **TURMERIC OLEORESIN**
MACARONI & CHEESE	**ARTIFICIAL COLOR; MODIFIED TAPIOCA STARCH;** MONOSODIUM GLUTAMATE; WHEY SOLIDS
NESTLÉ *LUNCH TIME*	
EGG NOODLES BEEF	**MODIFIED TAPIOCA STARCH;** MONOSODIUM GLUTAMATE
EGG NOODLES CHICKEN	**MODIFIED TAPIOCA STARCH;** MONOSODIUM GLUTAMATE; **SODIUM CASEINATE; TURMERIC**
EGG NOODLES TUNA & CELERY	**MODIFIED TAPIOCA STARCH;** MONOSODIUM GLUTAMATE; **SODIUM CASEINATE**
MACARONI CHEESE & HAM	**ARTIFICIAL COLOR; MODIFIED TAPIOCA STARCH;** MONOSODIUM GLUTAMATE

*Blend may contain one or more of these saturated oils.

BRANDS	ADDITIVES-OF-CONCERN **BOLD—CONCERN TO ALL** LIGHT—CONCERN TO SOME
TEMPO	
ITALIAN MEAT BALL; MEAT LOAF	NO ADDITIVES OF CONCERN
SWEDISH MEAT BALL	MONOSODIUM GLUTAMATE
SUPERMARKET PRIVATE BRANDS	
A & P—ANN PAGE	
MACARONI & CHEESE DINNER	**ARTIFICIAL COLOR;** BUTTERMILK SOLIDS; WHEY SOLIDS
LUCKY —LADY LEE	
MACARONI & CHEESE DINNER	BUTTERMILK SOLIDS; **CERTIFIED FOOD COLOR;** WHEY SOLIDS
SAFEWAY —TOWN HOUSE	
MACARONI & CHEESE DINNER	**ARTIFICIAL COLOR; MODIFIED CORN STARCH;** WHEY
WINN DIXIE —THRIFTY MAID	
DINNER	
MACARONI & CHEESE; SHELLS & CHEDDAR; TWISTS & CHEDDAR	**ARTIFICIAL COLOR;** BUTTERMILK SOLIDS; **MODIFIED FOOD STARCH;** WHEY SOLIDS
ITEMS IN HEALTH FOOD STORES	
DE BOLES *DINNER*	
MACARONI & CHEESE; WHOLE WHEAT HIGH PROTEIN MACARONI & CHEESE	EGG WHITE SOLIDS; GLUTEN (WHEAT)

BRANDS	ADDITIVES-OF-CONCERN BOLD—CONCERN TO ALL LIGHT—CONCERN TO SOME
EARTHWONDER *NATURAL DINNER-IN-A-BOX* MILLET STEW; MUSHROOM WHEAT PILAF; QUICK CHILI	NO ADDITIVES OF CONCERN
FANTASTIC FOODS *MIX* FANTASTIC FALAFIL; NATURE'S BURGER; TEMPURA BATTER	NO ADDITIVES OF CONCERN
FEARN *MIX* BREAKFAST PATTY; BRAZIL NUT BURGER	NO ADDITIVES OF CONCERN
SESAME BURGER	GLUTEN (WASHED WHEAT); KELP POWDER
FRITINI SWISS READY MIX FOR VEGETABLE PATTIES	NO ADDITIVES OF CONCERN
MANNA MEALS OF MARYLAND SLOPPY JOE	NO ADDITIVES OF CONCERN
NEAR EAST *MIX* WHEAT PILAF; WHEAT SALAD TABOULEH	NO ADDITIVES OF CONCERN

BRANDS	**ADDITIVES-OF-CONCERN** **BOLD—CONCERN TO ALL** LIGHT—CONCERN TO SOME
SOKEN-SHA *SOKEN RAMEN UNBLEACHED NOODLES*	
SEA VEGETABLES W/ MISO FLAVOR	WAKAME SEAWEED
W/ SOUP BASE; W/ WHOLE WHEAT/ BROWN RICE GERM	NO ADDITIVES OF CONCERN

FROZEN ONE-COURSE DISHES

By law, at this writing, it is not required of a manufacturer to state whether cheeses contain artificial colors, unless the color is FD & C Yellow No. 5 (Tartrazine). Their presence in frozen one-course dishes containing cheese therefore remains an uncertainty, unless it is voluntarily declared on the label that artificial colors have been used or have not. For a more detailed explanation of artificial color in cheese, refer to the beginning of the cheese section of "Dairy Products & Substitutes."

BRANDS	ADDITIVES-OF-CONCERN
BANQUET	
FRIED CHICKEN	MONOSODIUM GLUTAMATE
—BUFFET SUPPER	
CHICKEN & DUMPLINGS; GRAVY & SLICED TURKEY	MONOSODIUM GLUTAMATE; **PAPRIKA; TURMERIC**
GRAVY & SLICED BEEF	**BEEF FAT**
VEAL PARMIGIAN	**ARTIFICIAL COLORS; BHA;** MONOSODIUM GLUTAMATE; **PAPRIKA; SODIUM CASEINATE;** WHEY (DRY)

BRANDS	**ADDITIVES-OF-CONCERN** **BOLD—CONCERN TO ALL** LIGHT—CONCERN TO SOME
—COOKIN' BAG	
CHICKEN À LA KING	MONOSODIUM GLUTAMATE; **PAPRIKA**
CREAMED CHIPPED BEEF	**SODIUM NITRITE**
GRAVY & SLICED TURKEY	MONOSODIUM GLUTAMATE; **OLEORESIN OF TURMERIC; PAPRIKA**
MACARONI & CHEESE	**ARTIFICIALLY COLORED; COLORING; SODIUM CASEINATE;** WHEY
MEAT LOAF W/ TOMATO SAUCE	**CALCIUM CASEINATE; PAPRIKA;** WHEY
SALISBURY STEAK W/ GRAVY	**COLORING**
BUITONI	
BAKED ZITI; EGGPLANT PARMIGIANA	**MODIFIED FOOD STARCH; SODIUM/CALCIUM CASEINATE;** WHEY
CHEESE RAVIOLI; MANICOTTI	**FURCELLERAN; MODIFIED FOOD STARCH; SODIUM/CALCIUM CASEINATE;** WHEY
LASAGNE	**MODIFIED FOOD STARCH; SODIUM/CALCIUM CASEINATE; VEGETABLE GUM;** WHEY SOLIDS
MEAT RAVIOLI	NO ADDITIVES OF CONCERN

BRANDS	ADDITIVES-OF-CONCERN **BOLD—CONCERN TO ALL** LIGHT—CONCERN TO SOME
BUITONI *(Continued)*	
SAUSAGE & PEPPERS W/ MOSTACCIOLI RIGATI	**MODIFIED FOOD STARCH;** MONOSODIUM GLUTAMATE
VEAL PARMIGIANA W/ SPAGHETTI TWISTS	GLUTEN; **MODIFIED FOOD STARCH;** MONOSODIUM GLUTAMATE
CHUN KING CHICKEN CHOW MEIN; SHRIMP CHOW MEIN	**MODIFIED FOOD STARCH;** MONOSODIUM GLUTAMATE
FRIED RICE W/ PORK	NO ADDITIVES OF CONCERN
SHRIMP EGG ROLLS	**CALCIUM CASEINATE; EXTRACT OF PAPRIKA; MODIFIED FOOD STARCH;** MONOSODIUM GLUTAMATE; SODIUM ALGINATE; WHEY (DRIED)
FREEZER QUEEN BREADED VEAL PARMIGIANA	**MODIFIED FOOD STARCH;** MONOSODIUM GLUTAMATE; **PAPRIKA (& OLEORESIN)**
GRAVY & SALISBURY STEAK; GRAVY & SLICED BEEF; MUSHROOM GRAVY & CHAR BROILED BEEF PATTIES	**MODIFIED FOOD STARCH;** MONOSODIUM GLUTAMATE; **OLEORESIN PAPRIKA**
GRAVY & SLICED TURKEY	**MODIFIED FOOD STARCH;** MONOSODIUM GLUTAMATE; **OLEORESIN PAPRIKA; OLEORESIN TURMERIC**

BRANDS	**ADDITIVES-OF-CONCERN** **BOLD—CONCERN TO ALL** LIGHT—CONCERN TO SOME
SPAGHETTI W/ MEAT BALLS	**ARTIFICIAL COLOR; MODIFIED FOOD STARCH;** MONOSODIUM GLUTAMATE; **PAPRIKA (& OLEORESIN)**
GREEN GIANT	
CHICKEN & BISCUITS	GUAR GUM; **MODIFIED CORN STARCH;** MONOSODIUM GLUTAMATE; **TURMERIC**
LASAGNA	**MODIFIED FOOD STARCH;** MONOSODIUM GLUTAMATE
STUFFED GREEN PEPPERS	**MODIFIED FOOD STARCH;** MONOSODIUM GLUTAMATE; **PAPRIKA**
VEAL PARMIGIANA	**ARTIFICIAL COLOR; BHA; VEGETABLE SHORTENING**
HOWARD JOHNSON'S	
MACARONI & CHEESE	**PAPRIKA**
LA CHOY	
CHICKEN CHOW MEIN	**MODIFIED FOOD STARCH;** MONOSODIUM GLUTAMATE; **TURMERIC**
FRIED RICE W/ MEAT	MONOSODIUM GLUTAMATE
—EGG ROLLS	
CHICKEN; MEAT & SHRIMP; SHRIMP	**MODIFIED FOOD STARCH;** MONOSODIUM GLUTAMATE

BRANDS	**ADDITIVES-OF-CONCERN** **BOLD—CONCERN TO ALL** LIGHT—CONCERN TO SOME
MRS. PAUL'S EGGPLANT PARMESAN	NO ADDITIVES OF CONCERN
MORTON MACARONI & CHEESE	**ARTIFICIAL COLOR; MODIFIED FOOD STARCH;** WHEY SOLIDS
RONZONI FETTUCCINE ALFREDO	**BUTTER (U.S. GRADE A FRESH); MODIFIED FOOD STARCH**
STOUFFER'S BEEF STEW; MACARONI & CHEESE; ROAST BEEF HASH; TURKEY TETRAZZINI	MONOSODIUM GLUTAMATE
BEEF STROGANOFF W/ PARSLEY NOODLES	**MODIFIED CORN STARCH;** MILK SOLIDS (CULTURED, NONFAT); MONOSODIUM GLUTAMATE; **PAPRIKA**
CHICKEN À LA KING W/ RICE; CHICKEN DIVAN; CREAMED CHICKEN	**MODIFIED CORN STARCH;** MONOSODIUM GLUTAMATE; **TURMERIC**
CREAMED CHIPPED BEEF	**BHA; BUTTER; MODIFIED CORN STARCH; SODIUM NITRITE**
GREEN PEPPER STEAK W/ RICE; SHORT RIBS OF BEEF	ALGIN; **MODIFIED CORN STARCH;** MONOSODIUM GLUTAMATE
LASAGNA	**MODIFIED CORN STARCH**

BRANDS	**ADDITIVES-OF-CONCERN** BOLD—CONCERN TO ALL LIGHT—CONCERN TO SOME
MACARONI & BEEF W/ TOMATOES; STUFFED GREEN PEPPERS	**MODIFIED CORN STARCH;** MONOSODIUM GLUTAMATE
TUNA NOODLE CASSEROLE	**ARTIFICIAL COLOR;** MONOSODIUM GLUTAMATE
NOODLES ROMANOFF SIDE DISH	**CARRAGEENAN; COCONUT OIL; PAPRIKA; SODIUM CASEINATE;** WHEY SOLIDS
SWANSON	
FRIED CHICKEN	NO ADDITIVES OF CONCERN
MACARONI & CHEESE	**MODIFIED FOOD STARCH;** MONOSODIUM GLUTAMATE; **OLEORESIN PAPRIKA**
WEAVER	
—BATTER DIPPED	
FRIED CHICKEN BREASTS; FRIED CHICKEN THIGHS & DRUMSTICKS	NO ADDITIVES OF CONCERN
—DUTCH ENTREE	
CHICKEN AU GRATIN	**ARTIFICIAL COLOR;** BUTTERMILK SOLIDS; **MODIFIED FOOD STARCH;** WHEY & MODIFIED WHEY
CHICKEN CROQUETTES	EGG WHITE SOLIDS; **MODIFIED FOOD STARCH;** MONOSODIUM GLUTAMATE; **OLEORESIN OF TURMERIC; PAPRIKA (& OLEORESIN OF);** SODIUM ALGINATE; WHEY

BRANDS	ADDITIVES-OF-CONCERN **BOLD—CONCERN TO ALL** LIGHT—CONCERN TO SOME
WEAVER *(Continued)*	
—DUTCH FRYE	
FRIED CHICKEN BREASTS	**MODIFIED FOOD STARCH; OLEORESIN PAPRIKA;** SODIUM ALGINATE
FRIED CHICKEN DRUMSTICKS	ALGIN GUM; **PAPRIKA;** WHEY (DRIED)
—TOUCH-O-HONEY	
FRIED CHICKEN BREAST, THIGHS, DRUMSTICKS	**MODIFIED FOOD STARCH;** MONOSODIUM GLUTAMATE; **OLEORESIN PAPRIKA;** SODIUM ALGINATE
WEIGHT WATCHERS	
CHEESE & TOMATO PIES	NO ADDITIVES OF CONCERN
CHICKEN CREOLE	**VEGETABLE STABILIZERS**
EGGPLANT PARMIGIANA; LASAGNA W/ CHEESE, VEAL & SAUCE; ZITI MACARONI W/ VEAL, CHEESE & SAUCE	CAROB BEAN GUM; GUAR GUM; GUM TRAGACANTH
SUPERMARKET PRIVATE BRANDS	
A & P	
—A & P MACARONI & CHEESE	**FOOD COLOR;** GUAR GUM; **MODIFIED STARCH;** MONOSODIUM GLUTAMATE; WHEY (SWEET DAIRY)
—ANN PAGE	
GRAVY & 6 CHICKEN CROQUETTES	**MODIFIED FOOD STARCH;** MONOSODIUM GLUTAMATE; **PAPRIKA; TURMERIC**

BRANDS	ADDITIVES-OF-CONCERN **BOLD—CONCERN TO ALL** LIGHT—CONCERN TO SOME
ONION GRAVY & 4 CHAR-BROILED PATTIES	**MODIFIED FOOD STARCH;** MONOSODIUM GLUTAMATE
TOMATO SAUCE & MEAT LOAF	**MODIFIED FOOD STARCH**
FIRST NATIONAL —FINAST MACARONI & CHEESE	BUTTERMILK SOLIDS; **MODIFIED FOOD STARCH;** WHEY SOLIDS (MODIFIED)
—FINAST *BOIL-IN-BAG* BREADED VEAL PARMIGIANA	**MODIFIED FOOD STARCH;** MONOSODIUM GLUTAMATE; **OIL EXTRACTIVES OF PAPRIKA**
CHICKEN A LA KING	**MODIFIED FOOD STARCH;** MONOSODIUM GLUTAMATE
GRAVY & SLICED BEEF	NO ADDITIVES OF CONCERN
SAFEWAY —SAFEWAY BEEF & BEAN GREEN CHILI BURRITO; RED HOT BEEF BURRITO	**MODIFIED FOOD STARCH**
—BEL-AIR BARBEQUE SAUCE & SLICED BEEF	**PAPRIKA; SMOKE FLAVORING; SMOKED YEAST;* TURMERIC**

*Although "smoked" does not refer to a specific additive, it represents a process utilizing wood smoke, as "smoke flavoring" does, and therefore the food may contain some cancer-causing benzopyrene chemicals. For this reason, when a food or any of its constituents has been smoked, this has been noted in the additives-of-concern column.

BRANDS	**ADDITIVES-OF-CONCERN** **BOLD—CONCERN TO ALL** LIGHT—CONCERN TO SOME
SAFEWAY—BEL-AIR *(Continued)*	
SALISBURY STEAK & GRAVY	NO ADDITIVES OF CONCERN
GRAVY & SLICED BEEF	**RENDERED BEEF FAT**
GRAVY & SLICED TURKEY	MONOSODIUM GLUTAMATE; **OLEORESIN OF TURMERIC; PAPRIKA**
STOP & SHOP —STOP & SHOP	
CHEESE RAVIOLI	NO ADDITIVES OF CONCERN
MEAT TORTELLINI	**BHA; BHT; MODIFIED CORN STARCH;** MONOSODIUM GLUTAMATE; **OLEORESINS OF PAPRIKA & TURMERIC**
ITEMS IN HEALTH FOOD STORES	
HEALTH VALLEY	
CHEDDAR CHICKEN; CHEESE EGGPLANT; EGG ROLLS; LOBSTER ROLLS; NUT ROLLS; SHRIMP ROLLS; STUFFED PEPPERS	NO ADDITIVES OF CONCERN
PIZZA NATURALLY	
EGGPLANT PARMIGIANA; WHOLE WHEAT BAKED ZITI; WHOLE WHEAT LASAGNE	NO ADDITIVES OF CONCERN

BRANDS	ADDITIVES-OF-CONCERN **BOLD—CONCERN TO ALL** LIGHT—CONCERN TO SOME
TUMARO'S	
TEYA'S STUFFED GOLDEN SOY BEAN CAKES; TAMALE; TAMALE W/ SOY PROTEIN ADDED	NO ADDITIVES OF CONCERN
WORTHINGTON	
EGG ROLLS	**ARTIFICIAL COLOR;** EGG WHITES; **MODIFIED CORN STARCH;** MONOSODIUM GLUTAMATE
STAKES AU SAUCE	**ARTIFICIAL COLOR; CALCIUM & POTASSIUM CASEINATE;** EGG WHITES; GLUTEN (WHEAT); **MODIFIED CORN & TAPIOCA STARCH;** MONOSODIUM GLUTAMATE; WHEY
VEELETS PARMESANO	**ARTIFICIAL COLOR; CALCIUM & POTASSIUM CASEINATE;** EGG WHITES; GLUTEN (WHEAT); **MODIFIED TAPIOCA STARCH;** MONOSODIUM GLUTAMATE; WHEY
VEGETARIAN LASAGNA	**CALCIUM CASEINATE; FERRIC ORTHOPHOSPHATE; MODIFIED CORN STARCH;** MONOSODIUM GLUTAMATE

XIX. PASTA; POTATOES (INSTANT); RICE

PASTA

BRANDS	ADDITIVES-OF-CONCERN **BOLD—CONCERN TO ALL** LIGHT—CONCERN TO SOME
MUELLER'S ELBOWS; LASAGNE; OLD FASHIONED EGG NOODLES; READY-CUT MACARONI; SEA SHELLS; THIN SPAGHETTI; VERMICELLI	NO ADDITIVES OF CONCERN
PARADE ELBOWS; RIGATONI; SHELLS MEDIUM; VERMICELLI; ZITI	**FERRIC ORTHOPHOSPHATE**
PENNSYLVANIA DUTCH BROAD EGG NOODLES	NO ADDITIVES OF CONCERN
PRINCE ALPHABETS; CURLY LASAGNE; EGG PASTINA; ELBOWS; MACARONI MACARONCELLI; MANICOTTI; ORZO; RIGATONI; SHELLS MEDIUM	**FERRIC ORTHOPHOSPHATE**

BRANDS	ADDITIVES-OF-CONCERN BOLD—CONCERN TO ALL LIGHT—CONCERN TO SOME
ENRICHED EGG NOODLES; LINGUINE; SPAGHETTI; VERMICELLI; ZITI	**SODIUM IRON PYROPHOSPHATE**
—*SUPERONI* ELBOW MACARONI	**FERRIC ORTHOPHOSPHATE;** GLUTEN (WHEAT)
RIGATONI; THIN SPAGHETTI; ZITI	**CALCIUM CASEINATE; FERRIC ORTHOPHOSPHATE;** GLUTEN (WHEAT)
RONZONI CURLY EDGE LASAGNE; FETTUCCINE; RIGATONI; SHELLS; VERMICELLI; ZITI	NO ADDITIVES OF CONCERN
SUPERMARKET PRIVATE BRANDS **A & P** **—ANN PAGE** EGG NOODLES; ELBOW MACARONI; LASAGNE; LINGUINE; SEA SHELLS; SPAGHETTI	NO ADDITIVES OF CONCERN
FIRST NATIONAL **—EDWARDS-FINAST** ELBOW MACARONI; LARGE SHELLS; LASAGNA; RIGATONI; SPAGHETTI; VERMICELLI; ZITI	NO ADDITIVES OF CONCERN

BRANDS	ADDITIVES-OF-CONCERN BOLD—CONCERN TO ALL LIGHT—CONCERN TO SOME
LUCKY **—LADY LEE** COIL VERMICELLI; CUT MACARONI; EGG BUTTERFLIES; FINE EGG NOODLES; LARGE SHELLS; LASAGNE; RIGATONI; SMALL ELBOWS; SPAGHETTI	NO ADDITIVES OF CONCERN
SAFEWAY **—TOWN HOUSE** CUT MACARONI; FINE EGG NOODLES; RIGATONI; SMALL ELBOWS; SMALL SHELLS; SPAGHETTI	NO ADDITIVES OF CONCERN
STOP & SHOP **—STOP & SHOP** SPAGHETTI; THIN SPAGHETTI; ZITI	**FERRIC ORTHOPHOSPHATE**
WINN DIXIE **—THRIFTY MAID** CURLY LASAGNA; ELBOWS; SPAGHETTI; VERMICELLI	**FERRIC ORTHOPHOSPHATE**
ITEMS IN HEALTH FOOD STORES **DE BOLES** CURLY LASAGNE; FETTUCCINE; RIGATONI; SPINACH FETTUCCINE	NO ADDITIVES OF CONCERN

BRANDS	ADDITIVES-OF-CONCERN **BOLD—CONCERN TO ALL** LIGHT—CONCERN TO SOME
—WHOLE WHEAT HIGH PROTEIN ELBOWS; LINGUINE; MEDIUM SHELLS; RICE SUBSTITUTE; THIN SPAGHETTI; ZITI MACARONI	EGG WHITE SOLIDS; GLUTEN (WHEAT)
EREWHON CHINESE PASTA ORIENTAL NOODLES W/ INSTANT BROTH	DRIED KOMBU (KELP)
JAPANESE PASTA	NO ADDITIVES OF CONCERN
—WHOLEWHEAT ELBOW MACARONI; ELBOW PASTA W/ MIXED VEGETABLE POWDERS; LASAGNA; SPAGHETTI W/ ARTICHOKE POWDER; SPINACH LASAGNA; SPIRAL PASTA W/ SESAME & BROWN RICE; THIN SPAGHETTI; ZITI	NO ADDITIVES OF CONCERN
HEALTH VALLEY *WHOLE WHEAT* ELBOW PASTA W/ WHEAT GERM & 4 VEGETABLES; LASAGNA W/ WHEAT GERM; PASTA W/ WHEAT GERM; SPINACH PASTA W/ WHEAT GERM	NO ADDITIVES OF CONCERN
WESTBRAE NATURAL EGG NOODLES; SPINACH LASAGNA	NO ADDITIVES OF CONCERN

BRANDS	**ADDITIVES-OF-CONCERN** **BOLD—CONCERN TO ALL** LIGHT—CONCERN TO SOME
WESTBRAE NATURAL *(Continued)*	
WHOLEWHEAT RAMEN	DRIED KOMBU
—WHOLE-WHEAT LASAGNE; SPAGHETTI; STUFFING SHELLS	NO ADDITIVES OF CONCERN

POTATOES (INSTANT)

BETTY CROCKER	
AU GRATIN	**ARTIFICIAL COLOR;** MONOSODIUM GLUTAMATE; WHEY
HASH BROWN; POTATO BUDS MASHED	**BHA**
SCALLOPED	**ARTIFICIAL COLOR; BHA;** MONOSODIUM GLUTAMATE; **SODIUM CASEINATE;** WHEY
FRENCH'S	
IDAHO MASHED	**BHA; BHT**
—BIG TATE HASH BROWN	**BHT; PAPRIKA**
MASHED	**BHA**
SCALLOPED	**SODIUM CASEINATE;** WHEY
IDAHOAN MASHED	**BHA**

BRANDS	ADDITIVES-OF-CONCERN **BOLD—CONCERN TO ALL** LIGHT—CONCERN TO SOME
PILLSBURY HUNGRY JACK MASHED	**BHA; BHT**
SUPERMARKET PRIVATE BRANDS	
FIRST NATIONAL	
—EDWARDS AU GRATIN	**ARTIFICIAL COLOR;** GUAR GUM; **MODIFIED CORN STARCH;** MONOSODIUM GLUTAMATE; WHEY
—EDWARDS-FINAST INSTANT MASHED	**BHA**
LUCKY	
—LADY LEE MASHED	**BHA**
SAFEWAY	
—TOWN HOUSE	
AU GRATIN	**ARTIFICIAL COLOR;** MILK SOLIDS (NONFAT); **MODIFIED CORN STARCH;** WHEY (ACID & SWEET)
MASHED	**BHA**
SCALLOPED	MILK SOLIDS (NONFAT); **MODIFIED CORN STARCH;** WHEY (ACID & SWEET)
STOP & SHOP	
—STOP & SHOP MASHED	**BHA**
WINN DIXIE	
—ASTOR AU GRATIN; SCALLOPED	**ARTIFICIAL COLOR; MODIFIED CORN STARCH;** MONOSODIUM GLUTAMATE; WHEY

BRANDS	ADDITIVES-OF-CONCERN BOLD—CONCERN TO ALL LIGHT—CONCERN TO SOME
WINN DIXIE—ASTOR *(Continued)*	
HASH BROWN	NO ADDITIVES OF CONCERN
IDAHO MASHED	**BHA**
ITEMS IN HEALTH FOOD STORES	
BARBARA'S ORGANIC MASHED POTATOES	NO ADDITIVES OF CONCERN
PANNI BAVARIAN POTATO DUMPLING MIX	NO ADDITIVES OF CONCERN

RICE & RICE DISHES

GOLDEN GRAIN *—RICE-A-RONI*	
BEEF FLAVOR	MONOSODIUM GLUTAMATE
CHICKEN FLAVOR; SAVORY RICE PILAF	**BHA;** MONOSODIUM GLUTAMATE; **TURMERIC**
FRIED RICE W/ ALMONDS	**BHA; BHT;** MONOSODIUM GLUTAMATE; **TURMERIC**
HERB & BUTTER	**BHA; BHT**
LONG GRAIN & WILD RICE	**BHT;** MONOSODIUM GLUTAMATE
STROGANOFF	MONOSODIUM GLUTAMATE; **PAPRIKA;** WHEY
MINUTE RICE	**FERRIC ORTHOPHOSPHATE**
—RICE MIX	
DRUMSTICK	**BHA; BHT;** MONOSODIUM GLUTAMATE; **TURMERIC**

BRANDS	ADDITIVES-OF-CONCERN **BOLD—CONCERN TO ALL** LIGHT—CONCERN TO SOME
FRIED; RIB ROAST	MONOSODIUM GLUTAMATE
NEAR EAST RICE PILAF	**TURMERIC**
SUCCESS RICE	NO ADDITIVES OF CONCERN
UNCLE BEN'S *RICE*	
BROWN	NO ADDITIVES OF CONCERN
CONVERTED; QUICK	**FERRIC ORTHOPHOSPHATE**
LONG GRAIN & WILD	**BHA; BHT; FERRIC ORTHOPHOSPHATE; HICKORY SMOKED TORULA YEAST;*** MONOSODIUM GLUTAMATE
FOR BEEF W/ MUSHROOMS	**BHA; BHT; FERRIC ORTHOPHOSPHATE; MODIFIED CORN STARCH;** MONOSODIUM GLUTAMATE; **PAPRIKA**
FOR CHICKEN W/ VEGETABLES	**BHA; BHT; FERRIC ORTHOPHOSPHATE;** MONOSODIUM GLUTAMATE; **SMOKED TORULA YEAST;* TURMERIC**
PILAF W/ ROSAMARINA & PEAS	**BHA; BHT; FERRIC ORTHOPHOSPHATE;** MONOSODIUM GLUTAMATE; **TURMERIC**

*Although "smoked" does not refer to a specific additive, it represents a process utilizing wood smoke, as "smoke flavoring" does, and therefore the food may contain some cancer-causing benzopyrene chemicals. For this reason, when a food or any of its constituents has been smoked, this has been noted in the additives-of-concern column.

BRANDS	ADDITIVES-OF-CONCERN **BOLD—CONCERN TO ALL** LIGHT—CONCERN TO SOME
SUPERMARKET PRIVATE BRANDS	
A & P	
—ANN PAGE	
RICE 'N EASY CHICKEN FLAVORED	**BHA; BHT;** MONOSODIUM GLUTAMATE; **TURMERIC**
LUCKY	
—LADY LEE	
ENRICHED LONG GRAIN RICE	NO ADDITIVES OF CONCERN
SAFEWAY	
—TOWN HOUSE *RICE*	
CALIFORNIA PEARL	NO ADDITIVES OF CONCERN
LONG GRAIN; MEDIUM GRAIN	**FERRIC ORTHOPHOSPHATE**
ITEMS IN HEALTH FOOD STORES	
NEAR EAST *PILAF MIX*	
LENTIL	NO ADDITIVES OF CONCERN
RICE	**TURMERIC**
SPANISH RICE	**PAPRIKA; TURMERIC**

XX. PICKLES, SALAD DRESSINGS, & OTHER CONDIMENTS

CATSUP

None of the items in this section contain any additives of concern, and all the items appear without an additives-of-concern column.

DEL MONTE TOMATO CATSUP

HEINZ TOMATO KETCHUP

HUNT'S TOMATO KETCHUP

SUPERMARKET PRIVATE BRANDS
A & P ANN PAGE TOMATO KETCHUP

LUCKY HARVEST DAY TOMATO CATSUP

LUCKY LADY LEE TOMATO CATSUP

SAFEWAY TOWN HOUSE TOMATO CATSUP

STOP & SHOP TOMATO KETCHUP

WINN DIXIE THRIFTY MAID TOMATO CATSUP

ITEMS IN HEALTH FOOD STORES

HAIN IMITATION CATSUP

BRANDS	ADDITIVES-OF-CONCERN **BOLD—CONCERN TO ALL** LIGHT—CONCERN TO SOME

MAYONNAISE & IMITATIONS

HELLMANN'S MAYONNAISE	NO ADDITIVES OF CONCERN
KRAFT MAYONNAISE	**PAPRIKA**
MIRACLE WHIP	**MODIFIED FOOD STARCH; PAPRIKA**
MRS FILBERTS IMITATION MAYONNAISE	**ARTIFICIAL COLOR; MODIFIED FOOD STARCH; NATURAL COLOR**
SUPERMARKET PRIVATE BRANDS **A & P** **—ANN PAGE** MAYONNAISE	NO ADDITIVES OF CONCERN
SALAD DRESSING; SANDWICH SPREAD	**MODIFIED CORN STARCH**
FIRST NATIONAL **—EDWARDS** SALAD DRESSING	**MODIFIED FOOD STARCH; OLEORESIN PAPRIKA**
—FINAST MAYONNAISE	NO ADDITIVES OF CONCERN
LUCKY **—LADY LEE** IMITATION MAYONNAISE	**ARTIFICIAL COLOR; MODIFIED FOOD STARCH; OLEORESIN OF PAPRIKA;** PROPYLENE GLYCOL ALGINATE
MAYONNAISE	**OLEORESIN PAPRIKA**

BRANDS	ADDITIVES-OF-CONCERN **BOLD—CONCERN TO ALL** LIGHT—CONCERN TO SOME
SALAD DRESSING	ALGIN DERIVATIVE; **MODIFIED FOOD STARCH; PAPRIKA**
PUBLIX	
—FLAVOR PERFECT	
MAYONNAISE	**OLEORESIN PAPRIKA**
SALAD DRESSING	**MODIFIED FOOD STARCH**
SAFEWAY	
—NUMADE	
REAL MAYONNAISE	NO ADDITIVES OF CONCERN
RELISH SANDWICH SPREAD; SALAD DRESSING	**MODIFIED FOOD STARCH; OLEORESIN OF PAPRIKA**
—SCOTCH BUY IMITATION MAYONNAISE	**ARTIFICIAL COLOR; MODIFIED FOOD STARCH; OLEORESIN PAPRIKA**
STOP & SHOP	
—STOP & SHOP	
IMITATION MAYONNAISE	**ARTIFICIAL COLOR; MODIFIED FOOD STARCH; OLEORESIN OF PAPRIKA**
MAYONNAISE	NO ADDITIVES OF CONCERN
SALAD DRESSING	**MODIFIED CORN STARCH**
WINN DIXIE	
—DEEP SOUTH	
MAYONNAISE	NO ADDITIVES OF CONCERN
SALAD DRESSING	**MODIFIED FOOD STARCH**
ITEMS IN HEALTH FOOD STORES	
HAIN *MAYONNAISE*	
EGGLESS IMITATION	ALGIN

BRANDS	ADDITIVES-OF-CONCERN **BOLD—CONCERN TO ALL** LIGHT—CONCERN TO SOME
HAIN *(Continued)*	
SAF-FLOWER; UNSALTED	NO ADDITIVES OF CONCERN
NORGANIC GOLDEN SOYA MAYONNAISE	NO ADDITIVES OF CONCERN
WALNUT ACRES MAYONNAISE	NO ADDITIVES OF CONCERN

MUSTARD

COLMAN'S HOT ENGLISH	**ARTIFICIAL COLORING**
FRANK'S MISTER	NO ADDITIVES OF CONCERN
FRENCH'S ONION BITS; REGULAR	**TURMERIC**
GREY-POUPON DIJON	NO ADDITIVES OF CONCERN
GULDEN'S DIABLO HOT; SPICY BROWN	**TURMERIC**
KRAFT REGULAR	**OLEORESINS OF PAPRIKA & TURMERIC; TURMERIC**
MAILLE DIJON	NO ADDITIVES OF CONCERN
NANCE'S REGULAR	PROPYLENE GLYCOL ALGINATE
PLOCHMAN'S REGULAR	**PAPRIKA; TURMERIC**

BRANDS	ADDITIVES-OF-CONCERN BOLD—CONCERN TO ALL LIGHT—CONCERN TO SOME
SUPERMARKET PRIVATE BRANDS	
A & P	
—ANN PAGE SALAD	**PAPRIKA; TURMERIC**
LUCKY	
—LADY LEE	
REGULAR	**TURMERIC**
SAFEWAY	
—TOWN HOUSE	
REGULAR	**PAPRIKA; TURMERIC**
STOP & SHOP	
—STOP & SHOP	
SALAD STYLE	**TURMERIC**
ITEMS IN HEALTH FOOD STORES	
EDEN HOT	NO ADDITIVES OF CONCERN
HAIN	NO ADDITIVES OF CONCERN
REINE DIJON NO SALT ADDED	NO ADDITIVES OF CONCERN

PICKLES

HEINZ	
GENUINE DILL	NO ADDITIVES OF CONCERN
HAMBURGER DILL SLICES	**ARTIFICIAL COLORING**
KOSHER DILL SPEARS; SWEET CUCUMBER SLICES; SWEET GHERKINS; SWEET MIXED	**TURMERIC OLEORESIN**

BRANDS	ADDITIVES-OF-CONCERN BOLD—CONCERN TO ALL LIGHT—CONCERN TO SOME
VLASIC KOSHER DILLS; KOSHER GHERKINS; POLISH SPEARS; SWEET BUTTER CHIPS; SWEET GHERKINS; SWEET MIX	**ARTIFICIAL COLOR**
SUPERMARKET PRIVATE BRANDS	
FIRST NATIONAL —EDWARDS HAMBURGER SLICES; KOSHER DILLS	**ARTIFICIAL COLOR**
SWEET GHERKINS; WHOLE SWEETS	**TURMERIC**
SAFEWAY —TOWN HOUSE HAMBURGER DILL CHIPS; SWEET; SWEET CUCUMBER CHIPS; WHOLE DILL; WHOLE KOSHER STYLE DILL	**ARTIFICIAL COLOR** OR **U.S. CERTIFIED COLOR; TURMERIC**
KOSHER STYLE DILL GHERKINS; POLISH STYLE WHOLE DILLS	**TURMERIC**
SWEET MIDGETS	FD & C YELLOW NO. 5
STOP & SHOP —STOP & SHOP CUCUMBER SLICES; HAMBURGER DILL SLICES;	**CERTIFIED FOOD COLOR**

(continues)

BRANDS	**ADDITIVES-OF-CONCERN** BOLD—CONCERN TO ALL LIGHT—CONCERN TO SOME
KOSHER DILLS; SWEET GHERKINS; SWEET MIXED	**CERTIFIED FOOD COLOR**
WINN DIXIE **—DEEP SOUTH**	
KOSHER DILL SPEARS; SWEET CUCUMBER CHIPS	FD & C YELLOW NO. 5
SWEET MIXED; WHOLE DILL; WHOLE SOUR	**ARTIFICIAL COLORS** W/ FD & C YELLOW NO. 5; **TURMERIC**
ITEMS IN HEALTH FOOD STORES	
FEATHERWEIGHT	
SLICED CUCUMBER; WHOLE DILL	**TURMERIC** OR **OIL OF TURMERIC**
HAIN NATURALS	
NATURAL KOSHER DILL CHIPS	NO ADDITIVES OF CONCERN
NEW ENGLAND ORGANIC PRODUCE CENTER *—THE PICKLE EATERS*	
CHIPS W/ HONEY; POOKIE SIZE	NO ADDITIVES OF CONCERN
WESTBRAE NATURAL	
KOSHER DILL	NO ADDITIVES OF CONCERN

RELISH

BENNETT'S CHILI SAUCE	NO ADDITIVES OF CONCERN

BRANDS	ADDITIVES-OF-CONCERN BOLD—CONCERN TO ALL LIGHT—CONCERN TO SOME
HEINZ HAMBURGER	NO ADDITIVES OF CONCERN
HOT DOG	[GUM] TRAGACANTH
SWEET	**TURMERIC OLEORESIN**
VLASIC HAMBURG; HOT DOG; SWEET	**ARTIFICIAL COLOR** OR **ARTIFICIALLY COLORED**
SUPERMARKET PRIVATE BRANDS	
A & P	
—ANN PAGE SWEET GARDEN	**TURMERIC**
FIRST NATIONAL	
—EDWARDS SWEET	**MODIFIED FOOD STARCH; TURMERIC**
SAFEWAY	
—TOWN HOUSE	
HAMBURGER	**MODIFIED FOOD STARCH; PAPRIKA; TURMERIC**
HOT DOG	**ARTIFICIAL COLOR; MODIFIED FOOD STARCH; PAPRIKA; TURMERIC**
STOP & SHOP	
—STOP & SHOP	
SWEET PICKLED	**TURMERIC**
ITEMS IN HEALTH FOOD STORES	
MRS. WOOD'S FARM	
SWEET RELISH	NO ADDITIVES OF CONCERN
NEW ENGLAND ORGANIC PRODUCE CENTER	
PICCALILLI RELISH	**IRISH MOSS**

BRANDS	ADDITIVES-OF-CONCERN **BOLD—CONCERN TO ALL** LIGHT—CONCERN TO SOME

SALAD DRESSINGS

By law, at this writing, it is not required of a manufacturer to state whether cheeses contain artificial colors, unless the color is FD & C Yellow No. 5 (Tartrazine). Their presence in salad dressings containing cheese therefore remains an uncertainty, unless it is voluntarily declared on the label that artificial colors have been used or have not. For a more detailed explanation of artificial color in cheese, refer to the beginning of the cheese section of "Dairy Products & Substitutes."

BRANDS	ADDITIVES-OF-CONCERN
GOOD SEASONS *SALAD DRESSING MIX*	
BLEU CHEESE	**ARTIFICIAL COLOR;** MONOSODIUM GLUTAMATE; PROPYLENE GLYCOL ALGINATE
CHEESE ITALIAN	**ARTIFICIAL COLOR;** GUAR GUM; PROPYLENE GLYCOL ALGINATE
GARLIC	MONOSODIUM GLUTAMATE; **PAPRIKA;** PROPYLENE GLYCOL ALGINATE
ITALIAN	**BHA; CALCIUM CARRAGEENAN;** MONOSODIUM GLUTAMATE
RIVIERA FRENCH	**ARTIFICIAL COLOR; BHA;** GUAR GUM
HIDDEN VALLEY RANCH	
MILK RECIPE ORIGINAL RANCH	BUTTERMILK SOLIDS; **CALCIUM STEARATE;** CASEIN; MONOSODIUM GLUTAMATE; WHEY SOLIDS

BRANDS	ADDITIVES-OF-CONCERN **BOLD—CONCERN TO ALL** LIGHT—CONCERN TO SOME
KRAFT	
CATALINA; MIRACLE FRENCH	**OLEORESIN PAPRIKA**
COLESLAW; CREAMY CUCUMBER; ROKA BLUE CHEESE; THOUSAND ISLAND	PROPYLENE GLYCOL ALGINATE
CREAMY RUSSIAN; OIL & VINEGAR	**OLEORESIN PAPRIKA;** PROPYLENE GLYCOL ALGINATE
RED WINE VINEGAR & OIL	CAROB BEAN GUM; **OLEORESIN PAPRIKA**
ZESTY ITALIAN	**MODIFIED FOOD STARCH; OLEORESIN PAPRIKA**
PFEIFFER	
CAESAR	**MODIFIED FOOD STARCH**
CHEF ITALIAN; RED WINE VINEGAR & OIL	**ARTIFICIAL COLOR** OR **ARTIFICIAL FOOD COLORS**
ROQUEFORT CHEESE; THOUSAND ISLAND	PROPYLENE GLYCOL ALGINATE
RUSSIAN	**COLORING;** PROPYLENE GLYCOL ALGINATE
SPRING GARDEN	NO ADDITIVES OF CONCERN

BRANDS	ADDITIVES-OF-CONCERN **BOLD—CONCERN TO ALL** LIGHT—CONCERN TO SOME
SEVEN SEAS	
GREEN GODDESS	**ARTIFICIAL COLOR; CARRAGEENAN;** GUAR GUM; LOCUST BEAN GUM
RUSSIAN	NO ADDITIVES OF CONCERN
VIVA CAESAR	**ARTIFICIAL COLOR;** MONOSODIUM GLUTAMATE; **OXYSTEARIN**
VIVA ITALIAN!; VIVA RED WINE VINEGAR & OIL	**ARTIFICIAL COLOR; OXYSTEARIN**
WISH-BONE	
CALIFORNIA ONION; CHUNKY BLUE CHEESE; CREAMY GARLIC; CREAMY ITALIAN	ALGIN DERIVATIVE
DELUXE FRENCH; SWEET 'N SPICY FRENCH	ALGIN DERIVATIVE; **OLEORESIN PAPRIKA**
ITALIAN; RUSSIAN	NO ADDITIVES OF CONCERN
SUPERMARKET PRIVATE BRANDS	
A & P	
—ANN PAGE	
CAESAR; THOUSAND ISLAND	NO ADDITIVES OF CONCERN
FRENCH; ITALIAN	ALGIN DERIVATIVE
RED WINE VINEGAR & OIL	**ARTIFICIAL COLOR**

BRANDS	ADDITIVES-OF-CONCERN **BOLD—CONCERN TO ALL** LIGHT—CONCERN TO SOME
LUCKY **—LADY LEE** BLUE CHEESE	PROPYLENE GLYCOL ALGINATE
ITALIAN	**PAPRIKA**
THOUSAND ISLAND	ALGIN DERIVATIVE; **MODIFIED FOOD STARCH**
(REFRIGERATED) BLEU CHEESE; ROQUEFORT	**COCONUT OIL;** LOCUST BEAN GUM; MILK SOLIDS (NONFAT); MONOSODIUM GLUTAMATE; **SODIUM CASEINATE;** WHEY SOLIDS
1000 ISLAND	MONOSODIUM GLUTAMATE
SAFEWAY **—NUMADE** GREEN GODDESS	**ARTIFICIAL COLORS**
1000 ISLAND	NO ADDITIVES OF CONCERN
STOP & SHOP **—STOP & SHOP** BLEU CHEESE	NO ADDITIVES OF CONCERN
CREAMY ITALIAN; RED WINE VINEGAR & OIL	**ARTIFICIAL COLOR** OR **CERTIFIED FOOD COLOR**
FRENCH; RUSSIAN; THOUSAND ISLAND	**OLEORESIN PAPRIKA;** PROPYLENE GLYCOL ALGINATE
ITEMS IN HEALTH FOOD STORES **HAIN** BLEU CHEESE;	**VEGETABLE GUM**

(continues)

BRANDS	**ADDITIVES-OF-CONCERN** **BOLD—CONCERN TO ALL** LIGHT—CONCERN TO SOME
CREAMY FRENCH; GREEN GODDESS; HONEY & SESAME; ITALIAN; LEMON N HERB; REAL BLEU CHEESE BUTTERMILK; RUSSIAN; THOUSAND ISLAND	**VEGETABLE GUM**
GARLIC 'N OIL; OIL 'N VINEGAR	NO ADDITIVES OF CONCERN
HERB (UNSALTED)	SEA KELP
OLD WORLD FRENCH	ALGIN; WHEY
—SALAD DRESSING MIX	
BLEU CHEESE	WHEY
CAESAR; THOUSAND ISLAND	ALGIN; WHEY
FRENCH; ITALIAN	ALGIN; **PAPRIKA**
HONEY & LEMON	ALGIN
HEALTH VALLEY	
DRESSING & DIP	
AVOCADO; GREEN GODDESS; RUSSIAN	NO ADDITIVES OF CONCERN
FRENCH	ALGIN DERIVATIVE FROM SEAWEED
NORGANIC FRENCH	**VEGETABLE GUM**

BRANDS	ADDITIVES-OF-CONCERN BOLD—CONCERN TO ALL LIGHT—CONCERN TO SOME
WALNUT ACRES	
CREAMY FRENCH	CAROB [BEAN] GUM; **PAPRIKA**
CREAMY ITALIAN; CREAMY WATERCRESS; THOUSAND ISLAND	NO ADDITIVES OF CONCERN

XXI. SNACK ITEMS

CORN, POTATO, & TORTILLA CHIPS

By law, at this writing, it is not required of a manufacturer to state whether cheeses contain artificial colors, unless the color is FD & C Yellow No. 5 (Tartrazine). Their presence in corn, potato, and tortilla chips containing cheese therefore remains an uncertainty, unless it is voluntarily declared on the label that artificial colors have been used or have not. For a more detailed explanation of artificial color in cheese, refer to the beginning of the cheese section of "Dairy Products & Substitutes."

BRANDS	**ADDITIVES-OF-CONCERN** BOLD—CONCERN TO ALL LIGHT—CONCERN TO SOME
CAINS	
BARBEQUE POTATO	**MODIFIED WHEAT STARCH;** MONOSODIUM GLUTAMATE; **NATURAL HICKORY SMOKE FLAVOR; PAPRIKA**
POTATO	NO ADDITIVES OF CONCERN
SOUR CREAM & ONION RIPPLED POTATO	**MODIFIED FOOD STARCH;** MONOSODIUM GLUTAMATE; **SODIUM CASEINATE;** SOUR CREAM SOLIDS; WHEY POWDER
DORITOS NACHO CHEESE TORTILLA	**ARTIFICIAL COLOR;** MONOSODIUM GLUTAMATE; WHEY
FRITOS CORN	NO ADDITIVES OF CONCERN

BRANDS	ADDITIVES-OF-CONCERN **BOLD—CONCERN TO ALL** LIGHT—CONCERN TO SOME
KEYSTONE SNACKS	
CORN	**BUTYLATED HYDROXYTOLUENE; (COCONUT OIL)***
NACHO CHEESE TORTILLA	**(COCONUT OIL);* EXTRACTIVES OF PAPRIKA;** MONOSODIUM GLUTAMATE; **U.S. CERTIFIED FOOD COLOR;** WHEY POWDER
LAY'S POTATO	NO ADDITIVES OF CONCERN
NABISCO	
NACHO CHEESE TORTILLA	MILK SOLIDS (NONFAT); MONOSODIUM GLUTAMATE; **PAPRIKA; TURMERIC OLEORESIN;** WHEY SOLIDS
POTATO CHIPSTERS	**BHA; COCONUT OIL; TURMERIC OLEORESIN**
PARADE POTATO	**VEGETABLE OIL**
PRINGLE'S	
COUNTRY STYLE POTATO; ORIGINAL STYLE POTATO; RIPPLED STYLE POTATO	NO ADDITIVES OF CONCERN
TOSTITOS NACHO CHEESE TORTILLA	**ARTIFICIAL COLOR;** MONOSODIUM GLUTAMATE; WHEY

*Blend may contain this saturated oil.

BRANDS	ADDITIVES-OF-CONCERN **BOLD—CONCERN TO ALL** LIGHT—CONCERN TO SOME
WISE BARBEQUE POTATO	**ARTIFICIAL COLOR;** MONOSODIUM GLUTAMATE; **PAPRIKA**
CORN; POTATO	NO ADDITIVES OF CONCERN
—*BRAVOS* SOUR CREAM & ONION TORTILLA	MILK SOLIDS (CULTURED NONFAT); MONOSODIUM GLUTAMATE; SOUR CREAM SOLIDS
—*RIDGIES* SOUR CREAM & ONION POTATO	MONOSODIUM GLUTAMATE; **SODIUM CASEINATE;** WHEY
SUPERMARKET PRIVATE BRANDS **A & P—ANN PAGE** CORN; POTATO; RIPPLED POTATO	NO ADDITIVES OF CONCERN
LUCKY—LADY LEE DIP POTATO; POTATO	NO ADDITIVES OF CONCERN
SAFEWAY —(NO BRAND NAME) JALAPEÑO TORTILLA HOT	**PAPRIKA; TURMERIC**
—PARTY PRIDE BARBEQUE POTATO	**(HYDROGENATED COTTONSEED, PALM, SAFFLOWER OILS);* NATURAL HICKORY SMOKE FLAVOR; PAPRIKA (& EXTRACTIVES OF)**

*Blend may contain one or more of these saturated oils.

BRANDS	ADDITIVES-OF-CONCERN **BOLD—CONCERN TO ALL** LIGHT—CONCERN TO SOME
SAFEWAY—PARTY PRIDE *(Continued)*	
POTATO	**(HYDROGENATED CORN, COTTONSEED, PALM, PEANUT, SOYBEAN, SUNFLOWER OILS)***
TACO FLAVORED POTATO	NO ADDITIVES OF CONCERN
STOP & SHOP —STOP & SHOP	
CORN	NO ADDITIVES OF CONCERN
WINN DIXIE —CRACKIN GOOD	
CORN PLAIN; TACOS TORTILLA; WAVY	NO ADDITIVES OF CONCERN
ITEMS IN HEALTH FOOD STORES	
DR. BRONNER'S	
CORN & SESAME SNACK	NO ADDITIVES OF CONCERN
EREWHON	
—AZTEC CORN	
ONION & GARLIC; TAMARI; UNSALTED	NO ADDITIVES OF CONCERN
TACO	**PAPRIKA**
—ALL NATURAL POTATO	
NO SALT ADDED; SALTED; TAMARI	NO ADDITIVES OF CONCERN

*Blend may contain one or more of these saturated oils.

BRANDS	ADDITIVES-OF-CONCERN **BOLD—CONCERN TO ALL** LIGHT—CONCERN TO SOME
TACO	**PAPRIKA**
HAIN	
CHEESE SESAME TORTILLA	**PAPRIKA; TURMERIC;** WHEY
NATURAL POTATO	NO ADDITIVES OF CONCERN
VEGETIZED SESAME TORTILLA	DULSE; KELP
—SEVEN GRAIN	
W/ CINNAMON & RAISINS; W/ ONION & GARLIC	NO ADDITIVES OF CONCERN
—YOGURT	
W/ PEANUTS; W/ WHEAT GERM	NO ADDITIVES OF CONCERN
HEALTH VALLEY	
DIP POTATO CHIPS UNSALTED	NO ADDITIVES OF CONCERN
—CORN CHIPS	
UNSALTED; W/ CHEESE; W/ SESAME SEEDS	NO ADDITIVES OF CONCERN
—TORTILLA STRIPS	
CHEDDAR CHEESE; YOGURT	NO ADDITIVES OF CONCERN
MOTHER EARTH	
TAMARI TORTILLA	NO ADDITIVES OF CONCERN

BRANDS	ADDITIVES-OF-CONCERN BOLD—CONCERN TO ALL LIGHT—CONCERN TO SOME

DIPS & DIP MIXES

By law, at this writing, it is not required of a manufacturer to state whether cheeses contain artificial colors, unless the color is FD & C Yellow No. 5 (Tartrazine). Their presence in dips and dip mixes containing cheese therefore remains an uncertainty, unless it is voluntarily declared on the label that artificial colors have been used or have not. For a more detailed explanation of artificial color in cheese, refer to the beginning of the cheese section of "Dairy Products & Substitutes."

BRANDS	ADDITIVES-OF-CONCERN
BORDEN CLAM & LOBSTER DIP	**ARTIFICIAL COLOR;** CAROB BEAN GUM; **MODIFIED FOOD STARCH;** MONOSODIUM GLUTAMATE; WHEY
BREAKSTONE'S *DIP*	
CLAM; FRENCH ONION	CAROB BEAN GUM; **MODIFIED FOOD STARCH;** MONOSODIUM GLUTAMATE
FRITO-LAY *DIP MIX*	
GREEN GODDESS; ONION-BACON	**ARTIFICIAL COLOR** OR **COLORING;** MONOSODIUM GLUTAMATE
TOASTED ONION	**MODIFIED FOOD STARCH;** MONOSODIUM GLUTAMATE; **PAPRIKA**
FRITOS *DIP*	
BEAN	**LARD; PAPRIKA**
ENCHILADA	**LARD; MODIFIED FOOD STARCH;** MONOSODIUM GLUTAMATE; **PAPRIKA**

BRANDS	ADDITIVES-OF-CONCERN BOLD—CONCERN TO ALL LIGHT—CONCERN TO SOME
KRAFT *READY TO SERVE DIP*	
BLUE CHEESE	CAROB BEAN GUM; GUAR GUM
CREAMY CUCUMBER	**ARTIFICIAL COLOR;** CAROB BEAN GUM; **MODIFIED FOOD STARCH;** MONOSODIUM GLUTAMATE
ONION	CAROB BEAN GUM; GUAR GUM; **TURMERIC;** WHEY
SUPERMARKET PRIVATE BRANDS	
LUCKY	
—LADY LEE *DIP*	
AVOCADO; AVOCADO HOT	**ARTIFICIAL COLOR; COCONUT OIL;** GUAR GUM; MILK SOLIDS (NONFAT); **SODIUM CASEINATE;** WHEY SOLIDS
BACON & ONION; CLAM; FRENCH ONION	**COCONUT OIL;** MILK SOLIDS (NONFAT); PROPYLENE GLYCOL ALGINATE; WHEY SOLIDS
TORTILLA	**ARTIFICIAL COLOR; COCONUT OIL;** MILK SOLIDS (NONFAT); PROPYLENE GLYCOL ALGINATE; WHEY SOLIDS
PUBLIX	
—DAIRI-FRESH	
FRENCH ONION CHIP 'N DIP	**CARRAGEENAN;** LOCUST BEAN GUM

BRANDS	**ADDITIVES-OF-CONCERN** **BOLD—CONCERN TO ALL** LIGHT—CONCERN TO SOME
SAFEWAY —(NO BRAND NAME) *PARTY DIP*	
AVOCADO GUACAMOLE	ALGIN; **ARTIFICIAL COLOR;** GUAR GUM; MONOSODIUM GLUTAMATE; SODIUM ALGINATE; **SODIUM CASEINATE**
BLEU TANG; GARLIC	GUAR GUM; **MODIFIED FOOD STARCH;** MONOSODIUM GLUTAMATE; **SODIUM CASEINATE**
CHEDDAR CHEESE	BUTTERMILK SOLIDS; GUAR GUM; MONOSODIUM GLUTAMATE; **SODIUM CASEINATE;** WHEY
CLAM	FD & C YELLOW NO. 5; GUAR GUM; **SODIUM CASEINATE**
ITEMS IN HEALTH FOOD STORES HAIN *DIP*	
JALAPENO BEAN	**PAPRIKA**
ONION BEAN	NO ADDITIVES OF CONCERN
SEELECT VEGETABLE DIP MIX	NO ADDITIVES OF CONCERN

BRANDS	ADDITIVES-OF-CONCERN **BOLD—CONCERN TO ALL** LIGHT—CONCERN TO SOME

NUTS & SEEDS

BLUE DIAMOND *ALMONDS* CHEESE FLAVORED*	MONOSODIUM GLUTAMATE; **NATURAL SMOKE FLAVOR; VEGETABLE OIL;** WHEY (CHEDDAR CHEESE)
ROASTED BLANCHED SALTED	NO ADDITIVES OF CONCERN
SMOKEHOUSE	**NATURAL HICKORY SMOKE FLAVOR**
MAUNA LOA MACADAMIA NUTS	**COCONUT OIL**
PILLSBURY WHEAT NUTS	**ARTIFICIAL COLOR; HYDROGENATED SOYBEAN OIL; SODIUM CASEINATE**
PLANTERS COCKTAIL PEANUTS; MIXED NUTS; SALTED CASHEWS; SESAME NUT MIX; SUNFLOWER NUTS	**(COCONUT OIL)†**

*By law, at this writing, it is not required of a manufacturer to state whether cheeses contain artificial colors, unless the color is FD & C Yellow No. 5 (Tartrazine). Their presence in nuts and seeds containing cheese therefore remains an uncertainty, unless it is voluntarily declared on the label that artificial colors have been used or have not. For a more detailed explanation of artificial color in cheese, refer to the beginning of the cheese section of "Dairy Products & Substitutes."

†Blend may contain this saturated oil.

BRANDS	ADDITIVES-OF-CONCERN BOLD—CONCERN TO ALL LIGHT—CONCERN TO SOME
PLANTERS *(Continued)*	
RED PISTACHIOS	**ARTIFICIAL COLOR**
TAVERN NUTS	**MODIFIED FOOD STARCH**
—DRY ROASTED	
ALMONDS; CASHEWS; MIXED NUTS; PEANUTS; SPANISH PEANUTS; SUNFLOWER NUTS	GUM ARABIC; **MODIFIED FOOD STARCH;** MONOSODIUM GLUTAMATE; **PAPRIKA**
UNSALTED PEANUTS	NO ADDITIVES OF CONCERN
—SOUTHERN BELLE	
CASHEW HALVES; SUNFLOWER KERNELS	**(COCONUT OIL)***
SUNFLOWER SEEDS	NO ADDITIVES OF CONCERN
RIVER QUEEN	
CASHEW HALVES; SALTED MIXED NUTS	NO ADDITIVES OF CONCERN
SUPERMARKET PRIVATE BRANDS	
A & P	
—ANN PAGE	
ALMONDS; FANCY SALTED CASHEWS; FILBERTS; PECAN MEATS	NO ADDITIVES OF CONCERN
—DRY ROASTED	
CASHEWS; MIXED NUTS; PEANUTS	**MODIFIED STARCH;** MONOSODIUM GLUTAMATE

BRANDS	ADDITIVES-OF-CONCERN **BOLD—CONCERN TO ALL** LIGHT—CONCERN TO SOME
—SALTED FANCY MIXED NUTS; MIXED NUTS; PARTY PEANUTS; SPANISH PEANUTS	NO ADDITIVES OF CONCERN
FIRST NATIONAL **—FINAST** *DRY ROASTED* MIXED NUTS; PEANUTS	**PAPRIKA**
LUCKY **—LADY LEE** CASHEWS; MIXED NUTS W/ PEANUTS	**(COCONUT OIL)***
TOFFEE BUTTER PEANUTS	**BUTTER**
—DRY ROASTED CASHEWS; MIXED NUTS; PEANUTS	MONOSODIUM GLUTAMATE
SAFEWAY **—PARTY PRIDE** CASHEWS; DELUXE MIXED NUTS; PEANUTS	**(COCONUT OIL)***
—DRY ROASTED CASHEWS; MIXED NUTS; PEANUTS	MONOSODIUM GLUTAMATE

*Blend may contain this saturated oil.

BRANDS	**ADDITIVES-OF-CONCERN** BOLD—CONCERN TO ALL LIGHT—CONCERN TO SOME
STOP & SHOP —STOP & SHOP	
MIXED NUTS NO PEANUTS; SALTED PEANUTS	NO ADDITIVES OF CONCERN
—*DRY ROASTED*	
CASHEWS; MIXED NUTS; PEANUTS	GUM ARABIC; MONOSODIUM GLUTAMATE
UNSALTED PEANUTS	NO ADDITIVES OF CONCERN
ITEMS IN HEALTH FOOD STORES	
FLAVOR TREE	
NUT & SNACK MIX; SUNFLOWER & SESAME MIX	**TURMERIC**
HUNZA	
ALMONDS; WALNUTS	NO ADDITIVES OF CONCERN
NIK'S SNAKS	
RAISINS & RAW WALNUTS; TRAIL MIX	NO ADDITIVES OF CONCERN
—*DRY ROASTED*	
CASHEWS; MIXED NUTS; SPANISH PEANUTS; SUNFLOWER SEEDS	NO ADDITIVES OF CONCERN
—*RAW*	
MIXED NUTS; SUNFLOWER SEEDS; SUNFLOWER SEEDS & RAISINS	NO ADDITIVES OF CONCERN

BRANDS	ADDITIVES-OF-CONCERN **BOLD—CONCERN TO ALL** LIGHT—CONCERN TO SOME

POPCORN & POPCORN SNACKS

BANG-O *POPCORN*	
WHITE HULLESS; YELLOW HYBRID	NO ADDITIVES OF CONCERN
CRACKER JACK	NO ADDITIVES OF CONCERN
FIDDLE FADDLE	WHEY
FRANKLIN CRUNCH 'N MUNCH	NO ADDITIVES OF CONCERN
JIFFY-POP *POPCORN*	
BUTTER FLAVOR	**ARTIFICIAL VEGETABLE COLOR; BHA; BHT**
NATURAL FLAVOR	NO ADDITIVES OF CONCERN
ORVILLE REDENBACHER'S	
GOURMET POPPING CORN	NO ADDITIVES OF CONCERN
POPPYCOCK THE ORIGINAL	NO ADDITIVES OF CONCERN
SCREAMING YELLOW ZONKERS	**ARTIFICIAL COLOR**
TV TIME POPCORN	**ARTIFICIAL COLOR; COCONUT OIL**
WISE CHEEZ POP CORN	**ARTIFICIAL COLORS; (COCONUT OIL);*** MONOSODIUM GLUTAMATE

*Blend may contain this saturated oil.

BRANDS	ADDITIVES-OF-CONCERN **BOLD—CONCERN TO ALL** LIGHT—CONCERN TO SOME
SUPERMARKET PRIVATE BRANDS	
A & P	
—ANN PAGE	
CHEESE CORN	**ARTIFICIAL COLORING;** BUTTERMILK SOLIDS; **(COCONUT OIL);*** MONOSODIUM GLUTAMATE; WHEY SOLIDS
POPCORN	**ARTIFICIAL COLORING; (COCONUT OIL)***
LUCKY	
—LADY LEE	
POPCORN	NO ADDITIVES OF CONCERN
SAFEWAY	
—PARTY PRIDE	
POPCORN	**ARTIFICIAL COLOR; (COCONUT OIL)***
STOP & SHOP	
—STOP & SHOP	
CARAMEL CORN	NO ADDITIVES OF CONCERN

PRETZELS & MISCELLANEOUS SNACKS

BRANDS	ADDITIVES-OF-CONCERN
DURKEE/O & C	
FRENCH FRIED ONIONS	**ARTIFICIAL COLOR; BHA;**
POTATO STICKS	**BHA**
GENERAL MILLS	
BUGLES	**BHA; BHT; COCONUT OIL**

*Blend may contain this saturated oil.

BRANDS	ADDITIVES-OF-CONCERN BOLD—CONCERN TO ALL LIGHT—CONCERN TO SOME
KEYSTONE SNACKS CHEESE CURLS	**ARTIFICIAL COLORING; (COCONUT OIL);*** WHEY
NABISCO CORN DIGGERS	**BHA; COCONUT OIL**
DOO-DADS	**ARTIFICIAL COLOR; BHA; (COCONUT OIL; LARD);*** MONOSODIUM GLUTAMATE; **PAPRIKA;** WHEY
—*MISTER SALTY* DUTCH PRETZELS	NO ADDITIVES OF CONCERN
VERI-THIN PRETZELS	**(LARD)***
WISE *CHEESE DOODLES* CRUNCHY	**ARTIFICIAL COLORS; (COCONUT OIL);* FERRIC ORTHOPHOSPHATE;** WHEY (ACID & SWEET)
PUFFED	**ARTIFICIAL COLOR; (COCONUT OIL);* FERRIC ORTHOPHOSPHATE; MODIFIED CORN STARCH;** MONOSODIUM GLUTAMATE; **SODIUM CASEINATE;** WHEY
SUPERMARKET PRIVATE BRANDS A & P —A & P BITE SIZE PRETZELS	NO ADDITIVES OF CONCERN

*Blend may contain one or more of these saturated fats and/or oil.

BRANDS	ADDITIVES-OF-CONCERN BOLD—CONCERN TO ALL LIGHT—CONCERN TO SOME
—ANN PAGE CHEESE TWISTS	**ARTIFICIAL COLORING;** BUTTERMILK SOLIDS; **(COCONUT OIL);*** MILK SOLIDS (NONFAT DRY); MONOSODIUM GLUTAMATE; WHEY SOLIDS
—ANN PAGE *PRETZELS* PETITE; RODS; STIX; TEENIES	NO ADDITIVES OF CONCERN
FIRST NATIONAL —FINAST *PRETZELS* RODS; STIX	NO ADDITIVES OF CONCERN
LUCKY —LADY LEE *PRETZELS* MINI-TWIST; ROD; STICK; TWIST	NO ADDITIVES OF CONCERN
SAFEWAY —PARTY PRIDE *PRETZELS* BAVARIAN; MINI TWIST; STICK	NO ADDITIVES OF CONCERN
STOP & SHOP —STOP & SHOP CORN Q'S	**ARTIFICIAL COLOR;** BUTTERMILK SOLIDS; **(COCONUT OIL);*** MONOSODIUM GLUTAMATE; WHEY SOLIDS
—STOP & SHOP *PRETZELS* RINGS; STIX; THINS	NO ADDITIVES OF CONCERN

*Blend may contain one or more of these saturated fats and/or oils.

BRANDS	ADDITIVES-OF-CONCERN **BOLD—CONCERN TO ALL** LIGHT—CONCERN TO SOME
WINN DIXIE —CRACKIN GOOD CHEESE BALLS	**ARTIFICIAL COLOR;** WHEY
PRETZELS RINGS; RODS	NO ADDITIVES OF CONCERN
ITEMS IN HEALTH FOOD STORES	
BARBARA'S PRETZELS	NO ADDITIVES OF CONCERN
SESAME BREADSTICKS	GLUTEN (WHEAT)
DR. BRONNER'S CHEEZON CORN-SNACK*	**COCONUT OIL;** MILK SOLIDS (NONFAT)
EREWHON PRETZELS	NO ADDITIVES OF CONCERN
FLAVOR TREE SESAME & BRAN STICKS; SESAME STICKS	**TURMERIC**
HEALTH VALLEY PRETZELS	NO ADDITIVES OF CONCERN

*By law, at this writing, it is not required of a manufacturer to state whether cheeses contain artificial colors, unless the color is FD & C Yellow No. 5 (Tartrazine). Their presence in snacks containing cheese therefore remains an uncertainty, unless it is voluntarily declared on the label that artificial colors have been used or have not. For a more detailed explanation of artificial color in cheese, refer to the beginning of the cheese section of "Dairy Products & Substitutes."

BRANDS	ADDITIVES-OF-CONCERN **BOLD—CONCERN TO ALL** LIGHT—CONCERN TO SOME

TOASTER PASTRIES

KELLOGG'S *—POP-TARTS*	
BLUEBERRY; CHERRY CHIP; STRAWBERRY	**ARTIFICIAL COLORING; BHA; (COCONUT OIL);*** WHEY
BROWN SUGAR-CINNAMON; CHOCOLATE CHIP	**BHA; (COCONUT OIL);*** EGG WHITES; WHEY
—FROSTED POP-TARTS	
CHOCOLATE FUDGE; CHOCOLATE-VANILLA CREME	**ARTIFICIAL COLORING; BHA; (COCONUT OIL);*** EGG WHITES; WHEY
NABISCO *TOASTETTES*	
BLUEBERRY; BROWN SUGAR CINNAMON; CHERRY; STRAWBERRY	**ARTIFICIAL COLOR; MODIFIED CORN STARCH; MODIFIED TAPIOCA STARCH;** WHEY
SUPERMARKET PRIVATE BRANDS FIRST NATIONAL —FINAST *TOASTER PASTRIES*	
BROWN SUGAR CINNAMON	**BHA; MODIFIED CORN STARCH;** WHEY SOLIDS
CHERRY; FROSTED STRAWBERRY; STRAWBERRY	**ARTIFICIAL COLORS; BHA;** WHEY SOLIDS

BRANDS	ADDITIVES-OF-CONCERN BOLD—CONCERN TO ALL LIGHT—CONCERN TO SOME
SAFEWAY —TOWN HOUSE *FROSTED TOASTER TARTS* FUDGE; STRAWBERRY	**ARTIFICIAL COLORS; BHA;** WHEY SOLIDS
STOP & SHOP —STOP & SHOP *TOASTER TARTS* BLUEBERRY; STRAWBERRY	**ARTIFICIAL COLORS; BHA;** WHEY SOLIDS
FROSTED TOASTER TARTS CINNAMON	**ARTIFICIAL COLORS; BHA; MODIFIED CORN STARCH;** WHEY SOLIDS
WINN DIXIE —CRACKIN GOOD *TOASTER PASTRIES* BLUEBERRY; STRAWBERRY	**ARTIFICIAL COLOR; (HYDROGENATED PALM, SOYBEAN OILS);*** WHEY SOLIDS
FROSTED TOASTER PASTRIES APPLE	**ARTIFICIAL COLOR; NUTMEG;** WHEY SOLIDS
BLUEBERRY; CHERRY; CINNAMON; GRAPE; STRAWBERRY	**ARTIFICIAL COLOR;** WHEY SOLIDS

*Blend may contain these saturated oils.

XXII. SOUPS

BOUILLON, KOJI, MISO

BRANDS	ADDITIVES-OF-CONCERN BOLD—CONCERN TO ALL LIGHT—CONCERN TO SOME
CAINS *BOUILLON CUBES*	
BEEF	NO ADDITIVES OF CONCERN
CHICKEN	**TURMERIC**
HERB-OX	
—*BOUILLON CUBES*	
BEEF; CHICKEN	NO ADDITIVES OF CONCERN
—*INSTANT BROTH*	
BEEF	NO ADDITIVES OF CONCERN
CHICKEN	**TURMERIC**
STEERO	
INSTANT BEEF BOUILLON	MONOSODIUM GLUTAMATE
—*BOUILLON CUBES*	
BEEF; CHICKEN	MONOSODIUM GLUTAMATE
WYLER'S	
—*BOUILLON CUBES*	
BEEF	**ARTIFICIAL COLOR;** MONOSODIUM GLUTAMATE
CHICKEN	**BHA; BHT;** MONOSODIUM GLUTAMATE; **TURMERIC**

BRANDS	ADDITIVES-OF-CONCERN BOLD—CONCERN TO ALL LIGHT—CONCERN TO SOME
—INSTANT BOUILLON	
BEEF	**ARTIFICIAL COLOR;** MONOSODIUM GLUTAMATE
CHICKEN	**BHA; BHT;** MONOSODIUM GLUTAMATE; **TURMERIC**
ITEMS IN HEALTH FOOD STORES	
COLD MOUNTAIN	
KOJI	**ASPERGILLUS ORYZAE MOLD**
—MISO	
MELLOW WHITE; RED MISO	NO ADDITIVES OF CONCERN
EREWHON *SOYBEAN PASTE*	
PLAIN; W/ BARLEY; W/ RICE	NO ADDITIVES OF CONCERN
HAUSER BROTH	NO ADDITIVES OF CONCERN
MARUSAN SOYBEAN PASTE MAME MISO	NO ADDITIVES OF CONCERN
MORGA VEGETABLE BOUILLON CUBES	NO ADDITIVES OF CONCERN
PLANTAFORCE VEGETABLE CONCENTRATE	KELP
VEGEX BOUILLON CUBES	NO ADDITIVES OF CONCERN
WESTBRAE NATURAL MELLOW WHITE MISO	NO ADDITIVES OF CONCERN

BRANDS	ADDITIVES-OF-CONCERN **BOLD—CONCERN TO ALL** LIGHT—CONCERN TO SOME

CANNED SOUP

CAMPBELL'S	
BEEF NOODLE; CHICKEN BARLEY; CHICKEN NOODLE; MEATBALL ALPHABET; MUSHROOM BARLEY; NEW ENGLAND CLAM CHOWDER; TURKEY NOODLE; VEGETABLE; VEGETARIAN VEGETABLE	MONOSODIUM GLUTAMATE
BEEFY MUSHROOM; CLAM CHOWDER MANHATTAN STYLE; CREAM OF SHRIMP	**MODIFIED FOOD STARCH;** MONOSODIUM GLUTAMATE
BLACK BEAN	MONOSODIUM GLUTAMATE; **OLEORESIN PAPRIKA**
CHEDDAR CHEESE*	**CALCIUM CASEINATE;** MONOSODIUM GLUTAMATE; **OLEORESIN PAPRIKA; SODIUM CASEINATE;** WHEY
CREAM OF ASPARAGUS	**CALCIUM CASEINATE;** MONOSODIUM GLUTAMATE; **SODIUM CASEINATE;** WHEY

*By law, at this writing, it is not required of a manufacturer to state whether cheeses contain artificial colors, unless the color is FD & C Yellow No. 5 (Tartrazine). Their presence in canned soup containing cheese therefore remains an uncertainty, unless it is voluntarily declared on the label that artificial colors have been used or have not. For a more detailed explanation of artificial color in cheese, refer to the beginning of the cheese section in "Dairy Products & Substitutes."

BRANDS	**ADDITIVES-OF-CONCERN** BOLD—CONCERN TO ALL LIGHT—CONCERN TO SOME
CREAM OF CELERY; CREAM OF MUSHROOM	**CALCIUM CASEINATE; MODIFIED FOOD STARCH;** MONOSODIUM GLUTAMATE; WHEY
CREAM OF CHICKEN	**CALCIUM CASEINATE;** MONOSODIUM GLUTAMATE; WHEY
GREEN PEA; PEPPER POT; TOMATO; TOMATO BISQUE	NO ADDITIVES OF CONCERN
OLD FASHIONED TOMATO RICE	**CALCIUM CASEINATE;** WHEY
—CHUNKY SOUP	
BEEF	NO ADDITIVES OF CONCERN
CHICKEN; CLAM CHOWDER MANHATTAN STYLE; TURKEY; VEGETABLE BEEF	MONOSODIUM GLUTAMATE
HAM 'N BUTTER BEAN; SPLIT PEA W/ HAM	MONOSODIUM GLUTAMATE; **SMOKE FLAVORING; SODIUM NITRITE**
—ONE OF THE LIGHT ONES	
BEEF BROTH; CHICKEN & STARS; CHICKEN BROTH; CHICKEN VEGETABLE; CHICKEN W/ RICE; CONSOMMÉ (BEEF); ONION; TURKEY VEGETABLE	MONOSODIUM GLUTAMATE

BRANDS	ADDITIVES-OF-CONCERN BOLD—CONCERN TO ALL LIGHT—CONCERN TO SOME
CAMPBELL'S *(Continued)*	
—ONE OF THE MANHANDLERS	
BEAN W/ BACON	**MODIFIED FOOD STARCH;** MONOSODIUM GLUTAMATE; **SMOKE FLAVORING**
BEEF W/ VEGETABLES & BARLEY; MINESTRONE; SCOTCH BROTH; VEGETABLE BEEF	MONOSODIUM GLUTAMATE
CHILI BEEF	**OLEORESIN PAPRIKA**
SPLIT PEA W/ HAM & BACON	**SMOKE FLAVORING**
—SOUP FOR ONE	
BURLY VEGETABLE BEEF	**MODIFIED FOOD STARCH;** MONOSODIUM GLUTAMATE
GOLDEN CHICKEN & NOODLES; OLD WORLD VEGETABLE	MONOSODIUM GLUTAMATE
SAVORY CREAM OF MUSHROOM	**CALCIUM CASEINATE; MODIFIED FOOD STARCH;** MONOSODIUM GLUTAMATE; **SODIUM CASEINATE;** WHEY
CROSSE & BLACKWELL	
BLACK BEAN; CONSOMMÉ; CREME MUSHROOM BISQUE; CURRIED CREAM OF CHICKEN; LOBSTER BISQUE; MINESTRONE	MONOSODIUM GLUTAMATE

BRANDS	ADDITIVES-OF-CONCERN **BOLD—CONCERN TO ALL** LIGHT—CONCERN TO SOME
CRAB	MONOSODIUM GLUTAMATE; **SMOKED YEAST FLAVOR**
GAZPACHO	**MODIFIED CORN STARCH;** MONOSODIUM GLUTAMATE
LENTIL	MONOSODIUM GLUTAMATE; **SMOKED HAM;* SMOKED YEAST;* SODIUM NITRITE**
ONION	NO ADDITIVES OF CONCERN
SPLIT GREEN PEA	MONOSODIUM GLUTAMATE; **NATURAL SMOKE FLAVORS; SMOKED HAM;* SODIUM NITRITE**
DOXSEE ***CLAM CHOWDER***	
MANHATTAN	**MODIFIED FOOD STARCH;** MONOSODIUM GLUTAMATE; **PAPRIKA**
NEW ENGLAND	**MODIFIED FOOD STARCH;** MONOSODIUM GLUTAMATE
PEPPERIDGE FARM	
BACON, LETTUCE & TOMATO; WATERCRESS	**MODIFIED FOOD STARCH;** MONOSODIUM GLUTAMATE
BLACK BEAN	MONOSODIUM GLUTAMATE

*Although "smoked" does not refer to a specific additive, it represents a process utilizing wood smoke, as "smoke flavoring" does, and therefore the food may contain some cancer-causing benzopyrene chemicals. For this reason, when a food or any of its constituents has been smoked, this has been noted in the additives-of-concern column.

BRANDS	**ADDITIVES-OF-CONCERN** **BOLD—CONCERN TO ALL** LIGHT—CONCERN TO SOME
PEPPERIDGE FARM *(Continued)*	
CHICKEN CURRY	MONOSODIUM GLUTAMATE; **TURMERIC**
PROGRESSO	
BEAN & HAM	HYDROLYZED MILK PROTEIN; **MODIFIED FOOD STARCH;** MONOSODIUM GLUTAMATE; **NATURAL SMOKE FLAVORING**
CHICKARINA	EGG WHITES; MONOSODIUM GLUTAMATE; **TURMERIC**
GREEN SPLIT PEA	**NATURAL SMOKE FLAVOR**
LENTIL; MACARONI & BEAN; MINESTRONE; TOMATO	NO ADDITIVES OF CONCERN
SNOW'S	
OYSTER STEW	**GRADE AA BUTTER; MODIFIED FOOD STARCH;** MONOSODIUM GLUTAMATE
—CHOWDER	
CORN; MANHATTAN CLAM	**ARTIFICIAL COLORING; MODIFIED FOOD STARCH;** MONOSODIUM GLUTAMATE
FISH; SEAFOOD	**MODIFIED FOOD STARCH;** MONOSODIUM GLUTAMATE
NEW ENGLAND CLAM	MONOSODIUM GLUTAMATE
SUPERMARKET PRIVATE BRANDS	
A & P—ANN PAGE	
BEAN W/ BACON	**MODIFIED FOOD STARCH;** MONOSODIUM GLUTAMATE; **OLEORESIN OF PAPRIKA**

BRANDS	ADDITIVES-OF-CONCERN BOLD—CONCERN TO ALL LIGHT—CONCERN TO SOME
CHICKEN NOODLE; CHICKEN VEGETABLE; CHICKEN W/ RICE	MONOSODIUM GLUTAMATE
CREAM OF CELERY; CREAM OF CHICKEN	**MODIFIED STARCH;** MONOSODIUM GLUTAMATE; WHEY (DRIED OR POWDER)
TOMATO	NO ADDITIVES OF CONCERN
VEGETABLE BEEF; VEGETARIAN VEGETABLE	MONOSODIUM GLUTAMATE; **OLEORESIN OF PAPRIKA**
LUCKY	
—LADY LEE	
BEAN W/ BACON	MONOSODIUM GLUTAMATE
CREAM OF CHICKEN	**MODIFIED FOOD STARCH;** MONOSODIUM GLUTAMATE; **PAPRIKA; SODIUM CASEINATE**
SPLIT PEA W/ HAM	MONOSODIUM GLUTAMATE; **SMOKED HAM;* SMOKED YEAST***
VEGETABLE; VEGETABLE BEEF	MONOSODIUM GLUTAMATE; **PAPRIKA OLEORESIN**
—*CHUNKY* CHICKEN	**MODIFIED FOOD STARCH;** MONOSODIUM GLUTAMATE

*Although "smoked" does not refer to a specific additive, it represents a process utilizing wood smoke, as "smoke flavoring" does, and therefore the food may contain some cancer-causing benzopyrene chemicals. For this reason, when a food or any of its constituents has been smoked, this has been noted in the additives-of-concern column.

BRANDS	ADDITIVES-OF-CONCERN **BOLD—CONCERN TO ALL** LIGHT—CONCERN TO SOME
—LADY LEE *CHUNKY (Continued)*	
SIRLOIN BURGER	**MODIFIED FOOD STARCH;** MONOSODIUM GLUTAMATE; **OLEORESIN PAPRIKA; SMOKED YEAST***
SPLIT PEA W/ HAM	MONOSODIUM GLUTAMATE; **SODIUM NITRITE**
SAFEWAY	
—TOWN HOUSE	
CHICKEN W/ RICE	**MODIFIED FOOD STARCH;** MONOSODIUM GLUTAMATE
CREAM OF MUSHROOM	**CALCIUM CASEINATE; MODIFIED FOOD STARCH;** MONOSODIUM GLUTAMATE; WHEY
MINESTRONE	MONOSODIUM GLUTAMATE; **PAPRIKA OLEORESIN**
TOMATO	**MODIFIED FOOD STARCH**
CHUNKY	
SPLIT PEA W/ HAM	MONOSODIUM GLUTAMATE; **SODIUM NITRITE**
STOP & SHOP	
—STOP & SHOP	
CHICKEN NOODLE	**MODIFIED FOOD STARCH;** MONOSODIUM GLUTAMATE; **PAPRIKA**
CHICKEN VEGETABLE; TURKEY NOODLE	MONOSODIUM GLUTAMATE

*Although "smoked" does not refer to a specific additive, it represents a process utilizing wood smoke, as "smoke flavoring" does, and therefore the food may contain some cancer-causing benzopyrene chemicals. For this reason, when a food or any of its constituents has been smoked, this has been noted in the additives-of-concern column.

BRANDS	ADDITIVES-OF-CONCERN BOLD—CONCERN TO ALL LIGHT—CONCERN TO SOME
CREAM OF CELERY; CREAM OF MUSHROOM	**MODIFIED FOOD STARCH;** MONOSODIUM GLUTAMATE; **SODIUM CASEINATE**
VEGETABLE BEEF; VEGETARIAN VEGETABLE	MONOSODIUM GLUTAMATE; **PAPRIKA OLEORESIN**
WINN DIXIE —THRIFTY MAID	
CHICKEN NOODLE; CHICKEN RICE	**MODIFIED FOOD STARCH;** MONOSODIUM GLUTAMATE; **TURMERIC**
CREAM OF CHICKEN; VEGETABLE BEEF	**MODIFIED FOOD STARCH;** MONOSODIUM GLUTAMATE
CHUNKY	
BEEF	**MODIFIED FOOD STARCH**
CHICKEN	MONOSODIUM GLUTAMATE; **TURMERIC**
VEGETABLE	MONOSODIUM GLUTAMATE
ITEMS IN HEALTH FOOD STORES	
HEALTH VALLEY	
BEAN; CLAM CHOWDER; LENTIL; MINESTRONE; SPLIT PEA; VEGETABLE	NO ADDITIVES OF CONCERN
CHICKEN BROTH	**SPICE EXTRACTIVES OF TURMERIC**

BRANDS	ADDITIVES-OF-CONCERN **BOLD—CONCERN TO ALL** LIGHT—CONCERN TO SOME
WALNUT ACRES	
BEEF STEW; BLACK BEAN; CHICKEN CORN; FISH CHOWDER; MANHATTAN CLAM CHOWDER; TOMATO; VEGETABLE	NO ADDITIVES OF CONCERN

DEHYDRATED SOUP

BRANDS	ADDITIVES-OF-CONCERN
CUP O' NOODLES	
BEEF; BEEF/ONION; SHRIMP	MONOSODIUM GLUTAMATE
PORK	MONOSODIUM GLUTAMATE; **NATURAL HICKORY SMOKE FLAVOR; VEGETABLE OIL**
KNORR SWISS *SOUPMIX*	
ASPARAGUS	**MODIFIED FOOD STARCH;** MONOSODIUM GLUTAMATE; **TURMERIC; VEGETABLE GUM**
ONION & DIP MIX	**MODIFIED FOOD STARCH;** MONOSODIUM GLUTAMATE
OXTAIL	**BEEF FAT;** MONOSODIUM GLUTAMATE; **VEGETABLE GUM**
LA CHOY *RAMEN NOODLES*	
BEEF	**BHA; BHT;** GUAR GUM; MONOSODIUM GLUTAMATE; **PAPRIKA**

BRANDS	ADDITIVES-OF-CONCERN **BOLD—CONCERN TO ALL** LIGHT—CONCERN TO SOME
CHICKEN	**BHA; BHT;** GUAR GUM; MONOSODIUM GLUTAMATE; **PAPRIKA; TURMERIC**
LIPTON	
—SOUP MIX	
BEEF FLAVOR MUSHROOM	BUTTERMILK SOLIDS; MONOSODIUM GLUTAMATE; WHEY SOLIDS
CHICKEN NOODLE	**COLORING;** MONOSODIUM GLUTAMATE
CHICKEN RICE	**COLORING;** EGG WHITE SOLIDS; MONOSODIUM GLUTAMATE
CHICKEN RIPPLE NOODLE	**BHA; COLORING;** MONOSODIUM GLUTAMATE
GIGGLE NOODLE	**(HYDROGENATED COTTONSEED, PALM, SOYBEAN OILS);*** MONOSODIUM GLUTAMATE; **OLEORESIN TURMERIC**
ONION	NO ADDITIVES OF CONCERN
VEGETABLE BEEF	MONOSODIUM GLUTAMATE; **PAPRIKA; VEGETABLE GUM**
—CUP-A-BROTH	
CHICKEN	MONOSODIUM GLUTAMATE; **OLEORESIN TURMERIC**

*Blend may contain these saturated oils.

BRANDS	ADDITIVES-OF-CONCERN **BOLD—CONCERN TO ALL** LIGHT—CONCERN TO SOME
LIPTON *(Continued)*	
—CUP-A-SOUP	
BEEF NOODLE; ONION	MONOSODIUM GLUTAMATE
CHICKEN NOODLE	**BHA; COLORING; MODIFIED CORN STARCH;** MONOSODIUM GLUTAMATE
CHICKEN VEGETABLE	**BHA; COLORING; MODIFIED FOOD STARCH;** MONOSODIUM GLUTAMATE
CREAM OF CHICKEN	BUTTERMILK SOLIDS; **COCONUT OIL;** MILK SOLIDS (NONFAT); **MODIFIED FOOD STARCH;** MONOSODIUM GLUTAMATE; **OLEORESIN TURMERIC; VEGETABLE GUM;** WHEY SOLIDS
GREEN PEA	**MODIFIED FOOD STARCH;** MONOSODIUM GLUTAMATE; **VEGETABLE GUM**
TOMATO	**MODIFIED FOOD STARCH;** MONOSODIUM GLUTAMATE; **PAPRIKA OLEORESIN**
NESTLÉ'S *SOUPTIME*	
CHICKEN NOODLE	**MODIFIED CORN STARCH;** MONOSODIUM GLUTAMATE; **TURMERIC**

BRANDS	**ADDITIVES-OF-CONCERN** **BOLD—CONCERN TO ALL** LIGHT—CONCERN TO SOME
CREAM OF CHICKEN; CREAM OF GARDEN VEGETABLE	**HYDROGENATED COTTONSEED & SOY OILS; MODIFIED CORN STARCH;** MONOSODIUM GLUTAMATE; **SODIUM CASEINATE; TURMERIC**
GREEN PEA	**ARTIFICIAL COLOR;** GUAR BEAN GUM; **HYDROGENATED COTTONSEED & SOY OILS; MODIFIED CORN STARCH; NATURAL SMOKE FLAVOR; SMOKED TORULA YEAST***
TOMATO	**MODIFIED TAPIOCA & CORN STARCHES;** WHEY
OODLES OF NOODLES	
BEEF; PORK	ALGINIC ACID; MONOSODIUM GLUTAMATE; **VEGETABLE OIL**
CHICKEN	ALGINIC ACID; GUM ARABIC; MONOSODIUM GLUTAMATE; **SOLUBLE TURMERIC; VEGETABLE OIL**

*Although "smoked" does not refer to a specific additive, it represents a process utilizing wood smoke, as "smoke flavoring" does, and therefore the food may contain some cancer-causing benzopyrene chemicals. For this reason, when a food or any of its constituents has been smoked, this has been noted in the additives-of-concern column.

BRANDS	ADDITIVES-OF-CONCERN **BOLD—CONCERN TO ALL** LIGHT—CONCERN TO SOME
BEEF BARLEY; BEEF NOODLE	**MODIFIED FOOD STARCH;** MONOSODIUM GLUTAMATE; **PAPRIKA; TURMERIC**
SWIFT *SOUP STARTER*	
CHICKEN RICE	**BHA; BHT; COLORING;** GUAR GUM; MONOSODIUM GLUTAMATE; **SODIUM CASEINATE; TURMERIC**
CHICKEN NOODLE	MONOSODIUM GLUTAMATE; **TURMERIC**
SUPERMARKET PRIVATE BRANDS	
A & P —ANN PAGE	
NOODLE SOUP MIX	MONOSODIUM GLUTAMATE; **TURMERIC**
ONION	**MODIFIED STARCH;** MONOSODIUM GLUTAMATE
STOP & SHOP —STOP & SHOP	
NOODLE SOUP MIX	**BHA; MODIFIED WHEAT STARCH;** MONOSODIUM GLUTAMATE; **TURMERIC (& EXTRACTIVES OF)**
ITEMS IN HEALTH FOOD STORES	
EDWARD & SONS *MISO-CUP*	
ORIGINAL GOLDEN LIGHT	NO ADDITIVES OF CONCERN

BRANDS	ADDITIVES-OF-CONCERN BOLD—CONCERN TO ALL LIGHT—CONCERN TO SOME
RICH, RED W/ SEAWEED	WAKAME
HAIN NATURALS *SOUP MIX*	
CHICKEN; CREAM OF MUSHROOM; SPLIT-PEA	MILK SOLIDS (NONFAT)
ONION	WHEY
TOMATO	MILK SOLIDS (NONFAT); WHEY
VEGETABLE	NO ADDITIVES OF CONCERN
HUGLI MINESTRONE SOUP MIX	NO ADDITIVES OF CONCERN
MARUSAN'S *MISO SOUP*	
RED MISO; WHITE MISO	NO ADDITIVES OF CONCERN

FROZEN SOUP

LA CHOY WON TON	MONOSODIUM GLUTAMATE
ITEMS IN HEALTH FOOD STORES	
HEALTH VALLEY	
CHINESE VEGETABLE; EGG DROP; WON TON	NO ADDITIVES OF CONCERN

XXIII. VEGETABLE JUICES & VEGETABLES

CANNED BEANS & VEGETABLES

Most of the items in this section do not contain any additives of concern, and these appear first, without an additives-of-concern column.

The items that do contain additives of concern follow, in the usual style.

B IN B MUSHROOM CROWNS; SLICED MUSHROOMS

CONTADINA SLICED BABY TOMATOES; STEWED TOMATOES

DEL MONTE ASPARAGUS SPEARS; ASPARAGUS TIPS; WHOLE BEETS; CUT CARROTS; SLICED CARROTS; WHOLE KERNEL CORN; GREEN BEANS; GREEN LIMA BEANS; FRENCH STYLE GREEN BEANS; ITALIAN BEANS; PEAS & CARROTS; SPINACH; SWEET PEAS; STEWED TOMATOES; TOMATO WEDGES; WAX BEANS

FRESHLIKE VEG-ALL

GREEN GIANT CUT SPEARS ASPARAGUS; NIBLETS GOLDEN CORN; WHITE CORN; FRENCH STYLE GREEN BEANS; KITCHEN SLICED GREEN BEANS; WHOLE GREEN BEANS; MEXICORN; WHOLE MUSHROOMS; LE SUEUR EARLY PEAS; SWEET PEAS

GREENWOOD SLICED PICKLED BEETS; SWEET-SOUR RED CABBAGE

HUNT'S PEAR SHAPED TOMATOES; STEWED TOMATOES; WHOLE TOMATOES

KOUNTY KIST CREAM STYLE CORN; WHOLE KERNEL CORN

LIBBY'S SLICED BEETS; SLICED CARROTS; WHOLE KERNEL CORN; CUT GREEN BEANS; FRENCH STYLE GREEN BEANS; PEAS & CARROTS; SUCCOTASH; SWEET PEAS

PROGRESSO CANNELLINI; CHICK PEAS; FAVA BEANS; KIDNEY BEANS; ITALIAN PEELED TOMATOES

REDPACK CALIFORNIA WHOLE TOMATOES; CRUSHED TOMATOES; ITALIAN STYLE TOMATOES; STEWED TOMATOES; WHOLE TOMATOES

RITTER ASPARAGUS SPEARS

ROYAL PRINCE YAMS

SILVER FLOSS SAUERKRAUT

STEWART'S DARK RED KIDNEY BEANS

STOKELY VAN CAMP'S CUT GREEN BEANS; DARK RED KIDNEY BEANS; LIMA BEANS; PEAS & CARROTS; SAUERKRAUT; CUT WAX BEANS

SUPERMARKET PRIVATE BRANDS
A & P ANN PAGE BUTTER BEANS; SLICED BEETS; SLICED CARROTS; WHOLE KERNEL CORN GOLDEN SWEET; CUT GREEN BEANS; GREEN BEANS FRENCH STYLE; RED KIDNEY BEANS; STEWED TOMATOES; TOMATOES

FIRST NATIONAL FINAST SLICED BEETS; GOLDEN CORN; WHOLE GREEN BEANS; MIXED VEGETABLES; NAVY BEANS; PEAS & CARROTS; SLICED WHITE POTATOES; SAUERKRAUT; SPINACH; MEDIUM SMALL SWEET PEAS; WHOLE TOMATOES

STOP & SHOP ASPARAGUS ALL GREEN SPEARS; SLICED BEETS; SLICED PICKLED BEETS; SLICED
(continues)

STOP & SHOP *(Continued)*

CARROTS; CREAM STYLE CORN; GOLDEN SWEET CORN WHOLE KERNEL; CUT GREEN BEANS; FRENCH STYLE SLICED GREEN BEANS; LIMA BEANS; MIXED VEGETABLES; WHOLE ONIONS; PEAS & CARROTS; EARLY PEAS; WHOLE POTATOES; SAUERKRAUT; SPINACH; PEAR SHAPED PEELED TOMATOES; STEWED TOMATOES; TOMATOES; CUT WAX BEANS; FRENCH STYLE SLICED WAX BEANS

STOP & SHOP SUN GLORY CUT SPEARS ASPARAGUS; SLICED CARROTS; WHOLE KERNEL SWEET CORN; CUT GREEN BEANS; SWEET PEAS; TOMATOES

LUCKY HARVEST DAY PICKLED SLICED BEETS; SLICED BEETS; CREAM STYLE-GOLDEN SWEET CORN; WHOLE KERNEL GOLDEN SWEET CORN; CUT GREEN BEANS; SWEET PEAS; PEELED TOMATOES

LUCKY LADY LEE CUT ALL GREEN ASPARAGUS; SLICED BEETS; BLACKEYE PEAS; BUTTER BEANS; CREAM STYLE-GOLDEN SWEET CORN; WHOLE KERNEL GOLDEN SWEET CORN; GARBANZOS; CUT GREEN BEANS; WHITE HOMINY; KIDNEY BEANS; PIECES & STEMS MUSHROOMS; PINTO BEANS; SLICED WHITE POTATOES; SAUERKRAUT; SPINACH; SWEET PEAS; STEWED TOMATOES; GOLDEN SWEET YAMS

SAFEWAY SCOTCH BUY GOLDEN CREAM STYLE SWEET CORN; CUT GREEN BEANS; JULIENNE (FRENCH STYLE) GREEN BEANS; SWEET PEAS; TOMATOES

SAFEWAY TOWN HOUSE ARTICHOKE HEARTS; ASPARAGUS SPEARS; PICKLED SLICED BEETS; SLICED BEETS; BLACKEYE PEAS; PICKLED CAULIFLOWER; CREAM STYLE SWEET CORN; WHOLE KERNEL SWEET CORN; GREAT NORTHERN BEANS; CUT GREEN BEANS; FRENCH STYLE GREEN BEANS; GOLDEN HOMINY; MILD GIARDINIERA MIXED VEGETABLES; BUTTONS MUSHROOMS; STEMS & PIECES MUSHROOMS; MUSTARD GREENS; WHOLE WHITE POTATOES

WINN DIXIE ASTOR SWEET PEAS

WINN DIXIE THRIFTY MAID ASPARAGUS; MEDIUM WHOLE BEETS; FRESH SHELLED BLACKEYE PEAS; COLLARD GREENS; CREAM STYLE GOLDEN SWEET CORN; WHOLE KERNEL GOLDEN SWEET CORN; GREAT NORTHERN BEANS; FRENCH STYLE GREEN BEANS; GREEN BEANS; RED KIDNEY BEANS; GREEN & WHITE LIMA BEANS; MIXED VEGETABLES; FANCY SLICED BUTTONS MUSHROOMS; NAVY BEANS; GREEN BOILED PEANUTS; LARGE SWEET PEAS; PINTO BEANS; SWEET POTATOES; WHITE POTATOES; SPINACH; YELLOW CUT SQUASH; STEWED TOMATOES; TOMATOES; CHOPPED TURNIP GREENS

ITEMS IN HEALTH FOOD STORES

DEL GAIZO CRUSHED PEELED TOMATOES; PEELED PLUM TOMATOES; PEELED TOMATOES

HAIN SOY BEANS

NA-ZDROWIE CARAWAY SAUERKRAUT

VITARROZ BLACK BEANS; BLACKEYE PEAS; CHICK PEAS; GARBANZOS; GREEN PIGEON PEAS; KIDNEY BEANS; PINK BEANS; PINTO BEANS; SMALL WHITE BEANS

WALNUT ACRES TOMATOES

The following items contain additives of concern.

BRANDS	ADDITIVES-OF-CONCERN BOLD—CONCERN TO ALL LIGHT—CONCERN TO SOME
DEL MONTE CREAM STYLE CORN; ZUCCHINI	**MODIFIED FOOD STARCH**
GREENWOOD HARVARD BEETS	**MODIFIED FOOD STARCH**

BRANDS	ADDITIVES-OF-CONCERN **BOLD—CONCERN TO ALL** LIGHT—CONCERN TO SOME
LIBBY'S CREAM STYLE CORN	**MODIFIED CORN STARCH**
RITTER GOLDEN BUTTER BEANS	COLOR ADDITIVE FD & C YELLOW NO. 5; MONOSODIUM GLUTAMATE
SUPERMARKET PRIVATE BRANDS	
A & P	
ANN PAGE CREAM STYLE CORN	**MODIFIED FOOD STARCH**

CANNED TOMATO & VEGETABLE JUICE*

CAMPBELL'S TOMATO

LIBBY'S TOMATO

RITTER TOMATO; VEG-CREST COCKTAIL

SACRAMENTO TOMATO

V-8 CLAM; (REGULAR); LOW SODIUM; SPICY-HOT

SUPERMARKET PRIVATE BRANDS
A & P ANN PAGE TOMATO

FIRST NATIONAL FINAST TOMATO

LUCKY LADY LEE TOMATO; VEGETABLE COCKTAIL

SAFEWAY SCOTCH BUY TOMATO

—TOWN HOUSE TOMATO; VEGETABLE COCKTAIL

*See note on page 432

WINN DIXIE THRIFTY MAID TOMATO

ITEMS IN HEALTH FOOD STORES
HAIN BEET; CABBAGE; CARROT; CELERY; NATURAL VEGETABLE COCKTAIL; TOMATO

LAKEWOOD BIG TEN VEGETABLE

The following items contain additives of concern.

BRANDS	ADDITIVES-OF-CONCERN **BOLD—CONCERN TO ALL** LIGHT—CONCERN TO SOME
MOTT'S BEEFAMATO; CLAMATO	**ARTIFICIAL COLOR;** MONOSODIUM GLUTAMATE

CANNED TOMATO PASTE, PUREE, SAUCE*

CONTADINA TOMATO PASTE; TOMATO SAUCE

DEL MONTE TOMATO SAUCE

HUNT'S TOMATO PASTE

—*TOMATO SAUCE* PLAIN; W/ MUSHROOMS; W/ ONIONS

PROGRESSO TOMATO PASTE; TOMATO PUREE

SUPERMARKET PRIVATE BRANDS
FIRST NATIONAL EDWARDS TOMATO PASTE; TOMATO PUREE

LUCKY LADY LEE SPANISH STYLE TOMATO SAUCE; TOMATO PASTE; TOMATO SAUCE

SAFEWAY TOWN HOUSE TOMATO PASTE; TOMATO SAUCE

*See note on page 432

WINN DIXIE THRIFTY MAID TOMATO SAUCE SPANISH STYLE

ITEMS IN HEALTH FOOD STORES
DEL GAIZO HEAVY TOMATO PUREE; TOMATO PASTE

HEALTH VALLEY *TOMATO SAUCE* PLAIN; W/ MUSHROOMS & CHEESE*

WALNUT ACRES TOMATO PUREE

The following items contain additives of concern.

BRANDS	ADDITIVES-OF-CONCERN **BOLD—CONCERN TO ALL** LIGHT—CONCERN TO SOME
HUNT'S *TOMATO SAUCE* HERB; SPECIAL; W/ TOMATO BITS	**MODIFIED FOOD STARCH**

FROZEN FRENCH FRIED POTATOES & VARIATIONS

BIRDS EYE CRINKLE CUTS	**ARTIFICIAL COLOR**
TASTI FRIES; TASTI PUFFS; TINY TATERS	**HYDROGENATED PALM** OR **SOYBEAN OIL**
HEINZ *DEEP FRIES* COUNTRY STYLE DINNER FRIES; CRINKLE CUTS	NO ADDITIVES OF CONCERN

*By law, at this writing, it is not required of a manufacturer to state whether cheeses contain artificial colors, unless the color is FD & C Yellow No. 5 (Tartrazine). Their presence in tomato sauce containing cheese therefore remains an uncertainty, unless it is voluntarily declared on the label that artificial colors have been used or have not. For a more detailed explanation of artificial color in cheese, refer to the beginning of the cheese section of "Dairy Products & Substitutes."

BRANDS	ADDITIVES-OF-CONCERN BOLD—CONCERN TO ALL LIGHT—CONCERN TO SOME
ORE IDA COUNTRY STYLE DINNER FRIES; GOLDEN CRINKLES; SHOESTRINGS; SHREDDED HASH BROWNS; SOUTHERN STYLE HASH BROWNS	NO ADDITIVES OF CONCERN
CRISPERS	**BHA**
TATER TOTS	MONOSODIUM GLUTAMATE
SUPERMARKET PRIVATE BRANDS	
A & P—A & P CRINKLE CUT COTTAGE FRIED; FRENCH FRIED; SHOESTRING	NO ADDITIVES OF CONCERN
POTATO MORSELS	**MODIFIED FOOD STARCH**
FIRST NATIONAL —EDWARDS FRENCH FRIED	NO ADDITIVES OF CONCERN
LUCKY—LADY LEE CRINKLE CUT FRENCH FRIED; FRENCH FRIED; HASH BROWN	NO ADDITIVES OF CONCERN

BRANDS	ADDITIVES-OF-CONCERN BOLD—CONCERN TO ALL LIGHT—CONCERN TO SOME
SAFEWAY—BEL-AIR	
FRENCH FRIED; HASH BROWNS; POTATOES O'BRIEN; SHOESTRING; SOUTHERN STYLE HASH BROWN	NO ADDITIVES OF CONCERN
TATER TREATS	**MODIFIED FOOD STARCH;** MONOSODIUM GLUTAMATE
STOP & SHOP —STOP & SHOP	
FRENCH FRIED; OVEN FRIES CRINKLE CUT; SHOESTRING; TATERS POTATO PUFFS	NO ADDITIVES OF CONCERN

FROZEN VEGETABLES (SINGLE, COMBINATIONS) & RICE MIXTURES

By law, at this writing, it is not required of a manufacturer to state whether cheeses contain artificial colors, unless the color is FD & C Yellow No. 5 (Tartrazine). Their presence in frozen vegetables and rice mixtures containing cheese therefore remains an uncertainty, unless it is voluntarily declared on the label that artificial colors have been used or have not. For a more detailed explanation of artificial color in cheese, refer to the beginning of the cheese section of "Dairy Products & Substitutes."

BRANDS	ADDITIVES-OF-CONCERN
BIRDS EYE	
ARTICHOKE HEARTS; BABY BROCCOLI SPEARS; BABY BRUSSELS SPROUTS; CHOPPED COLLARD GREENS; CUT CORN;	NO ADDITIVES OF CONCERN

(continues)

BRANDS	ADDITIVES-OF-CONCERN **BOLD—CONCERN TO ALL** LIGHT—CONCERN TO SOME
CORN ON THE COB; FORDHOOK LIMA BEANS; CHOPPED MUSTARD GREENS; CUT OKRA; WHOLE ONIONS; TENDER TINY PEAS; COOKED SQUASH; SUMMER SQUASH	NO ADDITIVES OF CONCERN
—*AMERICANA RECIPE*	
NEW ENGLAND STYLE VEGETABLES	**ARTIFICIAL COLOR; BHA; MODIFIED CORN STARCH;** MONOSODIUM GLUTAMATE
SAN FRANCISCO STYLE VEGETABLES	**ARTIFICIAL COLOR; MODIFIED CORN STARCH**
WISCONSIN COUNTRY STYLE VEGETABLES	**ARTIFICIAL COLOR; BHA;** BUTTERMILK SOLIDS; **MODIFIED CORN STARCH;** MONOSODIUM GLUTAMATE; WHEY
—*COMBINATIONS*	
BROCCOLI W/ CHEESE SAUCE; CAULIFLOWER W/ CHEESE SAUCE	**ARTIFICIAL COLOR; BHA; CARRAGEENAN;** MONOSODIUM GLUTAMATE; WHEY
FRENCH GREEN BEANS W/ TOASTED ALMONDS	**BHA**
FRENCH GREEN BEANS W/ SLICED MUSHROOMS	NO ADDITIVES OF CONCERN

BRANDS	ADDITIVES-OF-CONCERN **BOLD—CONCERN TO ALL** LIGHT—CONCERN TO SOME
BIRDS EYE—*COMBINATIONS (Continued)*	
GREEN PEAS W/ SLICED MUSHROOMS; GREEN PEAS & PEARL ONIONS	**MODIFIED CORN STARCH**
SMALL ONIONS W/ CREAM SAUCE; CREAMED SPINACH DOUBLE CHOPPED; MIXED VEGETABLES W/ ONION SAUCE	**ARTIFICIAL COLOR; MODIFIED CORN STARCH;** WHEY
—5 MINUTE	
ASPARAGUS SPEARS; BROCCOLI SPEARS; CAULIFLOWER; CORN; CUT GREEN BEANS; FRENCH GREEN BEANS; ITALIAN GREEN BEANS; BABY LIMA BEANS; PEAS & CARROTS; CHOPPED SPINACH; SUCCOTASH; CUT WAX BEANS; ZUCCHINI SQUASH	NO ADDITIVES OF CONCERN
—INTERNATIONAL RECIPES	
BAVARIAN STYLE BEANS & SPAETZLE; CHINESE STYLE VEGETABLES	**MODIFIED CORN STARCH;** MONOSODIUM GLUTAMATE
DANISH STYLE VEGETABLES	**BHA; MODIFIED CORN STARCH;** MONOSODIUM GLUTAMATE

BRANDS	ADDITIVES-OF-CONCERN **BOLD—CONCERN TO ALL** LIGHT—CONCERN TO SOME
HAWAIIAN STYLE VEGETABLES; ITALIAN STYLE VEGETABLES; PARISIAN STYLE VEGETABLES	**MODIFIED CORN STARCH**
—INTERNATIONAL RICE RECIPES	
FRENCH STYLE RICE; ITALIAN STYLE RICE	MONOSODIUM GLUTAMATE
SPANISH STYLE RICE	MONOSODIUM GLUTAMATE; **PAPRIKA**
—STIR-FRY VEGETABLES	
CHINESE STYLE; JAPANESE STYLE; MANDARIN STYLE	NO ADDITIVES OF CONCERN
GREEN GIANT	
CREAM STYLE CORN	**MODIFIED CORN STARCH**
NIBBLERS; NIBLET EARS CORN-ON-THE-COB; LITTLE BABY EARLY PEAS; SWEET PEAS	NO ADDITIVES OF CONCERN
—BAKE 'N SERVE	
CAULIFLOWER IN CHEESE SAUCE; CUT BROCCOLI IN CHEESE SAUCE	**ARTIFICIAL COLOR; MODIFIED CORN STARCH**

BRANDS	ADDITIVES-OF-CONCERN **BOLD—CONCERN TO ALL** LIGHT—CONCERN TO SOME
GREEN GIANT *(Continued)*	
—IN BUTTER SAUCE	
BABY BRUSSELS SPROUTS; CRINKLE CUT CARROTS; NIBLETS CORN; BABY LIMA BEANS; SWEET PEAS; CUT LEAF SPINACH; YOUNG TENDER MIXED VEGETABLES	**AA BUTTER; MODIFIED CORN STARCH**
CUT GREEN BEANS; FRENCH STYLE GREEN BEANS	**AA BUTTER;** MONOSODIUM GLUTAMATE
SLICED POTATOES	**BUTTER; MODIFIED CORN STARCH; TURMERIC**
—IN A FLAVORED CHEESE SAUCE	
CUT BROCCOLI; CAULIFLOWER	**ARTIFICIAL COLORS;** BUTTERMILK SOLIDS; **MODIFIED CORN STARCH;** MONOSODIUM GLUTAMATE; SODIUM ALGINATE; WHEY SOLIDS (MODIFIED)
SMALL ONIONS	**ARTIFICIAL COLOR; MODIFIED CORN STARCH;** MONOSODIUM GLUTAMATE
—LE SUEUR	
BABY EARLY PEAS	**AA BUTTER; MODIFIED CORN STARCH**
BABY PEAS PEARL ONIONS & CARROTS; BABY PEAS PEA PODS & WATER CHESTNUTS	**BUTTER; MODIFIED CORN STARCH; TURMERIC**

BRANDS	ADDITIVES-OF-CONCERN **BOLD—CONCERN TO ALL** LIGHT—CONCERN TO SOME
—ORIENTAL COMBINATIONS CHINESE VEGETABLES; JAPANESE VEGETABLES	**ARTIFICIAL COLOR; MODIFIED CORN STARCH; TURMERIC**
HAWAIIAN VEGETABLES	**MODIFIED CORN STARCH; TURMERIC**
—RICE ORIGINALS MEDLEY RICE W/ PEAS & MUSHROOMS; PILAF RICE W/ MUSHROOMS & ONIONS	**FERRIC ORTHOPHOSPHATE; MODIFIED CORN STARCH; TURMERIC**
WHITE & WILD LONG GRAIN RICE	**FERRIC ORTHOPHOSPHATE; MODIFIED CORN STARCH;** MONOSODIUM GLUTAMATE
MRS. PAUL'S CANDIED SWEET POTATOES; FRIED ONION RINGS	NO ADDITIVES OF CONCERN
OH BOY! STUFFED POTATOES W/ NATURAL CHEDDAR CHEESE	**PAPRIKA**

BRANDS	ADDITIVES-OF-CONCERN BOLD—CONCERN TO ALL LIGHT—CONCERN TO SOME
ORE IDA CHOPPED ONIONS	NO ADDITIVES OF CONCERN
ONION RINGERS	**MODIFIED FOOD STARCH; OLEORESIN OF PAPRIKA**
PENOBSCOT *BAKED STUFFED POTATOES* CHEESE FLAVOR	**ARTIFICIAL COLOR;** WHEY (DRIED SWEET)
SOUR CREAM & CHIVES FLAVOR	MILK SOLIDS (CULTURED NONFAT DRY); SOUR CREAM SOLIDS
STOUFFER'S *SIDE DISH* BROCCOLI AU GRATIN	NO ADDITIVES OF CONCERN
CORN SOUFFLE; SPINACH SOUFFLE	MONOSODIUM GLUTAMATE
GREEN BEAN MUSHROOM CASSEROLE	**ARTIFICIAL COLOR; BHA; MODIFIED CORN STARCH;** MONOSODIUM GLUTAMATE; **TURMERIC**
POTATOES AU GRATIN	**MODIFIED CORN STARCH;** MONOSODIUM GLUTAMATE
SCALLOPED POTATOES	**MODIFIED CORN STARCH; U.S. CERTIFIED FOOD COLORING**

BRANDS	ADDITIVES-OF-CONCERN BOLD—CONCERN TO ALL LIGHT—CONCERN TO SOME
SUPERMARKET PRIVATE BRANDS A & P —A & P ASPARAGUS SPEARS; BROCCOLI SPEARS; BRUSSELS SPROUTS; WHOLE BABY CARROTS; CAULIFLOWER; CORN ON THE COB; CUT GREEN BEANS; BABY GREEN LIMA BEANS; FORDHOOK LIMA BEANS; MIXED VEGETABLES; PEAS & CARROTS; CHOPPED SPINACH; LEAF SPINACH; STEW VEGETABLES; SWEET PEAS	NO ADDITIVES OF CONCERN
LUCKY —HARVEST DAY BLACKEYE PEAS; BROCCOLI CUTS; BRUSSELS SPROUTS; BABY WHOLE CARROTS; CORN ON THE COB; FRENCH CUT GREEN BEANS; GREEN PEAS; MIXED VEGETABLES; CHOPPED MUSTARD GREENS; CUT OKRA; PEAS & CARROTS	NO ADDITIVES OF CONCERN

BRANDS	ADDITIVES-OF-CONCERN **BOLD—CONCERN TO ALL** LIGHT—CONCERN TO SOME
—LADY LEE	
BROCCOLI SPEARS; CAULIFLOWER; CORN ON THE COB; CUT CORN; CUT GREEN BEANS; BABY LIMA BEANS; MIXED VEGETABLES; CHOPPED SPINACH; LEAF SPINACH	NO ADDITIVES OF CONCERN
SAFEWAY **—BEL-AIR**	
3 BEANS PLUS; BLACKEYE PEAS; COLLARD GREENS; CORN MONTEREY; COUNTRY STYLE VEGETABLES; DICED GREEN BELL PEPPERS; MEXICALI CORN; CUT OKRA; COOKED SQUASH; SUCCOTASH	NO ADDITIVES OF CONCERN
BROCCOLI CUTS W/ CHEESE SAUCE	**ARTIFICIAL COLOR;** FD & C YELLOW NO. 5; **MODIFIED CORN STARCH; POTASSIUM CASEINATE;** WHEY
CHINESE STYLE VEGETABLES; FRENCH CUT GREEN BEANS W/ TOASTED ALMONDS; ITALIAN STYLE VEGETABLES	**MODIFIED FOOD STARCH**

BRANDS	**ADDITIVES-OF-CONCERN** **BOLD—CONCERN TO ALL** LIGHT—CONCERN TO SOME
FRENCH FRIED ONION RINGS	**ARTIFICIAL COLOR** W/ FD & C YELLOW NO. 5; **MODIFIED FOOD STARCH;** MONOSODIUM GLUTAMATE; **OLEORESIN PAPRIKA**
WHIPPED POTATOES W/ CHEESE	**PAPRIKA;** WHEY
WINTER MIX	**MODIFIED FOOD STARCH;** WHEY (DAIRY)
STOP & SHOP —STOP & SHOP	
BROCCOLI SPEARS; CHOPPED BROCCOLI; CAULIFLOWER; CORN ON COB; LITTLE EARS; CUT GREEN BEANS; FRENCH CUT GREEN BEANS; GREEN PEAS; CHOPPED ONIONS; PEAS & CARROTS; DICED PEPPERS; CHOPPED SPINACH; COOKED SQUASH; WAX CUT BEANS	NO ADDITIVES OF CONCERN

BRANDS	ADDITIVES-OF-CONCERN BOLD—CONCERN TO ALL LIGHT—CONCERN TO SOME
WINN DIXIE —ASTOR CHOPPED BROCCOLI; BRUSSELS SPROUTS; SPECKLED BUTTER BEANS; CAULIFLOWER; SWEET CORN; BABY LIMA BEANS; MIXED VEGETABLES; PEAS & CARROTS; CHOPPED SPINACH; SUCCOTASH; SWEET PEAS	NO ADDITIVES OF CONCERN
ITEMS IN HEALTH FOOD STORES	
HEALTH VALLEY BROCCOLI SPEARS; CHOPPED SPINACH; LEAF SPINACH; MIXED VEGETABLES	NO ADDITIVES OF CONCERN
NEWTON ACRES BROCCOLI CUTS; BROCCOLI CUTS W/ CAULIFLOWER; BRUSSELS SPROUTS; BUTTERNUT SQUASH; CAULIFLOWER FLORETS; CUT CORN; FRENCH CUT GREEN BEANS; GREEN PEAS; MIXED VEGETABLES; PEAS & CARROTS; SPINACH; STEW VEGETABLES; TURNIPS; WHOLE BABY CARROTS	NO ADDITIVES OF CONCERN

BRANDS	**ADDITIVES-OF-CONCERN** **BOLD—CONCERN TO ALL** LIGHT—CONCERN TO SOME

SEAWEED

ITEMS IN HEALTH FOOD STORES	
ATLANTIC MARICULTURE DULSE	DULSE
EREWHON	
ARAME	ARAME
HIJIKI	HIJIKI
JAPANESE NORI	NORI
KOMBU	KOMBU

THE DICTIONARY OF FOOD ADDITIVES

WHAT YOU SHOULD KNOW ABOUT A FOOD ADDITIVE

What is it?
Why is it used in food?
What do tests and medical experience
  tell us about it?*
Is it safe?

The reader will find in this part of the book the facts that address themselves to these questions, when such facts exist. Information from scientific investigations is not always as complete as one would like, or as tidy.

It has been necessary to use a selective procedure in the choice of material that is presented in the pages that follow, as there were hundreds of substances to be dealt with in this single volume. Care, however, has been exercised to report not only the results believed to be the most significant for the determination of safety, but divergencies from these findings as well.

To insure that the reader who wishes it can have access to the more detailed information summarized by the authors, identification of their main sources has been provided at the end of each report. When "NTIS PB" and a number follow the title, it means that the document, identified in this way, can be obtained from National Technical Information Service, 5285 Port Royal Road, Springfield, VA 22161.

ASSESSMENTS OF SAFETY

These safety ratings have been used in the Dictionary.

S—safe for everyone: where the additive has been adequately tested and found to be free of hazard for the consumer. This rating has been used only when the scientific evidence assures reasonable certainty of no harm.

---

*Some of the substances that are used as additives in food also are used as medicines, and their effects on humans, when available and relevant, have been reported in the dictionary section. The findings can provide invaluable information on human tolerances, but it is solely for this purpose that they have been included, not to recommend use in treatment. It is suggested, if anyone has this in mind, to consult a qualified physician.

X—unsafe for everyone: where the scientific reports clearly point to a possible health hazard or risk from the additive for *consumers in general.*

?—uncertain about safety: where the scientific data are too inconclusive or incomplete to warrant a well-informed judgment of safety.

Also combinations of these ratings, such as:

S—for some people; X for others: where the evidence indicates a possible health hazard or risk for a particular segment of consumers, especially heavy consumers of foods containing the additive.

When a combination of safety ratings occurs, the individuals to whom each rating applies are identified. For example:

Albumins: S for most people; X for those allergic to milk or eggs.

Caffeine: S for some people; X for pregnant women, nursing mothers, young children, anyone with a gastrointestinal or cardiovascular ailment.

A caution has not been included in a rating when the adverse effect of an additive applies to a comparatively few people or when it is likely to harm someone probably under medical supervision or on a restricted diet for its avoidance. The iron supplements in fortified foods could be harmful to anyone afflicted with hemochromatosis, a rare, genetically transmitted disorder. Patients suffering from serious kidney ailments should avoid foods containing substantial amounts of aluminum salts. These conditions and others have been noted in the text dealing with the additive when they arise, but are not mentioned in the ratings. Only cautions of wider application are believed to be appropriate in a volume such as this one, which is directed at a general audience.

The authors have concurred with the policy of the Select Committee on GRAS substances stated in its SCOGS reports "that reasoned judgment is expected even in instances where the available information is qualitatively or quantitatively limited"; and ratings of safety have been expressed in this volume in these circumstances in spite of omissions in the data.* Usually the omissions have been identified in the text. The Committee also recognized that on occasion "there are insufficient data upon which to draw a conclusion." With this

---

*An excellent review of the experiences of FASEB/SCOGS in food safety evaluation even when scientific information was very limited can be found in an article entitled "Evaluation of health aspects of GRAS food ingredients: lessons learned and questions unanswered," *Federation Proceedings,* volume 36, pp. 2519–62 (October, 1977).

in mind, as earlier stated, a ?—*uncertain* rating has been provided for this reason.

The reader will find *both additives-of-concern and additives that are not of concern* reviewed in this section. The purpose is to identify each additive so that the consumer will have no doubts, especially when something not previously encountered or noted appears on a food label.

THREE WAYS TO LOCATE WHAT YOU WANT TO KNOW IN THIS DICTIONARY SECTION

Table of Contents. Because it made sense in terms of the way the research has been conducted, and the considerable saving in space, whenever possible, related additives have been dealt with in groups. The various ammonium salts will be found together, as will the glutamates, the phosphates, the vegetable oils, and so on. It has not been difficult to provide ratings of safety when one or more substances in such a group is of concern and others are not. In other cases, such as brominated vegetable oil, mannitol, and others, each additive is reviewed individually. The Table of Contents which follows gives the location of the material arranged in this manner.

To locate a single additive. It may be necessary at times to refer to the Additives Index at the end of the dictionary section in order to find a specific additive. It has not always been possible to present single additives in the dictionary in alphabetical order because related additives are reviewed together. A typical example is Calcium Pantothenate, the form in which Pantothenic Acid is added to food as a nutritional supplement. It is grouped under Pantothenic Acid, one of the vitamins of the vitamin B complex.

To quickly determine whether or not an additive is of concern. You need go no further than the additive index to do this. Each additive-of-concern is identified in the manner described below. If it is of general concern, you'll find two asterisks(**) alongside the additive. If it is of concern to some people, but is viewed here as safe for others, you'll find one asterisk (*) plus a letter symbol alongside. The letter tells you who the "some" people are who are cautioned. If neither asterisks nor letters are present, the additive may be regarded as safe. Examples follow:

Brominated Vegetable Oil** (general concern)
Casein * *A (m)* (*A(m)* indicates caution for anyone allergic to milk)
Citric Acid (safe)

TABLE OF CONTENTS

DICTIONARY

ACETIC ACID

SODIUM ACETATE; SODIUM DIACETATE

Acetic acid, its salt *(sodium acetate),* and *sodium diacetate* (a combination of acetic acid and sodium acetate) are present in small amounts in most plant and animal tissues. Acetic acid is the acid in vinegar, and it is found naturally in substantial amounts in some aged cheeses and wine. It is added to catsup, mayonnaise, and pickles for acidity and taste, and it is in food products such as pickled fruits, vegetables, and meats that are preserved in vinegar. Sodium acetate is used as a preservative, while sodium diacetate is preferred by some food processors as a microbe preventative in baked goods and other products where acetic acid would impart an undesirable flavor.

SAFETY: The amount of acetic acid and the acetates as additives consumed in food in 1975 has been estimated at 80 milligrams daily per person in the U.S., or 1.3 milligrams per kilogram of body weight (kg/bw) for a person weighing 60 kg (132 lbs.). Acetic acid accounted for 93 percent of it.

Aside from what is obtained in food, the body itself produces acetic acid, which it needs as an intermediary in the transformation of ingested substances into other compounds used in the body, such as glucose (see p. 518). A constituent of the gastric juices stimulated in this process is hydrochloric acid, which is likely to remain in the intestines after the acetic acid, which is absorbed more readily. The sequence could lead to damage to mucous membranes in the gastrointestinal tract by the hydrochloric acid.

Highly concentrated acetic acid itself can produce damaging effects due to corrosive action in the digestive tract. This has been experienced by people attempting suicide, or through accidental ingestion. However, the amount thus consumed in a single dose should not be compared with the same amount consumed at lower concentrations, particularly when mixed with other ingredients in the diet over a period of time. For example, normal individuals have long consumed vinegar, which is 5.6 percent acetic acid, without reporting adverse effects.

MAJOR REFERENCE: Evaluation of the health aspects of acetic acid, sodium acetate and sodium diacetate as food ingredients. FASEB/SCOGS Report 82 (NTIS PB 274-670) 1977.

Acetic acid added to drinking water of rats at 390 milligrams per kg/bw, almost 300 times human intake of this additive, adversely affected growth; but did not when half the dosage was administered. Tests indicate that neither acetic acid nor sodium acetate is likely to cause abnormalities in fetuses or newborn, nor does acetic acid affect maternal or fetal survival. Sodium acetate does not exhibit mutagenic activity (gene mutations).

Studies do not appear to have been conducted to determine any long-term effects by these substances, or their possibility as a cause of cancer. Clinical evidence exists that acetic acid can cause allergic reactions in some individuals, which subside following its avoidance. No reports of investigations on sodium diacetate have been located, but since it breaks down in the body to acetic acid and sodium acetate, what is known of these should apply.

ASSESSMENT: Acetic acid is essential to the metabolic processes in the body that transform ingested food constituents into nourishment. In the quantities and concentrations present as additives in food, neither the acid nor the acetates pose a hazard to human health.

RATING: S.

ADIPIC ACID

Adipic acid occurs naturally as a minor component in such food products as beet juice and butter. The commercial additive is produced synthetically. Adipic acid has multiple uses: food processors use it in beverages and candies to control acidity, enhance flavor, and impart tartness; to lighten the texture of baked products; as a gelling agent in imitation jams; and as a preservative that binds chemically with, and so deactivates, metal impurities that can cause rancidity or flavor changes in edible oils, sausages, and other foods.

SAFETY: Based on surveys of usage by food processors conducted by the National Research Council in 1975 and 1976, it was estimated that the daily intake of added adipic acid in the diet of a U.S. individual averaged 41 milligrams, or about 0.7

MAJOR REFERENCE: Evaluation of the health aspects of adipic acid as a food ingredient. FASEB/SCOGS Report 80 (NTIS PB 266-279) 1977.

milligram per kilogram of body weight for a person weighing 60 kg (132 lbs.). In a study of human subjects over a nine-day period using up to 140 times this amount, there were no adverse effects. Long-term experiments with rats fed 700 to over 3500 times the daily human intake (adjusted for body weight) produced no harmful effects. Dosages of adipic acid more than 290 times the daily human intake fed by tube to pregnant rats and mice, and orally to hamsters, caused no ill effects to them or to their offspring; and laboratory tests investigating birth defects and gene mutation did not give reason for concern. Biological analyses indicated that much of the absorbed adipic acid is rapidly excreted in the urine, and the body's processes deal with it without difficulty by the same metabolic routes that are employed for fatty acids (the useful part of fat).

ASSESSMENT: The available evidence demonstrates that adipic acid does not represent a hazard to the public when it is used at levels that are current or that might reasonably be expected in foods in the future.

RATING: S.

ALBUMINS

Egg Albumin (Ovalbumin); Lactalbumin; Lactalbumin Phosphate

Albumins are important constituents in nearly all animal tissues and fluids (including blood serum). They are proteins rich in the essential amino acids which the body is unable to make and must obtain from the diet. *Egg albumin* is obtained from egg whites (which are largely albumin), and *lactalbumin* is derived from whey (the liquid part of milk; see p. 677), of which albumin is a major protein. These are the protein foods that contribute to the excellent nutritive value of eggs and milk.

Lactalbumin phosphate is prepared by adding polyphosphates (see p. 600) to lactalbumin protein. The resulting complex is useful as a partial replacement for milk solids in baked goods, in some gelatin products, and in imitation dairy products. The albumins are used as food additives in diet supplements, and as stabilizers, thickeners, and texturizers in baked goods, breakfast cereals, meat products, candies, fruit drinks, frostings, and sweet sauces.

SAFETY: The albumins used as food additives in 1976 accounted for a dietary intake of less than 24 milligrams per person daily. This is a minuscule contribution (1/3000) to the recommended protein intake of 44 to 56 grams required to meet the protein needs of most people. However, egg or milk protein is the standard for protein quality because of a nearly ideal amino acid makeup. When one of the albumins provides the dietary protein, an optimum growth response or utilization is shown.

Very high protein intakes are considered safe; the amounts in excess of the body's direct needs for protein and their essential amino acids will serve as a source of energy. Some investigators have recommended as much as 260 grams of protein a day under some conditions of physiological stress. Human subjects have consumed 600 grams of protein daily for many weeks without adverse effects other than increased excretion of calcium.

Some people have shown allergic reactions to milk or, to a lesser extent, eggs. The albumin fractions of milk and eggs can cause strong reactions in those allergic to these foods. Children are more likely to be susceptible; the symptoms may resemble hay fever, and include headaches and skin rash.

Two potential hazards that are known for excessively high protein intake are shared by albumins. One relates to the demand placed on the kidney to excrete the nitrogenous products of protein metabolism. This requires considerable water to keep the urea, ammonia, and other end products in solution—thus, an increased volume of urine. The other possible drawback of a very high protein diet is the greatly increased urinary calcium excretion that accompanies it. Calcium loss from the body, even when the diet provided well over twice the recommended calcium intake, has been noted by different investigators when dietary protein exceeded 142 grams daily.

ASSESSMENT: Albumins are excellent and nutritious proteins. They are used as food additives to serve as dietary supplements as well as aids in texture of the food. They add only minimally to the dietary intake of protein, and pose no hazard to consumers who are not allergic to milk or eggs.

RATING: S for most people; X for those allergic to milk or eggs.

ALGAE (SEAWEED) AND EXTRACTIVES

Algae constitute a group of plants that include seaweed and many single-cell marine and freshwater plants. Many species have been used for livestock and human food since before the Christian era. In the Orient, seaweeds are accepted foods and sometimes account for as much as 25 percent of the diet. Various gums that are used as additives in food processing in the U.S. are extracted from species of algae.

They include agar-agar, the alginates, carrageenan and furcelleran, and dulse and kelp. These are evaluated on the following pages.

ALGAE: AGAR-AGAR

Agaroid

Agar-agar is a polysaccharide, a complex carbohydrate consisting of a chemical combination of simple sugars, in this case mainly galactose. It is extracted from several varieties of red algae (seaweed). Agar-agar has the capacity to swell and can form resilient gels; it is used by the food industry as a gelling agent. *Agaroid,* a derivative of agar, serves similar purposes.

SAFETY: Agar-agar has been used for years as a gelling and bulking agent in the diets of experimental animals for various types of feeding studies of other substances. These have included investigations of cancer and gene mutation, in which agar-agar was fed in amounts up to 68,000 times (adjusted for body weight) as great as the 2.6 milligrams estimated to be ingested daily by a U.S. human in 1975, without evidence that it causes adverse effects. It has been used as a laxative for many years in dosages as much as 5000 times as great as that contained in the daily diet. When fed to rats as 30 percent of their diet for 44 weeks, many thousands of times human consumption, agar-agar caused an increase in the weight of the intestine, presumably due to its low digestibility and gelling properties, but the condition did not prevent the normal absorption of nutrients.

One investigation revealed that agar-agar fed to pregnant mice and rabbits at levels 35,000 and 9000 times respectively as great as daily human usage (adjusted for body weight) caused a significant increase in maternal deaths and a decrease in births

---

MAJOR REFERENCE: Evaluation of the health aspects of agar-agar as a food ingredient. FASEB/SCOGS Report 23 (NTIS PB 265-502) 1974.

by the survivors; and the offspring of mice were retarded in maturation. This did not occur with rats or hamsters, nor did it with mice and rabbits when the dosage was lowered respectively to 7400 and 2600 times human consumption. Taking into consideration the high feeding levels used, one explanation suggested for this effect is that harmful concentrations of certain metals, such as mercury, may be accumulated in agar-agar if the algae used in its manufacture are harvested in waters that are contaminated.

ASSESSMENT: Agar-agar has little effect when added to the diets of laboratory animals, except during pregnancy; and the toxic effects observed in studies of pregnant mice and rabbits only appeared when agar-agar was administered at levels many thousands of times that of human consumption. It has been used safely by man as a laxative in amounts substantially greater than contained in food.

RATING: S.

ALGAE: ALGINATES

Algin; Algin Derivative; Algin Gum; Alginic Acid; Ammonium, Calcium, Potassium, Sodium Alginate; Propylene Glycol Alginate

The alginates are extracted from several species of red and brown algae (seaweed). They are used in dressings, sauces, and sweets to blend ingredients and to prevent their separation, and as a gel and thickener. *Propylene glycol alginate* accounts for half of the total amount of alginates used in food, and *ammonium* and *sodium alginate* for most of the remainder. *Propylene glycol alginate* is preferred as a thickener in foods high in acidity.

SAFETY: A 1976 survey of the use of alginates by the food industry conducted by the National Academy of Sciences-National Research Council indicated an average daily intake by the U.S. consumer of 21 milligrams. Sodium alginate and propylene glycol alginate administered orally to a variety of animal species for periods ranging from several weeks to a

---

MAJOR REFERENCE: Evaluation of the health aspects of alginates as food ingredients. FASEB/SCOGS Report 24 (NTIS PB 265-503) 1974.

year, in amounts greatly exceeding normal human intake (adjusted for body weight) revealed no adverse effects. Six human subjects were given 8 grams of sodium alginate a day for 7 days, approximately 380 times the total amount of alginates currently present in the daily diet, without causing harm; 4 others who received 45 grams of alginic acid a day for the same period experienced only a mild laxative effect.

However, when fed by tube to pregnant mice in dosages 3250 times as great as that in the human diet (adjusted for body weight), propylene glycol alginate caused a significant number of maternal and fetal deaths. The same investigators using identical test procedures determined that propylene glycol by itself did not cause this, an indication that the alginate in the compound most likely was responsible. Another study, this time with pregnant rats and hamsters employing dosages of propylene glycol alginate equivalent to over 2900 times human consumption, determined that it did not cause birth defects.

Investigations have not been carried out with ammonium, calcium, potassium, or sodium alginate to determine whether they can cause defective offspring, gene mutation, or cancer. Only one cancer study of an alginate has been conducted; several injections of alginic acid in mice did not cause cancer.

ASSESSMENT: In sufficiently large amounts, the alginates, as is true of related polysaccharides (carbohydrates containing a union of several simple sugars), may pose some hazard during pregnancy. Two studies of human subjects provide some assurance that immediate adverse effects of any consequence are unlikely; but because of the almost total absence of investigation of the possibility of long-term effects, uncertainties remain unresolved. The evidence at hand does not indicate that the alginates as additives in food are a health hazard, except during pregnancy.

RATING: S for most people; ? for pregnant women.

ALGAE: CARRAGEENAN (IRISH MOSS)

Ammonium Carrageenan; Calcium Carrageenan; Potassium Carrageenan; Sodium Carrageenan

FURCELLERAN

Carrageenan, also called Irish moss, is extracted from a variety of red marine algae. It is composed of ammonium, calcium, potassium, or sodium salts (or combinations of these), and perhaps other salts, and a sulfur-containing polysaccharide (a condensation of a number of simple sugars). As a food additive, it may be identified as the ammonium, or calcium, or potassium, or sodium salt of carrageenan. Two types exist, but only one, undegraded carrageenan, is permitted in food. This is used as an emulsifier (blender), and as a stabilizer that keeps mixtures from separating, and as a gelling agent and thickener. Carrageenan is particularly useful with milk protein, enabling suspension of cocoa or chocolate in milk without the occurrence of settling or thickening.

Furcelleran, which is derived from another variety of red seaweed, is used for similar purposes as an additive in foods.

SAFETY: A survey conducted by the National Academy of Sciences-National Research Council among food processors in 1976, based on their usage, indicated that the daily intake of carrageenan in the diet averaged 15 milligrams per person, or 0.25 milligram per kilogram of body weight (kg/bw) for an individual weighing 60 kilograms (132 pounds). Far less use is made of furcelleran; the daily diet on the average contains 0.05 milligram.

Pregnant mice and rats fed calcium or sodium carrageenan orally or by tube at levels of 600-900 milligrams per kg/bw, somewhat over 2000 times average human consumption (adjusted for body weight), suffered a decrease in the number of live births because of dissolution or death of fetuses; and some of the newborn had immature or retarded skeletal structures or were abnormal. These effects were not observed in hamsters and rabbits.

No adequate feeding studies covering more than half of the life span of an animal species have been conducted, and these are needed to determine long-term consequences. Some information is available from laboratory tests in which calcium carrageenan caused aberrant chromosomes in rat bone marrow. Evidence has been produced that carrageenan can inhibit complement (a crucial catalyst in the body's immunological warning system) and increase the permeability of blood vessels; this could be of significance if a sufficient quantity of carrageenan were absorbed during infectious illness or a dis-

---

MAJOR REFERENCE: Evaluation of the health aspects of carrageenan as a food ingredient. FASEB/SCOGS Report 6 (NTIS PB 266-877) 1973.

turbance of the body's metabolic processes. Generally, however, carrageenan is not absorbed by the body.

Furcelleran has not been tested, but, according to the UN Joint FAO/WHO Expert Committee on Food Additives, it is so similar in chemical structure to carrageenan that the biological data available for carrageenan may be taken to apply to furcelleran.

ASSESSMENT: Based on some of the questions raised from animal experiments and laboratory tests, consumption of carrageenan (and furcelleran, which has a similar chemical structure) may best be avoided during pregnancy until additional studies are conducted with it and other polysaccharides which have had adverse effects on pregnant animals. It should be avoided, on the same basis, during infectious illness or disturbances of the body's metabolic processes. Laboratory tests indicate that carrageenan may cause chromosomal aberrations, and emphasize the desirability of long-term animal experiments to investigate cancer and gene mutation, which are lacking. The UN Joint FAO/WHO Expert Committee on Food Additives apparently does not share these reservations, as it regards up to 500 milligrams per kg/bw of carrageenan as an acceptable level in the daily diet.

RATING (carrageenan and furcelleran): ?; additional studies are needed to resolve uncertainties.

ALGAE: DULSE (RED ALGAE)

Nori

KELP (BROWN ALGAE)

Arame; Hijiki; Kombu; Wakame

Substances derived from red algae (species *Porphyra* and *Rhodymenia palmata*), such as *dulse,* consist mainly of galactose, a simple sugar in collodial form (gelatinlike); while two other sugars, guluronic and mannuronic, are the principal com-

---

MAJOR REFERENCE: Evaluation of the health aspects of certain red and brown algae as food ingredients. FASEB/SCOGS Report 38 (NTIS PB 265-505) 1974.

ponents in collodial compounds that originate from brown algae (species *Laminaria* and *Nereocystis*), such as *kelp*. Some *nori* are *Porphyra* and therefore dulse. *Arame, hijiki, kombu,* and *wakame* are regarded as kelp, or as closely related. As additives, algae may be used as seasonings or flavorings in food, and they provide iodine when used in dietary foods.

SAFETY: Information is not available on the amounts of red and brown algae, dulse and kelp, as such or their products, which are consumed daily by humans in the U.S. In the past this is likely to have been small, for these have been restricted as food ingredients to use as spices, seasonings, and flavorings. Now, however, some people appear to be consuming seaweed as a vegetable, as is done in the Orient where it can amount to a substantial portion of the total diet.

Most of the studies of these substances, other than those used for animals, have been conducted with *Laminaria*. Oral administration has been conducted mainly with ruminants (cud-chewing animals such as sheep, cows, and horses) as supplements in their food, with rats, pigs, and poultry occasionally included. Massive amounts of algae meal or dried seaweed from several species of red and brown algae, at times up to 20 percent of the diet, have not affected growth or productivity of these animals, lactation, or their ability to utilize their food; but caution is advised against quantities much above 15 percent because of iodine and mineral salt content.

A test of two algae ingested daily in the diet of many Japanese determined that 7 to 16 grams (containing 0.3 percent of iodine) of a species of kelp reduced normal thyroid function. But 1.2 to 3.2 grams (0.03 percent iodine content) of a species of dulse did not affect thyroid function. These results can be explained on the basis of the difference in the iodine content of the dosages.

It has been shown that mercury is present in several species of algae, probably due to water pollution. Tests to determine whether dulse or kelp can cause cancer, gene mutations, or abnormal or defective infants, or mortality during pregnancy have not been conducted.

ASSESSMENT: Data are not available on the consumption of dulse and kelp in the U.S. diet; but considering the limitations imposed on their usage as additives, only minute amounts are likely to be present in food ingested by humans in this country.

But the possibility exists that substantial amounts may be present as a food in the diet of some individuals who regard seaweed as a vegetable. Viewed solely as an additive, which is the province of this volume, the finding that mercury may be present in algae would only be of concern, as would the adverse effects on thyroid, should these substances be present at much higher levels as food additives than they appear to be today. On the other hand, the lack of research in several areas, especially during pregnancy where other gums derived from seaweeds such as carrageenan and the alginates have proved questionable, leaves uncertainties that need resolution.

RATING: S for most people; ? during pregnancy.

ALUMINUM COMPOUNDS

Aluminum Ammonium Sulfate; Aluminum Chloride; Aluminum Hydroxide; Aluminum Oleate; Aluminum Palmitate; Aluminum Potassium Sulfate (Alum); Aluminum Sodium Sulfate; Aluminum Sulfate; Sodium Aluminate; Sodium Aluminosilicate; Sodium Aluminum Phosphate; Sodium Calcium Aluminosilicate; Sodium Phosphoaluminate

Aluminum, the most abundant metallic element in the earth's crust, is found naturally in varying amounts in nearly all food and water. Aluminum compounds are added to foods to adjust acidity, to make foods light in texture, to keep processed fruits and vegetables firm, to aid in congealing cheese, and for other purposes. *Sodium aluminum phosphate* constitutes about 90 percent of the aluminum compounds that are used in food; it is a common ingredient in baking powders and self-rising flours.

SAFETY: The daily intake of aluminum as additives in food in 1976, based on a survey conducted by the National Academy of Sciences-National Research Council of usage by the food industry, has been estimated at less than 6 milligrams daily per person in the U.S. Another estimate, which includes aluminum in water, from cooking vessels, etc., suggests that intake can be as high as 100 milligrams daily.

Experiments with a number of animal species fed aluminum compounds greatly in excess of this latter estimate (adjusted

MAJOR REFERENCE: Evaluation of the health aspects of aluminum compounds as food ingredients. FASEB/SCOGS Report 43 (NTIS PB 262-655) 1976.

for body weight) indicate that an interaction can occur between aluminum and phosphorus which interferes with the enzyme that incorporates phosphate in the body's metabolic processes. This may cause a decrease in the retention of phosphorus and result in a disturbance of normal bone formation, kidney damage, and interference with the storage of carbohydrates. In one study, a single dose equivalent to 400 times the average amount of aluminum humans may ingest each day in food and water caused a significant decrease in retention of phosphorus in rats.

A study conducted with 11 patients fed relatively small doses of aluminum-containing antacid medicine (but still 170 to 560 times as great as the aluminum intake from food) confirmed the inhibition of phosphorus absorption in the intestine, followed by an increase in calcium loss; and other studies with human subjects bear out this depletion. In some animal experiments, the adverse effects were reduced or controlled by an increase of phosphorus in the diet. The high intake of phosphorus in the American diet may provide a protective effect, especially for someone who consumes large quantities of antacid preparations. But this cannot be assured for specific individuals at all times, particularly for those with kidney disease, who must exercise caution in consuming food containing high levels of aluminum salts which can aggravate their condition.

Evidence from research indicates that these salts do not cause cancer, or birth deformities, or gene mutations.

ASSESSMENT: Aluminum compounds in amounts greatly in excess of human intake have been proven to be hazardous in experiments with animals because they inhibit the retention of phosphorus. Individuals suffering from kidney disease are particularly vulnerable (but they are likely to be under diet control); and anyone regularly taking antacid medication containing aluminum (such as aluminum hydroxide gel) should make certain of sufficient phosphorus in the diet to counteract the effect of the aluminum that is in it.* While the margin of safety between the intake of aluminum compounds by some individuals and the amount that causes adverse effects in animals is narrow, the American diet is sufficiently plentiful in phosphorus to provide protection from aluminum compounds as used in processed foods.

RATING: S.

AMINO ACIDS

Cysteine (Cysteine Hydrochloride); Glycine (Aminoacetic Acid); Lysine (L-Lysine Hydrochloride); Methionine

All proteins are made up of chemical combinations of simpler nitrogen-containing amino acids, which provide the materials needed by the body for replacement and repair of tissue. Several of these amino acids are used in foods, frequently to improve the nutritive value but also for possible contributions to flavor or preservation of the food product. Four will be considered here: cysteine (usually available as the monohydrochloride, which has a molecule of hydrochloric acid attached to the amino acid), glycine, lysine (also usually available commercially as the monohydrochloride), and methionine. (Another amino acid contributing to food flavors, monosodium glutamate, is discussed on page 555).

Cysteine, a sulfur-containing amino acid, is usually present in food proteins as cystine (two molecules of cysteine linked together); cysteine is made by splitting cystine obtained from hair or wool. It is useful as a flavoring ingredient and can improve the effectiveness of certain dough strengtheners or conditioners. *Glycine,* also called aminoacetic acid, is prepared commercially by chemical synthesis. It is an important constituent of the proteins in most foods; it is unusually high in gelatin since it is a major building block in collagen, the connective tissue in the body. This amino acid can improve flavor in beverages that contain saccharin as a sweetener, by masking its aftertaste.

Lysine is usually made commercially by a bacterial fermentation. Most meat and dairy products are rich in lysine, as are legumes such as beans, peas, and soybeans. Lysine is one of the nutritionally essential amino acids; it must be supplied from the diet because the body cannot itself produce it, as it can some amino acids. When added to cereals, which do not contain a sufficient quantity of it, lysine can make the protein more useful in the body's metabolism. *Methionine* is present in eggs in substantial amounts, and in fish and most other animal products. The commercial product, made by chemical synthesis, is also a sulfur-containing, nutritionally essential amino

---

MAJOR REFERENCE: Scientific literature review of amino acids in flavor usage, volume I. FEMA Report to FDA (NTIS PB 265-526) 1977.

acid. As the proteins in legumes such as soybeans tend to be low in methionine, the nutritional value of soy-based infant formulas and other foods can be significantly increased by being fortified with it.

SAFETY: All these amino acids have limited usage as food additives, and all are present as natural constituents and in appreciable amounts in the protein in U.S. diets; in 1972 the average diet per person provided some 101 grams of protein daily, or over 1600 milligrams per kilogram of body weight (kg/bw) for an individual weighing 60 kg (132 lbs). Much of it is from milk and meat, which are rich sources of all these amino acids. The total use of these four amino acid additives by the food industry in 1976 was such that together they provided less than 1 milligram per kg/bw daily to the average diet.

Some of the amino acids needed in the tissue can be synthesized from protein components during the metabolic processes which the human body carries out. A few, including methionine and lysine, cannot be made by humans and thus must be provided by the diet. The National Academy of Sciences-National Research Council has recommended a minimum intake for adults of 12 milligrams per kg/bw per day of lysine. At least 10 milligrams per kg/bw of methionine is needed for adults, and an infant will require up to 49 milligrams per kg/bw daily.

Cysteine has been fed to rats through four generations at a concentration of 0.3 percent of the total diet, or 100-150 milligrams per kg/bw. No effects were seen on litter size, weight, or organ weight, or on examination at autopsy. Generally, as much as 1.5 percent cysteine in the diet has not affected growth of rats, but higher levels did. Cysteine and methionine are closely related sulfur-containing amino acids, and it is known that cysteine can supplement or spare methionine needs when the diet is very low in methionine.

Glycine, aminoacetic acid, has been administered to human adults at daily doses of up to 40 grams and for as long as six months; no adverse effects were seen.

Lysine has been extensively studied to determine its requirement. In the early studies with human adults, an allowance of 1.6 grams per day was suggested; the growing child requires substantially more lysine than the adult.

Methionine is the one amino acid for which a particular form or structure (isomer) is specified when added to an infant food. L-methionine, the form in which the amino acid occurs in nature, is well metabolized, and the D form (also present in DL-methionine) seems to be equally effective for adults. However, the very young infant does not have the full complement of enzyme catalysts to restructure and metabolize some nutrients as effectively as when he is older. One study with infants attempted to establish how much methionine could be safely added to a formula already providing 360 milligrams of methionine. When 180 milligrams of methionine was added daily, growth was superior; but with 360 milligrams additional (a total in the formula of 720 milligrams, or probably about 120 milligrams per kg/bw) some growth depression and extra urine output resulted. Methionine has been used medically in treatment of liver disease, the usual dosage being 3 to 6 grams daily.

ASSESSMENT: Cysteine and glycine are protein constituents that have found uses as flavor additives in certain foods. Two additional amino acids, lysine and methionine, are necessary in the diet because the human body cannot synthesize them. They are used to fortify certain foods deficient in one or the other, thus improving the nutritional quality of the protein. Since they are normal food components, the body regularly absorbs and utilizes each of these amino acids. The small amounts of cysteine, glycine, lysine, and methionine added to processed foods hardly affect the usual dietary intake of these substances, and are far below any possible hazardous levels.

RATING: S.

AMMONIUM SALTS

Ammonium Bicarbonate; Ammonium Carbonate; Ammonium Chloride; Ammonium Hydroxide; Ammonium Isovalerate; Ammonium Phosphate (Mono- and Dibasic); Ammonium Sulfate; Ammonium Sulfide

Ammonia and several ammonium salts (ammonia combined with certain acids) are present naturally in living substances. In man, ammonia plays a vital role in the functioning of a variety of essential processes, including the kidneys and the urinary system, where it participates in the acid-alkali balance. Ammonium salts are added to foods to lighten texture; to help

---

MAJOR REFERENCE: Evaluation of the health aspects of certain ammonium salts as food ingredients. FASEB/SCOGS Report 34 (NTIS PB 254-532) 1974.

achieve greater uniformity and palatability as they increase dryness and extensibility, enabling a better mixture despite variations in the raw food and processing time; and to adjust acidity. A few of the salts are used in small amounts as flavor enhancers.

SAFETY: It has been estimated that, in 1975, 100 milligrams of ammonium salts as additives were contained in the daily diet of U.S. individuals, or 1.7 milligrams per kilogram of body weight (kg/bw) for a person weighing 60 kg (132 lbs.).

Very few animal experiments have been conducted expressly to assess the safety of ammonium salts as ingredients when mixed in food. Some information is available from studies directed at determining their effect on certain vital body processes, and most of these have employed ammonium chloride in pure form or in water rather than in the diet. Investigators using dosages of this compound over 400 times the amount of ammonium salts consumed as additives by humans (adjusted for body weight) have reported kidney damage in animals, probably due to an overaccumulation of acid from the chloride portion. Two studies, one with rats, the other with mice, demonstrated that ammonium chloride inhibited cancer; in another experiment, precancerous changes were observed in rat stomachs. Female rabbits fed ammonium carbonate, chloride, hydroxide, or sulfate did not develop tumors. The weight of evidence appears to indicate that the ammonium salts are not cancer-causing. Studies have not been found that dealt with gene mutation or birth defects.

The healthy liver in normal circumstances readily prevents concentrations of ammonia from rising to harmful levels, but people with substantial liver impairment are not adequately safeguarded and can become seriously ill from it. Ammonium chloride is used medically to correct alkalosis (insufficient acid) and to increase urine; however, humans, some of them patients, fed doses of this compound ranging from 6 to 260 times as great as the amount of ammonium salts consumed in the daily diet experienced a number of adverse effects, such as disturbance of menses, headaches, loss of energy, and acidosis (accumulation of acid). On the other hand, when administered within this dosage range to patients with rheumatoid arthritis, ammonium chloride resulted in a decrease in swelling of joints, relief from pain, and increased mobility.

ASSESSMENT: Research to assess the limits of safety for ammonium salts when present as additives in food has not been adequate. However, findings from animal and human experiments investigating their effect on vital body processes indicate that it is unlikely that the ammonium salts constitute a health hazard, even if the quantity consumed were considerably greater than is present in the daily diet of humans, except to someone seriously ill with a liver impairment. The body requires ammonia for certain essential functions, and the normal liver acts as a safeguard by preventing concentrations of ammonia from rising to an excessive level in the blood, which could be dangerous. In assessing the safety of ammonium salts, a caution for anyone ill with serious liver impairment has not been added, as diets of these people are likely to be under strict control.

RATING: S.

ANIMAL FATS

Beef Fat (Tallow); Butter (Butter Fat); Lard (Pork Fat); Marine (Fish) Oil; Mutton; Poultry Fat, Skin

STEARIC ACID; Calcium Stearate

Beef fat and *tallow* are animal fats usually obtained from cattle, and *lard* is fat rendered from swine. These fats can be converted or hardened by the chemical addition of hydrogen for use in shortening, margarine, and for other purposes in food, and are identified as hydrogenated or partially so, depending on the degree desired. Hydrogenation converts fats to a more solid form, which is more stable and less subject to rancidity and other flavor changes. But it increases the saturation of the fat; the more hydrogenated, the greater the saturation. Hydrogenated lard is also used as a mastic for chewing gum, and lard serves as a component of adhesives used with food packaging materials. *Butter* is used in processed foods as a shortening, and to provide a distinctive flavor. Occasionally *marine* or *fish oils* may be used in a food, frequently after

---

MAJOR REFERENCES: Evaluation of the health aspects of tallow, hydrogenated tallow, stearic acid, and calcium stearate as food ingredients. FASEB/SCOGS Report 54 (NTIS PB 262-661) 1976; Evaluation of the health aspects of lard and lard oil as they may migrate to foods from packaging materials. FASEB/SCOGS Report 91 (NTIS PB 270-368) 1977; Evaluation of the health aspects of hydrogenated fish oil as a food ingredient. FASEB/SCOGS Report 66 (NTIS PB 262-667) 1976.

partial hydrogenation; and *poultry fat* may be used for its flavor contribution. The fish oils are from menhaden, herring, or tuna; the poultry fat from chicken or turkey. These fats or oils differ from most animal fats because they are relatively rich in polyunsaturated fatty acids; by contrast, butter, tallow, and lard have 4 to 12 times as much saturated fatty acids as polyunsaturated.

Stearic acid is a saturated fatty acid in animal fats and vegetable oils, and is extracted from these substances by food processors for use as a lubricant, blender, and binder; as a foam inhibitor; and as a coating in dry food packaging. It is used in beverages, baked goods, candy, and chewing gum. *Calcium stearate* is a combination of calcium, stearic acid, and palmitic acid (a fatty acid constituent of fat), and is used in food the way stearic acid is.

SAFETY: In 1972 daily U.S. consumption of tallow used in the manufacture of margarine and shortening amounted to 30 grams per person (about an ounce), 4 grams of which was stearic acid, a component of tallow. But this represents only a small fraction of the tallow that is consumed in the daily diet, which is present as the fat in beef. About one-fifth of the fat in hamburger is stearic acid. The use of calcium stearate in food is small, but it tripled between 1960 and 1970, when daily intake amounted to 4 milligrams per capita. In 1973 the quantity of lard used in food totaled 6.6 grams daily per person in the U.S., a decline by half since 1960. In 1975 use of enzyme-modified (lipolyzed) butter fat to enhance flavors provided a daily average of somewhat under 3 milligrams per person in the U.S. Beyond this, no reliable statistics are available for butter, poultry fat, or fish oils, but their uses as food additives obviously would contribute only a small fraction of the total dietary intake from consumption of dairy products, poultry, or fish.

Stearic acid (a completely saturated fat) fed to rats as 3 to 6 percent of their diet caused blood clots and cholesterol deposition in arteries. Beef tallow (which consists of saturated and unsaturated fats) produced less of this effect. In comparison with safflower oil (composed largely of unsaturated fat), beef tallow at a 15 percent level in the diet significantly accelerated blood clotting time in minature pigs. When fed to day-old chicks, tallow improved food utilization or efficiency, but this did not occur when hydrogenated fat or stearic acid was substituted. Butter fat and other animal fats with a high proportion of saturated fatty acids have been shown to elevate blood cholesterol levels when substituting for vegetable oils in the diet.

Poultry fat has considerable amounts of the polyunsaturated fatty acids that are essential dietary constituents. This composition is favorable to lowering blood cholesterol levels and cholesterol deposition in arteries in experimental animals whose diets include saturated fats and cholesterol. The fish oils that are unhydrogenated also are abundant in polyunsaturated fatty acids (though not the essential fatty acids). Experimental studies with rats have shown that lower blood cholesterol levels result when tallow in the diet is accompanied by fish oil; on the other hand, the fish oil is prone to oxygen uptake and rancidity, which can lead to toxic effects.

Long-term studies do not appear to have been conducted on the effects of feeding any of these fats to animals. Nor have tests been located that investigated their possibilities as the causes of gene mutation, fetus or birth abnormalities, or maternal or newborn mortality. Although it has been determined that stearic acid is not a cause of cancer, similar information is not available for various animal fats.

Experiments to determine adverse effects caused by lard have employed dosages ranging from 2 to 25 percent of the diets of laboratory animals, many thousands of times as great as the amount added to a human diet when adjusted for body weight. Mortality increased and the life span was shortened for male mice fed a diet of 24 percent lard, from weaning to death. Female mice fed lard at 2 to 10 percent of their diets through four generations lost considerable weight during lactation, but regained it rapidly after they were separated from litters, whose growth was not affected. The addition of brewer's yeast (see p. 681) to the diet prevented the weight loss, an indication that the large amounts of fat interfered with the ingestion of other food constituents or that the diet lacked some essential constituent.

The incidence of osteoarthritis doubled among male mice fed a diet supplemented with 25 percent of lard, but this did not occur among male mice of another strain with a more rapid growth rate. Oral studies to test whether lard can cause cancer have not been reported. In one instance when it was administered weekly by injection to rats, tumors did occur at injection sites. This is not accepted as evidence of a similar effect when lard is ingested, and did not happen in a number of comparable studies with mice.

ASSESSMENT: The adverse effects observed in experimental animals fed very high levels of lard can be ascribed to excessive fat in the diet rather than to the specific effects of lard. The evidence does indicate that a high degree of saturation of the fat is related to adverse effects on blood clotting, cholesterol deposits in arteries, and somewhat poorer fat digestibility. None of these studies, it must be noted, involved the small quantities of saturated animal fats estimated to be consumed by humans as food additives, which usually are present in exceedingly small amounts, and which account for only a minute fraction of total intake in the human diet. In their own right, saturated fats as additives are unlikely to present a significant hazard for normal people. On the other hand, there is concern over the role of saturated fats in the total diet of people with artery and associated cardiovascular diseases, and in these instances physicians and nutritional scientists recommend curtailment.

The potential health problem that a saturated fat may pose to a consumer will depend on the total amount in the diet, not merely its presence as an additive. The ingredients label can alert one to the presence of a highly saturated fat in a food, although quantities are not given. However, since the ingredients are listed in descending order of their presence in the food, it is possible to estimate when the amount of a given fat may be of concern. Accordingly, a caution will be raised in the Inventory in this volume about the safety of a food when one of the first three listed ingredients (other than water), alone or as part of a shortening blend containing other fats and oils, is any of the animal fats other than poultry or fish, or is stearic acid.

RATING: S for most additive uses; ? if the animal fat (other than poultry or fish) or stearic acid or calcium stearate appears as one of the first three ingredients on a food label.*

---

*The accuracy of this procedure for identifying highly saturated ingredients was tested with fatty acid analyses of over 200 purchased foods of all types contained in the 1980 thesis of Mary Gertrude Enig, conducted in collaboration with the U.S. Department of Agriculture. All the foods listing animal fats on the ingredient label were correctly identified as having excessive saturated fat not balanced by polyunsaturated fats or oils, with the single exception of one margarine.

ASCORBIC ACID (VITAMIN C)

Sodium Ascorbate

ERYTHORBIC ACID (ISOASCORBIC ACID)

Sodium Erythorbate

Ascorbic acid and *sodium ascorbate,* its salt, have multiple uses as food additives. They have value as nutrient supplements because of the vitamin C activity they possess. Along with the closely related *erythorbic acid* and its *sodium erythorbate* salt, these additives have antioxidant properties (prevention of deterioration caused by oxygen), and can help preserve flavor, color, or aroma of foods. They may be used in curing meat products such as bacon to inhibit the tendency of nitrite in curing salts to form nitrosamines following heating (see p. 632). The erythorbates, however, do not possess appreciable vitamin activity and are not regarded as important sources of vitamin C.

Ascorbic acid is found naturally in many plant products, especially leafy vegetables, fruits, and tomatoes; citrus fruits are particularly rich in this vitamin. The ascorbic acid available commercially is synthesized chemically, however. Erythorbic acid is not found in nature, but is a synthetic chemical additive.

SAFETY: The average daily intake per person of the ascorbates as food additives in the diet was estimated in 1970 as totaling 34 milligrams. This represented one-third of the average daily intake from all dietary sources. An estimate made in 1975 indicated that the use of ascorbates as additives in food had increased by over 80 percent, to 62 milligrams. The Recommended Dietary Allowance (RDA) set by the National Academy of Sciences-National Research Council for ascorbic acid is 60 milligrams for adults, or 1 milligram per kilogram of body weight (kg/bw) for a person weighing 60 kg (132 lbs.).

The use of the erythorbates as food additives in the average daily diet in 1970 amounted to 13 milligrams, a little over a third as much as the ascorbates. The use of these, too, has increased some, to 17 milligrams in 1975.

A severe deficiency of ascorbic acid causes scurvy in hu-

MAJOR REFERENCE: Evaluation of the health aspects of ascorbic acid and various ascorbates as food ingredients. FASEB/SCOGS Report 59 (NTIS PB 80-128796) 1979.

mans. Among other roles, the vitamin is required for the synthesis and metabolism of collagen, the protein making up connective tissue, cartilage, or gristle. This explains why the deficiency symptoms of scurvy include failure of cartilage, bone, and teeth to develop normally. Humans, according to the RDA, require about 1 milligram of ascorbic acid per kg/bw daily to maintain their body pools of the vitamin and make up the losses from excretion and metabolism. Erythorbic acid has weak antiscorbutic (scurvy prevention) activity, being about 5 percent as effective as the vitamin itself; however, the two chemicals are about equal in their antioxidant properties, which account for a significant part of their use as food additives.

In short-term studies, no harmful effects were seen in rats fed ascorbic acid at a level of 6500 milligrams per kg/bw. Guinea pigs also tolerated these high levels. Humans have not shown adverse effects after receiving daily supplements of 1 gram of ascorbic acid (some 16 times the average daily consumption of ascorbates as additives) for three months; much larger daily supplements have been taken for a number of weeks, again without apparent harm. Some investigations with experimental animals and humans produced evidence that continued intake of very large amounts of ascorbic acid (3 grams or more daily for humans) may create a dependency because of the overdosage. The rate of metabolism and excretion is accelerated, while concentration of ascorbic acid in tissues remains only slightly higher than normal; and if the dosage is reduced this accelerated metabolism continues, and can result in depletion of the vitamin in the tissues and even a vitamin C deficiency.

Ascorbic acid may significantly improve the absorption of iron in the diet, may depress the absorption of copper, and may prevent adverse effects from some toxic metals such as lead, mercury, or cadmium. There has been speculation that high dosages of ascorbic acid may produce kidney stones. It has been determined that calcium oxalate, the mineral salt causing this condition, increases only negligibly when less than 4 grams of ascorbic acid is ingested daily, which is far greater than the amount contributed to the diet as an additive.

Supplements of 100 milligrams of erythorbic acid daily (six times their presence in the diet) have been ingested safely by adults. Dogs have been fed erythorbic acid for six months or more at levels of 5 to 7 grams daily without toxic effects. There is some evidence that erythorbic acid at high levels can interact and compete with ascorbic acid, reducing uptake of the vitamin and its biological effectiveness. The animal studies with guinea pigs showed this only when erythorbic acid amounted to 2½ to 10 times the ascorbic acid in the diet (the reverse of the current use of these two substances in foods, where erythorbates are present in substantially smaller amounts than ascorbic acid). As much as 300 milligrams of erythorbic acid (5 milligrams per kg/bw) did not deplete ascorbic acid levels in the blood cells of adult humans partially deficient in vitamin C, and there was no indication that erythorbic acid could produce scurvy; indeed, this additive is itself mildly effective as a vitamin C substitute in preventing scurvy.

There are no clear indications that ascorbic acid taken during pregnancy has any adverse effect on survival of the mother or the fetus, or leads to any abnormalities in the offspring. Erythorbic acid and ascorbic acid, however, did cause embryo mortality when injected into the air cells of chick eggs at levels of 40 milligrams per kg/bw. Neither chemical was responsible for any chromosome damage in several types of tests, nor were the sodium salts.

ASSESSMENT: Ascorbic acid (vitamin C) is essential in the diet, and is present as a normal constituent of many fruits and vegetables. Erythorbic acid is an effective antioxidant but has only weak vitamin C activity. Many studies have shown that large amounts of ascorbic acid or sodium ascorbate, as well as erythorbic acid, can be tolerated without harm; the levels used were frequently 100 times the average amount of the substance incorporated as a food additive. No hazard is posed to the consumer from these chemicals.*

RATING: S.

AZODICARBONAMIDE

This chemical has been used since 1962 as a flour-maturing

MAJOR REFERENCE: Studies of the safety of azodicarbonamide as a flour-maturing agent. B. L. Oser, M. Oser, K. Morgareidge, and S. S. Sternberg. *Toxicology and Applied Pharmacology* 7:445 (1965).

*It is not the purpose of this volume to comment on aspects of the use of ascorbic acid (or other substances) other than the effect it has on health as an additive in food. As a consequence, its therapeutic usefulness in high doses, its role as a preventative or cure for various ailments, has not been subject to review or assessment here.

agent, up to 45 parts of azodicarbonamide per million parts (ppm) of flour being permitted by FDA regulations. When liquid is added to it, azodicarbonamide will strengthen the dough.

SAFETY: A survey in 1977 by the National Academy of Sciences-National Research Council determined that the average daily dietary intake per person in the U.S. of azodicarbonamide was 1 milligram.

When flour treated with this chemical is made into a bread dough, during the process the azodicarbonamide reacts chemically, converting rapidly to biurea, a very inert compound with a low solubility, which is mostly excreted in the feces. This reaction does not significantly affect the vitamins or amino acids in the flour. The safety of azodicarbonamide has been tested using mice, rats, and dogs. Mice tolerated doses as much as 6 grams per kilogram of body weight without adverse effect, though some diarrhea was noted.

Rats and dogs consumed a diet in which the principal ingredient was flour treated with this food additive at 100 ppm, ten times the normal usage level. After two years on this diet, no adverse effects were seen in growth, blood composition, microscopic appearance of the cells (no tumor growth), or in reproduction or lactation. This was also true when biurea was added at 1000 times the level expected from conversion of azodicarbonamide in treated doughs. Extremely high levels of biurea (5 percent or more of daily diet, which is many thousands of times above the level usually present in bread treated with azodicarbonamide), fed to rats for a year, were without effect; but dogs showed some deposits of the biurea in the kidney and bladder, due to the very low solubility of this compound.

ASSESSMENT: Azodicarbonamide at low levels is used as an aging or strengthening agent for flour in making bread or rolls. It is readily converted to biurea when the dough is formed. Studies show that no hazard exists when either of these compounds is present in the diet at levels far exceeding the 45 ppm that is permitted to be used in foods.

RATING: S.

BENZOIC ACID

SODIUM BENZOATE (BENZOATE OF SODA)

Benzoic acid occurs naturally in most berries, in prunes, tea, and in spices like cinnamon, ripe cloves, and anise. *Sodium benzoate* is its sodium salt. Both are used in a wide range of processed foods to prevent spoilage by microorganisms.

SAFETY: A survey of these two additives by the food industry in 1975 indicated an average daily intake per person in the U.S. of 48 milligrams.

These substances are rapidly absorbed, then combine with glycine (a component of protein; see p. 475) in the liver, and are excreted without any remaining in the body. The success of this process depends on a healthy functioning liver and a sufficient nutritional supply of glycine. Humans fed 1000 milligrams a day of benzoic acid for 88 days, more than 20 times the average consumption of benzoate additives in food, showed no observable ill effects. At 132 times daily consumption, sodium benzoate employed as a test of liver function caused temporary distress due to gastrointestinal irritation, but did no harm when administered in larger doses as medication to rheumatic patients.

Rats appear to employ body processes and pathways like ours to deal with these substances. Laboratory experiments provide evidence that rats can tolerate many hundreds of times the amount of sodium benzoate usually present in the average diet of a human (adjusted for body weight). A study covering four generations of rats demonstrated that benzoic acid did not cause abnormalities in the newborn. Research investigating cancer provided evidence that sodium benzoate did not cause tumors. Studies dealing with mutation of genes have not been conducted with these additives.

ASSESSMENT: Experiments with humans and rats, along with clinical experience with patients, indicate that in the quantities presently consumed in the daily diet, or that might reasonably be expected in the future, neither benzoic acid nor sodium benzoate offers any hazard to health.

RATING: S.

MAJOR REFERENCE: Evaluation of the health aspects of benzoic acid and sodium benzoate as food ingredients. FASEB/SCOGS Report 7 (NTIS PB 223-837) 1973.

BENZOYL PEROXIDE

This synthetic chemical has been used for many years, principally as a bleaching agent for flour; to a lesser extent it may be used to bleach milk used in manufacture of some cheeses, and to bleach lecithin (see p. 580) used in fats and oils.

SAFETY: A survey conducted by the National Academy of Sciences-National Research Council indicated that food usage of benzoyl peroxide in 1975 was 9 milligrams per person in the U.S. daily diet, or 0.14 milligram per kilogram of body weight (kg/bw) for a person weighing 60 kg (132 lbs.). This is the amount added during food processing, but only a small fraction of the bleach is ingested, as it decomposes to benzoic acid (see p. 487) during processing; most of any trace remaining is converted in the intestine to benzoic acid or to a form that is readily excreted in the urine.

Rats and mice have been fed a diet to which benzoyl peroxide had been added at levels up to 280 milligrams per kg/bw, 2000 times the average human intake. After 104 weeks there were no adverse effects on mortality and no increase in cancer, although there was some reduction in weight gain, perhaps because the bleach can destroy some nutrients such as vitamin E and the yellow carotene pigment precursors of vitamin A. Because carotene (see p. 505) is likely to be destroyed by treatment of milk with benzoyl peroxide in the manufacture of cheese, government regulations require that these losses be compensated for by the addition of supplementary vitamin A. (See p. 665.) Some vitamin C (see p. 483) and vitamin E (see p. 657) are also probably destroyed during bleaching with benzoyl peroxide, but their amounts in flour and milk are too small to have nutritional significance.

From studies with tissues, microorganisms, and mice, it is clear that benzoyl peroxide does not alter the chromosomes; but its effect on the newborn has not been tested. There has been some allergic response to this chemical among workers exposed externally to high levels of benzoyl peroxide, but such sensitivity has not been reported from foods treated with this chemical.

ASSESSMENT: Benzoyl peroxide is used as a bleaching agent in a few foods. It degrades almost wholly to benzoic acid by

---

MAJOR REFERENCE: Evaluation of the health aspects of benzoyl peroxide as a food ingredient. FASEB/SCOGS Report II-2 (NTIS PB 81-127854) 1980.

the time the food is consumed, and benzoic acid is regarded safe as a food ingredient. Certain vitamins can be destroyed by the addition of benzoyl peroxide to food, but where it could be significant, like destruction of vitamin A in milk used in cheese making, government regulations require the addition of the vitamin as a replacement. There appear to be no hazards associated with the use of benzoyl peroxide in foods at current levels or any likely to be used in the future.

RATING: S.

BHA (BUTYLATED HYDROXYANISOLE); BHT (BUTYLATED HYDROXYTOLUENE)

BHA and BHT are synthetic compounds used as preservatives to prevent or delay fats, oils, and fat-containing foods from becoming rancid and developing objectionable tastes and odors. By this means BHA and BHT extend the shelf life of such foods as breakfast cereals, baked goods, vegetable oils, and potato chips.

SAFETY: The average daily intake of BHA in the diet of individuals in the U.S. in 1975 has been estimated at 4.3 milligrams, or less than 0.1 milligram per kilogram of body weight (kg/bw) for a person weighing 60 kg (132 lbs.). For BHT, daily intake amounts to 0.04 milligram per kg/bw; totaling for both is thus a little over 0.1 milligram. The UN Joint FAO/WHO Expert Committee on Food Additives has suggested that intake of both or either should not be more than 0.5 milligram per kg/bw.

Extremely high dosages of BHA and BHT have consistently resulted in considerable enlargement of livers of experimental animals, reflecting a rapid growth of endoplasmic reticulum (the central portion of a cell other than the nucleus), accompanied by a marked increase in the production of microsomal enzymes (catalysts that activate this process). This did not occur when the dosages were reduced to the equivalent of 500 times human consumption. The claim is made that the enlargement of the liver is an adaptive response to an increased demand on this organ and is reversible with the elimination of

---

MAJOR REFERENCES: Evaluation of the health aspects of butylated hydroxyanisole as a food ingredient. FASEB/SCOGS Report 55 (NTIS PB 285-496) 1978; Evaluation of the health aspects of butylated hydroxytoluene as a food ingredient. FASEB/SCOGS Report 2 (NTIS PB 259-917) 1973.

these substances from the diet. However, it has been demonstrated with other compounds that a point can be reached at which adaptation fails and injury begins. The circumstances may arise, then, of further challenge to the liver, not only by BHA and BHT but also by other compounds. In fact, many common drugs and oral contraceptives activate the microsomal enzymes, and the possibility exists of an increase of these enzymes in the intestines, lungs, kidneys, etc. There is evidence that these enzymes can play a role in increasing the vulnerability of tissues to cancer-causing and other toxic substances. BHA and BHT may also stimulate the activity of steroid enzymes and, in this way, may affect the functioning of steroid hormones with adverse effects on reproduction.

Animal species differ in their sensitivity to BHA and BHT. Generally, larger dosages of BHA than of BHT have been needed to produce harmful effects in rodents, dogs, and guinea pigs. An important exception is primates; less of BHA than BHT does injury to monkeys, the significance being that the metabolic processes of monkeys are believed to more closely resemble those in man.

In contrast to these concerns, there are data which indicate that BHA and BHT reduced the occurrence of certain tumors, and they do not have an adverse effect on reproduction or the ability of the fetus to survive. Long-term studies of both additives fed to rats and mice in amounts far in excess of human consumption determined that by themselves they did not cause cancer.

ASSESSMENT: The amount of BHA and BHT that is consumed daily is within the safety levels specified by FAO/WHO, and some data are available indicating that they can reduce the occurrence of tumors and fetal failure. There remain uncertainties concerning their possible role in increasing microsomal enzymes in tissues other than the liver, thereby weakening resistance to cancer-provoking substances or causing hormonal imbalance that may affect reproduction. These possibilities have yet to be investigated, especially in conjunction with such widely used compounds as oral contraceptives and hormonal supplements, each of which may further aggravate BHA's and BHT's effect on the liver.

RATING: Both BHA and BHT, ? (additional studies needed to resolve uncertainties).

BIOTIN

Biotin is one of the B-complex vitamins, water-soluble vitamins that play a vital role as coenzymes* in the body's metabolic processes. Biotin is present in all living cells as a necessary nutrient for growth. It is found in small quantities in most foods, and in larger quantities in liver and other organ meats, some fishes, yeast, egg yolk, peanuts, and dried peas. The body provides a source of biotin because many microorganisms produce it in the intestine.

The biotin that manufacturers add to food is chemically synthesized. Its principal use as a food additive is as a nutrient in soy-based infant formulas and other milk-free formulas for infants and other age groups.

SAFETY: The daily intake of biotin from all food sources will range from 0.1 to 0.3 milligram per person. An infant receiving a fortified formula may receive 0.1 milligram of added biotin per day.

A protein in unheated egg white, avidin, interacts with biotin, binding it so tightly as to render it inactive; in effect, this creates a biotin deficiency known as "egg white injury." The deficiency has been produced experimentally in rats and rhesus monkeys, with symptoms of dermatitis, loss of hair around the eyes, and hind-leg stiffness or paralysis. Human volunteers fed a restricted diet of low biotin content and containing dehydrated egg whites (see p. 465) developed comparable symptoms, which disappeared after injection of as little as 0.15 milligram of biotin. A deficiency of biotin is unlikely to occur in adult humans without the combination of avidin and a diet low in biotin, because the availability of biotin is well in excess of the amount required for normal maintenance and growth. In fact, the body excretes more biotin than it obtains from the diet because of the ability of microorganisms in the gastrointestinal tract to produce it.

Humans have been administered oral doses of 10 milligrams of biotin daily for up to a month, at least a hundred times their

---

MAJOR REFERENCE: Evaluation of the health aspects of biotin as a food ingredient. FASEB/SCOGS Report 92 (NTIS PB 281-421) 1978.

*Coenzymes are linked together with enzymes (the body's protein regulators), which speed up the metabolism of proteins, carbohydrates, and fats, breaking them down to simpler forms usable by the body for energy, or using them in the replacement and building of cells and storage materials.

total dietary intake, without adverse effects. Infants under three months of age suffering from dermatitis, a possible symptom of biotin deficiency, have ingested 2 to 6 milligrams daily for several weeks (20 to 60 times the amount in the usual infant diet), again without adverse reactions and with improvement in the condition. Long-term studies on biotin have not been conducted, nor have there been any to determine whether it causes gene alteration (mutagenicity) or cancer induction (carcinogenicity). One experiment explored the effect of biotin on tumors induced in rats by a known carcinogen, and found it accelerated their growth.

Extremely large doses of biotin were given by injection (not by diet) to rats to determine the effect on reproductive performance. This treatment caused some irregularity in the estrous (fertility) cycle and failure to maintain a normal pregnancy.

ASSESSMENT: Biotin is an essential nutrient for humans, but one where a deficiency is extremely rare. Its use as a food additive is principally as a nutrient supplement in milk-free infant formulas. Even a hundred times the usual amount of biotin added to the formulas has not led to any harmful effects in infants, nor have excessive amounts adversely affected adults. Although adverse effects of biotin on the reproduction changes of rats have been reported, the amounts administered were thousands of times as great as those to which humans conceivably could be exposed. Thus, there is no reason to expect a hazard from this food additive as currently used or as may be expected in the future.

RATING: S.

BROMINATED VEGETABLE OIL (BVO)

Brominated vegetable oil is a combination of bromine (a chemical element, a reddish brown caustic liquid) with a vegetable oil. When bromine is chemically incorporated into the molecule of an unsaturated fat, it substantially increases the relative weight of the oil. The result is that the flavoring oils dissolved

MAJOR REFERENCES: Biochemical and pathological changes in rats fed low dietary levels of brominated cottonseed oil. I. C. Munro et al. *Food and Cosmetics Toxicology* 9:631 (1971); Brominated maize oil. II. Storage of lipid-bound bromine in pigs fed brominated maize oil. I. F. Gaunt et al. *Food and Cosmetics Toxicology* 9:13 (1971).

in the BVO will not rise and separate from the liquid mixture and will not form a ring in the neck of the bottle. It thus "stabilizes" the flavoring oils in the beverage. Used primarily in various citrus soft drinks and fruit-flavored beverages, BVO also provides a cloudy appearance, enabling the drinks to resemble natural fruit juices. Various vegetable oils are used in preparing BVO, including corn, cottonseed, olive, sesame, and soybean oils. (See p. 660.) Presently the FDA permits 15 parts per million of BVO in the finished beverage, pending additional studies concerning its safety.

SAFETY: A survey of the poundage of BVO used by the food industry in 1976, as reported by the National Academy of Sciences-National Research Council, indicated that an average of less than 0.2 milligram per person in the U.S. was present in the daily diet, or 0.03 milligram per kilogram of body weight (kg/bw) for a person weighing 60 kg (132 lbs.).

Although BVO made from various oils has been added to foods for nearly 50 years, toxicology studies during the past decade have raised questions about its safety. In 1969 and 1971 a series of studies reported that BVO from cottonseed oil produced adverse effects in the heart and liver of rats when fed at 2500 milligrams per kg/bw, even when the feeding was for less than a week, and even 400 milligrams per kg/bw resulted in marked reduction in the rate of fatty acid metabolism by the heart tissues. There was enlargement and fat deposition in the liver and kidneys, and some in the heart, because the brominated fatty acid deposited in the tissues was not readily metabolized. At levels of 250 milligrams per kg/bw there were cardiac (heart) lesions, inflammation of the heart muscle, and abnormal increases in thyroid cell growth.

BVO prepared from corn oil and fed to rats and pigs at levels down to 20 milligrams per kg/bw also resulted in deposits of brominated fat in the liver and other tissues. BVO from cottonseed, corn, olive, and sesame oils all caused pathological changes in the hearts of rats fed at 250 milligrams per kg/bw for 15 weeks; some showed these changes when the diet provided 50 milligrams per kg/bw. BVO from sesame or soybean oils fed to pigs at levels of 500 milligrams per kg/bw caused pathological changes, wasting of the testes, reduced growth, and lethargy.

ASSESSMENT: It should be noted that the adverse effects obtained in a variety of animals at levels as low as 20 milligrams per kg/bw were probably hundreds of times as high as the ingestion of BVO by humans, even for heavy users of beverages containing BVO. Nevertheless, the UN Joint FAO/WHO Expert Committee on Food Additives in 1971 considered that the accumulation of brominated fat in the body tissues and the degenerative changes in the hearts of experimental animals suggest problems in safety of BVO. They recommended that BVO not be used as a food additive. The FDA was sufficiently worried about this substance that it deleted it from the GRAS list. (The substance may nevertheless be used lawfully unless the FDA takes further action.) Although many tests were done by the manufacturer of BVO in the early 1970s to establish the safety of BVO in response to the interim permission for its use in foods, the chronic tests and reproduction studies using rats, dogs, and other species were judged to be faulty and unacceptable by the FDA when their results were submitted. The evidence to date has demonstrated that BVO can cause harm to vital organs when administered to experimental animals; it is felt that the technological benefits of BVO do not justify the potential health risks to a heavy consumer of beverages containing brominated vegetable oil.

RATING: X.

CAFFEINE

Caffeine belongs to a group of alkaloids, which are naturally occurring substances in plants. Many of the alkaloids are drugs with known stimulatory effects when ingested; they can also be habit-forming. This is true of caffeine, which is present as a natural constituent found principally in coffee, tea, chocolate, and the kola nut (whose extracts are used in cola drinks). Less than 10 percent of the caffeine in a cola drink is provided by the kola nut extract; the other 90 percent is added, presumably to enhance flavor and stimulation.

Caffeine need not be printed on the labels of coffee, tea, cocoa, or chocolate drinks as it occurs in them naturally. FDA regulations require no label declaration for caffeine on cola or other drinks or on finished foods as long as the caffeine concentration does not exceed 0.02 percent of the product.

MAJOR REFERENCE: Evaluation of the health aspects of caffeine as a food ingredient. FASEB/SCOGS Report 89 (NTIS PB 283-441) 1978.

SAFETY: In the U.S., total exposure to caffeine from all sources daily is estimated as averaging from 140 milligrams for most consumers to 300 milligrams or more for heavy users of beverages and other foods containing it; this would come to 2.3 to 5.0 milligrams per kilogram of body weight (kg/bw) for a person weighing 60 kg (132 lbs.). Approximately one-fourth to one-third of the daily intake of caffeine by adolescents and children comes from the consumption of cola-type beverages. Caffeine intake from all sources for one- to five-year-old cola users was estimated in 1977 as averaging 1.4 milligrams per kg/bw daily, and 1.0 for six-to-11-year-olds. Cola beverages contribute perhaps less than one-tenth of the caffeine intake of adults, who consume more coffee and tea than children do.

Caffeine is retained longer in the body fluids of infants and in pregnant women than in the normal adult. It's distributed into most of the body tissues and into breast milk; it can cross the placenta barrier to the fetus, and, if the nursing mother has been consuming beverages containing this chemical, an infant may have an intake of over 1 milligram per kg/bw. Some children can have an undesirably high consumption of caffeine from cola-type drinks during the period of brain growth and development. It is during this period that the developing nervous system is most sensitive to the effects of all aspects of the environment. In animals and humans, even 2.5 milligrams per kg/bw of caffeine can cause at least temporary effects on the central nervous system, adversely affecting behavior, motor activity, or sleep.

There are suggestions that in humans large doses of caffeine can markedly stimulate gastric secretions, which can cause peptic ulcers. In an FDA laboratory study, pregnant rats were force-fed caffeine in amounts not greatly exceeding the upper limits of human consumption (adjusted for body weight). Preliminary findings indicated that caffeine caused partial or complete absence of the digits of the paws of some of their progeny.

There is no good evidence that caffeine taken in quantities up to 50 milligrams per kg/bw (several times the normal daily intake) has any mutagenic effects (altering chromosomes) or causes cancer. However, results from the administration of larger doses suggest the need for more rigorous studies in these areas.

Because some doubts remain as a consequence of conflicts in findings, further study is also needed on immediate and long-range effects of caffeine stimulation on human behavior and the cardiovascular system.

ASSESSMENT: Caffeine consumption appears to be at a level that can lead to some undesirable effects in certain consumers, especially children from infancy through adolescence. While its use as a food additive contributes only a minor part of the intake of this substance, even the caffeine from cola-type beverages may be sufficient, through the pregnant or nursing mother, to have some effect on the fetus or the infant. A recent experiment with rats provides evidence that caffeine may be a teratogen (cause of birth defects). Although there is not strong evidence pointing to caffeine's contribution to peptic ulcers or heart ailments, physicians do caution susceptible patients to avoid beverages containing caffeine. For other consumers, according to the available evidence, caffeine intake at the present level of consumption is regarded as safe.

RATING: S for some people; X for pregnant women, nursing mothers, young children, anyone with a gastrointestinal or cardiovascular ailment.

CALCIUM PEROXIDE (CALCIUM DIOXIDE, CALCIUM SUPEROXIDE)

This synthetic compound is added to flour which is used in white bread and rolls. It acts as a dough-strengthening agent, making the dough more extensible and more uniform, even when ingredients vary. Its bleaching action may extend shelf life of the baked goods.

SAFETY: The average daily intake per person of calcium peroxide in foods in 1976, according to a survey conducted by the National Academy of Sciences-National Research Council, was calculated as 0.6 milligram.

There appear to be no published data on the biological safety of this additive, alone or in bread or flour treated with calcium peroxide. Because of this, the UN Joint FAO/WHO Expert Committee on Food Additives has set no level for an acceptable intake of calcium peroxide by humans.

When the additive completes its work in the flour, it becomes a simple calcium salt (see p. 498), which is harmless. The following discussion of safety deals with the peroxide part

MAJOR REFERENCE: Evaluation of the health aspects of hydrogen peroxide as a food ingredient. FASEB/SCOGS Report 113 (NTIS PB 80-104607) 1979.

of the calcium peroxide molecule, which resembles hydrogen peroxide in its action. During metabolism, the body produces hydrogen peroxide as a by-product of the interaction of certain enzymes (catalysts that accelerate the breakdown of foods) and oxygen; and hydrogen peroxide itself decomposes to oxygen and water in the gastrointestinal tract. Cases of poisoning have been reported in humans who accidentally consumed heavy concentrations of hydrogen peroxide in amounts many thousands of times as great as its presence from additives in food.

Hydrogen peroxide has been studied in many short-term experiments with mice and rats. At dietary levels of up to 30 milligrams per kilogram of body weight (thousands of times human intake), no harmful effects were seen. Higher levels reduced the rate of growth, but pregnant female rats were able to produce normal, healthy litters.

Since hydrogen peroxide can be produced by ionizing radiation (X-rays and gamma rays), its possible role in radiation-induced mutagenesis (gene alteration) has been studied extensively. It has proven to be mutagenic; however, to produce this effect in laboratory animals and tissue required a concentration of hydrogen peroxide ten times the amount produced by radiation, and far greater than the small amount expected from calcium peroxide added to food.

ASSESSMENT: The lack of experiments specifically testing the safety of calcium peroxide does not permit an assessment of any possible hazard from its use in foods. But evidence does exist concerning the mutagenic effects of hydrogen peroxide, a related compound, when used in high concentrations, warranting some caution in the use of calcium peroxide until appropriate studies are conducted.

RATING: ?

CALCIUM SALTS*

Calcium Acetate, Chloride, Gluconate, Hydroxide, Oxide, Phytate

These salts consist of calcium—an important nutrient needed for bones and teeth, as well as for blood coagulation and many metabolic functions in the body—and certain normal constituents present in the diet. The acetate portion of *calcium acetate* is present in acetic acid (see p. 463). The chloride part of *calcium chloride* is a component of sodium chloride (see p. 626). *Calcium gluconate* is calcium and gluconic acid (see p. 553). *Calcium phytate* is calcium and phytic acid, a relatively inert constituent found in many plant foods such as cereals, nuts, legumes, and potatoes. *Calcium oxide* is calcium combined with oxygen. *Calcium hydroxide* is formed by adding water to calcium oxide. *Calcium carbonate* is calcium oxide with the addition of carbon dioxide.

Some of these calcium salts are used in foods as agents to deactivate mineral substances that otherwise could cause undesirable flavor, color, or texture changes in the food; to alter the alkali-acid (pH) balance; or to firm up the texture. Among the large variety of products containing calcium salt additives are baked goods and baking mixes, carbonated fruit beverages, beer and wine, cheese, and processed vegetables. The salts are also used as nutrients or dietary supplements.

SAFETY: In 1975 the average daily intake per person of these added calcium salts, based on reported usage by the food industry, was estimated at about 41 milligrams in total, or somewhat less than 0.7 milligram per kilogram of body weight (kg/bw) for a person weighing 60 kg (132 lbs.). Calcium chloride accounted for 70 percent of this.

The contribution of these six additives toward body calcium needs represented 3 to 5 percent of the 800 to 1200 milligrams of calcium recommended daily for children and adults by the National Academy of Sciences-National Research Council.

MAJOR REFERENCES: Evaluation of the health aspects of certain calcium salts as food ingredients. FASEB/SCOGS Report 45 (NTIS PB 254-539) 1975; Evaluation of the health aspects of calcium oxide and calcium hydroxide as food ingredients. FASEB/SCOGS Report 72 (NTIS PB 254-540) 1975.

*Additional calcium salts are discussed elsewhere: see p. 513 for calcium citrate, p. 503 for calcium carbonate, p. 600 for calcium phosphate, and p. 652 for calcium sulfate.

The safety of these salts relates to the other parts of the compound combined with the calcium. Acetate and gluconate, as acetic acid and gluconic acid, are made by the body regularly in large quantities in its metabolism of carbohydrates. Chloride is an essential nutrient that helps to regulate the blood's acid-alkaline balance and pressure. The oxide and hydroxide readily form calcium carbonate, or chalk, itself a useful nutrient source. These calcium salts thus are substances that the body manufactures or requires in order to meet its needs.

At the high level of about 1500-2000 milligrams per kg/bw of calcium chloride, thousands of times human intake of all of these calcium salts, rabbits experienced an "acidosis" and damage to the stomach lining due to the excessive concentration of the salt solution. However, even at the same high levels, calcium chloride fed to rats for several months had no effect on thyroid size, and microscopic investigation did not reveal any abnormal changes. No adverse effects were seen in the heart, kidney, and liver of rats fed calcium gluconate or calcium chloride at levels of 400 milligrams of calcium per kg/bw. Calcium phytate, at about 300 milligrams per kg/bw, fed to rats as a diet supplement, successfully provided calcium for bone deposition, and the animals remained healthy.

Humans have taken as much as 10 grams of calcium gluconate as a calcium source, and the only deleterious effect was some diarrhea if the salt was taken on an empty stomach. Relevant studies of calcium oxide and hydroxide have not been made. However, these compounds when used in food change from caustic, alkaline materials to simple calcium salts.

No long-term studies appear to have been made on any of these calcium salts, and none on possible effects on cancer or on chromosome damage. Tests indicate that calcium chloride and calcium gluconate do not appear to cause embryo malformation.

ASSESSMENT: Calcium is a nutrient needed in substantial quantities in the diet. The acid components of its salts are normal food constituents and participate in metabolic processes in humans. The alkaline compounds, calcium oxide or hydroxide, change over to calcium salts when incorporated into foods and are then metabolized normally. The use of these various calcium compounds as food additives poses no hazard to the consumer.

RATING: S.

CALCIUM STEAROYL-2-LACTYLATE; SODIUM STEAROYL-2-LACTYLATE

These two additives are prepared by combining lactic acid (see p. 576) and stearic acid (a saturated fatty acid; see p. 479), and converting the product to its calcium or sodium salt by replacing the acidic hydrogen in it with calcium or sodium. The stearoyl lactylates (also referred to as lactylic stearates) are used as conditioners to produce a more stable dough and help ensure good volume for bakery products, baking mixes, pancakes and waffles, gelatins and puddings, and dietetic foods. Stearoyl lactylates are also used as whipping agents for vegetable oil toppings and icings and as emulsifiers to help blend fat and water fractions in products such as coffee cream substitutes and salad dressings.

SAFETY: A survey conducted for the National Academy of Sciences on use of these two additives by the food industry in 1976 determined that, on the average, 26 milligrams of calcium and sodium stearoyl-2-lactylates was present as additives in the daily diet of a U.S. adult, or slightly more than 0.4 milligram per kilogram of body weight (kg/bw) for a person weighing 60 kg (132 lbs.).

Short-term studies have been conducted with rats, but none has been reported for the longer period necessary to observe the possibility of malignancies, effects on the offspring, and so on. Perhaps this is because the stearoyl lactylates are handled in the body in the same manner as equivalent amounts of lactic acid and stearic acid, which are normal ingredients present in the diet and are regularly dealt with by our metabolic processes. The lactylates are readily split by enzymes in the body to yield these two acidic compounds.

In short-term studies with rats, calcium stearoyl lactylate at a level of 2 percent or more in the diet (perhaps 1600 milligrams per kg/bw) caused some reduction in growth over a four- to six-week period, and the liver weight was increased. This dosage is many thousands of times as great as the average human intake. Still higher levels of calcium or sodium stearoyl lactylate in the diet caused increased weights of other organs, including the brain, stomach, spleen, heart, and testes; but the

---

MAJOR REFERENCE: Toxicological evaluation of some food additives including anticaking agents, antimicrobials, antioxidants, emulsifiers and thickening agents. WHO Food Additives Series No. 5 (Geneva, 1974).

histology (microscopic structure) remained normal, with no pathological changes. Kidney weights remained unchanged. With dogs, calcium stearoyl lactylate at levels of some 3000 milligrams per kg/bw in the daily diet (thousands of times as great as average human consumption) allowed normal growth and metabolism over a two-year period, and organs did not change in weight and showed no abnormal microscopic changes.

Changes in inflammation and "granulation" of fat tissues have been observed in studies of rats employing very high levels of these additives in the diet. The changes were determined to be related to excessive intake of stearic acid. Adding unsaturated fats such as corn oil gave a better dietary balance and prevented or reversed the condition.

ASSESSMENT: Calcium and sodium stearoyl lactylates are readily split by the body to stearic acid and lactic acid, which can be metabolized normally. The adverse effects that have been reported occurred in rat studies that employed quantities of these compounds that far exceeded their presence in the human diet. One of these, changes in fatty tissues, appears to be related to a dietary imbalance caused by excessive intake of saturated fatty acids. The UN Joint FAO/WHO Expert Committee on Food Additives indicates that the intake of stearic acid from all sources may need to be taken into account; however, they consider that 20 milligrams per kg/bw of the stearoyl lactylates (thus, up to 1200 milligrams daily for a person weighing 132 pounds) is an acceptable daily intake for humans. As the actual presence of stearoyl lactylates in the U.S. diet is a small fraction of this, a cautionary rating appears to be unnecessary.

RATING: S.

CARAMEL

Caramel color is manufactured for commercial use by controlled heating of various sugars along with small quantities of an acid or a basic (alkaline) salt. Frequently, starch-derived sugars such as dextrose are employed, as are sucrose (table sugar), molasses, malt syrup, and lactose (milk sugar). The

MAJOR REFERENCE: Evaluation of the health aspects of caramel as a food ingredient. FASEB/SCOGS Report 20 (NTIS PB 266-880) 1973.

brown, caramelized color is imparted to many beverages, such as cola and root beer, and to a wide variety of processed foods. Caramel is also used as a flavoring. It is produced for this purpose commercially (or in the home) by heating sugar.

SAFETY: In 1976, the average daily intake of caramel as an additive by persons in the U.S., based on reported usage by the food industry, was 694 milligrams, or 12 milligrams per kilogram of body weight (kg/bw) for a person weighing 60 kg (132 lbs.). Studies with recently weaned rats fed for 90 days at levels 750 times the amount consumed as additives by humans (adjusted for body weight) determined that the only adverse effect experienced was that a greater amount of feed was required to produce a weight gain. Dosages over 4500 times human consumption (adjusted for body weight) fed to dogs for the same period failed to cause any damage to organs or to various body functions.

One human subject consumed up to 100 times the average U.S. intake of added caramel for 20 days without experiencing any adverse effects. In a long-term study, two generations of rats showed no harmful effects from daily feedings equivalent to almost 1000 times human intake, suggesting that caramel will not affect reproduction or lead to cancer. However, investigations specifically directed to determine whether caramel can cause cancer, gene mutation, or abnormalities in fetuses or offspring have not been located.

Some feed supplements for cattle, treated in a manner similar to that used in one of the processes of caramel manufacture (the ammonia process), produced toxic effects due to the presence of 4-methylimidazole, a compound containing nitrogen which has been found to occur in some food-grade caramels, but at levels regarded as insignificant with regard to possible hazard.

ASSESSMENT: Caramel as presently manufactured appears to be safe. Since questions have been raised concerning nitrogen-containing impurities that could be present in caramel produced by the ammonia process, limitation of the presence of these in the specifications for the additive would be desirable. The studies with both animals and humans indicate that caramel poses no hazard to the consumer at levels now consumed or likely to be consumed in future years.

RATING: S.

CARBONATES & BICARBONATES

Calcium Carbonate; Potassium Carbonate; Potassium Bicarbonate; Sodium Bicarbonate (Bicarbonate of Soda); Sodium Sesquicarbonate

Carbonates and bicarbonates are natural constituents of many foods, and are present in the fluids and tissues of the body as a product of its normal metabolic processes, where they play an important role in the control of the acid-alkali balance. The compounds specified above are some of their salts.* They are used as additives in foods to neutralize excess acidity and as leavening agents to lighten or raise dough in bakery products. Calcium carbonate is also considered to be a nutrient and a diet supplement, because it is useful in providing mineral for bone and tooth formation.

SAFETY: These salts as additives in food in the U.S. were estimated as totaling 1013 milligrams daily per person in 1975, or 17 milligrams per kilogram of body weight (kg/bw) for a person weighing 60 kg (132 lbs.). Sodium bicarbonate accounted for more than half, and sodium carbonate and calcium carbonate for most of the remainder. Actual daily intake is less, for when used as leavening agents the added carbonates disappear before the food is consumed.

Few experiments have attempted to determine whether the carbonate compounds are toxic when ingested orally in dosages approximating their intake as additives. *Sodium bicarbonate* injected into rats over a seven-day period at levels averaging 30 times the amount of carbonate salts added in human food, and over 50 times that amount in rabbits, did not cause harm to tissues and organs. However, when fed to dogs at almost 600 times the amount in human diets (adjusted for body weight) for 30 to 114 days, these unusually large doses did result in serious kidney damage and increased mortality due to the increased acidity.

Treatment of patients with large amounts of these salts has been conducted for years, but only rarely have disturbed acid-alkali balances been reported. Imbalance did occur in patients

---

MAJOR REFERENCE: Evaluation of the health aspects of carbonates and bicarbonates as food ingredients. FASEB/SCOGS Report 26 (NTIS PB 254-535) 1975.

*See p. 477 for ammonium carbonate and bicarbonate, and p. 585 for magnesium carbonate.

with gastric or peptic ulcers who were given daily doses of up to 100 grams of sodium bicarbonate by tube for three weeks, the equivalent of 98 times the amount of all these added carbonates in human food. The patients retained large amounts of sodium and developed alkalosis (a condition related to acid-alkali imbalance), but their kidneys were not damaged.

Sodium bicarbonate in laboratory tests proved not to be mutagenic (a cause of gene mutation); nor did it or sodium carbonate cause birth abnormalities in mice or rats.

A one-to-one ratio of calcium and phosphorus is desirable. A marked excess of calcium salts or phosphate salts in the diet can inhibit iron absorption, causing impaired iron utilization and anemia. Female mice were bred after a week on diets supplemented with *calcium carbonate* at 220 to 880 times the human intake of this substance as an additive. At all dosage levels, the number and weight of the first and second litters of newly weaned mice were lowered, and mortality was increased. The highest level caused heart enlargement. Supplementing the maternal diet with iron prevented this, indicating that the effects were really attributable to a mineral imbalance induced by excessive calcium.

In humans, 500 milligrams per kg/bw of calcium carbonate was fed daily to peptic ulcer patients for three weeks, over 145 times the amount ingested as an additive, resulting, in some patients, in an excess of calcium in the blood, accompanied by nausea, weakness, and dizziness.

In laboratory tests *potassium carbonate* proved not to cause gene mutations. Experiments with mice and rats indicate that it does not adversely affect maternal or fetal survival or cause birth abnormalities.

No investigations appear to have been conducted with *sodium sesquicarbonate,* but its biochemical conversion and metabolism in the body is similar to that of sodium carbonate and bicarbonate. Nor have any studies been located that have investigated the cancer-causing possibilities of any of these carbonates.

ASSESSMENT: The findings from animal feeding experiments and therapy with patients are not readily translatable to determine the level at which added carbonate salts in food can be harmful to humans. This is because of the wide discrepancy between the amounts administered and human intake as additives. Large quantities of calcium carbonate can interfere with reproductive performance, but this appears to be traceable to nutritive deficiencies caused by excessive calcium intake, which seem reversible with iron supplementation.

The carbonates are normal products of the body's metabolic processes and are essential for maintenance of its acid-alkali balance. In the quantities currently present in the diet as additives, or that might reasonably be expected in the future, the carbonate salts evaluated here are not regarded as posing any hazard to health.

RATING: S.

CAROTENE

Beta-Carotene

Carotene is a yellow-orange pigment that occurs naturally in many fruits and vegetables and is responsible for much of the yellow, orange, or red coloration of edible plants. Vegetables high in carotene content are parsley, carrots, spinach, turnip and beet greens, broccoli, and watercress. Fruits are less so; however, cantaloupe, peaches, apricots, prunes, and papaya are rich sources. Carotene can be converted by the body to vitamin A (see p. 665). Because the body can convert as much as 50 percent of carotene to this vitamin, nutritionists recommend at least one serving daily of a leafy green or yellow vegetable.

As an additive, carotene often is used in foods as a nutritional supplement, but its principal use is as a food colorant because of its intense and oil-soluble orange-yellow color. While there are a number of carotenes, *beta-carotene,* the only one used, is the important one for man as it possesses the most vitamin A activity. It is produced synthetically for commercial use in foods and it is added to many products such as margarine, cheese, ice cream, cake mixes, orange beverage, and puddings.

SAFETY: The average daily intake of vitamin A and its precursor carotene per person in the U.S. has been estimated at 5100 International Units (IU),* two-thirds of which (about 2 milli-

MAJOR REFERENCE: Evaluation of the health aspects of carotenes as food ingredients. FASEB/SCOGS Report 111 (NTIS PB 80-119837) 1979.

*The quantity of vitamin A activity is expressed by nutritionists in International Units (IU). For beta-carotene, 1700 IU are the equivalent of 1 milligram.

grams) comes from carotene present in the diet; 5000 to 6000 IU (3 milligrams) on the average probably would be sufficient for the body's needs. A survey in 1976 found that the carotene added to foods came to 0.67 milligram per person daily, a modest contribution to total intake.

When unusually large doses of carotene are taken, such as an excessive intake of carrots or carrot juice, the storage of carotene in the body can lead to an orange pigmentation of the skin. The condition, termed carotenemia, appears to be harmless, because when such food is removed from the diet, the pigment gradually fades away. Carotene has been used therapeutically to minimize an abnormal sensitivity to light. Doses of up to 180 milligrams of carotene daily for adults and 90 milligrams for children for long periods of time were tolerated well and without adverse symptoms other than yellowish skin coloring. The indications are that carotene taken in large amounts is inefficiently converted to vitamin A (see p. 665); apparently even these large doses of carotene do not lead to hypervitaminosis A, an excessive level of vitamin A in the body which can cause nausea, severe headache, loss of hair, and other toxic symptoms.

Carotene fed at very high levels to laboratory animals did not adversely affect fertility and the normal production of offspring. It did not cause cancer. A high incidence of skeletal abnormalities in fetuses did occur in a single study when pregnant rats were fed thousands of times the amount of added carotene contained in the human diet. The observations were difficult to interpret and as yet have not been confirmed.

ASSESSMENT: Carotene is present in a large variety of colored fruits and vegetables, where it serves as a source of vitamin A when metabolized by the body. Doses administered to humans far greater than the amount normally added to foods have been proved to be nontoxic. Carotene as a food additive poses no hazard to the consumer at current levels or at levels that may be expected in the future.

RATING: S.

CASEIN

AMMONIUM, CALCIUM, MAGNESIUM, POTASSIUM, SODIUM CASEINATE

Casein, the major protein in milk and milk products, has been a component in man's diet for centuries. The casein added to food is prepared from skim milk by processes that accelerate the formation of the casein curd and wash and dry it. The caseinates listed above are produced by dissolving casein in the desired alkaline solution, then spray or roller drying with the appropriate compound.

These substances are used for different purposes in food; some are added to serve as binders and extenders in imitation sausage, soups, and stews; as clarifying agents in wine; certain ones are permitted in specified frozen desserts; and calcium caseinate qualifies as a dietary supplement.

SAFETY: It is estimated that in 1970 the daily consumption of added casein and caseinates in food averaged 200 milligrams per person in the U.S. Sodium caseinate, the most widely used, accounted for two-thirds of this. About 100 times this amount of casein is consumed in the daily diet as a natural component of milk, cheese, and other dairy products.

The heating of casein, along with the alkali-treating process, results in the formation of some lysinoalanine (LAL). In a number of feeding studies with rats, LAL has caused kidney damage when present in the diet at levels of 100 parts per million (0.01 percent). Sensitivity to LAL appears to differ among animal species; when fed to mice, hamsters, rabbits, quail, dogs, and monkeys, LAL did no harm, possibly because of differences in metabolism. LAL has been detected in samplings of casein, calcium caseinate, and sodium caseinate used in food, at levels ranging up to 0.6 percent, although in some instances none at all was present.

Clinical sensitivity to cow's milk occurs in a small percentage of children. One study of subjects allergic to milk showed that 60 percent of them reacted allergically when they ingested casein. Sensitivity may increase or spread when added casein interacts during digestion with pepsin, the enzyme catalyst in gastric juice, or with other ingredients in processed foods.

MAJOR REFERENCE: Evaluation of the health aspects of casein and caseinates as food ingredients. FASEB/SCOGS Report 96 (NTIS PB 301-401) 1979.

Fortunately, only a small number of infants and children who suffer from this affliction carry it beyond age six.

The occurrence of nitrite (see p. 631) in spray-dried soy protein suggests that it may be present in spray-dried caseinates, which are manufactured by a similar process, although data do not exist to confirm this. Tests to determine whether casein or caseinates can cause birth defects or gene mutations have not been conducted, but the presence of LAL apparently does not produce abnormalities in the offspring of experimental animals.

ASSESSMENT: Casein as a natural protein component in cow's milk does not represent a hazard to health except to the few who are allergic to it. While its addition to food as an additive amounts to only a small fraction of its total presence in the diet of U.S. humans, some of the processes by means of which it is extracted and converted to caseinate can form lysinoalanine (LAL), a substance which has been found to cause kidney damage in experiments with rats, though not to several other species of animals, including subhuman primates, which is reassuring. For those allergic to milk, the possibility exists that casein as an additive can cause at least mild sensitive reactions.

RATING: Casein S for most people; X for anyone allergic to cow's milk. Caseinates ? because of uncertainties about the safety of LAL in the human diet.

CELLULOSE DERIVATIVES

Hydroxypropyl Cellulose; Hydroxypropylmethyl Cellulose; Methyl Cellulose; Methyl Ethyl Cellulose; Pure and Unregenerated Cellulose (Including Microcrystalline Cellulose); Sodium Carboxymethyl Cellulose (Carboxymethyl Cellulose, Cellulose Gum)

Cellulose is a natural fibrous carbohydrate present in the cell walls of green plants; it provides most of their structural strength. The cellulose used in food additives is obtained from wood pulp and cotton lint and is converted by chemical synthesis into forms or derivatives suitable for incorporation into

---

MAJOR REFERENCE: Evaluation of the health aspects of cellulose and certain cellulose derivatives as food ingredients. FASEB/SCOGS Report 25 (NTIS PB 274-667) 1974.

foods, or for use as surfacing in food packaging materials. The six cellulose derivatives listed above are permitted in food products and are assessed here for their safety. Three more that are allowed solely in packaging (and need not be listed on labels) are not covered here: carboxymethyl cellulose, cellulose acetate, and ethyl cellulose. It is unlikely that any of these could migrate to food in sufficient quantity to be of consequence; one, carboxymethyl cellulose, converts to sodium carboxymethyl cellulose (which is covered in this report).

Cellulose compounds are used in foods as thickeners. They add bulk, act as stabilizers that prevent separation and control texture, aid in blending of ingredients, help convert liquids to granular form or smooth-spreading gels, and prevent caking. They are used in such sweets as candy, ice cream, fillings, icings, and jellies, and they provide satisfying "body" to diet foods and breads by adding volume while increasing the "dietary fiber" content.

SAFETY: The amount of cellulose derivatives added to food in 1976, based on reported use by the food industry, totaled 40 milligrams per person per day, or 0.67 milligram per kilogram of body weight (kg/bw) for a human weighing 60 kg (132 lbs.). It has been suggested, since cellulose is contained naturally in many foods (cereals, vegetables, fruits), that the total amount of cellulose fiber in the daily diet may be 100 to 200 times the quantity added to processed foods.

The evidence from many metabolic studies indicates that ingested cellulose derivatives are not digestible, pass rapidly through the intestinal tracts of rats and man, and are excreted almost intact with little likelihood of being absorbed and stored in the body's tissues. It is not surprising under the circumstances that a diet consisting of 30 percent *microcrystalline cellulose* fed to rats for 72 weeks did not affect their appearance, behavior, or survival rate, although it did cause lower body weight in females and in some organs. The identical diet fed to three generations of rats lowered the reproduction rate, although it did not result in deformities in the offspring. The investigators attributed the adverse effects on reproduction to nutritional deficiencies caused by the high proportion of cellulose in the diet rather than to the derivative itself. Humans, when fed 30 grams of microcrystalline cellulose daily—more than 700 times the added cellulose in the average diet—for six weeks, did not experience any adverse effects on body tissues and organs.

Methyl cellulose, fed to a variety of animal species in

amounts far exceeding the added cellulose in the human diet for periods up to 95 days, did not affect growth rate, body weight gain, or behavior; nor did the animals suffer other toxic effects, tumor growths, or undesirable pathological changes, even after two years on diets consisting of 5 percent of this additive. Six normal humans taking up to 12 grams of methyl cellulose daily for a week or more, and 30 patients taking half again as much for as long as eight months, reported relief from constipation, and the continued use did not produce evidence of harm.

The survival rate during pregnancy was not affected in mice fed daily doses of methyl cellulose by tube up to 345 milligrams per kg/bw (over 100,000 times the average human consumption of added methyl cellulose, adjusted for body weight). Fertility and fetal survival were unaffected, and the incidence of abnormal young did not differ from that of the control group. However, when the dosage was raised to 1600 milligrams, it caused a significant increase in the mortality of dams, fewer fetuses survived, and these were smaller and retarded in maturation, although this did not occur with rats or rabbits.

Hydroxypropylmethyl cellulose, which is prepared from methyl cellulose, when fed to rats at 10 percent of their diet for three months, did not cause harm except for slight growth retardation. A dog ingested 25 grams daily for a month without adverse effects; at 50 grams daily (25,000 times its presence in the human diet) the substance caused anemia. Studies of effects when it is administered during pregnancy do not appear to have been conducted.

Sodium carboxymethyl cellulose, when fed to a variety of experimental animals at 20 percent of the diet, or up to 1 gram per kg/bw daily, for six months or longer, did not demonstrate ill effects. It has been used as a laxative in a study of 250 adult humans over a three-year period, given 2 to 18 grams twice daily, without harm. A reproduction study covering three generations of rats placed on diets containing 1 gram of sodium carboxymethyl cellulose per kg/bw did not find any undue effect on weight gain or body tissues or organs, or other evidences of hazard.

ASSESSMENT: Little or no breakdown and absorption of the cellulose derivatives considered in this review occurs in animal and human digestive tracts. Large amounts appear to have little effect other than providing "dietary fiber" bulk and thereby reducing the nutritive value of the diet, and possibly exerting a mild laxative effect. The increase in mortality of dams and retardation of unborn and developing young observed when extremely high dosages of methyl cellulose were administered to pregnant mice did not occur at lower dosages that were still substantially greater than the presence of cellulose derivatives in the human diet. Data are not available for hydroxypropyl cellulose or methyl ethyl cellulose, and until appropriate research is conducted on these a caution is indicated.

RATING: S for cellulose derivatives other than hydroxypropyl cellulose and methyl ethyl cellulose; for these ? (because of lack of information).

CHEWING-GUM BASE

The Food and Drugs section of the Code of Federal Regulations for chewing-gum base specifies 18 masticatory substances of vegetable origin consisting of coagulated or concentrated latex (frequently chicle), 8 synthetic masticatory materials, 13 plasticizing materials (softeners), a synthetic and a natural resin, and 5 chemicals used as antioxidants and for other preserving needs. These are permitted to be used separately or in combinations as a chewing-gum base. Some are reviewed elsewhere in this book. In addition, any food additive designated as GRAS* (many reviewed in this book) may also be used for this purpose, and these number in the hundreds. The regulation requires only that "chewing-gum base," if used, be listed on the label, but that the substance (or substances) comprising it need not be identified. It serves no purpose to provide assessments here of each of these, since there is no way of determining which one (or ones) have been used in a particular gum; and while some of the substances used for this purpose are viewed as of health concern when present in food which is ingested, it is doubtful that much gum base is eaten or swallowed, or even if it is, that much of the substances comprising it would "migrate" out of the digestive tract and into the bloodstream.

MAJOR REFERENCE: Code of Federal Regulations, Food & Drugs, *21*, Part 172.615, revised April 1, 1979.

*The Food Additives Amendments of 1958 exempted from FDA approval of safety substances that had been added to food prior to this legislation, which, based on the knowledge and experience of experts, were "generally recognized as safe." These additives are designated as GRAS.

CHOLINE BITARTRATE; CHOLINE CHLORIDE

These two choline salts are chemically synthesized for use as food additives. Choline is universally found in plant and animal products because it is a component of the fatty material, lecithin (see p. 580), in cell membranes. Egg yolk, meat, fish, milk, cereals, and legumes are particularly rich in choline. The average daily mixed diet may contain 150 milligrams of choline. Choline is added to milk products or infant formulas as a nutrient or dietary supplement to bring them up to the choline level normally found in milk.

SAFETY: Most of the choline salts used will be consumed by infants in formulas. Infants less than six months old on milk-free formulas will receive 70 to 80 milligrams of added choline chloride daily (15 milligrams per kilogram of body weight, kg/bw); when milk-based formulas are used, choline bitartrate intake will amount to about 30 milligrams per kg/bw. Older children and adults consuming choline as a supplement or on special diets will average less than this amount.

Choline helps the body's metabolism to prevent excessive fat deposition in the liver, and as a constituent of lecithin it is useful in transport of fat throughout the body. It is not an essential ingredient in the diet, like a vitamin, since it can be made within an animal or human if the diet contains adequate protein.

Rats and mice receiving 500 to 1000 milligrams choline chloride per kg/bw in the diet showed normal growth. Some growth depression resulted from higher amounts. However, no significant tissue changes due to the choline were found. No studies have been reported on any effects of choline on cancer, gene mutation, or embryo development during pregnancy; or on any long-term effects.

ASSESSMENT: Choline is an important constituent in metabolic functions in the body, including fat transport. However, its nutritional significance for humans still needs to be clarified. Data on possible harmful effects are scarce but, as noted by the UN Joint FAO/WHO Expert Committee on Food Addi-

---

MAJOR REFERENCE: Evaluation of the health aspects of choline chloride and choline bitartrate as food ingredients. FASEB/SCOGS Report 42 (NTIS PB 262-654) 1976.

tives, it appears to be without adverse effect when taken orally. Present evidence indicates that choline bitartrate and choline chloride as additives in foods pose no hazard even to the infant consumer.

RATING: S.

CITRIC ACID AND CITRATES

Ammonium Citrate; Calcium Citrate; Isopropyl Citrate; Mono-, Di-, Tripotassium Citrate; Mono-, Di-, Trisodium Citrate; Stearyl Citrate; Triethyl Citrate

Citric acid and its salts (ammonium, calcium, potassium, and sodium citrates) are natural constituents of plants and animals. The salts readily dissolve in water, releasing the free acid. Other derivatives of citric acid that are used as food additives are the isopropyl, stearyl, and triethyl citrates, which chemically are esters. (An ester is a compound formed from an acid and an alcohol by the removal of water.)

The acid and its salts are widely distributed in foodstuffs and are in especially high concentrations in citrus fruits and products. Citric acid is the dominant organic acid in apricots, many berries, pineapple, and tomatoes. By contrast, the esters occur naturally in foods only in extremely low quantities.

These substances have multiple uses as food additives. Citric acid and some of the salts are good sequestrants, collecting and deactivating metal contaminants, and thus may increase the effectiveness of preservatives used to maintain freshness and prevent rancidity. As flavor enhancers, citrates are used to impart tartness and to control acidity in foods and beverages. Calcium citrate is used as a firming agent in canned vegetables and as a nutrient in baby foods. The citrate esters are useful because of their solubility in fatty portions of foods such as margarines; triethyl citrate in particular performs functions as an additive somewhat similar to the citrate salts.

SAFETY: The use of citrates as food additives adds a relatively small part of the total consumed in human diets. The per capita intake in 1970 was estimated at approximately 500 milligrams. A resurvey in 1975 indicated an increase to more than 750

MAJOR REFERENCE: Evaluation of the health aspects of citric acid and citrates as food ingredients. FASEB/SCOGS Report 84 (NTIS PB 280-954) 1977.

milligrams in the daily diet of the average U.S. adult. This is the equivalent of the citrate intake from 3 ounces of orange juice.

It is well established that citrates participate in body metabolism as key intermediates in the process of deriving energy from carbohydrates, proteins, and fats, and can serve as energy-producing nutrients themselves. The esters, like the salts, are digested and metabolized normally. Understandably, citric acid and its salts and esters have very low levels of toxicity.

In adult human subjects, ammonium citrate added to the diet has been demonstrated useful as a nitrogen source and has a sparing effect on the protein requirement. Long-term studies with rats have shown that citric acid, even at 150 times the added amount found in human diets (adjusted for body weight), had no effect on survival or on the various organs and tissues. The citrate esters showed no adverse effects in rats in two-year feeding studies at even higher dosage levels. Because of its ability to bind calcium, citric acid may interfere with excess calcium absorption if the diet is deficient in phosphorus. Citric acid, its salts, and triethyl citrate neither alter the development of the fetus nor produce chromosome alterations. (Similar tests have not been conducted with the other esters.)

ASSESSMENT: Citric acid, its salts, and its esters are metabolized normally by the body. They are normal constituents of many foods, and are present naturally at far higher levels than the amount added to food in processed products. These compounds appear to pose no hazard to the consumer in amounts likely to be present in the diet.

RATING: S.

COCOA PROCESSED WITH ALKALI (DUTCHED)

DIOCTYL SODIUM SULFOSUCCINATE

Cocoa products may be processed from the cacao bean through heating with various mild alkalies, such as sodium, ammonium, or potassium bicarbonate, carbonate, or with the oxide or hydroxide. Magnesium carbonate or oxide also may be used. This alkali treatment is called the Dutch process.

Powdered cocoa is a hard-to-wet food; for some uses, a dispersing and solubilizing chemical, dioctyl sodium sulfosuccinate (a synthetic compound), is added to it (and other dry drink mixes) to serve as a wetting agent to make it easier to blend the dry powder with liquids.

SAFETY: The effects of the various means of processing or dispersing of cocoa come from either the alkali or the wetting agent. The alkalies used in the Dutch process are discussed and evaluated S on pages 477, 503, and 604.

Dioctyl sodium sulfosuccinate (DSS) as a dispersing agent in cocoa is not permitted to exceed 0.4 percent by weight. This compound has been studied in a few experiments with rats. Three successive generations received up to 1 percent of their diet as DSS. For the first mating, measures of fertility and gestation remained high. The ability of the young to survive and lactation by the mother were somewhat depressed in subsequent matings, but still similar to those of control animals. However, the mean weight of the young decreased with increased amounts of DSS in the diets of mothers. No significant changes due to the chemical showed up on autopsy. Another group of rats, which consumed these diets for two years, displayed no gross pathological changes or changes in the various organs and tissues.

Monkeys administered 125 milligrams per kilogram of body weight (kg/bw) of DSS by tube into the stomach for 24 weeks experienced gastrointestinal irritation, but autopsy did not reveal any organ abnormalities. Dogs given up to 0.25 milligram per kg/bw of DSS in their diets for the same period had no adverse effects; but higher levels caused gastrointestinal irrita-

---

MAJOR REFERENCE: Toxicological evaluation of some food colours, enzymes, flavour enhancers, thickening agents, and certain other food additives. WHO Food Additives Series No. 6 (Geneva, 1975).

tion. The high level of 500 milligrams of DSS per kg/bw fed by tube to rabbits caused severe diarrhea, loss of appetite, and death.

For many years DSS has been used as a bowel softener during chronic constipation for infants, children, and adults, involving doses up to 100 milligrams per day. Patch tests on humans show DSS to be nonirritating and not an allergic compound. Laboratory tests have not been conducted to assess whether DSS can induce cancer or gene mutation.

ASSESSMENT: The alkalies used in the Dutch process of cocoa demonstrate no hazard to the consumer. The additive dioctyl sodium sulfosuccinate, however, has been tested only minimally, and even with rats some suggestions of depressing effects on lactation and on the very young were noted. It also has caused gastrointestinal irritation in monkeys and dogs. More complete long-term studies are needed to properly evaluate possible hazard to the consumer.

RATING: S for cocoa processed with alkali; ? for dioctyl sodium sulfosuccinate.

COPPER SALTS

Basic Copper Carbonate (Copper Carbonate and Copper Hydroxide); Copper (Cupric) Gluconate; Copper (Cupric) Sulfate; Cuprous Iodide

Copper is a mineral trace element that is essential for man and other animals, and for most plants. Trace elements are necessary for the activity of various enzymes, hormones, and vitamins. Humans depend on obtaining copper almost entirely from food, water at times making a contribution. Shellfish, liver, nuts, wheat germ, legumes, and cocoa are rich sources of copper.

Basic copper carbonate, a combination of copper carbonate and copper hydroxide, and *copper (cupric) gluconate* are added to food as nutrients and as diet supplements. *Copper (cupric) sulfate* is used as a supplement in special dietary foods and in infant formulas. *Cuprous iodide* when present in table

---

MAJOR REFERENCE: Evaluation of the health aspects of copper gluconate as a food ingredient and copper sulfate as it may migrate to foods from packaging materials. FASEB/SCOGS Report 98 (NTIS PB 301-400) 1979.

salt provides iodine. The copper compounds are also used to clarify and stabilize wines and as preservatives in packaging materials that contact foods.

SAFETY: Most estimates of the total copper intake of an average adult for a single day's diet range from 2 to 4 milligrams, or 0.03 to 0.06 milligram per kilogram of body weight (kg/bw) for a person weighing 60 kg (132 lbs.). The National Academy of Sciences-National Research Council's Recommended Dietary Allowances advise a daily intake of 2 to 3 milligrams of copper for adults. A satisfactory balance in the body stores of the normal adult can be maintained on an intake of 0.03 milligram per kg/bw of copper per day, and somewhat larger amounts for infants and children.

The contribution of copper to the diet by copper salts used as additives is trivial compared to what is naturally present in food that is eaten. It was estimated that in 1975-76 only 0.05 milligram of copper gluconate and 0.02 milligram of copper sulfate were consumed daily by the average U.S. adult, and only a fraction of these weights consists of copper. Similar information for basic copper carbonate is not available, but its use in food is believed to be even less; and when last reported, cuprous iodide was not being used as an iodizing agent.

True copper deficiency is relatively rare in man because of the widespread distribution of copper in the ecosystem. However, copper deficiency has been reported in infants who are severely malnourished or who suffer from a rare genetic disease. Some scientists, on the other hand, believe that the usual copper intake is somewhat less than it should be. Copper intoxication also is rare, as man's intestinal capacity for absorption of copper that is ingested is limited, and much of it remains unabsorbed and is excreted. But cases of copper poisoning have been reported: people ingesting 5 to 32 milligrams of copper from cocktails mixed in a copper-lined container exhibited weakness, abdominal cramps, dizziness, and headache. A woman administered 15 milligrams per kg/bw of copper sulfate as an emetic, several thousand times the average daily intake of copper as an additive, died.

Adverse effects in experiments with animals were observed only when the dosages of copper salts that were administered far exceeded human consumption. Sheep appear to be particularly sensitive to copper salts; when fed 30 milligrams of copper sulfate per kg/bw daily for 26 to 73 days (some 30,000 times human consumption of copper salts), they suffered injury to liver, kidneys, and heart. Yet twice the amount produced no ill effect on ponies, and an even greater intake did not harm pigs. Rats fed diets containing 53 milligrams per kg/bw of copper in the form of copper gluconate or copper sulfate, thousands of times as much as the copper intake by humans from their total diet, did not suffer adverse effects in comparison with controls.

Microbial test-tube experiments indicated that copper gluconate, copper sulfate, and cuprous iodide will not cause gene mutations. Nor did copper gluconate or copper sulfate prove to be carcinogenic (cause of cancer) when administered by mouth. These tests were not conducted for cuprous iodide. Copper gluconate produced birth defects when injected into the egg in chicken embryos, but did not produce the same effects in mice or rats when given by stomach tube to pregnant animals. Copper sulfate was toxic to embryos and caused defective births when injected in large amounts in pregnant hamsters.

Research on basic copper carbonate has not been located.

ASSESSMENT: Copper is an essential trace element for man. The customary adult diet provides adequate copper to prevent a deficiency, as it is present in many foods. As the body can absorb only a fraction of the intake, excreting the rest, copper intoxication is rare. The amount contributed by copper salts in the average diet is negligible.

Although in humans adverse effects from copper have been reported, they have occurred only from ingestion of amounts far exceeding normal consumption. This is confirmed by the large quantity needed to cause harm in experiments with a variety of animal species. At the levels at which copper salts are present as additives in food, or that might reasonably be expected in the future, they do not pose a hazard.

RATING: S.

CORN SWEETENERS

Corn Syrup (Glucose Syrup); Dextrose (Corn Sugar, Glucose); Fructose (Fruit Sugar, Levulose); High-Fructose Corn Syrup; Invert Sugar; Invert Syrup; Maltodextrin

These carbohydrates are discussed together because they are interrelated in composition and in many properties. When corn starch is partially split chemically by heating in the presence of an acid (hydrolysis), *corn syrups* with varying proportions of dextrins (partially hydrolyzed starch; see p. 521) and free dextrose are obtained. Complete splitting by hydrolysis of the starch can lead to the simple sugar *dextrose* or glucose, or its syrup. *Maltodextrins* are manufactured by a similar procedure except that the hydrolysis is not as complete.

Glucose syrup may also be treated with an enzyme (a protein catalyst) that can convert glucose to its close relative, *fructose*, which is sweeter; such converted syrups are termed *high-fructose corn syrups*. They may closely approximate the composition of an invert syrup, which also contains equal parts of dextrose and fructose. *Invert sugar* or *invert syrup* is prepared by the splitting of sucrose (or common table sugar; see p. 648).

All these additives are used in foods as sweeteners. Dextrose has about three-fourths the sweetness of sucrose (cane or beet sugar), while invert sugar and fructose are somewhat sweeter than sucrose.

SAFETY: The presence of corn sweeteners in the daily diet of a person in the U.S. in 1976, based on usage reported by food manufacturers, was estimated at 18 grams of corn syrup, 22 to 25 milligrams each of dextrose and fructose, and close to 300 milligrams of invert sugar. High-fructose corn syrup has been replacing an appreciable amount of sucrose previously used in processed food as a sweetener.

Glucose and fructose are normal constituents of the body. Blood sugar is glucose. Fructose is largely converted to glucose when metabolized in the body. Corn syrup is readily broken down by digestive juices, and the resulting glucose is then absorbed and utilized as an energy source.

Studies have shown that high levels of fructose sugars in the diet, which may come from fructose itself or from sucrose, invert sugar, or high-fructose corn syrup, can increase the blood triglyceride (neutral fat) level in humans, though it does not have this effect on cholesterol levels. These studies involved some 30 times the average intake, and the triglyceride fraction in serum lipids is not a prime consideration in risk of coronary heart disease.

Studies have shown that corn syrup and corn sugar (dextrose) are free of allergenic properties, even for patients who are sensitive to corn starch. There are no confirmed reports of

---

MAJOR REFERENCE: Evaluation of the health aspects of corn sugar (dextrose), corn syrup, and invert sugar as food ingredients. FASEB/SCOGS Report 50 (NTIS PB 262-659) 1976.

cancer-producing responses to these sweeteners, and no known studies of effects on the developing embryo or genetic effects.

Sugars (and other carbohydrates) can be a serious hazard for some diabetics who are unable to produce a sufficient supply of insulin and cannot transform the glucose in sweeteners into energy; instead, glucose rises in the blood. The principal health concern for people is dental caries (cavities). All of these sugar products are readily fermentable and can support caries-producing organisms in the mouth. Sweet foods containing these additives may contribute to dental health problems, but more likely factors are the frequency and timing of eating, whether food remains in contact with the teeth, and dental hygiene habits.

ASSESSMENT: Dextrose, fructose, corn syrup, and invert sugar in the diet are all readily absorbed and utilized as energy sources by humans. Diabetics are an exception. Because of a metabolic disorder which interferes with utilization of sugar, they should avoid excessive amounts of it. A caution for them to avoid the substances under review here is not believed necessary, as they probably are, or should be, under medical supervision regarding their intake of all carbohydrates. There is no known hazard for nondiabetics at the present rate of consumption of these sweeteners as additives in the U.S. diet, other than their contribution to dental caries.*

RATING: S.

---

*Glucose, corn syrup, and related sweeteners have not been rated as hazardous because of their effect on cavities. While they play a role in tooth decay, they do so only in combination with other factors, such as improper dental hygiene, or between-meal snacks which contain fermentable carbohydrates, or water that is not fluoridated.

The rating procedure followed here does not consider an additive hazardous if it is harmful only when abused by the consumer, as almost anything can be. However, the sugar content in some dry breakfast cereals and instant breakfast bars and powders represents a sizable percentage of their total content. Because these percentages are not readily available, and because children consume much of these foods, they can be regarded as of concern to health. As a caution we have provided the percentage of sugar content in many of the brands in these categories, in the inventory section.

DEXTRIN

Dextrins are carbohydrates containing many molecules of glucose* chemically linked together, and are prepared commercially by dry-heating starch. This process changes some of its physical properties, including an increase in water solubility. Usually corn starch is the raw material, but potato starch, tapioca, and other starch sources may be used. The annual production and import of dextrins for direct food uses has been relatively stable for many years. In 1971 it was estimated at 30 million pounds, enough to provide a daily intake of 180 milligrams per person. Because of their water-holding properties, dextrins are used as food additives in baked goods, as thickeners in gravies and sauces, as coatings to prevent migration of oils from products containing nuts or from flavors in dry mixes. They are also employed as sugar substitutes.

SAFETY: The dextrins are only slightly poorer in digestibility than their parent starches. They too must be split to their constituent glucose units through digestive enzymes; the glucose is then absorbed and metabolized normally. Short-term and long-term studies with rats have not shown any adverse effects even where dextrin was a significant component of the diet (many thousandfold that of average human consumption, when adjusted for body weight).

There are no scientific studies on any effects of dextrin on such responses as allergy, cancer, embryo abnormalities, and genetic alterations.

ASSESSMENT: Dextrins behave in the body much like the starches from which they are derived. No toxic effects have been noted. Dextrins appear to present no hazard to the public at current levels in the diet or levels likely to be met in the future.

RATING: S.

MAJOR REFERENCE: Evaluation of the health aspects of dextrin and corn dextrin as food ingredients. FASEB/SCOGS Report 75 (NTIS PB 254-538) 1975.

*Glucose, also called dextrose, is a simple sugar found in plants and animals and is a source of energy in living organisms. See p. 518.

DIACETYL

STARTER CULTURE; STARTER DISTILLATE (BUTTER STARTER DISTILLATE)

When parts of milk are treated or cultured with selected harmless bacteria (*starter cultures* or bacterial starters), a mixture of flavor compounds is formed. The culture that results is distilled with steam to capture the flavor materials. The resulting *starter distillate,* also referred to as *butter starter distillate,* contains a large number of chemicals, but *diacetyl* is the major flavor component, comprising 80 to 90 percent of the mixture. The flavor chemicals in starter distillate account for less than 5 percent of the distillate, the remainder being water. Some 75 chemicals may be present, but obviously in very small amounts, since only a few hundredths of a percent of distillate will be incorporated into a food. Diacetyl, which occurs naturally in some fruits, berries, and cheese, usually is synthesized chemically rather than produced by natural bacterial fermentation. It is used alone as a flavoring ingredient in margarines, hard and chewy candies, and gum; it has a buttery odor and flavor. Starter distillate is also used in margarines, and to enhance the flavor of baked goods, fats and oils, dairy products and their analogs, beverages, and other processed foods.

SAFETY: The average daily consumption of starter distillate was 10 milligrams per person in 1978, or 0.25 milligram with the water removed. Use of diacetyl as a food additive, including its presence in starter distillate, was 0.3 milligram per person. The other starter distillate flavor components as a group contributed about 0.05 milligram to the daily diet.

The principal study on safety of the additive was to determine the amount of diacetyl that would cause any adverse effects in rats. Dosages of 30, 90, and 540 milligrams per kilogram of body weight (kg/bw) of the chemical were administered daily by stomach tube. After 90 days, some effects on blood and on organ weights were noted in animals receiving 540 milligrams per kg/bw, but no adverse effects were observed in those receiving 90. This amount is several thousand times the average dietary intake by humans when adjusted for body weight.

MAJOR REFERENCE: Evaluation of the health aspects of starter distillate and diacetyl as food ingredients. FASEB/SCOGS Report 94 (NTIS PB 80-178668) 1980.

Studies of possible chromosome alteration indicate lack of such activity from diacetyl or starter distillate. Starter distillate fed to pregnant rats, mice, hamsters, and rabbits did not harm either the fetus or the mother. The possibility that either of these compounds may cause cancer has not been investigated.

ASSESSMENT: Diacetyl accounts for 80 percent or more of the flavor ingredients in starter distillate. Less than a milligram is present in the average daily diet in the U.S. Several thousand times this amount has been administered to rats without adverse effects. The remaining components in starter distillate together contribute only hundredths of a milligram to the daily diet, any one of them representing a small fraction of this minute amount. While additional long-term studies on starter distillate of known composition and of diacetyl would be desirable, there is no reason at present to suspect any hazard to the consumer from the addition of either when used to enhance flavors of processed foods.

RATING: S.

DISODIUM 5′-RIBONUCLEOTIDES (RIBOTIDES)

Disodium 5′-Inosinate (IMP); Disodium 5′-Guanylate (GMP)

Disodium 5′-inosinate (referred to as *IMP*) and *disodium 5′-guanylate (GMP)* are prepared from ribonucleic acid (RNA) of yeast, or by a fermentation and partial chemical synthesis. RNA occurs in all cells, where it is responsible for the transfer of the genetic code and for protein synthesis, so it is understandable that the products derived from it occur widely in nature; meat, fish, and mushrooms are rich sources. IMP and GMP originally were produced in Japan from dried fish; they are principal flavor-enhancing components in this food.

The IMP and GMP ribonucleotides have the property of enhancing and improving food flavors without adding flavor of their own. Even at low concentrations, IMP and GMP help bring out "meaty" or "brothy" flavors in soups and bouillons, intensify the flavor of processed meat and fish, and can add to the "fresh" flavor of many canned products and mixes.

MAJOR REFERENCE: Toxicological evaluation of some food colours, enzymes, flavour enhancers, thickening agents, and certain other food additives. WHO Food Additives Series No. 6 (Geneva, 1975).

SAFETY: In 1976, usage of IMP and GMP as additives, according to a survey conducted by the National Academy of Sciences-National Research Council among food manufacturers, totaled approximately 0.5 milligram in the daily diet of the average U.S. adult, the intake being about equally divided between the two additives. This represented only a minuscule part of the nucleotide present as a natural ingredient in food in U.S. diets, where 2500 milligrams has been estimated as the average daily intake per person.

While IMP and GMP in the body are partially derived from the diet, they probably largely come from synthesis within the body, where they may be incorporated into various nitrogen-containing components of tissue, nucleic acids and genetic constituents, converted to the important energy mediator ATP in muscles, or broken down and metabolized to the excretory product uric acid.

Healthy humans have taken 2500 milligrams of IMP daily for a week with no signs of toxic effects, although serum and urinary uric acid levels increased sharply. Short-term and long-term diet studies have been made using mice, rats, and dogs. In rats, after 90 days of eating 1000 milligrams per kilogram of body weight (kg/bw) of IMP daily (equivalent to a human intake of 60 grams for a person weighing 60 kilograms), no adverse effects were observable on weight gain, organ size, or tissue and cellular structure. Similar negative findings were reported for animals receiving as much as 2 percent of these ribonucleotides in the diet (approximately 1000 milligrams per kg/bw daily) for up to two years.

Even at 8 percent of the diet fed to rats for a year, IMP did not cause toxicity, other than some calcification in the kidney due, it was reported, to a change in the concentration of the urine rather than the specific additive itself. Dogs also showed no significant adverse effects after two years on a diet containing 2 percent of a mixture of IMP and GMP (approximately 500 to 900 milligrams per kg/bw). IMP fed to dogs for two years at 2000 milligrams per kg/bw daily, more than double the amount in the previous study, again did not cause harm.

Neither of the additives has adverse effects on reproduction or on the offspring; nor do they produce malformed fetuses in mice, rats, monkeys, rabbits, or chicks.

ASSESSMENT: The safety of IMP and GMP has been demonstrated in many tests. The sole reservation is related to the metabolism of these and similar nitrogen-containing substances in the diet to uric acid. While this is normal and without complications in most people, those inflicted with gout accumulate an excess of uric acid which is deposited around their joints, causing a painful swelling and inflammation. The minuscule contribution of added IMP and GMP to the daily diet hardly warrants a caution to avoid these as additives in food. Usually people with gout are under diet restrictions and avoid foods rich in nucleotides such as organ meats, duck, and other "rich" foods, and 0.5 milligram is not likely to be of any consequence.

RATING: S.

EDTA (ETHYLENEDIAMINE TETRAACETIC ACID)

Calcium Disodium EDTA; Disodium EDTA

These two salts of ethylenediamine tetraacetic acid (EDTA) are capable of tightly binding with mineral elements present in foods (removing them from effective action), such as copper and iron. These metal contaminants may cause oxidation, rancidity, induce off-colors or off-flavors, and lead to texture changes. The EDTA salts thus may improve the stability of food. They are sometimes found in fruit drinks and beer, processed fruits, vegetables, vegetable juices, and margarines, salad dressings, mayonnaise, and condiments.

SAFETY: A survey conducted in 1976 for the National Academy of Sciences indicated that the average daily diet per person in the U.S. contained 1.3 milligrams of these EDTA salts, based on poundage of the additives used in processed foods. Only a small percentage of the EDTA is absorbed from normal diets containing one of its salts, and even this amount is rapidly excreted in the urine. At a level of 0.5 percent of the diet of dogs and rats for a period of two years, calcium disodium EDTA showed no toxic effects. This was true for four generations of offspring of rats receiving up to many thousands of times the amount present in the average human diet. When disodium EDTA was fed to rats in a two-year study at 5

MAJOR REFERENCE: Toxicological evaluation of some food additives including anticaking agents, antimicrobials, antioxidants, emulsifiers and thickening agents. WHO Food Additives Series No. 5 (Geneva, 1974).

percent of the diet—approximately 2500 milligrams per kilogram of body weight (kg/bw)—the only adverse symptom was diarrhea.

Disodium EDTA fed to pregnant rats at 3 percent of the diet (1500 milligrams per kg/bw) caused a reduction in litter size and the newborn were malformed. It is noteworthy that this did not occur if the zinc in the diet was raised. Since zinc deficiency is known to cause abnormalities in the newborn, the adverse effect noted above was ascribed to a deficiency due to the excess EDTA binding zinc in the tissues.

There are studies indicating that calcium disodium EDTA may cause an allergic reaction in some people, but this was after topical (surface) applications of pharmaceutical preparations containing the EDTA salt. This substance has proven to be useful medically in the treatment of metal poisoning and prevention of blood clotting; these clinical applications have demonstrated its safety for humans. Gross and microscopic findings in long-term studies with animals have uniformly been negative, indicating that the EDTA salts do not induce cancer.

The UN Joint FAO/WHO Expert Committee on Food Additives considers that 2.5 milligrams per kg/bw, or 150 milligrams for a person weighing 60 kg (132 lbs.), is an acceptable daily intake of these additives.

ASSESSMENT: The salts of EDTA are useful food additives because of their ability to bind and remove mineral contaminants that lower food quality and stability. Animal studies have demonstrated the safety of calcium disodium EDTA and disodium EDTA at feeding levels far higher than their presence in the human diet; and clinical applications substantiate that they are safe in humans. No health hazard to the consumer is posed by these additives even at levels in foods that may be expected in the future.

RATING: S.

ENZYMES

PROTEOLYTIC ENZYMES (PROTEASES, PROTEINASES)

Animal: Pepsin; Rennet (Rennin)
Microbial (Bacterial or Fungal): Fungal Protease; Fungal Enzyme derived from *Aspergillus (Aspergillus Oryzae);* Microbial Rennet
Plant: Bromelain (Bromelin); Ficin; Papain

CARBOHYDRASES

Animal: Alpha-Amylase
Microbial (Bacterial or Fungal): Alpha-Amylase; Glucoamylase (Amyloglucosidase)
Plant: Alpha-Amylase; Beta-Amylase

PHOSPHORYLASES

Enzymes are proteins and are present in all living matter. Most of them act as catalysts that accelerate the metabolic processes in which proteins, fats, and carbohydrates are broken down into simpler forms needed for the building and replacement of cells, and for energy required for essential activity. The human body contains many hundreds of varieties of enzymes; usually each produces only a single chemical change, and in a specific substance.

Man has known for centuries about the usefulness of enzymatic reactions in fermentation of yeast in beer and wine and of malted barley during the making of bread. Since the early part of this century food producers have made use of enzymes in the manufacture of cheese, baked goods, beer, and wine. Their enzyme sources have been *rennet* from mucosal tissues of young calves, *pepsin* from the stomach of pigs, *papain* from papaya fruit, *bromelain* from pineapple, *ficin* from figs, and *amylase* from malted barley. As it became known that various microorganisms (bacteria and fungi) could produce enzymes similar in their actions to those derived from animals and plants, and demand increased, production from these sources progressed rapidly. Microbial enzymes are less costly to produce and more readily available in the volume that food processors require.

MAJOR REFERENCES: Evaluation of the health aspects of rennet as a food ingredient. FASEB/SCOGS Report 76 (NTIS PB 274-668) 1977; Evaluation of the health aspects of papain as a food ingredient. FASEB/SCOGS Report 77 (NTIS PB 274-174) 1977; Toxicological evaluation of some food colours, enzymes, flavour enhancers, thickening agents, and certain other food additives. WHO Food Additives Series No. 6 (Geneva, 1975).

Three groups of enzymes are listed above. *Proteolytic enzymes,* also referred to as proteases or proteinases, are used when the catalytic action required is the splitting of the bonds or breakdown of certain proteins; *carbohydrases* are specific for starches or other carbohydrates; while *phosphorylases* are specific for inserting phosphorus (phosphate) into a molecule.

Individual enzymes are used in many ways to improve the quality of finished food. Papain and the other *plant proteases* are well-known meat tenderizers. They are also useful in splitting down large molecules in beer that otherwise would separate out and cause a haze when beer is chilled, and in the production of protein hydrolyzates (see p. 613). *Microbial proteases* have similar food uses, and may be used to modify the dough in cracker manufacture. *Fungal proteases* have been used for centuries in making soy sauce and in producing oriental foods. Animal and microbial rennet and pepsin, because of their milk-clotting activity, are of particular importance in cheese manufacture. Rennet, an extract which has been used for centuries in cheesemaking, contains the enzyme rennin, a catalyst which coagulates the protein in milk. Rennet is also used in frozen dairy desserts and mixes, puddings, gelatins, and fillings. Until recently it was obtained from the lining of stomachs of unweaned calves; a similar preparation called bovine rennet comes from the stomachs of older animals. An increase in the demand for cheese and a decrease in slaughtered calves led to the development of rennet substitutes. Pepsin, an enzyme from gastric juices, obtained from pigs, is used as an extender of rennet; microbial proteases may also be used as extenders.

The presence and balance of *alpha-* and *beta-amylases* are of significance in flour milling and baking. An important use of the amylases is in the production of sweet syrups from starch. FDA regulations permit addition of "a harmless preparation of enzymes of animal or plant origin capable of aiding in the curing or development of flavor" of many types of cheeses. Sometimes the specific enzyme doesn't appear on the ingredient label because the food—examples include various cheeses—has been "standardized" by FDA regulation. Thus a permitted "safe and suitable" microbial milk-clotting enzyme need not be separately mentioned on a label for cheese, or may be designated as "enzymes."

SAFETY: Enzymes extracted from animals and plants have been in use extensively for a long time, and questions have not arisen concerning the possibility of any hazard in their use in food processing. Little difference in chemical structure exists between them and microbial enzymes from bacterial and fungal sources; with few exceptions, all enzymes consist of a long, linear string of amino acids characteristic of many proteins. Because the absence of hazard has been assumed, few independent studies of bacterial and fungal enzymes have been published. Virtually all the research has been conducted by manufacturers petitioning the FDA for GRAS (generally recognized as safe) status, which would absolve them of the need for further proof of safety or for other use in foods. Generally, these studies have shown in animal experiments that large doses of purified microbial enzyme preparations are not toxic.

There is no information on the extent to which some of the enzymes are used as food additives, which limits knowledge of consumer exposure to them. It is certain to be less than the amount added to food during manufacture since often an enzyme is no longer present at the time the food is consumed. Much of the milk-clotting enzyme remains with the whey, and not with the curd from which cheese is made. Enzymes are readily deactivated by heat; thus if the food is pasteurized, baked, or otherwise heated after an enzyme treatment, enzyme activity will have been destroyed and the small amount of the added enzyme will behave merely as a food protein and be digested and metabolized like any other protein in the diet.

Amylases. A few studies of these enzymes have been reported in petitions by manufacturers to the FDA for GRAS status. Rats administered 3.5 to 7 grams orally per kilogram of body weight (kg/bw) of glucoamylase for 90 days showed no adverse effects. A single dose of 4 grams per kg/bw was fed by tube to ten mice without harm. No research has been located that employed alpha- or beta-amylase orally, though these enzymes are widely distributed in plants used as foods.

Bromelain is derived from the pineapple plant; it has been used for certain medical treatments, and this has enabled evaluation of effects of oral doses on humans. In separate studies, a total of 216 patients with inflammations, edema, bruises, or "black eye" purplish discolorations were given 2 to 8 bromelain tablets for two to ten days; generally, these conditions showed improvement. No side effects attributable to bromelain were reported. Doses of up to 10 grams of bromelain per kg/bw have been given orally to mice and rats without harm. Dogs given 750 milligrams per kg/bw of this substance in their daily diet for six months did not experience any toxic effects.

Tests conducted with rats receiving 1.5 grams of bromelain per kg/bw, as well as on their offspring, demonstrated this enzyme to be free of cancer-producing or teratogenic (deformed fetus) effects. Nor does it produce allergic reactions.

Ficin is obtained from the latex of the fig. Unfortunately, there is very little information concerning the safety or toxicity of this enzyme in experimental animals or humans, especially when the additive is a component of the diet rather than being injected. In 1959 a patent was issued on the addition of ficin or papain to poultry feed, which increased the weight of the birds over a seven-to-nine-week period. Amounts up to 10 grams of ficin or 40 grams of papain proved to be effective, the beneficial growth suggesting lack of toxicity.

Papain. The average intake of this enzyme in the U.S. is small, about 1 milligram per person in 1976. However, it is possible that some individuals who make frequent use of meat tenderizers in the home, or of condiments, seasonings, and gravies containing this enzyme, may consume as much as 25 milligrams daily, or 0.4 milligram per kg/bw. Preparations containing the pure enzyme have been studied in terms of possible usefulness as digestive aids or agents capable of controlling parasites in the digestive tract. Papain at high levels of 200-600 milligrams per kg/bw have reduced inflammation and pain in paws of experimental animals, and no adverse effects were seen. In rats and dogs, papain has been shown to increase the digestibility of protein in the meal. Secretions from the pancreas are generally needed for proper digestive action; dogs after removal of the pancreas had their digestive function restored by papain. In sheep, the enzyme proved effective in destroying the worms and eggs of parasites in the intestine. Humans have been successfully treated for roundworms by doses of 200 milligrams of papain per kg/bw, with no adverse effects in the subjects. However, it has been shown that even 45 milligrams of papain taken orally can somewhat reduce the clotting time of the blood.

Papain has not proved a safe method of digesting away meat that has lodged in the throat; where this has been attempted, using large doses of over a gram of the enzyme preparation, the lining of the throat has been damaged because it too was being digested by the enzyme.

If inhaled, papain can cause an allergic reaction; workers processing the enzyme have developed asthma. One investigator has concluded that papain is an antigen (a substance which, besides causing a specific allergic response, can broaden the range of sensitivity), and believes that allergic sensitization may possibly occur from ingestion. Routine allergy skin tests of 330 subjects found 7 (2 percent) who reacted to papain, an indication that not many are sensitive to it.

There appear to be no adverse effects of papain on maternal or fetal survival when administered orally to pregnant mice and rats. Studies of its possible cause of cancer apparently have not been made.

Pepsin, a normal constituent of the body, is the principal digestive enzyme in gastric juice for both humans and animals. It is usually derived from the stomachs of hogs when used as an additive. Pepsin has medicinal uses, including use as a digestive aid, particularly when stomach secretions are inadequate. A usual dosage for humans is 500 milligrams.

Phosphorylase. Research on the safety of this enzyme has not been located.

Fungal protease derived from *Aspergillus* (a genus of fungi) has had minimal short-term studies for toxicity. Rats fed for three months on a diet containing 7 grams per kg/bw daily showed no toxic symptoms and no microscopic changes in tissues. The enzyme has been given intravenously to dissolve blockages in the artery of the thigh or knee, 150 to 1050 milligrams of the fungal protease being administered during one to 11 days. Of nine patients thus treated, reopening of the artery was effected in four cases. A secondary effect of the treatment was local pain. The UN Joint FAO/WHO Expert Committee on Food Additives has pointed out that a product of metabolism of *Aspergillus* has been suspected of cancer-producing properties, and notes the need for broader long-term studies to investigate this matter.

Rennet. In 1975, a survey of the poundage of rennet added to foods by the industry as reported to the National Academy of Sciences-National Research Council indicated a daily intake per person in the U.S. of a little over 3 milligrams. The major source of rennet in the diet is from the consumption of cheese; about 40 percent of the milk coagulants used is rennet or a rennet-pepsin blend.

Few studies of rennet or rennin exist. One, an investigation of the effect of 1 gram of rennin on the digestion of milk which was administered by tube to human patients on three to five alternate days, found it had no effect on digestion, nor did it cause any harm. Improvements in weight gain and diet tolerance were noted among 130 hospitalized infants given cow's milk curdled with a rennin preparation.

Ulcers occurred in intestines exposed directly to solutions containing several proteolytic enzymes (including rennin and pepsin). This is unlikely to occur from these substances in the normal human diet, since they are ingested in low concentrations which are embedded in substantial amounts of bulk food (like cheese), and would be rapidly deactivated by digestion.

Laboratory tests to determine whether rennet can cause birth defects proved negative, but no tests as yet have been conducted to determine whether it can cause cancer or gene mutations. Nor are there any feeding experiments with animals available.

ASSESSMENT: Many of the enzymes used in food processing are extracted from edible plant and animal tissues, and have been in use by man for a considerable period without evidence of causing harm. Most frequently, they are destroyed (inactivated) by heat during processing or food cookery. Experimental studies of these, as well as clinical use with humans, confirm their safety. Enzymes of microbial origin are similar in chemical structure to those of animal origin; they are proteins which the body is accustomed to metabolize in a normal way. Research on these has been sparse, but the findings that do exist indicate lack of hazard. A caution is noted, however, about *Aspergillus,* a fungus source for some amylases and proteases. A suspicion exists that it may produce a carcinogen (cancer-causing substance) during metabolism, and further research is indicated. The FDA requires that when alpha-amylase obtained from *Aspergillus oryzae* is used in flour or other products, it must be declared in the list of ingredients by that name.

RATING: S for enzymes of animal, plant, and microbial origins. Exception: ? for enzymes derived from *Aspergillus*.

FOLIC ACID

Folacin

Folic acid is one of the vitamins of the vitamin B complex. It is needed for forming heme, the iron-containing pigment attached to the protein in hemoglobin, which is the oxygen-carrying constituent of red blood cells, and for nucleic acid,

MAJOR REFERENCE: A conspectus of research on folacin requirements of man. M. S. Rodriguez. *Journal of Nutrition* 108:1893 (1978).

which is essential to the growth and reproduction of the body's cells. Folic acid also participates with other vitamins (B_{12} and C) in the processes that break down or alter amino acids and proteins.

This vitamin occurs widely in nature; good sources are liver, leafy vegetables, eggs, and whole grain cereals. When found in foods, it frequently is as *folacin,* folic acid combined with additional molecules of glutamic acid (see p. 555), an amino acid. In this form it is less readily available to the body than free folic acid—a synthetic compound when used as an additive—which contains only one glutamic acid molecule. Folic acid is added as a nutrient in foods like breakfast cereals, frozen fruit juice, baby foods, and a variety of dietetic products.

SAFETY: A survey among food manufacturers, conducted by the National Academy of Sciences-National Research Council, determined that in 1976 the daily diet of an average U.S. adult contained 0.25 milligram of added folic acid. It is unlikely that this amount actually was ingested, as the vitamin is liable to destruction during cooking.

The Food and Nutrition Board of the National Academy of Sciences-National Research Council has proposed a Recommended Dietary Allowance for total dietary folacin of 50 micrograms (0.05 milligram) for small children, 400 micrograms (0.4 milligram) for most adults, and 800 micrograms (0.8 milligram) for pregnant or lactating women. They note that when the source is pure folic acid, the daily intake may need to be only a fourth as much, because folic acid is more readily absorbed and utilized than many of the combined forms of the vitamin occurring in foods.

One of the primary results of folic acid deficiency is megaloblastic anemia, where abnormal red blood cells with large nuclei appear in the blood circulation. This condition is also a characteristic of pernicious anemia resulting from a vitamin B_{12} (see p. 671) deficiency. Folic acid can bring about a temporary remission or alleviation of symptoms of the anemia in vitamin B_{12} deficiency. Thus one of the concerns about extra folic acid in foods or vitamin preparations is that it can mask some of the complications of the nervous system which can result from a vitamin B_{12} deficiency. Nevertheless, megaloblastic anemia is more likely to be caused by a folic acid deficiency than by inadequate vitamin B_{12}.

Folic acid is regarded as nontoxic. In earlier studies, no

adverse effects were seen in adults who were given the large doses of 400 milligrams of folic acid daily for five months. In 1970 a report, which was not confirmed, claimed that 15 milligrams of folic acid daily brought about mental changes and other adverse effects in healthy subjects after a month.

There are some apparent interactions of folic acid metabolism and some drugs. Aspirin may change the distribution of the folic acid in the body tissues. Both methotrexate (a folic acid inhibitor used in acute leukemia) and oral contraceptives can alter folic acid metabolism, but the possible effect on the dietary requirement is not clear. Alcohol can affect the body's use of dietary folacin; the Recommended Dietary Allowance is intended to cover such increased needs of the vitamin. Generally, the effect of drugs is likely to increase the need for folic acid.

ASSESSMENT: Folic acid is an essential vitamin; inadequate intake can cause a serious anemia. Its use as a food additive is limited by the Food and Drug Administration regulations to assure an intake of no more than the daily recommended allowance of the vitamin. In the quantities present as a nutrient in some fortified foods, folic acid poses no problems or hazard to the consumer.

RATING: S.

FOOD COLORS

Artificial colors are dyes and pigments that are synthesized from coal tar and petroleum. *U.S. certified colors (FD & C colors)* are colors that have met government specifications for composition and purity. Approval of a color for use in food

MAJOR REFERENCES: FOOD COLORS—Food colors. National Academy of Sciences-National Research Council Food Protection Committee (Washington, D.C., 1971); Handbook of food additives, 2nd ed. T. E. Furia, ed. (Cleveland: CRC Press, 1972); UN Joint FAO/WHO Expert Committee on Food Additives Reports 38B, 1966; 40 ABC, 1967; 46A, 1969; 48A, 1971; 54A, 1975. FOOD COLORS AND HYPERACTIVITY—G. T. Augustine and H. Levitan. *Science* 207:1489 (1980); C. K. Connors et al. *Pediatrics* 58:154 (1976); P. S. Cook and J. M. Woodhill. *Medical Journal of Australia* 2:85 (1976); Why your child is hyperactive. B. F. Feingold (New York: Random House, 1975); J. P. Harley et al. *Contemporary Nutrition* 3:No. 4 (April 1978), and *Pediatrics* 61:818, 62:975 (1978); S. Palmer et al. *Clinical Pediatrics* 14:956 (1975); L. K. Salzman. *Medical Journal of Australia* 2:248 (1976); J. A. Swanson and M. Kinsbourne. *Science* 207:1485 (1980).

requires documentation of safety based on research findings that are satisfactory to the FDA, guided by laws. Unfortunately, the judgment of this agency at times has been based on too meager evidence and is open to question. Thus a number of colors have been removed from the lists as evidence has come in of toxic effects. *Natural colors* are extracted from pigment naturally occurring in plants, animals, small organisms, and minerals; some of these can also be produced synthetically.

The obvious purpose of color additives is to make food look as acceptable and desirable as possible. Because of consumer expectation of what a food should look like, any marked deviation is likely to arouse suspicion that the product is inferior and to discourage sales. This expectation is reinforced by food grading and regulatory quality standards that take color into consideration. Besides some variation in the food to begin with, food can change color during manufacture and storage; dyes and pigments frequently are needed to achieve the desired and uniform hue, and for the color to remain stable afterward. By the same means, regrettably, foods can be made to look as if they contain ingredients that, in fact, they do not. A color rarely adds nutritive value to food; of the colors used in food, only the carotene pigments (see p. 505) and riboflavin (see p. 668) significantly contribute to nutrition. Most color additives are used in beverages, but they are also put into candy and confections, cereals, ice cream, butter and cheese, sausage casings and meats, baked foods, snack foods, gravies, jams and jellies, nuts, salad dressings, and much more.

ARTIFICIAL COLORS

Citrus Red No. 2; Orange B; FD & C Red No. 3 (Erythrosine); FD & C Red No. 40 (Allura AC); FD & C Yellow No. 5 (Tartrazine); FD & C Yellow No. 6; FD & C Blue No. 1; FD & C Blue No. 2; FD & C Green No. 3

SAFETY: There are two major aspects of concern regarding the health implications of these food additives. One relates to their possible effects on behavior of children, the other to their toxicology and experimental testing for safety.

Food Colors and Hyperactivity. A connection between diet and hyperactivity (hyperkinesis) in the child has been postulated and publicized by Dr. Benjamin Feingold, stating the belief that food additives such as artificial food colors and flavors, along with foods containing "salicylatelike" natural compounds* and medicines that contain aspirin, are responsible for many behavioral disturbances in some growing children. These may show up as shortened attention span, an easier distractibility, and various degrees of compulsive behavior or overactivity.

This provocative theory, implicating many food additives, has caused a flurry of experiments with children. Although there is now some scientific information on the effect of artificial colors in the diet of the hyperactive child, there are no final answers. Artificial flavors could not be tested because they number in the hundreds and because the children used in the experiment probably would know by the taste whether they were receiving the placebo (inactive substance) or the flavor.

Two types of investigations have evolved. In one, a group of hyperactive children are fed the Feingold diet (which omits foods that contain artificial colors and flavors, foods that contain salicylic acid, and certain preservatives), while another group are fed a normal diet which contains these foods. After a number of weeks, those fed the Feingold diet are shifted to the normal diet for a similar period, while those initially fed the normal diet are switched to the Feingold diet. Objective scientific tests are performed, and parents and teachers also subjectively evaluate the children's behavior in both conditions. Studies conducted in this manner at the universities of Pittsburgh and Wisconsin came up with similar findings; by parents' ratings, a minority of the children showed improvement (a quarter of the hyperactive children in the former study, a third in the latter one) when changed from the normal diet to the Feingold diet; teacher ratings showed much fewer differences. However, no behavioral change occurred when the shift was made from the Feingold diet to the normal diet.

A second procedure employed the Feingold diet throughout the experiment, and at intervals added to it a dose of artificial colors (a blend of eight FD & C certified colors) in a cracker, capsule, or soft drink. The dosage of artificial colors varied from 26 to 150 milligrams on the day of administration, the latter amount corresponding, according to FDA estimates, to the daily intake of the highest 10 percent of child consumers of these substances. Tests of learning performance, standardized scales for measurement of behavior, and other measures were

---

*These include fruits (apples, apricots, oranges, peaches, prunes, raisins, and many berries), vegetables (cucumbers, tomatoes), and flavorings (cloves, oil of wintergreen) that contain salicylic acid.

conducted to assess the effect of the addition of artificial color to the diet. When dosages of less than 100 milligrams were used, very few children appeared to be adversely affected; in one study, two of 22 children showed a response to 35 milligrams of artificial color (estimated as a daily average intake of these additives). When the artificial colors were upped to 100 milligrams or more in a Canadian study, the learning performance of 17 out of 20 hyperactive children deteriorated, though not to a considerable extent, but their social behavior did not seem to have been affected.

Two studies have been conducted in Australia. One used children who had shown allergic-type sensitivity to salicylates and artificial colors and flavors. When the rating of the mother was used to assess the diet effect, 93 percent indicated some improved behavior on the Feingold diet; but the average score on "overactivity" only slightly improved, from 5.8 to 4.6, which may not be significant by statistical testing. The other study used a parent questionnaire after the children had been placed on the diet recommended by Dr. Feingold, but adapted to Australian food habits; most of the children reportedly showed substantial improvement in behavior. A comparative study was made of the food consumption and dietary habits in the U.S. of hyperactive versus more normal children. Here no significant difference in the presence of food additives was found in the diets. It is possible that the hyperactive child may be more susceptible to certain food additives than other children.

An animal experiment conducted at the University of Maryland studied the effect of FD & C Red No. 3 (which is permitted in food) on the nerve impulses leading to a frog's muscle, and the findings may provide an explanation of the manner in which a food color might contribute to hyperactive behavior. In the experiment this food color was able independently to increase the frequency of the transmission of nerve impulses, an indication that it was able to affect nerve cells with the capability of altering behavior.

Individual Artificial Colors (U.S. Certified Colors). Nine artificial colors are permitted as additives in food. Two are restricted, each to a single product. The UN Joint FAO/WHO Expert Committee on Food Additives has cautioned that one of these is a health hazard, and the other is similar to Amaranth (FD & C No. 2), which has been banned.

Citrus Red No. 2 is restricted to coloring orange skins not used for processing. It has induced cancer in animals.

Orange B is restricted to casings and surfaces of frankfurters and sausages; it is similar in chemical structure to Amaranth, banned because of research linking it to cancer. The FDA in 1978 proposed that Orange B be banned because of a cancer-causing contaminant appearing during its manufacture. The sole manufacturer has discontinued production.

Some evidence exists that two other artificial colors may be harmful, but clear evidence is lacking. *FD & C Red No. 3* (Erythrosine) has demonstrated adverse effects on blood and may also cause gene mutation. FAO/WHO considers up to 2 milligrams per kilogram of body weight (kg/bw) as an acceptable daily intake, but has asked for metabolic studies with humans. *FD & C Red No. 40* (Allura AC) has been imputed by some to induce cancer in animals. A working group from the National Cancer Institute and the FDA, in a reappraisal of the data in 1978, concluded that evidence was not provided that it causes cancer.

A fifth color, *FD & C Yellow No. 5* (Tartrazine), has been found to cause allergic reactions, usually minor, but serious on occasion. The FDA has issued a regulation which requires it to be listed on labels of food that contain it after July 1, 1981 (but July 1, 1982, for ice cream and frozen custard). Most people allergic to it are also allergic to aspirin. It is used in candy, desserts, cereals, and dairy products.

No evidence of hazard has come to light concerning the four artificial colors that remain, but FAO/WHO is critical of the adequacy of the research that has been conducted on some of these. They are *FD & C Blue No. 1; FD & C Blue No. 2; FD & C Green No. 3;* and *FD & C Yellow No. 6.*

ASSESSMENT: Since food manufacturers are not required to, and usually do not, list each artificial color they use in a product, for the purposes of evaluation of safety the color of the product must be considered a clue to the artificial color used. On this basis, if the color is blue, green, or yellow it is likely to be one of the four colors listed immediately above, which according to present knowledge are safe. The one exception to this color judgment is FD & C Yellow No. 5, which can be dealt with separately because it must now be identified on the label. It causes allergic reactions in some people.

If the product is colored red, orange, or violet, a questionable artificial color additive probably is involved. There is no way at present to avoid the uncertain or the potentially hazardous ones except by not purchasing brands of foods of these hues that list artificial color(s) among their ingredients.

An issue of concern regarding the safety not only of artificial colors, but also of artificial food flavors and some natural components in foods, is a possibility of their connection with behavioral disturbances in growing children. The research to date on the relationship between food additives in the diet and hyperactivity of some children has not provided clear-cut answers. Many parents have observed benefits in the child's behavior when the diet is changed to eliminate additives; this is not clearly apparent when standardized, objective measures are employed.

RATING: If food is yellow, blue or green: S for adults; ? for children because of uncertainties about contribution to hyperactivity. If food is red, orange, or violet: ? for everyone. If FD & C Yellow No. 5 (Tartrazine) is listed: S for most adults; X for anyone susceptible to allergic reactions, especially if intolerant to aspirin; ? for children because of uncertainties about contribution to hyperactivity.

NATURAL COLORS

Annatto (Bixin); Beet Red (Dehydrated or Powdered Beets); Beta-Apo 8′-Carotenal; Beta-Carotene; Canthaxanthine; Caramel Color; Carrot Oil; Cochineal and Carmine; Ferrous Gluconate; Fruit Juice; Grape Skin Extract (Enocianina); Paprika; Paprika Oleoresin; Riboflavin; Saffron; Titanium Dioxide; Toasted Defatted Cottonseed Flour; Turmeric; Turmeric Oleoresin; Vegetable Juice

SAFETY: The substances above may appear as individual items on labels, or may just be identified as natural color(s). The FDA's exemption from further investigation of safety for these substances causes a problem for assessment of hazard, as at times research has been inadequate or nonexistent. The groupings below make an attempt at guidance. A number of these substances are also used as flavorings, and a few are used as vitamin sources. Most of those in the "safe" group can be produced synthetically. When a synthetic color is used in a food, it is regarded as an artificial color by the FDA, and in most products must be identified on the label as artificial coloring. On occasion, "artificial color (beta-carotene)," or another compound in the parentheses, may be found on the list of ingredients in order to meet this requirement and to specify the substance actually used.

The safety of the following natural colors is supported by research or clinical experience.

*Beta-carotene** and *Beta-apo 8′-carotenal* occur naturally in vegetables and citrus fruits and are a source of vitamin A (see p. 665). They impart orange and yellow colors to food.

Canthaxanthine is found in edible mushrooms, crustaceans, trout and salmon, and tropical birds. It produces a pink to red color when used in foods. FAO/WHO considers up to 25 milligrams per kilogram of body weight (kg/bw) as an acceptable intake.

Caramel color (see p. 501) provides a tan or brown color. The substances used to produce it are various carbohydrates (such as table sugar, corn sugar, or molasses) heated together with small amounts of acids, bases, or malts, and using a variety of temperature and pressure conditions.

Ferrous gluconate (see p. 573) is present in the body, and its components are essential for the body's chemical processes. As a color additive, it is used to provide a greenish yellow coloring.

Riboflavin (vitamin B_2; see p. 668) exists in plants and animals and as a coloring adds yellow with a greenish fluorescence.

Titanium dioxide is a white pigment which is present in minerals; and, as it is very insoluble, it is rapidly excreted. Studies in several species, including man, have shown neither significant absorption nor tissue storage following ingestion, nor did the pigment cause toxic effects.

A second group of natural colors are probably safe: research is incomplete, but further study is unwarranted.

Annatto (bixin) is extracted from the seeds of a tropical tree *(Bixa orellano L.)*. Research so far on this yellow pigment reports no adverse effects. FAO/WHO has established a temporary "acceptable daily intake" of 1.25 milligrams per kg/bw, but is withholding final judgment pending the completion of metabolic studies.

Beet red (dehydrated or powdered beets), a normal ingredient in the diet, is unlikely to be harmful. Nevertheless, research has been inadequate, and metabolic and long-term studies have been requested by FAO/WHO.

A third group of natural colors are also probably safe: research is nonexistent, but they are components of food in the normal diet. Included are *carrot oil; fruit juice; grape skin*

---

*This substance is also used for other purposes as a food additive. See p. 505.

extract; toasted defatted cottonseed flour; and vegetable juice.

A fourth group should be regarded with caution: there is some possibility of harm.

Cochineal is a red color derived from an insect found in the Canary Islands and South America, and *carmine* is the pigment that is extracted. A few short-term studies of carminic acid (the active agent in carmine) injected into veins or in the abdominal cavities of laboratory animals revealed an abnormal effect on spleen tissue. Research is unavailable on long-term effects, reproduction, and metabolism.

Paprika, a sweet orange-red pepper, contains capsaicin, which when injected in cats and mice has slowed heartbeat, caused abnormally low blood pressure, and stopped breathing. Administered by tube, it reduced body temperature and raised acid secretion because of irritation. Rats fed chili, which contained paprika, as 10 percent of their diet developed liver tumors. Paprika is used as well as a spicy flavoring, and the amount necessary to cause harm probably would make the food unpalatable. *Paprika oleoresin* is the fraction that is extracted by a percolation process, and is more concentrated, stronger in color, and less abrasive.

Turmeric is an herb whose coloring agent is curcumin, which gives a yellow color. Limited research on turmeric employing a single dosage level did not disclose adverse effects, but curcumin did affect the body liquids and livers of rats. *Turmeric oleoresin* is the active fraction extracted by a percolation process, and thus is more concentrated. FAO/WHO considers that adequate short-term studies indicate safety, but that long-term ones are needed to establish the safety level of turmeric in food. For curcumin the FDA has requested research on metabolic and long-term effects, on reproduction, and on the developing organism.

One natural color, *saffron,* does not fit in any of these groups, as further information is needed in order to provide a safety rating. It is extracted from the seed-bearing organ of a Near East crocus, and crocin is the substance in it that provides the orange-red powder used in food. FAO/WHO reports that available data are inadequate for assessing the possibility of harmful effects.

ASSESSMENT: An evaluation of hazard for a natural or uncertified food color can be provided for a specific substance if adequate information concerning it exists and if it is identified on the label. However, when only "natural color(s)" or "uncertified color(s)" are mentioned, this is not possible. The ratings that follow must reflect this limitation.

RATING: S for annatto (bixin), beet red (dehydrated or powdered beets), beta-apo 8′-carotenal, beta-carotene, canthaxanthine, caramel color, carrot oil, ferrous gluconate, fruit juice, grape skin extract (enocianina), riboflavin, titanium dioxide, and vegetable juice; ? for cochineal and carmine, paprika and paprika oleoresin, saffron, and turmeric and turmeric oleoresin (uncertainties remain, although the amount used in food is unlikely to cause harm).

FOOD FLAVORS

More flavor additives are used in foods (perhaps 1700 flavoring agents) than all other food additives combined. Some are derived from natural ingredients, while others are synthetic chemicals. Most often, complicated combinations of them are required to achieve the desired flavor, which the flavor chemist usually cannot create using a single chemical or ingredient, especially if a natural flavor is to be simulated. Because it is not required by law or of much help to the average consumer, only rarely are the individual chemical components that constitute the flavoring identified on the brand's label. The formulas are regarded as trade secrets and need only be specified in general terms like "artificial flavors" or "imitation flavors" (similar to artificial) or "natural flavors," and even this is not required for some products.

Natural flavors are often processed in ways that concentrate and enhance them, enabling the manufacturer to provide a better distribution and effect in the food. *Essential oil,* for example, is obtained from plants and usually retains its original taste and smell, but is as much as 100 times as concentrated. *Oleoresin* is usually the extractive of a spice and is more concentrated than the original material. *Extracts* are derived from certain fruits and berries, but they may be weak. Further

MAJOR REFERENCES: Evaluation of the health aspects as a food ingredient of caprylic acid. FASEB/SCOGS Report 29 (NTIS PB 254-530) 1975; . . . of dill. Report 22 (NTIS PB 234-906) 1973; . . . of garlic and oil of garlic. Report 17 (NTIS PB 223-838) 1973; . . . of mustard and oil of mustard. Report 16 (NTIS PB 254-528) 1975; . . . of nutmeg, mace and their essential oils. Report 18 (NTIS PB 266-878) 1973; . . . of oil of cloves. Report 19 (NTIS PB 238-792) 1973; Handbook of Food Additives, 2d ed. (Cleveland: CRC Press) 1972.

strength can be obtained by combining them with other natural flavors. An extract must consist of at least 51 percent of the original fruit's flavor to be identified as its extract.

Artificial flavors may consist of synthetic chemicals alone, or may combine them with natural substances. If a single artificial ingredient is contained in a flavoring, it must be labeled as an artificial flavoring. Occasionally a single ingredient is used as a flavoring; an example is vanillin or ethyl vanillin (see p. 659).

Flavor additives are used to achieve or assure an acceptable, desirable, or uniform flavor in the product. The flavor of food can deteriorate when it is subjected to the varying conditions imposed by processing, packaging, and storage over a period of time. Often the original ingredients themselves may have varied in flavor quality. To achieve a desirable and uniform flavor that meets the consumer's expectations, flavors are added to some food products to modify any undesirable changes, mask disagreeable ones, or even to wholly re-create the taste and odor of the food. Foods in which these flavor additives are found frequently are beverages, ice cream, cereals, candy, baked goods, meats, icings and toppings, gelatins, jellies, syrups, sauces, seasoning, margarine, and shortening; flavor additives are essential for simulated, tailor-made products such as artificial bacon bits, which are made from soybean protein.

SAFETY: The Food Additives Amendment of the Food, Drug, and Cosmetic Act, which initiated federal regulations in 1958 on safety-in-use of additives, permitted continuing use of food additives such as flavorings in existence at the time, provided that they qualified on the basis of experts' experience in the past, or of scientific data already at hand. Neither measure proved to be a very effective screen; normal use at best might reveal immediate acute reactions, but hardly any insidious but traceable long-term effects, and the research available was inadequate by today's standards. In effect, the vast majority of flavors went unchallenged. Since then laboratory studies have uncovered a few dangerous offenders; for example, safrole, a flavoring that had been used for years in root beer, can induce cancer. It no longer is permitted as a flavor additive. Coumarin, once present in imitation vanilla extract, is now banned because large amounts can cause hemorrhaging. The FDA has been aware of the inadequacy of the more permissive initial rulings, and in its own studies of some substances has found a few that could do serious harm to vital organs, cause malignancies, and increase fatal consequences. But only recently has there been a concerted effort by the agency to review the adequacy of safety information on many flavor ingredients. Flavorings that have come into existence since 1958 are required to have their safety determined by scientific procedures.

Since the 1958 Food Additives Amendment does not require the identification of the individual chemicals that comprise an added flavoring, it poses an insoluble problem for our assessment of safety. A typical artificial flavor may contain as many as a dozen or more chemicals, with no means of determining which ones of these are among the FDA list of 800 to 900 permitted synthetic flavoring substances.

The fact that "natural flavors" (no other identification) are added is no assurance of safety either. Natural flavors often contain the same chemicals as the artificial flavors that imitate them. A host of edible plants and animal products contain small amounts of natural components which can be toxic. If they were to be added to foods rather than introduced by nature, they would be prohibited by the FDA. (See Nutmeg, p. 548.) The National Academy of Sciences in a publication on toxicants occurring naturally in food flavors has noted that there are only a few of them; however, a concentration of some natural flavors, such as in essential oils, oleoresins, and extracts, in sufficient amounts can do harm.

It is argued that disclosure of the composition of food flavorings is not needed since only an extremely minute amount of each compound in a flavoring is used. (More would make the food unpalatable, a self-limiting safeguard.) Low dosages clearly lessen the hazard; while gram quantities of some chemicals and ingredients used in food flavors are required to cause serious illness or fatality, the daily intake of an individual flavor ingredient in the U.S. diet is likely to be only a small fraction of this amount. However, an important exception should be noted; the amount sufficient to cause cancer, if a substance should have this capability, remains an unknown.

The authors regret that an evaluation of safety is not possible here for flavors when their sole identification on a label is grouping as artificial, imitation, or natural flavors. Nor is it possible to assess the validity of the claim that artificial flavors can contribute to the hyperactivity of some children, because, unlike artificial colors, no research has been conducted to determine its truth.

Individual flavors at times are identified on labels, usually when used alone; most of these are reviewed elsewhere in this volume. The others are dealt with below.

ASSESSMENT: When unspecified flavors are grouped as artificial, imitation, or natural flavors, an assessment of safety is not possible.

See elsewhere in this volume for assessments of the following individual flavors.

Safe: acetic acid, adipic acid, caramel, citric acid, corn sweeteners (dextrose, fructose, corn syrup, invert sugar), formic acid and ethyl formate, fumaric acid, lactic acid, licorice, malic acid, malt, maltol and ethyl maltol, succinic acid, sucrose (cane or beet sugar), tannic acid, vanillin and ethyl vanillin.

Safe for some, not for others: caffeine, quinine.

Not safe or questionable: saccharin, smoke flavoring, xylitol.

Other individual flavors are assessed below:

Caprylic acid, a fatty acid (a component of fat) and a natural constituent in many foods, is nutritionally utilizable by man and animals. Less than 0.02 milligram was present in the daily diet of the average U.S. adult in 1976. Feeding experiments with rats using dosages approximating several hundred thousand times that amount (adjusted for body weight) did not cause injury.

RATING: S.

Cassia oil and extract are prepared from cassia, a variety of an aromatic Chinese cinnamon bark. It has a pungent flavor and is used to flavor liqueurs, baked goods, meats, confections, condiments, and beverages. The value of cassia oil depends on its content of cinnamaldehyde; it may contain 80 percent or more of this compound. Cassia oil has been used as a colic to help cleanse and remove excess gas from the intestinal tract. The use of cassia oil or extracts in 1976 as reported by the food industry indicated an average daily intake per person in the U.S. of just under 1 milligram. Cassia oil and cinnamaldehyde can cause inflammation and irritation of the intestinal lining if present in large amounts; they have a low order of toxicity, and have gained conditional acceptance as flavoring substances by the UN Joint FAO/WHO Expert Committee on Food Additives.

RATING: S.

Dill oil, an essential oil, is obtained by steam distillation of the freshly cut stalks, leaves, or seeds of dill, an herb. Daily consumption in the U.S. in 1970 averaged about 1 milligram per person. The most plentiful constituent, carvone, present to the extent of 60 percent of the oil, can cause toxic responses, but only when present in the diet of rats at thousands of times human consumption (adjusted for body weight).

RATING: S.

Ethyl acetate occurs naturally in many fruits, and is the principal volatile ester in pineapple flavor, where it is present along with closely related esters including *ethyl propionate* and *ethyl butyrate*. It is used as a flavor modifier in baked goods, milk products, frozen dairy desserts, jams, candy, gelatins, beverages, and chewing gum. In 1976 the usage of ethyl acetate in foods averaged a little over 5 milligrams per person in the daily diet. Ethyl acetate is considered relatively innocuous; FAO/WHO considers that there is adequate toxicological data to give an unconditional acceptance to use of ethyl acetate and ethyl butyrate as flavoring substances. The chemically related ester *ethyl heptanoate* (ethyl heptoate) also has conditional acceptance by FAO/WHO.

RATING: S.

Ethyl methyl phenylglycidate is used as a flavor agent in baked goods, frozen dairy desserts, candy, gelatins and puddings, beverages, and chewing gum. Its use by the food industry in 1976 amounted to an average daily intake per person in the U.S. of only 0.2 milligram. This chemical has been shown to produce adverse neurological effects in experimental animals, and more studies, with rats, are in progress.

RATING: ?.

Extractives of fenugreek: Fenugreek is a leguminous Asian herb with aromatic seeds; the seeds have been used externally and internally as a softening and soothing agent for inflammation. Extractives of fenugreek are used in curry powders and in some artificial vanilla flavorings, candies, frozen dairy desserts, sausages, baked goods, gelatins, and toppings. In 1976 use of these extractives in the daily diet averaged 2 milligrams per person. The limited information on its use by humans does not suggest any adverse effects.

RATING: S.

Garlic and *oil of garlic* are obtained from *Allium,* a genus of the lily family which includes onions, leeks, shallots, and chives. Garlic oil is obtained by steam distillation of fresh garlic, but garlic yields very little of the oil; it takes almost 500 milligrams of garlic to produce 1 milligram of garlic oil. The long history of use of garlic in food, and animal feeding studies (which are limited), reveal no credible adverse biological effects at concentrations many thousands of times as great as the 1 to 5 milligrams of garlic and garlic oil contained in the daily diet of the average U.S. adult in 1970.

RATING: S.

Ionone is an intermediate compound in the chemical synthesis of vitamin A (see p. 665). It has the drawback of being able to cause allergic reactions; susceptible individuals should avoid products containing ionone.

RATING: S for most people; X for anyone susceptible to allergies.

Kola nut extractives: The seeds of the kola nut may be extracted to give a flavoring ingredient used in cola-type beverages. While kola nut extractives contain some caffeine (see p. 494), this does not pose a hazard, as the amount in the processed extract is small and contributes less than a tenth of the caffeine in cola drinks.

RATING: S.

Mustard and *mustard oil:* The brown and yellow mustards commonly used in the U.S. are derived from two *Brassica* species that are quite different in the major constituents of their essential oils. The characteristic flavor of brown mustard, described as "horseradish bite," comes from allyl isothiocyanate. The distilled, commonly used mustard oil consists of more than 90 percent allyl isothiocyanate. The active flavor constituent of yellow mustard apparently is quite different; other constituents of the seed and added spice are responsible.

Daily consumption of these mustards and their oils, based on industry usage in 1970, totaled 305 milligrams per person, about 94 percent of it being the yellow variety. The available information from research findings indicates that the major constituents of these mustards require dosages far exceeding their human consumption to cause harm to experimental animals.

RATING: S.

Nutmeg, mace, and their essential oils, and mace oleoresin: Nutmeg comes from the ripe seed and mace from the dried shell that contains the seed of trees of the *Myristica* species cultivated in the East and West Indies. Oil of nutmeg and oil of mace are the essential oils obtained by steam distillation of nutmeg and mace, respectively. Mace oleoresin is a butterlike product obtained by pressing; 73 percent of it is the fat trimyristin, and about 12 percent is essential oil. It has been estimated, based on usage for food purposes reported by manufacturers in 1970, that the average daily intake per person in the U.S. was 11 milligrams of nutmeg and 3 milligrams of mace.

Nutmeg can have a modifying effect on mental activity. Single oral doses of 10 to 15 grams may produce acute intoxication. In doses of 5 grams or more, nutmeg can produce euphoria and hallucination, followed by abdominal pain, nausea, giddiness, drowsiness, and stupor; rapid heartbeat and respiration, effects on vision, and fever may also occur. A constituent in nutmeg, safrole (it is present in the East Indian variety but not the West Indian), is believed to be a weak hepatocarcinogen (cause of liver cancer). Safrole itself is prohibited as an additive.

The amount of nutmeg required to elicit its psychoactive effect is over 450 times as great as the amount contained in the daily diet of the average U.S. adult, and for this reason may not warrant caution in ingesting food containing it. However, the presence of safrole, a possible cause of liver cancer, in East Indian nutmeg does merit concern.

RATING: X.

Oil of clove, also called clove bud oil, clove stem oil, and clove leaf oil, is the essential oil obtained by steam distillation of these products of the clove tree. Two other clove products are used in food, *clove bud extract* (obtained by solvent extraction from clove buds) and *clove bud oleoresin*. It is estimated, based on the average annual import of cloves and stems and clove oils, that the average intake of these flavorings totals 23 milligrams per person daily. The available research has been conducted with eugenol, the chief constituent of clove oil (70 to 90 percent of clove bud oil consists of eugenol). Only when administered at far greater levels than those occurring in foods did eugenol produce irritation in the gastrointestinal tracts of laboratory animals.

RATING: S.

FORMIC ACID

ETHYL FORMATE

Formic acid is a fatty acid, one of several that function in the body as intermediaries in metabolic processes that transform ingested substances into useful compounds. Formic acid acts with components of proteins and is incorporated into DNA and RNA, constituents that carry and transmit the genetic code. The acid occurs naturally in honey, some fruits and berries, coffee, rum, wine, and milk and cheese. *Ethyl formate*, an ester of formic acid, is a component in certain plant oils, fruits, honey, and wine. As additives, both are used principally to enhance flavors or as flavorings themselves.

SAFETY: The average intake per person in 1970 of added formic acid and ethyl formate has been estimated at 1 milligram per kilogram of body weight (kg/bw). The UN Joint FAO/WHO Expert Committee on Food Additives has conditionally proposed an acceptable intake from additives of five times this amount.

The tolerance of the body for these substances is high. 160 milligrams of formic acid per kg/bw, at least 160 times human consumption, administered orally to rats did not harm them. Nor did it harm human subjects who over a four-week period ingested eight times the amount they would have normally in their diet. Calcium formate (not used in food, but containing formic acid) in dosages 150 to 200 times human consumption,

MAJOR REFERENCE: Evaluation of the health aspects of formic acid, sodium formate and ethyl formate as food ingredients. FASEB/SCOGS Report 71 (NTIS PB 266-282) 1977.

adjusted for body weight, fed in drinking water to rats through five successive generations, had no adverse effect on mortality rate, fertility, or fetal development. The amount administered was doubled for two years, again without adverse effect.

Human deaths have been recorded after accidental or intentional overdoses of 50 grams or more. But considering the quantities present as additives in food, such a finding is not regarded as relevant.

Results of in vitro (test-tube) experiments investigating gene mutation activity of formic acid are in conflict. Some have indicated it to be moderately mutagenic, while others have not. The possibility of formic acid or ethyl formate as a cause of cancer does appear to have been studied.

ASSESSMENT: Formic acid and its ester ethyl formate are natural constituents in a number of foods. Formic acid is a normal intermediary in the metabolic processes of the body. Ingestion of it by animals and man far exceeding human consumption as an additive provides evidence that at present levels in food, or at levels likely in the future, it does not pose a hazard to health.

RATING: S.

FUMARIC ACID

This compound is a normal constituent of the body, since it is one of the acids produced in tissues during metabolism of carbohydrates. It has been found in blood, brain tissue, kidney, liver, muscle, and bones. As fumaric acid is essential in all plants and animals, it is naturally present in foods in the normal diet. For the quantities required for commercial use as an additive, it is produced synthetically by chemicals or by bacterial fermentation.

Fumaric acid is used in foods to impart an acid flavor, as a flavor enhancer, and to extend the shelf life of powdered food products such as beverages, puddings, and gelatin desserts, because it absorbs moisture very slowly.

SAFETY: A survey conducted by the National Academy of Sciences-National Research Council on usage of fumaric acid

---

MAJOR REFERENCE: Toxicological evaluation of some food colours, enzymes, flavour enhancers, thickening agents, and certain other food additives. WHO Food Additives Series No. 6 (Geneva, 1975).

as an additive in foods in 1976 indicated an average daily intake per person of 31 milligrams. This amounts to about 0.5 milligram per kilogram of body weight (kg/bw) for a person weighing 60 kg (132 lbs.).

Fumaric acid has been fed to rats and to guinea pigs for a year or two at levels several hundred times the average human intake (adjusted for body weight) with no adverse effects being observed. Indeed, rabbits have been fed for 150 days at a rate of 1500 milligrams per kg/bw daily without harm to appetite, body weight gain, blood counts, mortality rate, or tissue structure. Some 75 human patients have taken fumaric acid at daily doses of 500 milligrams (8 milligrams per kg/bw) for a year, again with no toxic effects. This is not surprising, since fumaric acid is regularly converted to citric acid and other intermediates in the body's metabolism of carbohydrates. At unusually high levels (oral doses of 5 to 30 grams of the sodium salt of fumaric acid) a laxative effect has been observed.

The UN Joint FAO/WHO Expert Committee on Food Additives considers 6 milligrams per kg/bw an acceptable and safe daily intake for humans.

ASSESSMENT: Fumaric acid is a regular constituent in body tissues and is found in the normal metabolism of carbohydrates by humans. At levels hundreds of times the average intake of the food additive, no adverse effects have been seen in experimental animals or humans. At levels now used in foods or expected to be used in the future, fumaric acid appears without hazard.

RATING: S.

GELATIN

Gelatin is extracted from collagen, the main protein component in the connective tissue of the animal body. Collagen occurs to some extent in all tissues and organs, and is concentrated in skin, skeletal bones, and tendons that attach muscles to the skeleton. Two types of gelatin are produced commercially: one converts collagen in pig skins by an acid process, while the other uses cattle hides and bones and an alkaline or lime procedure.

MAJOR REFERENCE: Evaluation of the health aspects of gelatin as a food ingredient. FASEB/SCOGS Report 58 (NTIS PB 254-527) 1975.

Gelatin functions as a thickener and a stabilizer (prevents separation) in food. It is found most often in gelatin desserts, meat products, consommés, candies, bakery and dairy products, marshmallows, and ice cream.

SAFETY: In the U.S. in 1972 the average daily intake of gelatin was estimated as 262 milligrams per person, about 4 milligrams per kilogram of body weight (kg/bw) for a human weighing 60 kg (132 lbs.).

A good deal of the research on gelatin, which has been directed at discovering its nutritional properties rather than its toxicity, has demonstrated its low nutritional value. Although gelatin is derived from a protein, it lacks tryptophan, which is necessary for human metabolism, and it is deficient in several other biologically essential amino acids (components of proteins). Rats fed a diet containing 35 grams per kg/bw of gelatin daily, supplemented with some but not all of its amino acid deficiencies, suffered retarded growth, and half of the animals died within 48 days. Thus gelatin is a poor-quality protein which, without adequate supplementation, will not sustain life.

Allergic reactions to gelatin derived from bone have been reported in human patients sensitive to beef, but the same patients did not repeat this response to gelatin originating from a pork source. Among allergic individuals, 1 in 150 tested gave positive skin reactions to gelatin, but only 1 in 500 produced clinical symptoms, an indication that an allergic reaction is likely to occur only rarely.

Tests do not appear to have been conducted investigating the possibilities of gelatin as a cause of birth defects or gene mutations. A number of studies provide evidence that it does not cause tumors in mice.

ASSESSMENT: Gelatin is derived from collagen, a natural ingredient in commonly consumed foods of animal origin. There is no evidence that it is harmful to humans when ingested, except for a rare allergic response. It lacks one essential amino acid, tryptophan, and is deficient in several others. While by itself it would be insufficient for maintaining health, gelatin is unlikely to be used as the sole source of food by humans, and considering its current level of use in food, or levels that might reasonably be expected in the future, it appears to be without hazard.

RATING: S.

GLUCONIC ACID

MAGNESIUM, POTASSIUM, SODIUM, ZINC GLUCONATE

A number of the gluconates—mineral salts of *gluconic acid*—are used in a variety of ways in nutrition and in food processing. Easily soluble in water and readily utilized by the body, these food additives provide an excellent way in which essential minerals that they contain may be introduced as nutrients or dietary supplements.* *Magnesium, potassium,* and *sodium gluconate* act as buffering agents in soda water that help retain carbonation. The gluconates may also be used to neutralize some mineral elements in foods such as gelatins and puddings, helping to maintain the consistency and appearance of the product.

SAFETY: The daily intake of sodium and zinc gluconate per person is 0.25 milligram, estimated on total poundage reported as used by the food industry in 1976. No usage figures for magnesium and potassium gluconates were reported at that time, indicating that these salts were used in minimal amounts.

Gluconate is a normal product in the metabolism of glucose in mammals. The daily production of gluconate during metabolism in the body is about 450 milligrams per kilogram of body weight for a person weighing 60 kg (132 lbs.). This level is so much higher than the amount of gluconate supplied by the diet that the amounts added to foods are not likely to be of significance.

The information available from acute (high-dosage) toxicity studies that have used a variety of animals and various gluconates indicates that any adverse effects observed are to be ascribed to the mineral portion of the salt and not to the gluconate part. Evidence from clinical medicine supports the usefulness of *zinc gluconate* administered in substantial doses to patients for such disturbances as skin inflammation and leukemia; the dosages were well tolerated.

MAJOR REFERENCE: Evaluation of the health aspects of gluconates as food ingredients. FASEB/SCOGS Report 78 (NTIS PB 288-675) 1978; Potassium gluconate, supplemental review and evaluation. FASEB/SCOGS Report II-18 (NTIS PB 81-12786) 1980.

*This review and evaluation is limited to the gluconates of magnesium, potassium, sodium, and zinc. Calcium, copper, iron, and manganese gluconate, and glucono-delta-lactone, are covered elsewhere in this volume.

Tests have shown that sodium and zinc gluconates do not produce gene mutations, but no studies have been reported in this area for either magnesium or potassium gluconate. None of the gluconates reviewed here appears to have been investigated for possibilities as causes of cancer. Tests using one species (chicks) gave no evidence that zinc gluconate could cause birth abnormalities.

ASSESSMENT: These gluconate salts are useful as nutritional supplements, since their high solubility allows rapid absorption of their mineral constituents. The body's normal metabolic processes produce gluconic acid from glucose, in amounts many times as great as are likely to be consumed from food additives. The minerals in these salts appear to bear the responsibility for any adverse effects that may occur from extremely high intakes. At the present low level of use as additives, or levels that may be expected in the future, there is no evidence of hazard from the gluconic salts in review here.

RATING: S.

GLUCONO-DELTA-LACTONE

Glucono-delta-lactone is a form of the sugar-acid gluconic acid (see p. 553) which is obtained by the removal of the elements of water from the acid molecule; in water, glucono-delta-lactone readily reverts to gluconic acid. Humans (and other mammals) produce gluconic acid as an intermediate in the metabolic processes that break down the simple sugar glucose, and other carbohydrates, into sources of energy. The acid is a natural constituent in many foods because of its presence in living cells and blood. Glucono-delta-lactone is used as an additive when a dry acid is required, increasing the acidity in foods like jelly powders and soft-drink mixes; it is used also as a leavening agent, improving the characteristics of a dough.

SAFETY: The total amount of glucono-delta-lactone estimated to be added to foods averaged slightly over 3 milligrams in the daily diet of U.S. adults in 1975.

MAJOR REFERENCES: Toxicological evaluation of some food colours, enzymes, flour enhancers, thickening agents, and certain other food additives. WHO Food Additives Series No. 6 (Geneva, 1975); Evaluation of the health aspects of glucono-delta-lactone as a food ingredient. FASEB/SCOGS Report II-11 (NTIS PB 82-108663) 1981.

Human patients have been given doses of 5 to 25 grams of glucono-delta-lactone daily, several thousand times its presence as a food additive, with no change in urinary constituents and no discernible effects other than some diarrhea.

Rats have been fed glucono-delta-lactone at a level of 0.4 percent of the diet for 29 months, with no differences being noted compared with control animals receiving no added glucono-delta-lactone. Fed during pregnancy, this compound had no effect on maternal or fetal survival or on abnormalities in offspring in mice, rats, hamsters, or rabbits. In the long-term experiment with rats, glucono-delta-lactone showed no evidence of causing cancer.

ASSESSMENT: The dietary intake of glucono-delta-lactone as an additive represents a rather minute fraction compared to the body's exposure to gluconic acid during metabolism of carbohydrates. According to experimental and clinical evidence, glucono-delta-lactone does not present a hazard to the consumer as a food additive at levels currently in use or levels likely to be used in the future.

RATING: S.

GLUTAMATES

Glutamic Acid; Glutamic Acid Hydrochloride; Monoammonium, Monopotassium, Monosodium Glutamate (MSG)

Glutamic acid is one of the amino acid compounds that form proteins; it is present naturally in appreciable concentrations in fish, vegetables, cereals and grains, cow's milk, and meats. On the average, 20 percent of food proteins consist of glutamic acid. The *acid,* its *hydrochloride,* and its ammonium, potassium, and sodium salts are produced in commercial quantities by fermentation of glucose or by extraction from hydrolyzates (products obtained by a chemical splitting involving the addition of water) of wheat, corn, soybean, and sugar beet proteins. The glutamates are added to foods to improve or enhance flavor, particularly to high-protein foods. *Monopotassium* and *monoammonium glutamate* are also used as components of sodium-free salt substitutes. *Monosodium glutamate*

MAJOR REFERENCE: Evaluation of the health aspects of certain glutamates as food ingredients. FASEB/SCOGS Report 37a (NTIS PB 283-475) 1978, Report 37a Supplement (NTIS PB 80-178635) 1980.

(MSG), a component found in seasoning salts, accounts for 99 percent of the glutamates used as food additives.

SAFETY: It has been estimated, based on usage reported by the food industry, that the average daily intake of MSG as an additive in food in 1976 was approximately 170 milligrams per person (other glutamates may have contributed an additional 1 or 2 milligrams), or slightly more than 3 milligrams per kilogram of body weight (kg/bw) for a person weighing 60 kg (132 lbs.). The intake of naturally occurring glutamic acid from foods is almost 70 times as great, the daily diet contributing 200 milligrams per kg/bw.

Glutamic acid concentrates in the brain, and the mature brain appears to be protected by mechanisms that maintain a steady level of brain glutamate. While it is possible to overload these mechanisms, this occurs only at very high levels of glutamic acid intake, and the presence of other foods in the diet appears to restrict its spread and modify the effect. However, anyone sensitive to sugar beets, corn, or wheat may suffer an allergic reaction if MSG is added, since it is derived from these. If food containing large amounts of MSG is consumed on an empty stomach, the "Chinese restaurant syndrome" of burning sensations, facial and chest pressure, and headache will be experienced by some people.

For infants and very young children, the adverse effects of high glutamate levels can be more serious. Several investigators have demonstrated that glutamates caused damage in the central nervous system in a number of animal species during the first ten days of life, such as lesions in the retina of the eye and the hypothalamus of the brain. While in most of these studies the dosages were administered by injection and involved levels more than 600 times human dietary intake (adjusted for body weight), in one experiment 160 times such intake force-fed to 10-to-12-day-old mice also produced brain lesions in half of the animals. When the dosage was halved, this did not occur. In mice or rats, a brain barrier develops by the twelfth day, reducing the hazard. These adverse effects have not been observed by a number of other researchers. Significant decreases in learning performance have been observed in rats fed high dosages of MSG; however, clinical studies have not convincingly demonstrated that this extends to intelligence in humans.

Data from animal experiments suggest that the blood-brain protective barrier in adults may be weakened by acute hypertension and illnesses affecting the vascular system, brain edema (excessive fluid), eye inflammation, and eye surgery. Damage from these illnesses may be increased by the entrance of glutamic acid.

Glutamates administered to pregnant animals do not cause birth defects, or adversely affect growth or development in the newborn; nor do they alter genes or cause cancer. In one experiment a glutamate inhibited growth of malignant tumors.

ASSESSMENT: Glutamic acid, its salts, and its hydrochloride should not be added to infant foods (the practice has been discontinued by manufacturers). Moderation in intake of foods containing these substances is advisable for younger children. MSG might best be avoided by anyone sensitive to sugar beets, corn, or wheat, and particularly on an empty stomach. It would be prudent for anyone afflicted with vascular illness, brain edema, or eye inflammation, undergoing eye surgery, or suffering from acute hypertension to restrict intake of glutamates. With these exceptions, it is believed that these substances offer no health hazard at their present levels as additives in food.

RATING: S for some people; X for infants and very young children, persons with vascular and eye ailments, and persons allergic to corn, wheat, or sugar beets.

GLUTEN (CORN OR WHEAT)

Vital Gluten

Gluten is the principal protein component of cereal grains. Wheat and corn gluten used as food additives are prepared from flours by washing out the starch, leaving a doughlike gluten residue which is then dried. Wheat and corn gluten are mixtures of proteins. One part of wheat gluten helps trap gas bubbles during yeast fermentation, while another protein component provides elasticity to the dough. Corn gluten has a different protein mixture and is not used in baking. Instead, it is added to extend the protein in a food. As a nutritional

---

MAJOR REFERENCES: Nutrition in preventive medicine. G. H. Beaton and J. M. Bengoa, eds. WHO Monograph Series No. 62 (Geneva, 1976); Evaluation of the health aspects of wheat gluten, corn gluten, and zein as food ingredients. FASEB/SCOGS Report II-12 (NTIS PB 82–108671) 1981.

supplement, gluten is used in protein-fortified cereals. When gluten has been dried without altering the protein by heat (prepared in this way it is called *vital gluten*), it improves the texture in baked goods.

SAFETY: Gluten is a useful protein source, but is deficient in the nutritionally essential amino acid lysine (see p. 475). However, the U.S. diet is high in proteins, especially animal protein, and almost always will provide adequate lysine to balance out the body's needs.

The sole concern about the safety of gluten in the diet is gluten sensitivity in some individuals; about 1 person in 3000 in the U.S. is believed to have this malabsorption disease. Its cause is not known, but may be of genetic origin. In such individuals, gluten may irritate the lining of the intestine and interfere with the normal absorption of nutrients in the diet. The symptoms at times can be severe. There may be distress, diarrhea, weakness, damage to the small intestine, and weight loss, particularly in infants. The condition has been called celiac sprue (abdominal malabsorption), gluten-sensitive enteropathy, idiopathic steatorrhea, and nontropical sprue. The only successful treatment is to go onto a gluten-free diet, which usually means elimination of cereals, especially wheat and rye, since these grains contain gluten; corn seems to be acceptable in not having this toxicity, however.

ASSESSMENT: Gluten is the edible protein of grains, and is useful as a diet supplement; wheat gluten also is used as a dough-strengthener in baked goods. A few people have a gluten sensitivity and cannot adequately digest this protein. For them, use of wheat gluten as an added food ingredient poses a hazard due to the distress and diarrhea associated with this celiac disease.

RATING: S for most people; X for wheat gluten for those with a gluten sensitivity (celiac sprue).

GLYCERIDES

SIMPLE GLYCERIDES: Monoglycerides (Glyceryl Monostearate, Monostearin, Monoglyceryl Stearate); Diglycerides; Triglycerides

CHEMICALLY-MODIFIED GLYCERIDES: Acetylated Glycerides; Glyceryl Triacetate (Triacetin), Diacetyl Tartaric Esters of Mono- and Diglycerides
Ethoxylated Mono- and Diglycerides (Polyglycerates; Polyoxyethylene Stearates)
Glyceryl-Lacto Esters of Fatty Acids (Glyceryl Lacto-Palmitate, Lactopalmitate)
Lactylated Fatty Acid Esters of Glycerol and Propylene Glycol (Lacto Esters of Propylene Glycol)
Oxystearin
Polyglycerol Esters of Fatty Acids
Succinylated Mono- and Diglycerides, Succistearin

Glycerin or glycerol (see p. 563), a type of alcohol and a component of fat, can be combined chemically with various acids, including fatty acids, to form glyceryl esters, which are called glycerides. There are three types of the more simple, natural glycerides: monoglycerides, diglycerides, and triglycerides, depending on the number of alcohol groups (hydroxyl groups) that are combined in the molecule.

Many natural or simple glycerides occur in foods, such as vegetable oils (see p. 660), animal fats (see p. 479; these are all triglycerides), and lecithin (see p. 580). Mono- and diglycerides are formed in the body as the first step in digestion of edible fats and oils. They have been added to food for over 60 years, and are used to maintain softness in bakery products. *Monostearin,* a monoglyceride, has only one fatty acid, stearic acid, attached to glycerol.

Various chemically modified glycerides are now synthesized to serve as emulsifiers (to help blend oil and water mixtures, such as shortening), as crystallization inhibitors in fatty food products (to prevent clouding when food is chilled), and to help dissolve antioxidant additives (which inhibit the undesirable changes caused by oxygen, such as rancidity). Sometimes acids in addition to fatty acids are incorporated into the

---

MAJOR REFERENCE: Evaluation of the health aspects of glycerin and glycerides as food ingredients. FASEB/SCOGS Report 30 (NTIS PB 254-536) 1975.

glyceryl ester. Acetic acid (see p. 463), lactic acid (see p. 576), succinic acid (see p. 648), and tartaric acid (see p. 655) may be used; a mixture of related compounds usually results.

When acetic acid is incorporated, an *acetylated glyceride* (acetoglyceride) may be formed. *Glyceryl triacetate* (triacetin, a triglyceride ester with acetic acid, the ingredient of vinegar substituting for the fatty acid of a fat) may be used as a flavoring agent.

Ethoxylated mono- and *diglycerides* (sometimes referred to as polyglycerates or polyoxyethylene stearates) are manufactured by reacting an edible glyceride with ethylene oxide, a gas often used to sterilize biological preparations. The resulting emulsifier is used in bakery products, toppings and icings, frozen desserts, and coffee cream substitutes.

Lactic acid incorporated into the glyceride (lactylated, giving *glyceryl-lacto esters of fatty acids*) provides an additive useful as an emulsifier in baked goods. Another emulsifier used in baked goods, identified on food labels as *lactylated fatty acid esters of glycerol and propylene glycol* (or lacto esters of propylene glycol), includes lactic acid and propylene glycol (see p. 611) in the modified glyceride additive.

Oxystearin is a glyceride combined with oxygen that is prepared from hydrogenated soybean or cottonseed oil. It is a cloud inhibitor in salad oils.

Polyglycerol esters of fatty acids are prepared from a variety of vegetable oils or fatty acids, and also from tallow. They are made up of chemical combinations of glycerol that may involve as many as ten glycerol molecules linked together. They may be used as food emulsifiers, or as cloud inhibitors in salad oils.

When succinic acid replaces one or more fatty acids in a fat, *succinylated mono-* and *diglycerides* are formed. (A glyceride that includes the term "succinylated" indicates the presence of this acid.) The resulting additive may be used as an emulsifier or as a dough conditioner in baked products. In the case of the emulsifier *succistearin,* propylene glycol, fully hydrogenated vegetable oil, and succinic anhydride (a version of succinic acid) are used in the manufacture, incorporating succinic acid in various glycerin and propylene glycol esters of the fatty acids.

SAFETY: The daily diet of simple as well as chemically modified glycerides as additives in food in the diet, based on reported usage by food manufacturers in 1976, has been estimated at somewhat over 600 milligrams per person, or 10 milligrams per kilogram of body weight (kg/bw) for a person weighing 60 kg (132 lbs.). The modified glycerides represented less than 10 percent of the total, the principal ones used being ethoxylated mono- and diglycerides and the glyceryl-lacto esters of fatty acids, each accounting for about a third of these chemically altered glyceride additives. Most of the consumer exposure to these additives thus is in the form of the mono- or diglycerides, which are natural constituents obtained from fats during digestion and absorption of fatty foods in humans. As additives they are readily used and metabolized by the body, serving as an energy source for the body, just as do animal and vegetable fats and oils in the diet.

Monoglycerides of saturated fatty acids are more poorly absorbed than the more unsaturated fatty acids. Generally, however, the diet will include enough fats containing oleic acid (having some unsaturated linkages) and polyunsaturated fatty acids to help bring the saturated fatty acid into solution, increasing the absorption and assuring availability to the body of an additive such as monostearin.

The high degree of saturation of the fatty acid is of concern only in its presumed role in human heart disease. A high-fat diet, especially when the fat contains considerable amounts of saturated fatty acids, is considered by many authorities as a risk factor in arteriosclerosis and heart disease, based on population studies and experimental studies with both animals and humans. The fatty foods of the diet obviously contribute most of the fatty acids; food additives such as monostearin add a tiny amount to the total fat intake, which may amount to over 150 grams daily.

When *triacetin* (glyceryl triacetate) was fed to rats at unusually high levels (at 55 percent of the diet for two months, and for nearly two years at a level of slightly over 3 percent of the diet, some 1500 milligrams per kg/bw), no differences in growth from that of control animals were observed, and no toxic effects were seen. When *acetylated glycerides* or *diacetyl tartaric acid esters of mono- and diglycerides* made up 20 percent of the diet of rats (10 grams per kg/bw) over a two-year period, no adverse effects were seen on growth, survival, or tissue abnormalities, if there was adequate vitamin E in the diet; the same lack of effect was noted in dogs over a 25-month period.

Two *ethoxylated monoglycerides* (polyoxyethylene stearates, or polyglycerates) have been examined for safety in the diet. In humans, as much as 6 grams daily (several hundred times the usual dietary intake) for up to two months had no effect; only small amounts of it are absorbed. Similar findings of lack of effect were seen in studies of the *glyceryl-lacto esters of fatty acids* added to the diet of experimental animals.

Ninety-day studies on the safety of *lacto esters of propylene glycol* (propylene glycol lactostearate) have used rats and dogs as experimental animals. No significant abnormalities were noted in the studies when the additive constituted 10 percent of the diet, an enormously high level when compared with use in human foods.

A two-year feeding study with *oxystearin* at dietary levels up to 7½ grams per kg/bw (hundreds of thousands of times the usual human intake of oxystearin) again resulted in normal growth of the animals and no toxic effects, with the exception of an observation of some tumors in the testes, which the investigators doubted were related to the oxystearin fed; the UN Joint FAO/WHO Expert Committee on Food Additives questions the validity of such an excessive level of feeding.

Polyglycerol esters of fatty acids have been tested with mice, rats, and dogs. A two-year study with mice (receiving amounts much larger than those ingested by humans) showed that the substance does not induce cancer or other harmful effects. Rats fed at levels as high as 5 grams per kg/bw (thousands of times the usual human intake of polyglycerol esters) showed no adverse effects beyond a slight enlargement of the liver at the highest dosage. Dogs eating the additive for 90 days at 5 grams per kg/bw had no abnormalities in any of many examinations that were made.

Succistearin at the level of 10 percent of the diet for rats (5 grams per kg/bw) provided no evidence of toxicity or harm.

There appear to be no studies on whether any of these glycerides can cause allergy, malformation of the fetus, or chromosome damage.

ASSESSMENT: The mono-, di-, and triglycerides used as food additives are handled by the body in the same manner as the food fats. They will add to the caloric level of the food, and in some cases will add to the saturated fatty acid intake. Long-term studies at levels many hundreds of times as high as exposure through diets of consumers have demonstrated, with a single exception, that the glyceride additives are probably free from hazard other than the small contribution they make to the normal daily intake of saturated fatty acids. Since these additives represent only a minute fraction of the total intake of saturated fat in the human diet, they pose little additional risk and do not represent a significant hazard for people with heart ailments.

A possible exception to the otherwise risk-free modified glycerides is oxystearin; the only adverse effect noted for it was the presence of some tumors after rats had been fed for two years with the additive at 7.5 grams per kg/bw (equivalent to 450 grams daily for a human weighing 60 kilograms). Despite the extremely high level fed, caution is suggested regarding intake of oxystearin until more detailed research becomes available.

RATING: For oxystearin, ? pending clarification of tumor effects; for all other glycerides or modified glycerides, S.

GLYCERIN (GLYCEROL)

Glycerin (or glycerol), an alcohol, is a natural constituent of all fats; about 10 percent by weight of animal and vegetable fats consists of it. Glycerin is a slightly sweet food additive that has a variety of uses. It readily absorbs moisture and thus serves to help retain moistness of a food. As a processing aid, it serves as a crystallization modifier and a plasticizer to maintain consistent texture in candy creams. It may be found in marshmallows, various candies, baked goods, some meat products, and many other foods. For commercial use by food manufacturers, it is produced in a number of ways, among them chemical synthesis, bacterial fermentation of sugar, and as a by-product of soap manufacture.

SAFETY: A survey of food usage by processors in 1976, conducted by the National Academy of Sciences-National Research Council, established that the average daily dietary intake in the U.S. was 106 milligrams per person, or close to 2 milligrams per kilogram of body weight (kg/bw) for a person weighing 60 kg (132 lbs.).

The body is accustomed to dealing with glycerin usefully. It forms glycerin from ingested carbohydrates, from glycogen (the body's carbohydrate reserve of energy), and during the breakdown of fats. Along with the fatty acids, glycerin aids in the metabolic processes that build and reconstitute cells.

MAJOR REFERENCE: Evaluation of the health aspects of glycerin and glycerides as food ingredients. FASEB/SCOGS Report 30 (NTIS PB 254-536) 1975.

Glycerin has proven useful medically in the treatment of humans suffering from cerebral pressure. Oral administration of 1.5 grams per kg/bw (some 800 times the presence of this substance as an additive in the diet) for four days resulted in improvement, without toxic effects. Dosages of up to 2 grams per kg/bw have lowered intraocular pressure in patients afflicted with glaucoma, and also in healthy subjects.

Long-term studies with rats and dogs fed glycerin at levels of up to 5 grams per kg/bw showed no adverse effects from the chemical. Similarly, glycerin did not affect reproductive capabilities, growth, maternal or fetal survival, or result in malformations of the offspring. One investigation conducted through seven generations of rats reared on a diet providing 15 grams per kg/bw of this compound found even this massive dosage to be without influence on growth and reproduction.

Research has not been conducted to determine whether glycerin can induce cancer or gene mutations. Glycerin does not appear to cause allergic reactions in humans.

ASSESSMENT: Glycerin is a normal constituent of the body, accounting for some 10 percent of the body fats or dietary fat. Many studies with this material conducted with humans and animals, and employing dosages far in excess of its consumption as an additive, demonstrate it to be free from hazard to the consumer.

RATING: S.

GLYCEROL ESTERS OF WOOD ROSIN (TALL OIL ROSIN ESTERS)

ESTER GUM; GLYCERYL ABIETATE; SOFTENERS

Rosin is obtained from the wood of pine trees. When the acid fraction of rosin is chemically combined with the alcohol groups of glycerol (glycerin, see p. 563), *glycerol esters of wood rosin* are produced. The rosin may be obtained from chips of pine wood in paper manufacturing; its acid fraction when combined with glycerol is called *tall oil rosin esters*. These rosin esters are a very complex mixture and usually are purified by treating with steam. Abietic acid is one of the

MAJOR REFERENCE: Evaluation of certain food additives. WHO Technical Report Series No. 557 (Geneva, 1974).

components of pine rosin; its ester is *glyceryl abietate*. As food additives, sometimes called *ester gum*, they are used in beverages to modify the tendency of citrus oils to separate, or as cloud crystal aids or inhibitors; for serving as a *softener* in chewing gum to impart plasticity to chewing-gum base materials;* as an emulsifier or blender; and as a flavor agent. Frozen desserts and gelatins make use of them.

SAFETY: In a 1976 survey of the use of these additives by the food industry, it was estimated that on the average approximately 130 milligrams of these glycerol esters of rosin were contained in food in the daily diet of a person in the U.S. Most of this is in chewing gum, which usually is not ingested and thus not truly part of the diet. The esters of wood rosin are essentially insoluble in water and are likely to be excreted intact.

There apparently have been no published studies on the safety of these materials in foods. The composition of the rosin is not controlled, and is known to vary depending on the starting material, species of pine trees used, and manufacturing process for preparing the rosin acids.

ASSESSMENT: A considerable portion of the use made of glycerol esters of wood rosin as an additive is as a softener in chewing gum, which usually is not part of the diet. They also are insoluble, and there is little likelihood of absorption by the body of these esters. Nevertheless, the lack of toxicity studies prevents any basis for rating the safety of the rosin esters. Long-term studies to characterize these additives for any possible effects on cancer, reproduction, or other metabolic changes are needed.

RATING: ? (inadequate data on safety are available).

---

*See p. 511. There are softeners used as chewing-gum base materials other than the rosin esters.

GUAR GUM

Guar gum is extracted from the seed of the guar plant, a legume resembling the soybean plant. India and Pakistan are major producers. The gum is a complex sugar, a condensation of the simple sugars mannose (found in many plants) and galactose (a constituent of milk sugar). It absorbs cold water readily, forming thick, semifluid mixtures. This property makes guar gum very useful in the food industry to stabilize the consistency of ice cream and as a texture modifier or thickening agent in various food products.

SAFETY: The intake of guar gum in food in the U.S. in 1975, calculated from figures reported on weight used in food, averaged 50 milligrams per person per day, or 0.8 milligram per kilogram of body weight (kg/bw) for a person weighing 60 kg (132 lbs.).

No apparent adverse effects were found in studies with monkeys fed guar gum at levels 250 times as high as the estimated daily human intake (adjusted for body weight); rats were free of any harmful effects, even when consuming guar gum at several thousand times human intake. Tests using several animal species have shown that guar gum does not cause physical defects in the embryo during pregnancy. In vitro (laboratory test-tube) investigations indicate that it apparently does not produce gene mutation or chromosome damage, though some cell division effects were seen in one test at high levels. Force-feeding relatively high dosages (about 1000 times the estimated daily human intake) to pregnant mice did result in the loss of some animals. While human consumption of guar gum is unlikely to approach this amount, a greater mortality rate has occurred in a variety of pregnant animals when fed very large doses of other vegetable gums (carrageenan, gum arabic, gum tragacanth, locust bean gum, and others); this suggests the advisability of conducting additional feeding studies to help explain this maternal toxicity. No evidence of cancer-producing or allergenic properties of guar gum has been reported.

The UN Joint FAO/WHO Expert Committee on Food Additives considers the acceptable daily intake of guar gum for man as ranging up to 125 milligrams per kg/bw. This amount is about 150 times the U.S. average daily intake.

MAJOR REFERENCE: Evaluation of the health aspects of guar gum as a food ingredient. FASEB/SCOGS Report 13 (NTIS PB 223-836) 1973.

ASSESSMENT: Long-term feeding studies are needed, but the available information indicates that guar gum is not hazardous to human health at the present consumption level. However, when fed at high levels it caused some mortality in pregnant mice, and similar findings have occurred with other vegetable gums, suggesting caution in its use by pregnant women pending further investigation.

RATING: S for most people; ? for pregnant women.

GUM ARABIC (ACACIA GUM)

Gum arabic is a vegetable gum obtained from the discharge of a number of species of acacia, a Middle Eastern tree. The U.S. supply comes mostly from the *Acacia senegal* in the Sudan. Gum arabic is a complex polysaccharide, a carbohydrate consisting of several simple sugars. It contains calcium, magnesium, and potassium as well and can be almost completely dissolved in water—an unusual characteristic for a gum—making it useful in a range of concentrations and ways for food processors. As an additive, gum arabic enables mixtures of ingredients to blend together successfully, prevents separation, and can improve "body feel" of a food and act as a thickener.

SAFETY: The daily intake of gum arabic in food by U.S. individuals in 1975 averaged less than 50 milligrams per person. A study of hospital patients determined that even when administered intravenously at least 180 times this amount was required to cause harm. Under normal conditions, gum arabic breaks down into simple sugars in the body and is completely digested by animals and, in all likelihood, by man.

Investigations conducted among a number of animal species during pregnancy did not reveal that gum arabic caused defects in the newborn; but among rabbits it did result in the death of the majority of the mothers at a dosage over 1000 times the average daily human intake (adjusted for body weight). Laboratory tests failed to disclose gene mutation properties, and short-term studies of rats and mice revealed no evidence that this substance can cause cancer. However, gum arabic is a proven cause of allergic reactions in animals. Human sensitivity has been reported among people who work with it, and this

---

MAJOR REFERENCE: Evaluation of the health aspects of gum arabic as a food ingredient. FASEB/SCOGS Report 1 (NTIS PB 234-904) 1973.

effect has been confirmed in research among patients prone to allergies.

ASSESSMENT: Gum arabic appears to be an antigen, a substance capable of stimulating a reaction in a specific antibody of the body's immunological warning system and resulting in an allergic response, such as an attack of asthma or a rash. As an antigen, it may spread its effect to other closely related antibodies, increasing the severity of the attack or broadening the range of allergic sensitivity. The extent to which this occurs has not been determined. People prone to allergies will be well advised to avoid gum arabic.

Other polysaccharides have caused fatalities among pregnant animals when fed at very high levels, and similar findings among rabbits fed gum arabic merit a caution about its safety for women during pregnancy. Further research is needed. Aside from these areas, the evidence indicates that this substance is not hazardous to human health in the quantities presently in food, or even if these quantities substantially increase.

RATING: S for most people; X for anyone susceptible to allergies; ? for pregnant women (additional studies are needed to resolve uncertainties).

GUM GHATTI

Gum ghatti is the dried sap that oozes from taps in a tree native to India and Sri Lanka. This gum is a complex carbohydrate containing a number of simple sugars; its exact chemical structure has not been determined. The gum is used in the food industry only to a limited extent, primarily in frozen dairy products and nonalcoholic beverages as a stabilizer for oil and water emulsions.

SAFETY: The intake of gum ghatti in food in the U.S. in 1970 averaged slightly less than 0.05 milligram per person per day.

Minimal biological and toxicological information is available on gum ghatti in animals or humans. Laboratory tests on rats and mice, as well as on human and animal cells, have shown that gum ghatti does not produce gene mutation. Experiments

MAJOR REFERENCE: Evaluation of the health aspects of gum ghatti as a food ingredient. FASEB/SCOGS Report 12 (NTIS PB 223-841) 1973.

with four species of pregnant animals confirm that this substance does not cause physical defects in the embryo. However, it has been noted that force-feeding pregnant rats an extremely high dose of 1700 milligrams per kilogram of body weight (kg/bw), equivalent to 2 million times the daily estimated human intake, resulted in the loss of some of the animals. Similar findings did not occur with hamsters or mice, nor did they in rats when the dosage was reduced to 370 milligrams. There were losses among pregnant rabbits fed at a lower dose level (33 milligrams per kg/bw), which still is the equivalent of 180,000 times the estimated daily human intake.

No evidence that gum ghatti has cancer-producing or allergy-producing properties has been reported.

ASSESSMENT: The very limited information available on gum ghatti presents no evidence to indicate that this substance is hazardous to human health at current levels of consumption. The sole adverse effect was with pregnant rats and rabbits, but this occurred only when force-fed at levels 180,000 times as great as the presence of gum ghatti in human diets.

RATING: S.

GUM TRAGACANTH

Gum tragacanth, a vegetable gum, is the dried discharge obtained from several species of *Astragalus,* a wild shrub found in the Middle East. A complex mixture of sugars and sugar acids that contains calcium, magnesium, and potassium salts, gum tragacanth swells in cold water to produce very viscous solutions. It is used as a stabilizer or thickening agent in a wide variety of food products.

SAFETY: The intake of gum tragacanth in food in the U.S. in 1975 averaged a little under 4 milligrams per person per day. Very few reports have been published on the biological activity of this substance.

Gum tragacanth can induce an allergic response in sensitive people by ingestion, contact, or inhalation. With respect to incidence or severity of these reactions, it resembles many allergens commonly encountered in the diet. Cases of sensitiv-

---

MAJOR REFERENCE: Evaluation of the health aspects of gum tragacanth as a food ingredient. FASEB/SCOGS Report 4 (NTIS PB 223-835) 1973.

ity to gum tragacanth have been reported. Seven times the average daily consumption of 4 milligrams administered for one week caused severe symptoms in susceptible people. Other vegetable gums that may cause an allergic response are arabic (acacia) and sterculia.

A study published in 1972, using several species of animals, reported that gum tragacanth did not cause physical defects in the developing embryo. However, feeding very high dosages (adjusted for weight differences, equivalent to 2200 to 18,000 times the estimated daily human intake) did result in the loss of a significant number of pregnant rats and rabbits. This observation suggests the advisability of conducting additional feeding studies with pregnant animals that would include other dosage levels equivalent to and exceeding the current estimated daily human intake of gum tragacanth.

No reports of studies of gene mutation or cancer-producing properties of gum tragacanth have been found thus far, although there is evidence that it can inhibit some tumors in mice.

ASSESSMENT: Since gum tragacanth can cause an allergic response, persons sensitive to such allergens should avoid this substance. While the dosages of this substance that caused fatalities in pregnant animals were far in excess of human consumption, similar findings employing other polysaccharides suggest caution in its use by pregnant women pending further investigation.

Excluding these areas, the evidence indicates that gum tragacanth is not hazardous to human health at the present consumption level.

RATING: S for most people; ? for pregnant women and for anyone with allergic reactions.

INOSITOL

Inositol occurs naturally in plant and animal tissues, where it is one of the constituents of phosphorus-containing fatty substances essential for maintenance of life. The organ meats of animals and cereals are rich sources of inositol. It is not con-

---

MAJOR REFERENCE: Evaluation of the health aspects of inositol as a food ingredient. FASEB/SCOGS Report 51 (NTIS PB 262-660) 1976.

sidered to be a vitamin (a substance which the body needs, cannot make, and must obtain from food), because man is able to synthesize it from other substances. When inositol is added to foods, it generally is as a nutrient for milk-free infant formulas or special dietary foods.

SAFETY: The information on daily human intake is confined to infants, and only to those 10 percent who are on milk-free infant formulas. These are infants under six months of age, and their added intake of inositol averages 65 milligrams daily. This is within the same range as the normal inositol intake received by babies from cow's milk, and about a fourth of the intake from breast feeding. Adults average an intake of 1 gram of inositol daily from the normal foods in their diet.

Inositol has some capability to facilitate the distribution of fat throughout the body, and helps in metabolism to prevent excessive fat deposition in the liver. Added intakes up to 2 grams daily for several weeks produced no harmful effects in humans. A variety of tests have shown that inositol does not cause cancer, abnormalities in the newborn, or gene alterations.

ASSESSMENT: Inositol is widely distributed in foods and is readily synthesized in humans. It participates in fat metabolism, but its role in nutrition is not fully understood. The rationale for adding inositol to milk-free infant foods is the assumption that the greater intake ensures against possible deficiency of this substance during early growth and development, when the need might be maximum and the body's ability to synthesize it might not be fully developed. It has shown no adverse effects even when appreciable amounts have been taken by adults. There is no likelihood of any hazard from its use in foods.

RATING: S.

IODINE SALTS*

Calcium Iodate; Potassium Iodate; Potassium Iodide

Iodine is found naturally in some drinking waters and in various plants, but especially in seafoods; by contrast, freshwater fish are rather poor sources of this element. Iodine is a nutrient essential for production of thyroxine in the body, which serves to monitor or regulate the body's metabolism. Because the food and water in many parts of the country are deficient in this nutrient, a situation that can lead to goiter, *iodine salts* are added to some foods to supplement human diets. Iodized salt is the major fortified product; it contains *potassium iodide*. In bread manufacture, *calcium* or *potassium iodate* may be used as a dough conditioner to improve texture; generally the iodates are converted to iodide salts during the baking process.

SAFETY: A survey conducted by the National Academy of Sciences-National Research Council (NAS-NRC) to determine the intake in the diet of added potassium iodide, based on reported usage by food processors in 1976, indicated it to be 395 micrograms (0.395 milligram) daily, or 302 micrograms calculated as iodine. The iodates used as food additives in bakery products would be in addition to this. While the recommended daily intake of iodine of 150 micrograms proposed for optimum health by NAS-NRC is considerably lower, this group considers 1000 micrograms of iodine in the daily diet to be safe.

A deficiency of iodine will cause changes in the thyroid gland leading to goiter. An excess of dietary iodine also can cause adverse effects, including thyroid enlargement, but inhibition of iodine uptake by the thyroid gland in humans occurs only with doses of potassium iodide at least 100 times as high as the average dietary intake. With experimental animals, toxic effects have been seen only when iodine intake was many hundred times human intake, adjusted for body weight differences; at these high iodine levels, occasional effects on lactation, body weight and appetite, blood components, inflammation of the gastrointestinal tract, and mortality among the newborn have been seen in various studies.

MAJOR REFERENCE: Evaluation of the health aspects of potassium iodide, potassium iodate, and calcium iodate as food ingredients. FASEB/SCOGS Report 39 (NTIS PB 254-533) 1975.

*See p. 516 for cuprous iodide.

Potassium iodide supplementation has no apparent effect on thyroid cancer, and shows no adverse effect on the chromosomes.

ASSESSMENT: Iodine is essential in the diet, and is frequently added to some foods because many diets otherwise would be deficient. The iodine-containing salts that are used as food additives or in seasoning salts more than meet human requirements; their toxic effects are evident only after an intake hundreds of times as high as normal consumption. These additives pose no hazard to the consumer at current levels in the diet.

RATING: S.

IRON

Electrolytic Iron; Reduced or Elemental Iron

IRON SALTS

Ferric Ammonium Citrate; Ferric Phosphate (Iron Phosphate, Ferric Orthophosphate); Ferric Pyrophosphate; Ferric Sodium Pyrophosphate (Sodium Iron Pyrophosphate); Ferrous Fumarate; Ferrous Gluconate; Ferrous Lactate; Ferrous Sulphate

Iron is an essential mineral element. A deficiency of iron leads to anemia. Iron occurs widely in foods, especially in organ meats such as liver, red meats, poultry, and leafy vegetables. Nevertheless, iron deficiency anemia is extensive, and iron fortification of some foods is recommended by nutritionists. Women of childbearing age are especially in need of iron; their recommended dietary intake is 18 milligrams daily.

The principal foods to which iron or iron salts (compounds resulting from interaction of iron with an acid) have been added are enriched cereal products and some beverages, including milk. In addition, some iron salts may have special uses as food additives; ferrous gluconate is used for coloring ripe olives.

Iron may occur in two chemical forms in its salts—ferrous (iron with two chemical linkages available for forming a salt) and ferric iron (the more stable form, with three linkages available).

MAJOR REFERENCE: Evaluation of the health aspects of iron and iron salts as food ingredients. FASEB/SCOGS Report 35 (NTIS PB 80-178676) 1980.

SAFETY: A survey of the reported use of iron salts as food additives conducted in 1976 by the National Academy of Sciences-National Research Council indicated an average daily intake per person in the U.S. of 11 milligrams.

The form of iron used in food enrichment affects its usefulness as a nutrient for humans. There is great variability in the solubility and the body's ability to absorb the various iron compounds that are used. Iron powder (reduced or elemental iron) and the iron phosphates, for example, are so insoluble that they are poorly absorbed. By contrast, the ferrous salts are quite readily absorbed and thus have higher biological availability in replenishing the body's stores. The iron in heme (a part of hemoglobin, and the major form in which iron occurs in red meats) is the most readily useful form of iron in the diet, although it is not used as a food additive. However, either heme or ascorbic acid (vitamin C) in the diet can appreciably raise the absorption of iron from foods that are eaten at the same time.

The excessive accumulation of iron in the body can lead to a toxic clinical condition called hemochromatosis, which is due to a faulty metabolism occurring in only a small part of the population. Feeding studies with animals using diets containing iron in amounts of 200 to 1000 milligrams per kilogram of body weight daily (1000 to 5000 times as high as average human intake) have failed to reproduce this condition.

Elemental iron has been shown to be most effective in maintaining normal blood hemoglobin when the particle size of the iron powder is extremely fine. Even massive doses (up to 10 grams) have been fed to animals and humans without apparent harm; it is the least toxic form of iron compounds used in foods.

Ferrous fumarate is widely used as an iron supplement in medication, but apparently has very limited use as a food additive. It is well absorbed and utilized by humans, and has been recommended for use in infant formulas. High doses caused less gastric irritation than ferrous sulfate and ferrous gluconate.

Ferrous gluconate also has high biological availability. Patients have been given this iron salt at levels of up to 180 milligrams iron daily, with some individuals showing gastrointestinal side effects (common to excessive doses of any of the iron salts). Since the body normally produces large amounts of gluconate (see p. 553), this should not in itself pose any hazard. The additive has shown chromosome alteration activity in some tests but not in others.

Ferrous lactate dissolves readily and is absorbed quite efficiently by the body. It did not affect the developing embryo when injected into chick eggs. However, there are no long-term studies on this compound, and the one limited test of its effect on cancer when injected suggests need for a more thorough study to discover whether it might cause tumors when present in the diet.

Ferrous sulfate has been quite thoroughly investigated, since it is the recognized standard in comparing iron compounds for their anemia prevention usefulness. Adults have taken 300 milligrams three times daily for years with no ill effects; however, higher doses are to be avoided, since an excess of any soluble iron salt may be toxic. Ferrous sulfate has been shown to be free from any adverse effects on fetal development, and most studies indicate that it does not damage chromosomes or cause tumors.

There is more limited information on which to assess the safety of ferric salts as food additives. While experimental studies indicate the relative biological availability of the iron (its ability to be absorbed and used in metabolism), there are few studies with animals to examine any possible long-term hazards.

Ferric ammonium citrate has limited use as a food additive, principally in dairy products, since it is one of the few iron salts that are soluble and do not induce off-flavors in such foods. It is less useful for hemoglobin regeneration in anemia than the ferrous salts. In treating patients with anemia, dosages equivalent to 400 milligrams of iron have been administered daily with good tolerance.

Ferric phosphate (ferric orthophosphate) and *ferric sodium pyrophosphate* have been used in the past as a major source of iron in fortified cereals because of their chemical inertness. Unfortunately this inertness also means very poor absorption by the body. Both are used in foods much less today. Neither chemical caused gene alteration in a variety of tests, and ferric sodium pyrophosphate induced no abnormal fetal growth when administered to pregnant animals.

Ferric pyrophosphate has only limited use as a food additive. It, too, is relatively insoluble and thus is poorly absorbed; it has low biological availability for replenishing the blood cells. It does not cause chromosome damage. Long-term studies of its effect on animals or humans have not been reported.

ASSESSMENT: The addition of some form of iron to selected foods is sometimes necessary to prevent anemia from an iron deficiency, since many consumers, and especially women, get less than recommended amounts in their usual diets. For nearly all people, the body mechanisms prevent accumulation of too much iron; for a very few with a metabolic defect, excessive stores may arise from dietary iron and lead to a hemachromatosis toxicity. Ferrous salts are preferred additives (when compared to iron powder or ferric salts) because of their greater biological availability in red-blood-cell regeneration. There is only sparse experimental evidence of the safety of many of these forms of iron, particularly in long-term studies or studies of cancer-induction. It is possible to conclude that elemental iron powders (reduced iron), ferrous fumarate, ferrous gluconate, or ferrous sulfate pose no hazard to the consumer at levels currently used in foods. In view of the very poor biological availability of the ferric salts used as food additives, they may not be acceptable ingredients in food fortification.

RATING:* S for iron and most ferrous iron salts; ? for ferrous lactate (more studies needed) and ferric iron salts (ineffective for fortification).

LACTIC ACID

CALCIUM LACTATE

Lactic acid is widely distributed in living cells, and is contained in blood. The body produces lactic acid during certain metabolic processes, primarily during the breakdown of carbohydrates into simpler compounds. The acid occurs naturally in many foods. For the quantities required in commercial food processing, lactic acid can be produced by chemical synthesis or from a bacterial fermentation of a carbohydrate such as corn sugar. To form *calcium lactate,* the acid is treated with calcium carbonate (chalk).

Calcium lactate has several food uses. By providing calcium, it can help firm the texture of some processed foods; it also is used to improve the properties of baked products and dry milk powders, and to inhibit discoloration in processed

MAJOR REFERENCE: Evaluation of the health aspects of lactic acid and calcium lactate as food ingredients. FASEB/SCOGS Report 116 (NTIS PB 283-713) 1978.

*Iron salts are not additionally rated hazardous for individuals known to have hemachromatosis, since diets of these people are likely to be under strict control.

fruits and vegetables. Lactic acid is used to impart flavor and tartness to some desserts and carbonated juices. It provides acidity, thus inhibiting spoilage by fermentation in some processed foods, including cheeses and Spanish olives. It conditions dough and stabilizes certain wines.

SAFETY: Average daily intake has been estimated at 11 milligrams for lactic acid and less than 1 milligram for calcium lactate, based on use reported by the food industry in 1975. Both substances are readily absorbed by the body and either converted back to sugar or utilized as energy sources in metabolism. The normal adult may make and convert as much as 140 grams of lactate daily in the course of carbohydrate metabolism.

A single dose of 10 grams of calcium lactate caused vomiting, diarrhea, and abdominal distress in humans, but no such violent reactions occurred with a 5-gram dose. (Such doses have been used to supply calcium for medical reasons.) Infants have been fed formulas acidified with lactic acid to reduce curd tension (this no longer is done in the U.S.). When the added lactic acid consisted of the L and D forms present in the usual food additives, premature infants usually showed a disturbed acid-base metabolism, a condition which could inhibit growth of certain gastrointestinal microorganisms in the infant. This was not observed when the formula contained only the L form normally produced in the body's metabolism. It should be noted that the daily intake of lactic acid in these cases was at least 2000 times the amount used as additives in the average U.S. diet.

Lactic acid and calcium lactate have been shown to be devoid of gene-alteration effects.

ASSESSMENT: The additional lactic acid provided by its use as a food additive is probably less than 0.1 percent of the lactate produced and metabolized normally in the body. This poses no hazard to the consumer beyond infancy. The UN Joint FAO/WHO Expert Committee on Food Additives has recommended that lactic acid and calcium lactate should not be added to formulas for infants less than three months of age, since there is evidence that high dosages as additives may cause health problems with young infants. These additives are not used in infant formulas in the U.S.

RATING: S.

LACTOBACILLUS CULTURE (LACTIC CULTURE, CHEESE CULTURE, VIABLE YOGURT CULTURE)

The *Lactobacillus* microorganism occurs widely in nature. It enables milk to sour (lactic acid—see p. 576—has been formed from the milk sugar), and shredded cabbage to ferment and form sauerkraut. Since the organism is killed by heating, for instance in pasteurization, *Lactobacillus cultures* (lactic cultures) are sometimes added for acidity during food processing. For example, if pasteurized milk is used to make cheese, the FDA states that it may be "subjected to the action of harmless lactic acid bacteria, present in such milk or added thereto" in order to make food products such as blue cheese or cheddar cheese. For some products, such as cottage cheese, the use of these bacterial cultures may appear on the ingredient label as "cultured" or "made from cultured skim milk"; for others, there may be reference to *"cheese culture"* or *"viable yogurt culture."*

SAFETY: Because this group of microorganisms is so widespread, and is present in the gastrointestinal tract as well as in unprocessed foods, direct tests of possible effects on health are rare. There are population studies that have related longevity of peasants in Bulgaria to their consumption of milk soured with *Lactobacillus bulgaricus*. Many articles in medical journals have recommended cultured dairy foods, such as yogurt, to help gastrointestinal disorders. Infants have had some relief from diarrhea when fed either yogurt or milk containing *Lactobacillus acidophilus,* the culture used to make "acidophilus buttermilk."

ASSESSMENT: The lactic acid-producing organisms used in preparing certain foods are harmless and ubiquitous. Use of these cultures in processed foods poses no hazard to the consumer.

RATING: S.

MAJOR REFERENCE: Cultured dairy foods. *Dairy Council Digest* 43:No. 4 (July-August 1972).

LACTOSE

Lactose, or milk sugar, is a major carbohydrate constituent of milk and occurs naturally in all dairy products; human milk contains 7 percent lactose, and cow's milk approximately 5 percent. Lactose is a sugar which, when split, will yield a molecule of glucose and one of galactose. It has only about a sixth the sweetness of sucrose (cane or beet sugar). It is used in foods as a carrier of other flavors, aromas, and color, and can contribute to the texture, flavor, and toasting qualities of baked goods.

SAFETY: About 29 grams of lactose is contained in the daily diet of the average consumer in the U.S., but only a tiny fraction, less than 1 milligram, was contributed to this intake by the lactose used as a food additive in 1976. A breast-fed infant consumes 50 grams daily, or 35 grams if fed cow's milk. As much as a fifth of ice cream consists of lactose, and over 50 percent of dry skim milk.

In most people, lactose is split into its constituent sugars by an enzyme secreted in the intestines; only then can it be absorbed and utilized by the body. This enzyme, lactase, is at its maximum during the early years of childhood, but begins to decrease after weaning. Among adolescents or adults, some individuals do not secrete adequate amounts of lactase. This leads to "lactose intolerance," characterized by a bloated feeling, cramps, gassiness, and a watery diarrhea, due to the bacterial fermentation of the undigested lactose in the lower intestine. Lactose intolerance varies among individuals, and even among races. It is most frequent among Africans and Orientals. It is believed present in 70 percent of U.S. blacks, but only 10 to 15 percent of U.S. whites appear to be afflicted. The symptoms are not to be confused with the allergic reaction of those with "milk intolerance"; lactose-intolerant individuals usually can consume a glass or two of milk daily (containing perhaps 10 grams of lactose) with no difficulty, since some residual lactase activity is usually present. When lactose is accompanied by protein and other food components present in a mixed diet, tolerance for it by susceptible individuals is increased, and this holds true when the lactose is fermented, as it is in buttermilk and yogurt.

Galactose, one of the constituent sugars of lactose, or lactose itself can cause cataracts in the eyes of rats fed excessively large quantities (over 30 percent of the diet). In rats, inability to efficiently metabolize such large quantities of galactose apparently is a cause of this cataract development. In humans, a rare "inborn error of metabolism" or genetic defect (galactosemia) can lead to cataract formation when galactose or lactose is ingested, since these people do not have the right enzymes in the liver to properly metabolize the sugar. With this exception, lactose has not been shown to induce cataracts in humans.

Lactose is known to improve the absorption of some essential minerals in the diet, especially calcium. No reports were found to assess whether lactose has any adverse effect on reproduction, or whether it can cause chromosome damage, or cancer.

ASSESSMENT: Lactose is the principal carbohydrate of milk, the earliest food for infants. Its presence in food is not hazardous except for the individuals unable to fully metabolize high amounts of lactose in the diet. The intestinal distress of lactose-intolerant individuals may arise from consumption of dairy products, but is unlikely to occur from the minute contribution to the total diet of the lactose used as a food additive.

RATING: S.

LECITHIN (SOY LECITHIN)

HYDROXYLATED LECITHIN

The *lecithin* that is added to food by manufacturers consists of a complex mixture of fatty substances derived from the processing of soybeans, for the most part, and to a lesser extent from corn and eggs. The main constituents of lecithin are choline (see p. 512), phosphoric acid (see p. 600), glycerin (see p. 563), and fatty acids (which, in combination with glycerin, form fats). None of them is regarded as a health hazard. They are present in all living organisms as chief components of all cell membranes and also in bile and blood. *Hydroxylated lecithin,* which has been modified by reacting with hydrogen peroxide or benzoyl peroxide, acetic or lactic acid, and sodium hydroxide, disperses more easily in water and is superior in its emulsifying (blending) properties.

MAJOR REFERENCE: Evaluation of the health aspects of lecithin as a food ingredient. FASEB/SCOGS Report 106 (NTIS PB 301-405) 1979.

Lecithin as an additive is used principally as an emulsifier and stabilizer of components in products like oils and margarines, frozen desserts, chocolate, and baked goods. It serves also as an antioxidant (one of its functions in living organisms), preventing destruction of fats by oxygen, which otherwise would cause rancidity, objectionable odors, and destruction of flavor.

SAFETY: The UN Joint FAO/WHO Expert Committee on Food Additives has estimated that the average daily diet contains between 1 and 5 grams of lecithin. Based on reported usage by food manufacturers in 1970, its presence as an additive averaged 96 milligrams daily, or 2 to 10 percent of its total presence in the diet.

A two-year feeding study of rats fed soy lecithin in dosages 15 to 80 times as great (in proportion to their body weight) as the amount consumed by humans in their total diets did not result in any adverse effects accountable to lecithin. An unconfirmed experiment conducted with mice, investigating separately the effects of lecithin diet and a cholesterol diet, reported a high incidence of brain tumors in both. No such result occurred in two rat studies.

Human volunteers ingesting 20 grams or more daily for 6 to 12 weeks did not experience harm.

In a test to determine the advisability of adding lecithin to bacon, in which conditions simulated smokehouse temperatures, a nitrosamine known to cause gene alterations was reportedly formed by the reaction of lecithin with sodium nitrite (see p. 631). Tests of lecithin by itself indicated that it is not mutagenic. Experiments with pregnant mice and rats determined that lecithin had no adverse effect on fertility, or maternal or fetal survival, and does not cause deformed offspring. Chick embryos showed no abnormalities after treatment with benzoyl peroxide-bleached lecithin.

One rat study with hydroxylated lecithin has been located. For a year rats were fed dosages comparable to thousands of times its presence in the average diet of U.S. humans, without ill effects. Studies are available of animals fed compounds which could correspond to the action of hydrogen peroxide or benzoyl peroxide on unsaturated fatty acids. They indicate that neither bleaching agent acting on fatty acids causes cancer, and that the products are toxic only in amounts thousands of times as great as the presence of hydroxylated lecithin or bleached lecithin in food.

ASSESSMENT: Lecithin is a component in membranes, blood, and other parts of the body. It is a natural constituent in foods consumed in the average diet, since it is present in all living plants and animals. This natural lecithin accounts for 90 to 98 percent of human intake of this substance, only a fraction of the total being added by food processors. Humans have ingested amounts considerably in excess of what is contained in their food without difficulty, and animal feeding studies support the conclusion that there is no cause for concern. The one investigation that reported the occurrence of brain tumors in mice was not confirmed in two other studies. The evidence at hand on bleached or hydroxylated lecithin treated with hydrogen peroxide or benzoyl peroxide indicates that neither will introduce a hazard when employed as an additive.

RATING: S.

LICORICE (GLYCYRRHIZIN)

AMMONIATED GLYCYRRHIZIN

Licorice is an extract prepared commercially from the roots and root stems of a shrub, *Glycyrrhizin glabra L.*, which is found in moderate and semitropical regions of Europe and Asia. Its biologically active component, glycyrrhizin, which constitutes about 20 percent of the extract, is a compound of sugars and a complex acid. *Ammoniated glycyrrhizin,* obtained from the licorice root by a hot-water process involving sulfuric acid, and then by neutralization by dilute ammonia, is 50 times as sweet as sucrose (cane or beet sugar), and can greatly enhance flavor. In foods, licorice is used as a spice and seasoning, as well as a flavoring. Also, it has long been valued as a medicinal remedy. Ninety percent of licorice production is used in tobacco products.

SAFETY: The amount of licorice present as an additive in food has been estimated at 3 milligrams daily in the diet of a U.S. adult in 1976, based on a survey of usage reported by the food industry conducted by the National Academy of Sciences-National Research Council; this is 0.05 milligram per kilogram

MAJOR REFERENCE: Evaluation of the health aspects of licorice, glycyrrhiza and ammoniated glycyrrhizin as food ingredients. FASEB/SCOGS Report 28 (NTIS PB 254-529) 1975.

of body weight (kg/bw) for a human weighing 60 kg (132 lbs.).

Ammoniated glycyrrhizin is only slightly absorbed in the body, with the remainder, for the most part, excreted in the feces. It (like licorice), when fed to or injected in a variety of animal species, caused suppression of urine. This occurred in rats fed 125 milligrams per kg/bw, about 2500 times average human consumption (adjusted for body weight), while 160 milligrams of glycyrrhizin caused a 25 percent rise in blood pressure.

On the other hand, a smaller amount, 100 milligrams per kg/bw, showed antiarthritic and anti-inflammatory effects in rats. Ammoniated glycyrrhizin in tests conducted with different animals did not adversely affect maternal or fetus survival or cause birth defects or gene mutations. Studies to determine possible cancer-causing properties have not been located.

Observations of the effect of licorice in humans record disparate findings. Individuals who overindulged in licorice candy, some ingesting as much as 35 to 75 grams in a day (12,000 to 25,000 times the amount in the average diet), experienced elevated blood pressure, hypertension with unpleasant cardiac sensations, severe muscle and nerve discomfort, and many related symptoms. A controlled experiment with ten people, to whom 20 to 45 grams of licorice extract were administered daily for periods up to three weeks, resulted in decreases in hemoglobin and serum protein and a considerable rise in blood and pulse pressure. Four grams of ammonium glycyrrhizin fed for five to ten days to the same group caused inhibition of hormone output of the pituitary-adrenal system. On a positive note, 100 grams of licorice extract mixed in water and given three times daily to 45 patients with gastric ulcers resulted in the disappearance of the ulcers in nearly two-thirds of the cases. This treatment, however, was not as effective for patients with duodenal ulcers.

ASSESSMENT: While licorice and licorice derivatives elicit a variety of adverse effects in humans as well as in laboratory animals, these are at levels thousands of times as great as likely to occur in usual diets. The substances provide some remedial benefits for arthritic inflammation and peptic ulcers. At the present level of licorice consumption in food as an additive, the evidence does not indicate a health hazard; but overindulgence, particularly in licorice candy, is inadvisable.

RATING: S.

LOCUST BEAN GUM (CAROB BEAN GUM, CAROB SEED GUM)

Locust bean gum, also called *carob bean gum* or *carob seed gum,* is extracted from the seed of the carob tree, an evergreen widely cultivated in the Mediterranean area. The gum, a carbohydrate, is a compound that contains D-mannose (a simple sugar found in many plants), and D-galactose (a component of milk sugar). It is used to blend ingredients and to prevent separation in foods such as ice creams, sauces, and salad dressings. It also acts as a binder in sausages and will improve texture.

SAFETY: Research on the possibility of adverse effects from locust bean gum has not been adequate. Feeding studies exceeding half the life span of any animal species have not been conducted, and these are required to determine the likelihood of long-term consequences. Studies covering shorter periods produced evidence that this additive depressed growth in chicks, but not in rats. The relevance of this finding to human beings is doubtful, as the dosages employed were thousands of times as great (adjusted for body weight) as the 17 milligrams per person estimated as present in the U.S. daily diet in 1976. Information on the absorption, digestion, metabolism, and excretion of locust bean gum is scanty. A single dosage amounting to over 800 times the daily human intake was followed through the intestinal tracts of eight adults by means of X ray, and stools were examined, with no evidence of interference with normal digestion.

Test-tube experiments in laboratories using living matter such as rat bone marrow and human lung cells found no evidence of gene mutation properties, nor did the feeding of locust bean gum by tube in massive doses to four species of pregnant animals result in abnormalities in their newborn. However, administered in amounts more than 4000 times human consumption (adjusted for body weight), the gum did result in a significant number of maternal deaths among rabbits and mice, although this effect did not occur with doses lowered to 600 times (for rabbits) and 900 times (for mice) the amount consumed by humans.

The evidence at hand does not indicate that this substance can cause cancer or allergic activity.

MAJOR REFERENCE: Evaluation of the health aspects of carob bean gum as a food ingredient. FASEB/SCOGS Report 3 (NTIS PB 221-952) 1972.

ASSESSMENT: While the available evidence does not suggest that locust bean gum offers a hazard at current levels of consumption, further research is needed, including long-term feeding studies and ones with pregnant animals, to determine whether its increase in use in food could pose a health hazard. The high incidence of maternal deaths in two animal species when fed locust bean gum in amounts far greater than it is likely that humans will ever consume, has been repeated in identical tests with a number of substances which, like locust bean gum, contain a condensation of simple sugars. Meanwhile, it is viewed as wise to avoid this additive during pregnancy until the animal deaths are satisfactorily explained.

RATING: S for most people; ? during pregnancy, as additional studies are needed to resolve uncertainties.

MAGNESIUM SALTS*

Magnesium Carbonate; Magnesium Chloride; Magnesium Hydroxide (Milk of Magnesia); Magnesium Oxide; Magnesium Phosphate; Magnesium Stearate; Magnesium Sulfate (Epsom Salts)

Magnesium is an essential nutrient needed in human diets. It is necessary for many of the body's processes, including the production and transfer of energy, fat and protein synthesis, contractability in muscle, excitability of nerves, and enzyme activity. An adult body may contain about 24 grams of magnesium, much of it in bone.

Salts of magnesium find a variety of uses as general purpose food additives. They may serve as binders and firming agents, as anticaking agents, as alkalies to adjust the acidity of foods, as flavor enhancers and color retention agents, and as nutrients or dietary supplements.

SAFETY: The average American diet has been estimated to contain 300 milligrams of magnesium daily, whereas 7 to 10 milligrams per kilogram of body weight (kg/bw) are recommended (420 to 600 milligrams for a person weighing 60 kg, or 132 lbs.), an indication that average U.S. consumption is not sufficient. The magnesium in the additives in food in the daily

---

MAJOR REFERENCE: Evaluation of the health aspects of magnesium salts as food ingredients. FASEB/SCOGS Report 60 (NTIS PB 265-509) 1977.

*See p. 622 for magnesium silicate and p. 553 for magnesium gluconate.

diet of an average U.S. individual amounted to only approximately 2 milligrams in 1976, according to a survey of usage by food processors conducted by the National Academy of Sciences-National Research Council.

Very large doses of magnesium sulfate may cause toxic symptoms and kidney problems, but only at levels several thousand times the normal intake. Magnesium salts have been beneficial in minimizing deposits of cholesterol and other deposits in coronary arteries (rabbits) and kidney stones (humans).

Studies of possible effects of magnesium salts on cancer do not provide a clear picture. One study suggests that high levels of magnesium (320 milligrams per kg/bw) may increase the effect of a known carcinogen, the chemical urethan. A second study, in contrast, showed that a twentieth of this dosage led to a marked reduction in tumors induced by two carcinogenic benzanthracene compounds. High levels of magnesium sulfate given by injection to female rats before and during pregnancy had adverse effects on their newborn, but these levels were thousands of times normal human intake, and also were not taken orally as foods are.

ASSESSMENT: Magnesium is an essential nutrient in the diet. The contribution of magnesium salts as food additives is so small that they rarely will bring the dietary intake even up to recommended levels. The few observed toxic effects from magnesium salts are at levels far beyond the intake from foods. Magnesium salts as added to foods appear to pose no hazard to the consumer, either now or at levels likely to be found in the future.

RATING: S.

MALIC ACID

Malic acid occurs naturally in a wide variety of fruits and vegetables. It is the major acid in rhubarb and in many fruits such as apples, bananas, cherries, pears, plums, and many berries. In the normal metabolism of carbohydrates in the human body, malic acid is formed as one of the key intermediates in a series of energy-supplying reactions. Malic acid is

MAJOR REFERENCE: Evaluation of the health aspects of malic acid as a food ingredient. FASEB/SCOGS Report 56 (NTIS PB 262-662) 1976.

used in foods as a flavoring agent as well as to acidify the product, especially by imparting a tart taste for sweets and fruit-flavored foods such as jellies, preserves, some candies, and sherbets.

SAFETY: The average daily intake of malic acid added to processed foods has been estimated at 112 milligrams per person (1975 data), or less than 2 milligrams per kilogram body weight (kg/bw). The malic acid added to food is all in the form of the natural L-isomer normally found in nature. Usual daily food consumption provides up to 3 grams of malic acid from natural sources, 27 times as much as from the food additive.

Because of its key role in metabolism of carbohydrates and organic acids in the body, L-malic acid has been studied thoroughly. Another form of malic acid, the D-isomer, apparently is not used in metabolism by rabbits and dogs, being excreted unchanged. Malic acid has a very low toxicity when ingested. No significant effects were seen in dogs or rats over a two-year period when they were fed as much as 200 milligrams (rats) and 1400 milligrams (dogs) per kg/bw of the additive in the daily ration. These levels are 100 to 700 times the intake for humans. No effects of malic acid on reproduction were observed when it was fed to rats at 20 times the usual human intake (adjusted for body weight). Similarly, malic acid was not found to cause abnormal fetal growth when fed to experimental animals or injected into eggs. Some concern has been expressed by the UN Joint FAO/WHO Expert Committee on Food Additives that D-malic acid might be toxic in early infancy, since the enzyme responsible for conversion of this substance is relatively deficient in very young infants. Fortunately, D-malic acid, which normally is not found in nature, no longer is added to infant foods.

ASSESSMENT: Malic acid is normally produced by the body and is a common constituent of many foods used in human diets. The many studies with experimental animals demonstrate that malic acid as a food additive poses no hazard to the public at current levels of consumption or levels that may be expected in the future.

RATING: S.

MALT (DIASTATIC, NONDIASTATIC)

Malt Extract; Malt Syrup

Malt is a product obtained from germinated barley. *Malt* (or nondiastatic malt) *syrup,* a thick concentrate, is extracted from the malt with water, and the dried *malt extract,* a powder, is obtained by drying the syrup. Malt products are used to flavor meat and poultry products and a variety of other foods and beverages, including vanilla ice cream, flavored milk, sour cream, chocolate syrup, candy, cough drops, condiments and dressings, and breakfast cereals. Malt syrup may be used in preparation of caramel coloring. Because it contains active enzymes (proteins that can hasten a chemical reaction) which split or hydrolyze starch, malt (or *diastatic malt*) is used in bread and cereal products and in beer and whiskey production. (*Nondiastatic malt* does not contain these enzymes.)

SAFETY: Derivatives of malt have been used in foods for many centuries. In a 1976 survey of the use of malt extracts by the food industry, the average intake in the U.S. per person was estimated as about 68 milligrams, or slightly over 1 milligram per kilogram of body weight (kg/bw) daily.

Many of the substances present in malt extract and malt syrup may be used as food additives. Among those assessed elsewhere in this dictionary are dextrins, glucose, fructose, starch, sucrose, various fatty acid components of animal fats and vegetable oils, mineral salts (calcium, iron, magnesium, phosphorus, potassium, silicon), and several vitamins (biotin, niacin, pantothenic acid, riboflavin, thiamin). When used as food additives, these individual compounds in malt have not been found to cause harm, providing support for the suggestion that this is likely to be true of the malt products under review.

Few studies on malt syrup or extract have evaluated the substances for short- or long-term toxicity, or for the possibility that they may cause cancer. However, malt extracts have been used in human diets throughout historical times with no evidence that they have adverse effects. Tests of malt extract have determined that it did not alter chromosomes, and, when

---

MAJOR REFERENCE: Evaluation of the health aspects of malt syrup and malt extract as food ingredients. FASEB/SCOGS Report II-13 (NTIS PB 81-121402) 1980.

fed to rats, actually improved growth and produced no adverse effects. Malt syrup did not induce abnormalities in the offspring when introduced into chicken eggs at a dosage of 200 milligrams per kg/bw. Infrequent cases of allergic reactions to malt have been reported; only two reports of allergy (in children) have been cited in medical literature.

Humans have consumed this product for a month to improve constipation, with no undesirable side effects. Again, no harmful effects were seen when athletes took 100 grams of malt extract (approximately 1.5 grams per kg/bw) daily for six days.

ASSESSMENT: Malt extract and syrup, which are derived from barley, are used in many foods for the associated flavor and texture changes they induce. The few controlled studies on their safety are augmented by the many studies on individual components and the centuries of use by humans with no indication of hazard. Apparently the instances of anyone being allergic to malt products are extremely rare and thus pose no hazard.

RATING: S.

MALTOL; ETHYL MALTOL

Maltol, a natural flavor that occurs in the bark of larch trees, pine needles, chicory, and wood tars and oils, is also found in many heated and roasted foods such as coffee, cocoa, bread, and milk products. *Ethyl maltol* is a synthetic compound not present naturally in unprocessed foods. Both substances are often used as flavorings and flavor enhancers in baked goods, gelatin desserts, frozen dairy products, soft drinks, and other carbohydrate-rich or sugary foods. They bring out sweetness in a food and thus permit a reduction in the sugar content while maintaining the desired degree of sweetness. They also can impart a fresh-baked or browned odor to bakery products, and act as synthetic berry and citrus fruit flavorings.

SAFETY: Flavors and flavor enhancers are added to foods in very small amounts because their intense flavor in larger amounts may be unpalatable. As a consequence they are con-

---

MAJOR REFERENCE: Toxicological evaluation of some food colours, enzymes, flavour enhancers, thickening agents, and certain other food additives. WHO Food Additives Series No. 6 (Geneva, 1975).

sumed only in minute quantities. The average daily intake per person in the U.S. in 1970 has been estimated as 0.4 milligram of maltol and 0.3 milligram for ethyl maltol as additives in foods. Combined, this amounts to about 0.01 milligram per kilogram of body weight (kg/bw) for a person weighing 60 kg (132 lbs.).

In short-term studies, rats and dogs were fed ethyl maltol for 90 days at levels of up to 500 milligrams per kg/bw, many thousands of times human intake when adjusted for body size. No abnormalities or other effects were observed other than a mild anemia and some vomiting at the highest level of 500 milligrams. Maltol in the diet was somewhat more toxic, and the 500 milligrams per kg/bw dose proved excessive and caused tissue damage and even death, although this did not occur at 250 milligrams per kg/bw. Long-term studies, where ethyl maltol was fed daily to rats and dogs for as long as two years, showed that at 200 milligrams per kg/bw, the flavor enhancer was without adverse effect. The animals were mated to investigate any effects on fertility, survival and size of the offspring, and development of abnormalities of the fetus, and there were none. No adequate long-term feeding studies using maltol appear to have been made.

Neither of these additives produces any allergic reaction or sensitization. No studies directed toward possible cancer-producing properties have been conducted.

ASSESSMENT: Maltol and ethyl maltol are used as ingredients to accentuate natural flavors and as flavorings in foods. Both additives have been without hazard at levels in the diet many thousands of times the average intake by humans. The UN Joint FAO/WHO Expert Committee on Food Additives has indicated that up to 2 milligrams per kg/bw (or 120 milligrams for a person weighing 132 pounds) is an acceptable human intake of ethyl maltol, which is well over 2000 times present average consumption of both compounds. The low level of these additives used in processed foods indicates that harm to the consumer is unlikely from their use, now or in the future.

RATING: S.

MANGANOUS SALTS

Manganous Chloride; Manganous Glycerophosphate; Manganous Hypophosphite; Manganous Sulfate

Manganese, a mineral element, is nutritionally essential for man. It participates as an activator of enzyme systems in metabolism and is necessary in the normal development of bones and the nervous system, and in sex-hormone production. Manganese occurs in many foods of plant and animal origin and is especially abundant in nuts and whole grains. Any of the salts *manganous chloride, manganous glycerophosphate, manganous hypophosphite,* and *manganous sulfate* may be added as a nutrient or diet supplement to foods such as baked goods and baking mixes, infant formulas, and dairy product substitutes.

SAFETY: The average daily adult intake of manganese from the amounts occurring naturally in foods is estimated to range from 2 to 9 milligrams. In 1975, approximately 0.2 milligram of manganous sulfate was added to foods, and a much smaller amount of manganous chloride (almost entirely in infant formulas). The Food and Nutrition Board of the National Academy of Sciences-National Research Council has recommended that the daily manganese intake for an adult be in the range of 2.5 to 5 milligrams. The varied nature of U.S. diets indicates that a deficiency of this mineral is unlikely to occur with any frequency in the U.S.

In experimental animals, a deficiency of manganese in the diet impaired bone formation and produced diabeteslike symptoms. When extremely large doses (thousands of times human intake) of manganous chloride or manganous sulfate were given to rats, rabbits, and farm animals, they developed anemia because of an interaction of manganese and iron in the body. Liver damage has been observed in hamsters and rats when the salt is fed at these high levels.

Manganous sulfate fed to rabbits in drinking water for several months, again in amounts thousands of times as high as its presence as an additive in the diet, caused transient paralysis in hind paws and some effects on the nerves. Several cases of poisoning of humans have occurred from drinking well water

---

MAJOR REFERENCE: Evaluation of the health aspects of manganous salts as food ingredients. FASEB/SCOGS Report 67 (NTIS PB 301-404) 1979.

contaminated with manganese, with similar neurological effects. Large dosages of manganous sulfate fed for four weeks to rats adversely affected formation of blood cells.

No adverse effects have been seen from manganous sulfate fed to pregnant animals when studied for possible abnormal fetal growth. Both manganous salts have been found to cause chromosome damage in some tests, but not in all. Long-term feeding studies to determine whether they cause cancer have not been made.

ASSESSMENT: Manganese is an essential mineral in the human diet. It has been demonstrated that a deficiency can have serious effects. It has also been shown that amounts administered far in excess of human intake have been extremely harmful; and disputed findings exist about whether the salts can cause chromosomal alterations. Considering the amounts added to foods, a minute fraction of the total dietary intake of manganese, these salts are hardly likely to be a hazard to health; they are used as nutrients in fortifying foods that otherwise would be low in manganese and thus are likely to be of benefit.

RATING: S.

MANNITOL

Mannitol is prepared commercially by the chemical addition of hydrogen to glucose (corn sugar). It is found naturally in such plant foods as beets, celery, and olives. Mannitol has about half the sweetness of sucrose (cane sugar), and is used in some sugarless dietary foods. Because this substance does not readily take up moisture, it is useful in powdered products, such as the dusting on chewing gum. It also can help blend ingredients and improve texture. It is present in amounts as high as 20 percent in chewing gum and 32.5 percent in soft candy. It may be used in breakfast cereals and frostings.

SAFETY: The average daily intake of mannitol, based on reported use by food processors in 1975, was estimated at about 36 milligrams per person. Mannitol is only moderately utilized by the body. Adult humans have ingested up to 100 grams of

---

MAJOR REFERENCE: Evaluation of the health aspects of mannitol as a food ingredient. FASEB/SCOGS Report 10 (NTIS PB 221-953) 1973.

mannitol in one dose, with two-thirds of it being absorbed and either excreted in the urine or metabolized to more readily useful substances such as sugar and other carbohydrates. Mannitol may have an energy value of about 2 kilocalories per gram. (The energy value of sugar is about 4 kilocalories per gram.)

Mannitol shows a laxative effect at lower total intakes (10 to 20 grams) than does the closely related compound sorbitol (which requires 50 grams). Short-term studies with humans and experimental animals have shown mannitol to have very low toxicity and to have no adverse effect on the fetus when added to the diet of pregnant animals at levels several thousand times the average intake for humans (adjusted for body weight). There are no reported long-term animal feeding studies (extending more than half the life spans of the species), and a lack of experimental data prevents assessment of any carcinogenic or mutagenic effects of mannitol, or of its effects on reproduction.

ASSESSMENT: The available evidence reveals no short-term adverse consequences in a variety of animal species or in man when mannitol is fed in amounts exceeding those currently consumed in the U.S. diet, and it has been used as a food additive for three decades without known adverse effects. The absence of information to determine whether mannitol is free of cancer-producing or chromosome-damaging properties or can affect reproduction is of concern. Mannitol exerts a laxative effect at levels well above the probable average adult intake. If diets are high in mannitol, it is possible that infants and children less than two years of age could consume amounts close to those capable of causing a laxative effect. Current evidence indicates that the adult consumer has no hazard from current levels of mannitol in the diet, or ones likely to be encountered in the future.

RATING: S.

NIACIN

Nicotinic Acid; Nicotinamide (Niacinamide)

Niacin is a vitamin of the B-complex which prevents pellagra. It is a nutrient that is essential for many of the body's crucial functions: it participates as a coenzyme in metabolic processes that break down foods for their utilization as energy and also for fat synthesis in the body. It is needed to maintain the body's protection from infection, to avoid certain changes in the nervous system, to improve circulation, and to aid in the formation of sex hormones. *Nicotinic acid* is the active substance in niacin. It is converted to the physiologically active *nicotinamide,* also referred to as *niacinamide.*

Niacin normally has to be obtained by humans from their diets, although the body can make the vitamin from an excess of the essential amino acid tryptophan. However, 60 milligrams of tryptophan are required to produce the equivalent of 1 milligram of niacin. The recommended daily intake of niacin to prevent a deficiency is 6.6 milligrams per 1000 calories in the diet for the adult, with not less ingested than 13 milligrams—somewhat less for children or infants, and more for pregnant and lactating women. These levels are normally exceeded because lean meat, poultry, cereals, milk, legumes, and other high-protein foods contain not only tryptophan but appreciable amounts of niacin. As a food additive, niacin is used solely for nutritional purposes, and FDA regulations require it to be added to enrich foods such as bakery, cereal, and pasta products.

SAFETY: A nationwide nutritional survey in 1974 determined that the total niacin/niacinamide intake of adults averaged 20 milligrams daily (about 0.33 milligram per kilogram of body weight, kg/bw). Less than half of this probably came from the vitamin being added to the food.

Humans have taken doses of this vitamin as large as 10 grams daily, for periods of several years, often to reduce elevated blood cholesterol levels or to help in schizophrenia. When niacin by itself is fed, a flushing of the skin is common; the symptoms include redness of the face and neck, a sensation of warmth, some sweating, and occasional nausea. In

---

MAJOR REFERENCE: Evaluation of the health aspects of niacin and niacinamide as food ingredients. FASEB/SCOGS Report 108 (NTIS PB 80-112030) 1979.

addition to flushing reactions, abnormal liver function has been observed in tests where gram quantities of niacin were taken. Liver function returned to normal once the increased niacin intake was discontinued, but there have been reports of liver disease and jaundice after administration of ¾ to 3 grams daily of niacin. Many coronary patients receiving 3 grams of niacin daily developed abnormal skin pigmentation, scaliness, rash, and inflammatory diseases of the skin.

Laboratory animals fed diets containing 1 to 2 grams per kg/bw per day have demonstrated growth depression in some studies, but not in all. Fatty livers may occur at a level of 1 gram of niacin per kg/bw, reflecting an induced choline deficiency. The addition of choline (see p. 512), one of the components of lecithin (see p. 580) in various fats, prevents this.

There are no animal studies on the effects of large doses of niacin or niacinamide covering several generations, or reports on possible effects on normal embryo growth or on chromosome damage. Niacinamide does not appear to cause cancer. It did not alter the effect of a known chemical carcinogen applied to the skin of mice, but it did promote the appearance of tumors induced by another carcinogen in rats.

ASSESSMENT: Niacin is a vitamin that is essential for humans and is added for fortification or enrichment of certain foods. When large doses (more than 100 times the amount in the average U.S. diet) are administered to patients, particularly to reduce high blood cholesterol concentrations, undesirable side effects frequently are experienced. Generally these are temporary. Greater dosages (when adjusted for body weight) have produced similar effects in laboratory animals. The quantity of niacin present naturally and as an additive in the diet of individuals in the U.S. does not pose a health hazard, but rather the reverse, for it is a necessity for good health.

RATING: S.

PANTOTHENIC ACID

CALCIUM PANTOTHENATE

Pantothenic acid, one of the vitamins of the vitamin B complex, is a necessity in human diets. It is a constituent of coenzyme A, an essential enzyme factor involved in metabolism of fats, proteins, and carbohydrates. It also participates in a number of other vital body functions, among them the production of cortisone and other adrenal hormones responsible for the health of nerves and skin. This vitamin is naturally present in many foods; rich sources are liver, meats, whole-grain cereals, legumes, and many fresh vegetables. When added to some processed foods, it is as a nutritional or dietary supplement, almost always in the form of *calcium pantothenate.*

SAFETY: The daily per capita consumption of pantothenates in the average U.S. diet from all food sources is estimated at 5 to 19 milligrams. Based on poundage used by the food industry, the amount of calcium pantothenate added to foods in 1976 amounted to 0.3 milligram per person in the daily diet, a minute fraction of the total intake of this vitamin. The Food and Nutrition Board of the National Academy of Sciences has stated that a daily intake of 4 to 7 milligrams is probably adequate for adults, with a higher intake suggested for pregnant and lactating women. It is known that the vitamin will readily pass through the placenta and be taken up by the fetus.

Adult patients suffering from a disease characterized by superficial inflammation of the skin have ingested oral dosages of 1 gram or more daily for several months with improvement in their condition and no evidence of toxic effects. Excess amounts of the vitamin taken orally are rapidly excreted in the urine.

Experimental animals (mice, rats, dogs, monkeys) have not suffered adverse effects over a period of six months or longer from daily intakes of calcium pantothenate many thousands of times normal human consumption. Nor have investigations produced evidence of harmful effect on the normal development of the fetus, on reproduction, or on alteration of chromosomes.

MAJOR REFERENCE: Evaluation of the health aspects of calcium and sodium pantothenate and d-pantothenyl alcohol as food ingredients. FASEB/SCOGS Report 93 (NTIS PB 288-672) 1978.

ASSESSMENT: Pantothenic acid is an essential vitamin needed in the daily diet. Its use as a food additive provides only a small fraction of what is needed daily by humans and provided by naturally occurring food sources. There is no evidence, either from clinical treatment of humans or experiments with animals, that this vitamin used as an additive to supplement the diet poses any hazard to the consumer.

RATING: S.

PARABENS

Methyl Paraben; Propyl Paraben

Methyl and *propyl paraben* are esters of *paraben* (parahydroxybenzoic acid). Neither of them occurs in nature; they are synthetic compounds produced for food, cosmetic, and pharmaceutical purposes. In food, these parabens function as preservatives that prevent microbial activity, particularly the growth of yeasts and molds. The parabens are closely related to benzoic acid (see p. 487) and its salt sodium benzoate and when used in combination with them extend their effective range of antimicrobial activity to foods high in pH (high in alkalinity, low in acidity). Methyl and propyl paraben are used in some baked goods such as fruit cakes, sweet rolls, and cookies, in sugar substitutes and in products that use them—for example, artificially sweetened jams and jellies and low-calorie foods and beverages, in fats and oils, and in frozen dairy desserts and many other milk products.

SAFETY: A survey conducted by the National Academy of Sciences-National Research Council, based on usage by food processors in 1976, determined that the daily diet of the average U.S. adult contained only 0.1 milligram of methyl and propyl paraben. Far more extensive use is made of benzoic acid and sodium benzoate by the food industry for similar purposes.

Experiments with a variety of animal species indicate that even when administered in their diets in amounts thousands of times human consumption, these esters of paraben are harmless. Dogs fed 1000 milligrams per kilogram of body weight

---

MAJOR REFERENCE: Evaluation of the health aspects of methyl paraben and propyl paraben as food ingredients. FASEB/SCOGS Report 8 (NTIS PB 221-950) 1973.

(kg/bw) six days a week for one year were unaffected. Evidence of toxicity did occur when dosages of methyl paraben were doubled, and those of propyl paraben were tripled.

Rabbits fed 500 milligrams per kg/bw per day for six days of either compound showed no ill effects, but these did appear when the dosages were increased to 3000 milligrams. 500 milligrams per kg/bw in animals is the equivalent of over 300,000 times human consumption when adjusted for body weight.

Rats fed up to 1200 milligrams per kg/bw daily of methyl or propyl paraben for almost two years were unaffected. In another experiment, a decrease in growth rate was experienced by rats at a level of 1500 milligrams per kg/bw of propyl paraben for a year and a half. A human volunteer who ingested 2000 milligrams of methyl paraben daily for a month was not adversely affected, nor was another volunteer who ingested the same amount of propyl paraben each day for the same period.

Methyl paraben did not cause birth defects in offspring of mice and rats fed 550 milligrams per kg/bw daily during pregnancy, and in hamsters fed 300 milligrams. Studies to determine whether these paraben esters can cause cancer when taken orally have not been conducted.

ASSESSMENT: The available information reveals that methyl and propyl paraben do not cause harmful consequences in various species of animals or humans when ingested in amounts greatly exceeding the minute quantity currently present in the diet of the average U.S. human, or likely to be so in the future.

RATING: S.

PECTIN

Amidated Low-Ester Pectin; Low-Ester Pectin; High-Ester Pectin

Pectin is present in most plants, where it provides a cementlike strength to cell walls. Fruits and vegetables are highest in pectin content. The main sources for production of this substance for the food industry are citrus peels and apple pomace

---

MAJOR REFERENCE: Evaluation of the health aspects of pectin and pectinates as food ingredients. FASEB/SCOGS Report 81 (NTIS PB 274-477) 1977.

(the residue from apple pressings). Pectin is a complex polysaccharide, a carbohydrate that consists of a number of simple sugars and sugar-acids.

Pectin is added to foods for its gelling, thickening, and blending properties and to prevent separation of ingredients. It is useful in ice creams and ices; in processed fruits and juices; in fruit jellies, jams, and preserves; and in soft candies. The presence of sugar is usually required to make pectin gel, but its molecular structure can be altered to enable it to gel in products with low-sugar content. These are termed *low-ester pectins* (or *amidated low-ester pectins,* when ammonia is employed). *High-ester pectins* require sugar.

SAFETY: It has been estimated, based on a survey of usage by food processors in 1976 conducted by the National Academy of Sciences-National Research Council, that 23 milligrams of pectin as additives are consumed daily in the diet of an average individual in the U.S., or 0.4 milligram per kilogram of body weight (kg/bw) for a person weighing 60 kg (132 lbs.). Less than a fifth of this is low-ester pectin. The UN Joint FAO/WHO Expert Committee on Food Additives considers pectins and their salts to be normal constituents of the human diet and places a limitation on acceptable daily intake only on amidated low-ester pectin, their suggested limitation being 25 milligrams per kg/bw, or over 350 times the average consumption of all low-ester pectins.

Extensive studies of pectins have demonstrated that they are decomposed by organisms in the intestinal tract and, for the most part, are excreted in the feces. They do not appear to enter into the metabolic processes, and they are not absorbed into tissues to any extent. Thus, unlike some carbohydrates, pectins do not serve as a source for the body's energy.

Animal studies have failed to disclose adverse effects when these compounds, including amidated pectins, are fed in amounts far greater than their estimated human intake as additives. Amidated pectin administered in diets of pregnant rats did not cause birth defects in their newborn, or adversely affect maternal health, fertility, or fetuses. However, pectins have yet to be investigated as possible causes of cancer and gene mutation.

A study conducted on 24 men provided evidence that pectin can lower blood cholesterol level somewhat. Medically, it is used in combination with other ingredients for treatment of diarrhea.

ASSESSMENT: Pectin is a natural constituent in many edible plants, and, as such, is consumed in the normal diet in quantities far in excess of the amount that is added to foods. Animal feeding studies confirm that pectins do not present a hazard even in quantities far exceeding their consumption as additives by humans, and, since little if any is absorbed in the body's tissues, it appears unlikely for pectins to be of health concern.

RATING: S.

PHOSPHATES*

Ammonium, Calcium, Potassium, Sodium Phosphate; Phosphoric Acid; Sodium Hexametaphosphate; Sodium Polyphosphate; Sodium Pyrophosphate

A large number of phosphates (which are phosphorus-containing chemicals) are used as food additives. They are prepared commercially from phosphoric acid. In foods they are used as flavoring agents, as nutritional supplements, to keep mixtures of ingredients from separating, to change or control the degree of acidity or alkalinity, and for other purposes. Many of the phosphates occur as natural constituents of both plants and animals, since phosphorus is an essential nutrient in the diet; phosphates are found in high concentrations in muscle and organ meats, cereals, legumes, nuts, cheese, and eggs.

MAJOR REFERENCE: Evaluation of the health aspects of phosphates as food ingredients. FASEB/SCOGS Report 32 (NTIS PB 262-651) 1976.

*Many phosphates may appear on food labels, and food processors have different ways of naming them. In forming phosphates, one to three of the acid groups of phosphoric acid may be combined with a base, or alkali, rather than with acidic hydrogen. Thus in the case of sodium phosphate, there can be monosodium (or monobasic sodium, or sodium acid, or monosodium acid) phosphate, disodium (dibasic sodium) phosphate, or trisodium (tribasic sodium) phosphate as permissible food additives. The phosphates listed above provide the primary terms and can vary according to the base, or alkali, and the number of phosphoric acid units as well as the different ways of identifying each one. When two phosphoric acid units are chemically linked together, a pyrophosphate results, with four acid groups available on it; a salt of this may be a disodium pyrophosphate, or a tetrasodium pyrophosphate (sometimes referred to merely as tetrasodium phosphate). Polyphosphates (metaphosphates) generally occur with linkages of four or more phosphoric acid units; a hexametaphosphate, for example, would have six such phosphoric acid units, and eight acidic groups on which mineral salts such as sodium can replace the acidic hydrogen.

SAFETY: A survey of the use of phosphates by food manufacturers in 1976, conducted by the National Academy of Sciences-National Research Council, determined that the daily diet of the average individual in the U.S. contained 210 milligrams of added phosphate, or 3.5 milligrams per kilogram of body weight (kg/bw) for a person weighing 60 kg (132 lbs.). As phosphorus is necessary for life, 800 milligrams per day are recommended in the diet for an adult. Generally, however, this is met by the phosphorus present naturally in food.

The safety of phosphates depends on their amount in relation to calcium in the diet; a calcium-phosphorus ratio of at least one to one is desirable in humans. Appreciably raising the level of phosphate without an accompanying increase in calcium may have undesirable consequences; it can result in some loss of calcium from bones or teeth and in damage such as calcification in the kidneys.

Rats on a diet containing 1000 milligrams of added phosphorus per kg/bw as phosphoric acid, sodium or potassium phosphate, or sodium pyrophosphate (285 times the average additive intake by humans) showed some kidney damage after several weeks. Appreciably higher levels of phosphate caused gradual bone decalcification. In long-term studies with rats, 630 milligrams of phosphorus per kg/bw added as sodium tripolyphosphate or sodium phosphate led to some growth retardation, anemia, and kidney damage, and thigh bones were reduced in size. The level at which phosphates did not cause harm when added to the laboratory diet of rats was 500 milligrams phosphorus per kg/bw.

Brief studies with humans consuming 450 to 600 milligrams of calcium and at least 2000 milligrams of phosphorus daily did not disclose an adverse effect on calcium retention. However, in another investigation, 1000 milligrams of added phosphorus resulted among adult humans in a marked increase in serum parathyroid hormones (which influence calcium and phosphorus metabolism and bone formation) and a slight decrease in serum calcium, which could lead to adverse effects on bones and kidneys.

A large number of studies have demonstrated the lack of effect of added phosphate on either the developing fetus or on change in the chromosomes. Extremely high levels of phosphate additives did not cause an increase in the incidence of tumors in rats, but long-term experiments specifically designed to test for ability to cause cancer have not been conducted.

ASSESSMENT: Phosphorus is an essential dietary constituent that is closely interrelated with calcium in its effect; the calcium-phosphorus ratio should not vary substantially from one to one. Despite some uncertainties about the amount of consumer exposure to phosphates in the diet in relation to calcium intake, phosphate additives appear to pose no hazard when used at current levels. Nutritionists consider that the ratio of calcium to phosphorus in the usual diets of man would be more favorable if calcium were higher, but phosphorus levels generally appear adequate. The very high levels of added phosphate that can lead to some deterioration in bones and teeth, or cause kidney disturbances, are unlikely from customary food intakes.

RATING: S.

POTASSIUM BROMATE

BROMATED FLOUR

Potassium bromate is a chemical that has been used for 60 years as a conditioner or maturing agent in flour. It reacts with and changes some undesirable flour constituents and in this way helps improve baking quality and produce loaves that have better volume and keeping quality. *Bromated flour* is flour that contains added potassium bromate.

SAFETY: The daily intake of potassium bromate as an additive in food averaged about one milligram in the diet of a U.S. adult in 1976, according to a survey conducted by the National Academy of Sciences-National Research Council. Potassium bromate converts completely to bromide when the dough is mixed and the product is baked, if the concentration in the flour does not exceed 75 parts per million (ppm), the limit permitted by the FDA. A pound loaf of bread contributes about 20 milligrams of potassium to the diet. Potassium bromide has been used therapeutically in humans, and doses as high as 15 milligrams per kilogram of body weight (900 milligrams for a 130-pound person) have been administered without adverse effects. Many foods have a natural bromide content, ranging up to 10 ppm.

MAJOR REFERENCE: Specifications for the identity and purity of food additives and their toxicological evaluation: emulsifiers, stabilizers, bleaching and maturing agents. WHO Technical Report Series No. 281 (Geneva, 1964).

Some 20 years ago studies were made using rats, dogs, mice, and monkeys to determine any possible effect of bromated flour in the diet. Both flour containing potassium bromate (in concentrations up to 627 ppm) and bread from flour containing potassium bromate (in concentrations up to 200 ppm) were used, and at times the bread or flour constituted 84 percent of the total diet of the animals. The rat studies covered up to five generations, and those of mice over eight generations. No adverse effects were noted in any of these, or in short-term studies with dogs and monkeys. Reproductive performance of rodents also was normal.

Recent studies conducted with rats for two years and mice for 80 weeks have examined the possibility of long-term effects on, for example, cancer rate and survival rate, of bread-based diets when the bread was made from flour containing up to the maximum limit (75 ppm) permitted by FDA. No adverse effects were observed.

The treatment of flour with potassium bromate at a concentration of 45 ppm does not affect its nutritive value, since it does not cause a decrease in its content of thiamin, riboflavin, or niacin. Flour treated with this chemical at a concentration of 25 ppm does not show any greater decrease in tocopherol content (35 to 50 percent) than untreated flour when stored for 12 months.

The UN Joint FAO/WHO Expert Committee on Food Additives has estimated that up to 20 ppm of potassium bromate in flour, the average level used, is unconditionally acceptable as presenting no significant hazard, and 20 to 75 ppm of potassium bromate is conditionally acceptable for special purposes, such as certain biscuit flours.

ASSESSMENT: Potassium bromate is used as a food additive to improve certain breads, rolls, buns, and pasta. It has been without adverse effect in feeding studies of a variety of animals. Bromate converts to bromide in the baking process, and bromide has been used therapeutically with humans in concentrations 100 times as great as the total intake of bromate in the daily diet. The low level permitted in flours and breads poses no hazard to the consumer.

RATING: S.

POTASSIUM CHLORIDE; POTASSIUM HYDROXIDE

Natural deposits of mineral sylvinite, crystalline masses like rock salt, are a commercial source of *potassium chloride,* which occurs naturally in substantial quantities in foods in the daily diet: in dairy products; meat, fish, and poultry; cereals, grains, and potatoes; and in lesser amounts in fruits, vegetables, and beverages. Potassium is essential to animals including man for a number of the body's vital functions, among them the regulation of the acid-base balance, for it is a principal base in tissues and blood. It also functions as an intermediary in the metabolic processes that transform carbohydrates into glycogen for storage of energy in the liver and muscles and as activator of transmission of nerve impulses. And potassium provides a protective effect against the hypertensive action of high salt intake.

Potassium chloride is used as an aid to fermentation in brewing. Food processors add it as a flavoring agent and flavor enhancer, and as a nutrient. Medicinally, it is administered orally or intravenously for potassium depletion caused by treatment of kidney, liver or heart failure, hypertension, and other cardiac ailments. *Potassium hydroxide* is an alkali used as an agent to neutralize acids in food products, including cocoa (see p. 515) and instant coffee and tea.

SAFETY: The average diet of a U.S. adult contains 4 to 8 grams of potassium chloride daily. In a survey of food manufacturers in 1975, it was determined that about 20 milligrams of potassium chloride and 16 milligrams of potassium hydroxide per person were being added to foods. The National Academy of Sciences-National Research Council in 1980 suggested that between 1875 and 5625 milligrams of potassium is desirable to meet an adult's daily nutritional needs.

Potassium hydroxide during processing reacts with the acids in the foods, and changes to potassium chloride or another potassium salt. Serious toxic reactions by man to potassium chloride have occurred rarely, and only when it was adminis-

---

MAJOR REFERENCES: Evaluation of the health aspects of potassium chloride and sodium chloride as food ingredients. FASEB/SCOGS Report 102 (NTIS PB 298-139) 1979; Evaluation of the health aspects of sodium hydroxide and potassium hydroxide as food ingredients. FASEB/SCOGS Report 85 (NTIS PB 265-507) 1977.

tered for medical reasons. It is difficult to administer excessive amounts of potassium salts to healthy people without causing nausea and elimination by vomiting. Elimination of excess potassium may be impaired by diseases of the kidney, heart, and liver, and special precautions are indicated here to avoid overdosage. However, oral dosages of up to 10 grams of potassium chloride daily, when administered to patients with kidney impairment, did not cause adverse effects, leading the investigators to conclude that even in these instances the intestinal absorption of potassium and its mode of distribution in the body tended to prevent high concentrations in the blood.

Occasional occurrence of ulcers has been observed in patients receiving potassium chloride tablets; in most cases the potassium chloride had been incorporated in a thiazide drug prescribed for control of hypertension. It occurred only in a small percentage of the patients, 1 in 35,000. Rhesus monkeys were administered similar tablets, with and without thiazide. Intestinal ulcers developed in those receiving the potassium chloride (equivalent to 1 to 2 grams for a human), whether or not it contained thiazide.

In test-tube experiments, neither potassium hydroxide nor potassium chloride proved to be a cause of gene mutations. Tests conducted on pregnant mice and rats did not disclose any discernible adverse effects on fertility or on maternal or fetal survival; nor did it result in any malformations of their offspring.

ASSESSMENT: Potassium is a mineral that is essential for many of the body's metabolic and chemical processes. The body appears well equipped to dispose of excessive oral intake without difficulty. Ulcers have occurred rarely in patients administered potassium chloride for treatment of hypertension, but the low frequency as well as the minuscule amounts of potassium chloride used as an additive in food—perhaps 0.25 percent of the amount naturally present in food in the average diet—does not warrant concern about its contribution as a health hazard.

RATING: S.

PROPELLANT GASES

Carbon Dioxide; Chloropentafluorethane; Isobutane; Nitrogen; Nitrous Oxide; Propane

These additives are relatively inert or chemically nonreacting gases. They are used to aerate and dispense foamed or sprayed food preparations from pressurized or aerosol containers. They may be used in dairy products such as whipped cream, vegetable fat toppings, and dairy product substitutes. Some, such as nitrogen, also may be used to displace air or oxygen in a food package such as canned coffee, to prevent oxidation or chemical deterioration of the food. Carbon dioxide is the gas used in preparing carbonated or sparkling soft drinks, where it also imparts a sharp flavor to the beverage because it forms a weak acid when dissolved in water.

Carbon dioxide and *nitrogen* are important components of the air that we breathe. (Carbon dioxide in solid form is "dry ice.") *Chloropentafluorethane* is related to chemicals developed for use as refrigerants, the fluorinated hydrocarbons. Their use is being limited because of environmental concerns that their presence may reduce the protective ozone layer in the upper atmosphere. *Isobutane* and *propane* are present in natural gas and petroleum; propane is a major ingredient of bottled gas fuel. The atmosphere contains *nitrous oxide,* a gas which occurs from bacterial action on ammonium or nitrate salts in the soil. It is the most widely used anesthetic in the U.S., one of low potency that is used for the relief of pain without loss of consciousness. It is sometimes referred to as "laughing gas."

SAFETY: Consumer exposure to any of these gases from their presence in food products is minimal. The gases are relatively unreactive with the food and are designed to be released to the atmosphere when the food product is dispensed. Thus the amount used in preparing processed foods is no indication of the exposure of the user. Obviously, the principal exposure would be by inhalation, since little of the gas will remain in the

---

MAJOR REFERENCES: Evaluation of the health aspects of butane, helium, nitrogen, nitrous oxide, and propane as food ingredients. FASEB/SCOGS Report 112 (NTIS PB 275-750) 1979; Evaluation of the health aspects of carbon dioxide as a food ingredient. FASEB/SCOGS Report 117 (NTIS PB 80-104615) 1979.

product—though the solubility of carbon dioxide means it is still present in a beverage when it is drunk. People drinking many bottles of carbonated beverages daily may ingest gram quantities of carbon dioxide.

Carbon dioxide is the major respiratory product or end product when the body metabolizes food to produce energy. Even with little physical activity, a person may produce and excrete (largely through the lungs) several hundred grams of carbon dioxide each day. The blood in the veins contains about 55 to 60 milliliters of the gas per 100 milliliters of blood; an average carbonated drink may contain several times this concentration, much of which is dispersed to the air after the pressure is released. Humans have been exposed in a submarine for six weeks to air containing 1.5 percent carbon dioxide (the air we normally breathe contains 0.03 percent), with an initial acidosis (increase in the acidity of body tissues) that later was compensated for by the body. The acid-base balance returned to normal and no other lasting effects were noted. Carbon dioxide has had some adverse effects on reproduction and fetal malformation and has prevented successful pregnancies in some experimental animals, but only after exposure to levels far higher than any obtained from foods or beverages.

Chloropentafluorethane is judged relatively nontoxic in the Underwriters Laboratory ratings, 1 part per 1000 being allowable when inhaled. Increased fluorination of the hydrocarbon usually decreases its pharmacological action so that the effect of this highly-fluorinated gas on metabolism is minimized. Chronic inhalation of 20 percent of chloropentafluorethane has no effect on rats, guinea pigs, dogs, and cats. It produced no central nervous system or behavioral effects when inhaled at a concentration of 60 percent.

Isobutane exposure to humans at levels far higher than can be expected from its use in foods was studied for a two-week period. No adverse effects were noted from these high levels of inhalation.

Nitrogen, which makes up nearly four-fifths of the air, is also present as a dissolved gas in body tissues and blood. The only known effects from it are a feeling of euphoria and decrease in motor performance when retained nitrogen has been markedly increased by pressure, as in deep-sea diving.

Nitrous oxide can produce a light-headed feeling if one inhales 2 to 3 grams over a period of several minutes. For general anesthesia in dentistry or surgery, two-thirds of the gas inhaled may be nitrous oxide, and many hundreds of grams may be inhaled. Some short-term effects on blood pressure and heart rate have been observed in humans, but the exposure was thousands of times what could be derived from the gas used as a propellant in foods. Nurses exposed to nitrous oxide in operating rooms apparently suffered no chromosome damage. A 1980 study indicated that this gas can cause damage to the fetus when pregnant rats are exposed to it.

Exposure of mice or monkeys to 20 percent concentrations of *propane* did not affect heart rhythms or blood pressure. Propane has been inhaled by humans for up to eight hours per day for one to two weeks, and at a level as high as 100 parts per million in air, with no adverse effects. Because of its nontoxicity, industrial hygienists have set a threshold limit of propane in the air of workspace at 1000 parts per million.

ASSESSMENT: Several gases, selected in part because of their chemical inertness, have been used as propellants in aerosol-type containers to expel food products. In addition, carbon dioxide is used in carbonated beverages. Nearly all the gas will be released to the atmosphere when the food is dispensed, and only a tiny fraction will be exposed to the consumer. Each gas is considered relatively nontoxic and has been shown to be free of adverse effects when inhaled at concentrations far higher than could be expected from its use as a food additive. The present or anticipated usage in foods of carbon dioxide, chloropentafluorethane, isobutane, nitrogen, nitrous oxide, and propane appears to pose no hazard to the consumer.

RATING: S.

PROPIONIC ACID

CALCIUM PROPIONATE; SODIUM PROPIONATE

Propionic acid, along with its mineral salts *calcium propionate* and *sodium propionate,* is a natural constituent in dairy products, such as butter and Swiss cheese. These substances also function as products of metabolism in the body, participating as intermediaries during the processes that break down ingested food into usable, accessible forms. They are usually

---

MAJOR REFERENCE: Evaluation of the health aspects of propionates as food ingredients. FASEB/SCOGS Report 79 (NTIS PB 80-104599) 1979.

produced by chemical synthesis for food processors, who use them as additives in food to inhibit mold growth and as preservatives in some cheeses, and in bread and other baked goods.

SAFETY: A survey by the National Academy of Sciences-National Research Council, based on usage of the propionates as food additives by food processors in 1976, indicated an average of 24 milligrams per person added to the daily diet, about 90 percent as calcium propionate and less than 1 percent as propionic acid. This combined intake would be about 0.4 milligram per kilogram of body weight (kg/bw) for a person weighing 60 kg (132 lbs.).

Rats have been fed sodium or calcium propionate at levels of 1 to 3 percent of the diet (up to 1200 milligrams per kg/bw) for several weeks with no effect on growth. Administered orally to rats in even larger amounts and for a longer period (4000 milligrams per kg/bw for one year), sodium propionate caused growth depression for the first few weeks, but resulted in no other adverse effects. Some stomach lesions did occur in rats fed 5 grams of propionic acid per kg/bw daily for 110 days, many thousands of times human consumption of all added propionates—but without evidence of any malignancy. As much as 6 grams of sodium propionate taken orally by an adult human for a number of days caused no adverse effects, only a faintly alkaline urine.

When pregnant hamsters, rats, and mice consumed calcium propionate at levels up to 400 milligrams per kg/bw, it did not affect survival of the mother or young, or cause abnormalities in their offspring. Neither calcium nor sodium propionate was found to cause chromosome damage in a number of laboratory tests.

ASSESSMENT: The propionates are effective inhibitors of fungi and some bacteria, and thus are particularly useful in baked goods and cheese. They occur naturally in dairy products, and as such, are handled without difficulty metabolically. As metabolites themselves, they facilitate this process. The feeding studies demonstrate that there is no hazard to the consumer from calcium propionate, sodium propionate, or propionic acid at levels currently used in foods or levels that may be expected in the future.

RATING: S.

PROPYL GALLATE

Propyl gallate, a synthetic chemical compound, is one of several that are added to foods containing fats and oils. It prevents rancidity and the development of objectionable tastes and odors caused by the reaction of fatty ingredients to oxygen. These antioxidants are frequently used in combinations, such as BHA and BHT (see p. 489) and/or propyl gallate, as they become more effective together and less of them is required in the food. They act as "chain breakers" by blocking or retarding deterioration of fats under the usual conditions of processing, storage, and use of foods. Propyl gallate is found in such items as shortenings and vegetable oils, meats, candies and gum, some snack foods, baked goods, nuts, and frozen dairy products.

SAFETY: The daily intake of propyl gallate in the diet of the average individual in the U.S., based on reported usage by food manufacturers in 1975, was 1 milligram. This amounts to 0.02 milligram per kilogram of body weight (kg/bw) for a person weighing 60 kg (132 lbs.), which is well under the 0.2 milligram per kg/bw regarded unconditionally as a safe and acceptable daily intake by the UN Joint FAO/WHO Expert Committee on Food Additives.

Interpretation of some of the research on propyl gallate is difficult, since it often has been studied in mixtures with other antioxidants. However, feeding studies with propyl gallate alone have been conducted with a number of animal species (rat, mouse, guinea pig, dog), and have established 100 milligrams per kg/bw as the level at which it has no adverse effect on them. This is more than 5000 times its presence in the daily diet of U.S. humans (adjusted for body weight).

Fed at five times this amount (500 milligrams per kg/bw) to pregnant rats, propyl gallate reduced fertility, but did not do so when the dosage was reduced to 202 milligrams per kg/bw. In another experiment, excessively high levels of propyl gallate fed to rats for one to two years caused some growth retardation and kidney damage.

A man ingested 500 milligrams of propyl gallate daily for six consecutive days. Examination of his urine failed to reveal any presence of blood constituents (such as albumin), abnormal sedimentation, or evidence of kidney damage.

MAJOR REFERENCE: Evaluation of the health aspects of propyl gallate as a food ingredient. FASEB/SCOGS Report 11 (NTIS PB 223-840) 1973.

Tests have yet to be conducted to determine whether this compound can cause cancer or alteration of genes.

ASSESSMENT: The "no-adverse-effect" level of propyl gallate fed to experimental animals has been established at several thousand times its presence in the daily diet of the average adult. Although adverse effects have been reported at materially higher dosages, these are not regarded as relevant considering the minute quantities (in comparison) of propyl gallate present at current levels in food, or at levels that might reasonably be expected in the future. The available information indicates that this compound as an additive in food does not constitute a hazard to the health of the consumer.

RATING: S.

PROPYLENE GLYCOL

PROPYLENE GLYCOL MONOSTEARATE (PROPYLENE GLYCOL MONO- AND DIESTERS)

The *propylene glycol* used in food processing is derived from propylene, a gaseous by-product in petroleum refining, or from glycerol, the alcohol component in fats. *Propylene glycol monostearate,* a mixture of propylene glycol mono- and diesters of stearic and palmitic acids (see p. 479), is the product of the chemical interaction between propylene glycol and a hydrogenated vegetable oil (an oil chemically bonded with hydrogen; see p. 660). Propylene glycol monoester contains one molecule of the fatty acid from the vegetable oil; the diester has two. Another food additive, propylene glycol alginate, is discussed under alginates (see p. 468).

Food manufacturers find propylene glycol useful in blending ingredients, and in increasing their flexibility and spreadability. It is soluble in both watery and dry mixtures and readily absorbs water. Food processors employ it in confections, ice creams, beverages, baked goods, and some meat products. Propylene glycol monostearate is used to improve texture, softness, and to preserve the quality of several foods, including baked goods, puddings, and toppings.

MAJOR REFERENCE: Evaluation of the health aspects of propylene glycol and propylene glycol monostearate as food ingredients. FASEB/SCOGS Report 27 (NTIS PB 265-504) 1974.

SAFETY: The daily intake per person of the propylene compounds in the U.S., based on a survey conducted by the National Research Council-National Academy of Sciences of actual poundage used by food processors in 1976, averaged 51 milligrams, almost all of it propylene glycol monostearate. This amounted to slightly less than 1 milligram per kilogram of body weight (kg/bw) for a person weighing 60 kg (132 lbs.). The UN Joint FAO/WHO Expert Committee on Food Additives has indicated that an acceptable daily intake for man is 25 milligrams per kg/bw, many times the amount normally present in the diet of the average U.S. adult.

During digestion, propylene glycol mono- and diesters are split into fatty acid plus the propylene glycol. These substances are absorbed in the body, where the propylene glycol is converted (by the same metabolic processes that handle carbohydrates) into glucose, which is the chief source of energy for living organisms. Propylene glycol can be ingested in substantial quantities by experimental animals for long periods of time, without adverse effects. Dogs have tolerated the equivalent of 2000 milligrams per kg/bw (over 200 times estimated human consumption). A long-term study with rats showed that up to 2500 milligrams per kg/bw had no deleterious effect on reproduction through three generations. Investigations employing a variety of animal species provided evidence that propylene glycol does not cause cancer, nor does it adversely affect maternal or fetal survival.

Some investigators, but not all, have found that at extremely high doses, 6000 milligrams per kg/bw or more, propylene glycol can cause kidney damage. However, considering the amount normally consumed by humans, this finding hardly indicates that the substance is a threat.

ASSESSMENT: Propylene glycol and propylene glycol monostearate are metabolized and used by the body like carbohydrates. Experimental evidence indicates that propylene glycol can be ingested in substantial quantities over long periods by laboratory animals far in excess of man's present consumption without causing harm. These additives do not pose a hazard at levels now current in foods, or likely to be used in them in the future.

RATING: S.

PROTEIN HYDROLYZATES

Autolyzed Yeast Extract; Hydrolyzed Casein (Hydrolyzed Milk Protein); Hydrolyzed Cereal Solids; Hydrolyzed Plant Protein (HPP) and Hydrolyzed Vegetable Protein (HVP); Soy Sauce (Fermented Soy Sauce, Shoyu, Tamari); Hydrolyzed Soy Sauce

Protein hydrolyzates are mixtures of naturally occurring amino acids, the chief constituents of proteins. Their composition varies according to the source materials from which they are derived, and they are extracted from these by a variety of chemical and manufacturing procedures. *Hydrolyzed plant protein* (HPP) and *hydrolyzed vegetable protein* (HVP), interchangeable terms, are obtained either by hydrolysis (chemical splitting involving the use of water) of soybean and peanut meals or of crude protein recovered from the wet milling of grains such as wheat and corn. At times they are identified on labels as *hydrolyzed cereal solids*. Soybeans are the base of most manufactured HPP and HVP.

Autolyzed yeast extract, a concentration of the soluble components of hydrolyzed brewer's or baker's yeast (a by-product of brewing), provides a potent source of B-vitamin complex in addition to the nitrogen-containing materials derived from the yeast protein. A starting mixture of soybeans or soy grits and wheat is the base of *fermented soy sauce*. The Japanese have produced such sauces for hundreds of years. *Hydrolyzed soy sauce* is produced in the same manner and from the same material as HPP and HVP. *Hydrolyzed casein*, used as a nutritional supplement and amino acid source in infant formulas, is prepared from casein, the principal protein in milk, by use of an enzyme (a catalyst) to hydrolyze the protein.

When hydrolysis is brought about by an acid rather than by an enzyme or natural distintegration, the final product after neutralizing with alkali is likely to have a high content of salt. This is true of HPP and HVP, and hydrolyzed soy sauce. Fermented soy sauce will also be high in salt content, since the fermentation is carried out in the presence of up to 18 percent salt.

MAJOR REFERENCES: Evaluation of the health aspects of certain protein hydrolyzates as food ingredients. FASEB/SCOGS Report 37b (NTIS PB 283-440) 1978; Supplemental review and evaluation. FASEB/SCOGS Report 37b Supplement (NTIS PB 80-178643) 1980.

The major uses of protein hydrolyzates (such as hydrolyzed cereal solids) are as flavor enhancers and sources of "meat-like" flavor when added to foods like soup mixes, gravies, and chili. Soy sauce is used solely as a flavor ingredient. Some protein hydrolyzates are also included as nutritional supplements in special diet products; for example, those used in rapid weight reduction and, specifically, hydrolyzed casein for infants with health problems.

SAFETY: In 1976, an estimated 313 milligrams of hydrolyzed proteins was consumed daily on the average by U.S. individuals in processed foods and in sauces used by restaurants and other retailers, or 5 milligrams per kilogram of body weight (kg/bw) for a person weighing 60 kg (132 lbs.).

Protein and its amino acid components are essential nutrients and energy sources in the diet, and are routinely metabolized by the body. Thus it is not surprising that animal feeding studies experimenting with a variety of hydrolyzates, and employing amounts far exceeding average human consumption, did not provide evidence for concern at the levels currently present in the human diet.

However, the components of protein hydrolyzates contain a substantial presence of glutamic acid, ranging from 5 to 25 percent depending on the original source material; and some concern does exist about glutamic acid's possible effect on the central nervous system and the brain of infants, and on adults afflicted with vascular and brain-related ailments. Brain lesions in ten-day-old mice occurred after the injection of doses of hydrolyzed casein 300 times average human intake (adjusted for body weight). Similar findings with young mice have been reported by two other investigators using monosodium glutamate equivalent to that contained in 2 grams per kg/bw of a protein hydrolyzate, some 400 times human intake. It has been pointed out that hydrolyzed casein has been used for decades in feeding infants, with no reports of any abnormalities. (For health hazards that may arise from the consumption of glutamic acid, see p. 555.)

A major problem associated with the addition of an amino acid mixture from hydrolyzed proteins as a supplement to diet is the possibility of producing an imbalance of amino acids. The efficiency with which proteins are utilized is dependent on the amino acids contained in the mixture. Experiments with young rats have demonstrated that an unbalanced mixture, which could come from an improperly selected diet, can depress food intake and retard growth. The U.S. Public Health Service has had a special task force investigating reports of deaths associated with rapid reduction diets consisting solely or primarily of protein hydrolyzates.

Long-term studies of animals fed protein hydrolyzates such as acid-hydrolyzed soy protein or wheat gluten have shown that they do not affect reproduction, maternal or fetal survival, or the normal development of offspring. Nor do they produce pathological changes or tumors.

ASSESSMENT: With the exception of enzymatically hydrolyzed casein, protein hydrolyzates are used exclusively as flavor enhancers. This, together with their high salt content, restricts their consumption as food additives to about 3 milligrams per kg/bw. Casein hydrolyzates may also be used as a nutrient supplement and thus can be a more significant part of the diet. In digesting protein, the body regularly metabolizes protein hydrolyzates, but the reservation that experimental studies have raised is the potential hazard from an imbalance of the constituent amino acids or an excess of glutamic acid. Studies associated with the administration of hydrolyzed casein (about 20 percent of which can be glutamic acid) to newborn mice that resulted in brain lesions were by injection, but more recent research has established 400 milligrams per kg/bw as the minimum oral intake of glutamic acid required to cause brain lesions immediately after birth (shortly thereafter, the minimum required to cause an effect becomes much higher). The other possible health hazard is the use of hydrolyzed casein exclusively or as the main part of the diet; this is not advised until the deaths associated with the "liquid protein diet" are otherwise explained.

RATING: S for all protein hydrolyzates except hydrolyzed casein; for hydrolyzed casein, S for some people, ? for infants, X for individuals on weight-reducing diets without adequate supplementation.

QUININE

Quinine Hydrochloride; Quinine Sulfate

The hydrochloride or sulfate salts of quinine are used to impart a refreshing bitter flavor to beverages, especially tonic water. Use in carbonated beverages is limited to a maximum of 83 parts per million, equivalent to about 24 milligrams per 10-fluid-ounce bottle. Quinine is an alkaloid* extracted from the bark of the *Cinchona* tree, and for several hundred years has been a therapeutic agent for malaria.

SAFETY: In 1976 a survey of the usage of quinine salts as an additive, based on reports by the food industry, indicated an average daily intake for individuals consuming products containing these quinine salts of 0.6 milligram, or a daily intake of 0.01 milligram per kilogram of body weight (kg/bw) on the average for a consumer of these products weighing 60 kg (132 lbs.). A gin-and-tonic highball may provide 13 to 20 milligrams of quinine. Medical use in controlling malaria may call for daily use of over 1 gram of quinine, 50 times this amount.

Quinine does not accumulate in the body. It usually disappears from the blood circulation in less than 24 hours, since it and the products obtained from its breakdown metabolically are excreted in the urine. Although repeated use of doses of 1 gram or more of quinine can cause adverse or toxic effects in humans, the only symptoms reported from use of quinine salts in carbonated beverages is an allergic type of hypersensitivity. In susceptible people, temporary small hemorrhages beneath the skin may cause purplish blotches to appear.

Hearing loss and effects on equilibrum have been observed in guinea pigs consuming quinine hydrochloride daily, at 67 milligrams per kg/bw or more, for several months. (This is many hundreds of times the usual human consumption, adjusted for body weight.) The fetuses of pregnant rabbits have also shown hearing-nerve damage when the mother received approximately 125 milligrams per kg/bw of quinine sulfate daily for ten days. There are studies with monkeys and humans suggesting that they do not show such effects. It should be

---

MAJOR REFERENCE: Scientific literature review of quinine salts in flavor usage. FEMA Report to FDA (NTIS PB 296-017) 1979.

*Alkaloids are bitter substances of vegetable origin with a nitrogen base, such as caffeine, morphine, nicotine, quinine, and strychnine, which often have a powerful effect on animals and humans.

noted, however, that among the toxic effects of large doses of quinine in humans is the temporary impairment of hearing, and there are claims associating deafness of offspring with quinine usage.

Quinine did not cause malformation of the offspring when fed to pregnant rats or monkeys. When quinine sulfate was injected into the vaginal tissue of mice twice weekly for 40 weeks, no evidence of cancers or tumors in the vaginal area or elsewhere was found.

ASSESSMENT: Quinine salts are used as flavor ingredients in beverages, notably tonics. Quinine has been used therapeutically in areas where malaria is present; its use as a food additive involves substantially lower levels of intake. Some evidence that hearing may be impaired from high intakes of quinine, and that the fetus may also be subject to toxic effect, suggests that pregnant women should avoid quinine. Obviously, the very few people who show an unusual sensitivity, manifested by a flushing caused by temporary hemorrhages, should also avoid beverages containing quinine.

RATING: S for most people; X for pregnant women and for hypersensitive individuals.

SACCHARIN

Sodium Saccharin

The one noncaloric sweetener most commonly used today* and approved for use in dietetic foods is saccharin, usually as the sodium salt, which is more water-soluble. Saccharin is several hundred times as sweet as ordinary sugar. It has a bitter aftertaste, which usually is masked by the addition of the amino acid glycine, a protein component. Saccharin is widely used as an additive to provide sweetness in low-calorie soft

---

MAJOR REFERENCES: Artificial sweeteners and human bladder cancer. R. N. Hoover and P. H. Strasser. *Lancet* (April 19, 1980), p. 837; Artificial sweetener use and bladder cancer: a case-control study. E. Wynder and S. Stellman. *Science* 207:1214 (1980); Artificial sweeteners and cancer of the lower urinary tract. A. Morrison and J. Buring. *New England Journal of Medicine* (March 6, 1980).

*Aspartame, a chemical combination of two amino acids, aspartic acid and phenylalanine, was approved in July 1981 for use as a low-calorie sweetener in cold cereals, powdered beverages, puddings, gelatins, chewing gum, dessert toppings, and instant coffee and tea.

drinks (not to exceed 12 milligrams per fluid ounce), dietetic ice creams, and various other low-calorie processed foods (not to exceed 30 milligrams per serving), and as a replacement for table sugar. It is a synthetic sweetener which has been used for calorie reduction in foods throughout this century, and, since it does not convert to glucose in the body, diabetics use it instead of sugar.

SAFETY: The National Academy of Sciences-National Research Council (NAS-NRC) has estimated that saccharin consumption in 1977-78 by users of low-calorie or dietetic foods and beverages (roughly a third of all consumers) averaged 30 milligrams daily. By far the greatest amount is consumed in diet soft drinks; some consumers averaged as many as five bottles daily, containing a total of some 365 milligrams of saccharin. For a person weighing 60 kg (132 lbs.), this would amount to 6 milligrams per kilogram of body weight (kg/bw).

Saccharin has been the subject of a seemingly endless series of safety scrutinies. President Theodore Roosevelt had a group of scientists review it in 1912; NAS-NRC reviewed toxicity data on saccharin in 1955, 1968, 1974, 1978, and 1979. Additional reviews by the federal government have come from the Food and Drug Administration (FDA), the National Cancer Institute, and the U.S. Congress Office of Technology Assessment.

In two studies, one conducted by the FDA, there was an indication of a higher incidence of bladder cancer among rats consuming saccharin, but whether saccharin or the impurities it contained was the cause remained uncertain. In 1977 researchers for Canada's Health Protection Branch (similar to our FDA), who had been investigating this matter for 25 years, produced findings substantiating that saccharin without impurities was the sole culprit. The levels used in the initial studies in Canada were 5 percent saccharin in the rat diet, or approximately 2500 milligrams per kg/bw. This is many hundred times the upper level of human consumption, but such an amount is not uncommon in tests of possible cancer effect of a chemical when using a necessarily limited number of experimental animals.*

---

*Nobody knows with any assurance the level at which a known carcinogen (cancer-causing substance), or a substance suspected of being one, causes the malignancy. The type of cancer under investigation may occur in only one of a thousand people exposed to the substance. To test it experimentally while

Since the initial report in 1977, additional studies have confirmed that some bladder tumors may develop in male rats fed saccharin throughout their lives, especially if their mothers, during pregnancy, ate a similar diet. Some evidence exists that saccharin is also a cocarcinogen, promoting the cancer-causing effects of other carcinogens.

The findings led to a proposal by the FDA to ban use of saccharin as a food additive, since the 1958 Delaney Clause in the Food and Drug Act bans use of any chemical in food at any level if the chemical is found to induce cancer in man or animals. The widespread opposition to such a ban among the public, health professionals, and legislators led to the Saccharin Study and Labeling Law, which came into effect in November 1977, prohibiting the FDA from banning this additive for eighteen months; in August 1979, the moratorium was extended to June 30, 1981; and on expiration, the moratorium was extended for an additional two years.

The saccharin moratorium had been recommended by the American Medical Association, the American Society of Internal Medicine, the saccharin committee of the National Academy of Sciences, the American Diabetes Association, and the American Society of Bariatric Physicians. But there is opposition to a "unilateral modification of present laws which would exempt saccharin from complying with specific regulations."*

That saccharin is a "weak carcinogen" seems evident. That the dosages used were unusually huge is also recognized; indeed, the findings of a tumor increase were only observed at a dosage of 5 percent of total diet or more. The linkage between such studies and probable risks to humans is uncertain.

---

using low levels of the chemical would require groups of 50,000 animals. Instead, many scientists believe detection is possible by employing large doses administered to a few hundred animals.

The evidence has not supported the contention that any substance can cause cancer, if the dosage administered is sufficiently large. One hundred and twenty chemicals, including pesticides, many of which were suspected of being able to cause malignancies, were tested in the same way as saccharin. The findings were reported in 1969 in the *Journal of the National Cancer Institute:* only 11 of these chemicals caused cancers. It appears likely that the substance must be a carcinogen to begin with, and that the dosage by itself, however large or small, is not the responsible agent. And while it is not certain that all substances causing cancer in rats will also do so in humans, it is known that of 30 which cause cancer in man, 29 have produced cancer in rats.

*The saccharin question re-examined: an A.D.A. statement. *Journal of the American Dietetic Association* (May 1979), pp. 574-581.

One study in Canada did suggest a connection, but for males only. However, at least six studies of various population groups (epidemiologic studies), which included diabetic patients who are large users of saccharin, have been conducted and failed to give statistical evidence of any association between bladder cancer in humans and intake of saccharin in the diet. Three of these studies were reported early in 1980. One was conducted by epidemiologists of the National Cancer Institute and involved 9000 people, about a third of whom were bladder cancer patients. There was little or no increased risk of bladder cancer for average users of saccharin. The researchers inferred a somewhat greater risk among those consuming eight or more diet soft drinks and tabletop artificial sweeteners daily, but this was based on only seven cases of 9000 people studied.

A second population survey by the American Health Foundation focused on 367 bladder cancer patients, comparing their usage of artificial sweeteners with 367 others from a control population of 5597 patients with no bladder cancer. These investigators found no indication of increased cancer with saccharin intake, no response to increasing dosages or duration, and no evidence that the artificial sweeteners promoted the tumor-producing effects of smoking.

The third large-scale population study was made by scientists at Harvard University. More than half of the 1128 people studied had a bladder tumor. This cancer bore no relationship to a history of use of dietetic beverages or of sugar substitutes. Again, increasing frequency or duration of use did not appear to be associated with greater relative risk of bladder cancer.

Saccharin appears not to cause malformations of the fetus, and does not affect reproduction or normal growth.

ASSESSMENT: To many, the scientific facts on saccharin are based on dosages given at such excessive levels that it is easy to belittle any danger to a human eating a diet even where saccharin is present in appreciable amounts, as in consumption of many diet soft drinks daily. Yet the concern that saccharin in the diet may pose an unnecessary risk to some consumers is less easily dismissed. At high levels, saccharin has caused bladder cancer in rats. The risk that it could bring about a similar result in humans seems minimal, based on population studies. Nevertheless, that there is some risk suggests that overriding benefits are not likely to fully offset any potential hazards. Some diabetics who need to drastically restrict carbohydrate intake may find the risk worth taking. Overweight but sweet-toothed persons truly needing to restrict calories despite large consumption of beverages may concur. For the public in general, even the small risk from saccharin in food suggests that this food additive should be avoided.

RATING: ? for persons on medically restricted low-calorie diets; X for all others.

SHELLAC

Confectioner's Glaze

Shellac, the only commercial resin secreted by an insect* (the other resins occur in plants), is harvested principally in India and Thailand. It is a familiar material for finishing furniture because of its protective, glossy, hard finish. The shellac used in food ("food-grade" shellac) is produced by refining and bleaching regular shellac. Refined food-grade shellac is used as a food glaze, especially in coating candy *(confectioner's glaze)* and some fruits and vegetables, such as citrus and avocados, and ice-cream cones and fruit cakes.

SAFETY: The bulk of food-grade shellac is on skins of citrus fruit and avocados that are peeled off and discarded; this shellac is not likely to be ingested. The remainder, which has very limited use in food, contributes on the average approximately 0.25 milligram to the daily diet of a U.S. adult.

There are no studies on the biological effects in humans or in experimental animals from ingestion of shellac, although its food use is "generally recognized as safe" by FDA. It has been tested in laboratory tests for chromosome damage and been found free of such activity.

ASSESSMENT: Because there are no studies on which to base an evaluation of the safety of shellac in food use, no meaningful assessment of its safety is possible.

RATING: ?

MAJOR REFERENCE: Evaluation of the health aspects of shellac and shellac wax as food ingredients. FASEB/SCOGS Report II-19 1980.

**Kerria lacca*, a scale insect which lives on twigs of various trees.

SILICATES

Calcium Silicate (Calcium Trisilicate); Magnesium Silicate (Magnesium Trisilicate); Methyl Polysilicone (Dimethyl Polysiloxane, Methyl Silicone); Phenylmethyl Cyclosiloxane (Cyclophenyl Methylsilicone); Silicon Dioxide (Silica, Silica Aerogel); Sodium Aluminosilicate (Aluminum Sodium Silicate, Sodium Silicoaluminate); Sodium Calcium Aluminosilicate; Sodium Silicate; Talc

Silicates are compounds of silica, a principal constituent of rocks and sand, and other minerals. They are present in practically all natural waters (city water may contain 7 milligrams per quart), plants, animals, and humans. Silicon in trace amounts is an essential mineral element for humans; it is believed to participate in bone calcification, and it may be a component of the body's connective tissues.

As additives, the silicates contribute only a minor portion of the total intake of silica in the normal diet, because many foods and liquids naturally contain substantial amounts. The compounds listed above are usually employed in foods for only a few purposes; several are used for preventing salt and dry mixes from caking; methyl polysilicone and silicon dioxide are used to reduce foam and reduce sticking; talc provides a base for chewing gum and a coating for rice.

SAFETY:* Estimates of the amount of silicates as additives in the daily diet of U.S. adults, based on a survey conducted by the National Academy of Sciences-National Research Council of usage by food processors in 1976, are available for silicon dioxide (60 milligrams); talc (25); sodium aluminum silicate (12); calcium silicate (3); dimethyl polysiloxane (2); and magnesium silicate (1)—for a total of 103 milligrams, or less than 2 milligrams per kilogram of body weight (kg/bw) for a person weighing 60 kg (132 lbs.).

The ability of the body to absorb silicates depends on their solubility in water. The silicates that are used as direct food

---

MAJOR REFERENCES: Evaluation of the health aspects of silicates as food ingredients. FASEB/SCOGS Report 61 (NTIS PB 301-402) 1979; Evaluation of the health aspects of methylpolysilicones as food ingredients. FASEB/SCOGS Report II-14 (NTIS PB 81-229239) 1981.

*For safety findings on sodium aluminosilicate and sodium calcium aluminosilicate, see also p. 473.

additives are essentially insoluble, and are not likely to be absorbed; instead, they pass through the body inert and are excreted. Studies of the effects of feeding various silicon compounds to laboratory animals have generally shown them not to be harmful under the test conditions, but some adverse effects have been recorded:

Magnesium trisilicate at extremely high levels in the diets of young dogs resulted in kidney damage. Contrary to these results, the substance has been administered without serious harm to patients as an antacid in the treatment of peptic ulcers and related illnesses, some patients taking several grams daily for a number of years.

Silicon dioxide fed at 6 milligrams per kg/bw in one experiment caused kidney damage in guinea pigs, but did not in other experiments when dogs and rats were used. It is possible that this compound is more noxious for one species of animal than another.

Experiments with pregnant animals indicated that neither silicon dioxide, *sodium silicoaluminate,* nor *calcium silicate* caused birth defects, and laboratory tests determined that they did not induce mutations or other genetic changes.

Talc may contain asbestos fibers, but the FDA requires that when it is used in food it be free of this substance. At this writing, no practical means exists for determining absence of asbestos. Inhalation of talcum powder has resulted in fatalities, and injections of crushed drug tablets containing talc have caused serious illness. Evidence exists of inflammatory tissue reaction caused by contact with talc dust from surgical gloves and from its intentional use to induce lesions in the treatment of lung diseases. But these effects have not been found when talc has been administered orally; and laboratory tests indicate that it does not cause mutations, birth defects, or genetic changes.

Biologic effects and safety data are not available for *sodium calcium aluminosilicate,* but its toxicity should not differ from closely related compounds for which there are data that indicate their safety.

Several short-term feeding studies using rats and dogs have shown that *methyl polysilicone* does not adversely affect their health when doses of up to 10 grams per kg/bw were administered. Monkeys fed as much as 60 grams daily, five days a week for eight months, did not demonstrate abnormalities in behavior, blood, body weight, or urine, but did exhibit occasional diarrhea. A life-span study with mice given 5.8 grams per kg/bw of methyl polysilicone in their daily diets found it caused no adverse health effects, nor were any observed in rats fed silicone through three generations. The dosages employed in the experiments cited above were many thousands of times as great as the 2 milligrams of methyl polysilicone present in the daily diet of the average U.S. adult. This compound has been used therapeutically in man at dosages of up to 200 milligrams for excess gas in the digestive tract. It has been shown to be free of properties that can lead to cancer, malformed offspring, or chromosome (gene) alterations.

Methyl polysilicone is a high-molecular-weight (large-molecule) compound. A variation that in part contains small-molecule silicon polymers has caused some damage to livers of dogs fed 3 grams daily per kg/bw for six months, and to kidneys of rabbits, an effect not seen in mice and rats. It was identified in these experiments as Dow-Corning Antifoam A. When small molecules are present, they may be absorbed in limited amounts and then may accumulate in tissues. *Phenylmethyl cyclosiloxane* (which includes the phenyl group as well as small molecules, with cyclic structure rather than linear) caused testicle atrophy and reduction in sperm in a variety of animals, including monkeys who developed this condition when given as little as 50 milligrams per kg/bw by mouth.

ASSESSMENT: Except for talc and phenylmethyl cyclosiloxane, the evidence indicates that the silicates can be regarded as safe when used as additives in food at levels now current or that reasonably can be expected in the future. Silicates added in this manner represent only a minute portion of total intake by humans, because of their presence in most natural foods. When talc is used in food, uncertainty remains until a means is devised of assuring the absence of asbestos fibers; phenylmethyl cyclosiloxanes have caused damage to reproductive capacities, livers, and kidneys in a number of experimental animals.

RATING: S for silicates other than talc and phenylmethyl cyclosiloxane; ? for talc until absence of asbestos fibers can be determined; X for phenylmethyl cyclosiloxane.

SMOKE FLAVORING (LIQUID SMOKE, CHAR-SMOKE FLAVOR)

These solutions are condensed fractions or extracts of wood smoke that has been trapped from wood burned with a limited amount of air and then further treated to remove some of the tars. Hickory and maple are preferred woods. Smoke flavoring is used in meats, cheese, barbecue sauces, baked beans, pizza, fabricated snack foods and vegetables, and other food products. Smoked flavorings have an advantage over conventional, direct smoking of foods (meat, poultry, fish) in that most of the resinous tar materials present in the smoke, which can contain some cancer-producing benzopyrene chemicals, have been removed in manufacturing.

SAFETY: In 1978 the food usage of smoke-flavoring solutions indicated a daily consumption on the average of about 30 milligrams per person, or 0.5 milligram per kilogram of body weight (kg/bw) for a person weighing 60 kg (132 lbs.).

In a study of approximately three months' duration, rats consuming meals containing up to 2 grams per kg/bw of two liquid smoke preparations, an amount 4000 times as great as average human intake, showed no significant abnormalities. In a study of similar duration conducted in Poland, pigs fed up to 300 milligrams per kg/bw of a Polish wood-smoke extract in their daily diet did not develop abnormalities, and results of blood and urine tests were in the normal range.

The majority of long-term toxicity studies on smoked products have been primarily concerned with the presence of benzopyrene chemicals, suspected carcinogens which are associated with the tars and resinous materials developing from wood smoke. This concern has been substantiated by investigations in which a significant number of experimental animals fed meats smoked directly over wood fire developed malignancies. The smoked flavorings reviewed here, while derived from wood smoke, employed processing methods that appear to have made them essentially free of benzopyrene. Tests have indicated that remaining traces, if any, are likely to be less than 0.5 parts per billion. The only study of sufficient duration to determine whether a hazard remains, a two-year experiment

---

MAJOR REFERENCE: Evaluation of the health aspects of smoke flavoring solutions and smoked yeast flavoring as food ingredients. FASEB/SCOGS Report II-7 1980.

on rats fed sausage containing a Polish smoke extract estimated to be the equivalent of 100 times the amount eaten by the average Polish consumer (adjusted for body weight), found that it did not cause a higher incidence of tumors among the experimental animals who ingested it than among the controls who did not. However, the Polish extract differs from those available in the U.S. in its components.

Smoke-flavoring products have proved to be free of chromosome-altering properties in several types of tests.

ASSESSMENT: The composition of smoke flavoring varies, depending on the kind of wood used, combustion conditions, and method of entrapment and extraction of the smoke and purification of the extract. Although the procedures used commercially remove nearly all benzopyrene chemicals, there is inadequate evidence from long-term studies with experimental animals to assess the absolute safety in the food supply of liquid smoke as prepared and used in the U.S. Clearly, liquid smoke is less hazardous to a consumer than the simple smoking of foods or charcoal broiling of meats. But while the existing evidence suggests no hazard from consuming products containing smoke flavoring, the unknowns and uncertainties call for further research, and especially long-term studies, to properly assess safety of this additive.

RATING: ?

SODIUM CHLORIDE

Sodium chloride is common table salt. It is abundant in nature and is found in many natural waters and in underground deposits. Commercial quantities for use in food are obtained by mining, by evaporation from natural brines, and from seawater. The food industry finds it useful as a seasoning, preservative, and curing agent; in formulating and processing; as a nutritional supplement; and as a dough conditioner. Solutions of sodium chloride are used in medicine for replacement of fluid following trauma or operations, usually in intravenous preparations, and as tablets for rapid replacement of body salt loss.

MAJOR REFERENCE: Evaluation of the health aspects of potassium chloride and sodium chloride as food ingredients. FASEB/SCOGS Report 102 (NTIS PB 298-139) 1979.

SAFETY: The diet of the U.S. adult, on the average, contains 10 to 12 grams of sodium chloride each day. While salt is essential to the body, the minimum requirement is substantially less than 1 gram; this amount is already exceeded by the 3 grams that are naturally present in the foods we eat. Four to 6 grams are added to foods by manufacturers, baked goods and meat products accounting for more than half, making it difficult to restrict salt intake. Salting foods by the consumer is responsible for the remainder of the salt in the daily diet.

Investigations of many of the chronic effects of ingesting sodium chloride in amounts far exceeding requirements have been focused on its role in high blood pressure. The U.S. Health and Nutrition Examination Survey has estimated that 23 million persons 12 to 74 years of age suffer from hypertension. Clinical and epidemiological studies suggest a relationship between salt intake and the onset of hypertension. A low sodium intake is characteristic of populations that do not exhibit much if any hypertension; while comparable populations with high sodium intake experience a higher incidence.

A critical determinant appears to be a genetic predisposition. Data on 1000 New York City residents, 563 normotensive (normal blood pressure) individuals and 437 hypertensive (who did not know it), revealed little difference in sodium intake, except that higher intakes were found more frequently among the "normals." A separate study comparing 717 hypertensive and 819 normotensive patients, however, found a significant correlation between family history of high blood pressure and hypertension in the patients. Consistent excess consumption of sodium chloride seems to play a primary role in the prevalence of hypertension among "responsive" individuals, and a number of clinical studies confirm the therapeutic effectiveness of low-salt intake for this condition.

Animal experimentation with strains of mice and rats bred either to be resistant or susceptible to salt toxicity impressively demonstrates the significance of heredity. Both resistant and susceptible mice were given the same dosages of sodium chloride in their drinking water, comparable to thousands of times that of human intake. All the susceptible strain died from salt toxicity within three months, while a significant number of the resistant variety survived for more than a year. Two strains of rats, differing genetically in their response to chronic ingestion of excess salt, were studied for its effect on systolic (period of heart contraction) blood pressure. The sensitive strain developed hypertension, while the resistant strain did not. Neither strain developed high blood pressure on a normal diet.

A National Academy of Sciences committee has expressed concern about the capacity of an infant to excrete sodium chloride daily in excess of 23 to 58 milligrams per kilogram of body weight. The committee found no nutritional justification for the addition of salt to the normal infant's diet, since commonly used unsalted foods supply more than does human breast milk, and by four months of age most infants are receiving cow's milk, which supplies three times as much sodium as human milk.

ASSESSMENT: The amount of sodium chloride in the daily diet of the average U.S. individual is many times in excess of a human's requirements. (However, extra salt—and water—may be needed where there is salt loss by sweating, diarrhea, or vomiting.) A more than sufficient amount usually is present naturally in food before what is added by the food manufacturer and by consumer salting. While this excess is not likely to be harmful for many people, it does represent a risk to the estimated 23 million who exhibit tendencies to develop high blood pressure. The addition of salt to the foods of most infants is not warranted nutritionally, and may be beyond their disposal capacities.

RATING: S for many people; X for anyone with high blood pressure, or with family history of this tendency, or for normal infants in their prepared foods.

SODIUM FERROCYANIDE (YELLOW PRUSSIATE OF SODA)

Sodium ferrocyanide is an effective anticaking agent. It is used in small amounts in table salt to prevent the formation of clumps and keep it free-flowing by causing the salt to form into rough crystals during crystallization. The additive is produced synthetically by heating sodium carbonate and iron together with organic materials. FDA regulations permit a maximum of 13 parts of sodium ferrocyanide per million parts of salt (sodium chloride; see p. 626). As salt is added to many pro-

---

MAJOR REFERENCE: Monograph on ferrocyanide salts. Informatics, Inc., Report to FDA (NTIS PB 289-591) 1978.

cessed foods, sodium ferrocyanide may be present in a variety of products, such as breads, breakfast cereals, natural cheeses, meat products, and salad dressings.

SAFETY: The average daily diet in the U.S., based on reported usage in food in 1976 by manufacturers, contained 0.6 milligram of sodium ferrocyanide per person.

The cyanide part of sodium ferrocyanide is a deadly poison, but when a strong chemical bond between iron and the cyanide part of the molecule is formed, the toxic action is neutralized, and sodium ferrocyanide shows a low toxicity. There was no evidence of urinary disturbance in infants given 0.1 percent of sodium ferrocyanide administered intravenously, or of harm in kidney function tests in which 550 milligrams, over 800 times the presence of this chemical in the diet, was injected into adults. Indeed, as much as 6 grams of sodium ferrocyanide has been injected into humans to study excretion problems, the only adverse finding being the presence of blood albumin and granulation in the urine, which disappeared in two weeks.

There are no long-term studies to assess toxicity of sodium ferrocyanide, but 13-week studies have included rats and dogs. Rats fed 25 milligrams of the additive per kilogram of body weight, 2500 times the average human consumption when adjusted for body weight, did not experience any adverse effects. However, at ten times this dosage there was evidence of minimal damage or inflammation in the kidney tubules. Dogs were not harmed when fed this compound at a level of 0.1 percent of their diet, an amount thousands of times as high as is present in human diets.

Tests with bacteria showed potassium ferrocyanide to be free of chromosome alteration effects; sodium ferrocyanide should be similar.

ASSESSMENT: Sodium ferrocyanide has a very low toxicity, and its use solely as an anticaking agent in salt limits consumer exposure to minimal amounts. The UN Joint FAO/WHO Expert Committee on Food Additives considers 1.5 milligrams daily an acceptable and safe intake for a 132-pound human (2½ times the amount in the U.S. diet). There is no evidence of hazard to the consumer from the present or probable future usage of sodium ferrocyanide as a food additive.

RATING: S.

SODIUM LAURYL SULFATE

This additive is a mixture prepared from coconut oil fatty acids (components of its fat), chiefly lauric acid, which are chemically converted to fatty alcohols and then combined with sulfuric acid and made into sodium salts. Sodium lauryl sulfate is a surface-active agent (one that modifies the surface properties of liquid food components because of detergent properties); it acts as a cleanser of fresh fruits and vegetables and improves whipping or emulsion formation with egg whites or gelatin. It also is used as a finishing agent in hard and soft candies, and as a wetting agent in dry beverage mixes containing fumaric acid.

SAFETY: The presence of sodium lauryl sulfate as an additive in food, according to 1976 usage reported by food processors, amounted to less than 0.2 milligram daily in the average diet of a U.S. adult.

Many studies of the possible toxicity of sodium lauryl sulfate have been conducted during the past 40 years. When the compound was added in amounts of 0.1 percent in the drinking water of rats, it did not produce any harmful effects, but above this level it did cause some depressed growth and gastrointestinal irritation. An intake of 0.1 percent approximates 130 milligrams per kilogram of body weight (kg/bw) daily. In a 132-pound person it is the equivalent of well over 7 grams per day, many thousands of times its presence in the daily diet. Other studies have confirmed that levels of this chemical in the diet of experimental animals providing over 100 milligrams per kg/bw of sodium lauryl sulfate daily are without adverse effect.

Sodium lauryl sulfate has been used therapeutically in human patients. Such use has demonstrated that even gram quantities do not produce toxic effects. Studies of this additive as a possible cause of cancer or gene mutation, or of harmful effects on reproduction, have not been found.

ASSESSMENT: Sodium lauryl sulfate added to the diet of experimental animals to provide an intake of 100 milligrams per kg/bw has had no apparent adverse effect, nor have gram quantities used medicinally in humans. Any hazard to the

---

MAJOR REFERENCE: Toxicity of sodium lauryl sulphate, sodium lauryl ethoxysulphate and corresponding surfactants derived from synthetic alcohols. A. I. T. Walker, V. K. H. Brown, L. W. Ferrigan, R. G. Pickering, and D. A. Williams. *Food and Cosmetics Toxicology* 5:763 (1967).

consumer is unlikely from the minute quantity of this additive presently in processed foods, or likely to be expected in the future.

RATING: S.

SODIUM NITRATE, POTASSIUM NITRATE; SODIUM NITRITE

Nitrate and nitrite are closely alike in chemical composition, but nitrite has less oxygen in the molecule and is less stable when exposed to air. Nitrate can convert to nitrite by the loss of oxygen. Nitrates are present in tap water and in many foods, in some measure because of the widespread use of nitrogen fertilizers. Spinach, celery, other leafy vegetables, and their juices contain substantial quantities. About 80 percent of the human dietary exposure to nitrites comes from nitrates in drinking water and in foods that have not had any added nitrate or nitrite; about 20 percent comes from cured food products.

The nitrate in food before or during the digestive process can degrade to nitrite by enzymatic or bacterial action. Nitrate is a regular constituent of saliva and also can be converted into nitrite, entering the stomach in amounts estimated at as much as 10 times the quantity actually present in cured meats and other foods in the diet. Further formation of nitrite in the intestines raises it to levels many times as great as the nitrite added to foods, and probably thousands of times as great as the nitrite remaining in processed foods at the time they are consumed.

For decades bacon, ham, corned beef, frankfurters, sausages, and fish products have been prepared by curing with sodium and potassium nitrate and sodium nitrite (always in combination with sodium chloride, or table salt; see p. 626). These two additives contribute to the characteristic flavor, produce a pink color in the meat, and serve as meat preservatives, greatly extending the shelf life of the product. Nitrite has the property of inhibiting the botulinus microorganism, thus preventing development of botulism toxin, a deadly poison. Many years ago it was discovered that nitrate was effective

---

MAJOR REFERENCE: Toxicological evaluation of some food additives including anticaking agents, antimicrobials, emulsifiers and thickening agents. WHO Food Additives Series No. 5 (Geneva, 1974).

only after it was converted to nitrite by bacterial action; today sodium nitrite is used directly in meat curing.

SAFETY: A survey conducted for the National Academy of Sciences-National Research Council determined the total poundage in 1976 of sodium nitrate and nitrite used in food processing. These figures would suggest an average daily intake per person of approximately 11 milligrams, most of it added to food as sodium nitrite (though it is recognized that much of the additive will be further altered during processing and storage and thus end up as nitrate when consumed). In comparison, a person's exposure to salivary nitrite has been estimated at 15 milligrams daily; and an additional 90 milligrams are produced in the intestine. Calculated in terms of body weight, the intake of nitrate-nitrite used as food additives would be less than 0.2 milligram per kilogram of body weight (kg/bw) for a person weighing 60 kg (132 lbs.).

There are two major and opposing factors in considering possible hazards to the consumer through use of these additives. Botulinus toxin is a potent and deadly food poisoning agent. It has been found that at least 150 to 200 parts per million of nitrite are needed to prevent growth of the botulinus organism, which is not eliminated from cured meats because they are not sterilized by a high heat treatment. The other side of the coin is that nitrite can, under certain conditions such as the frying of bacon, chemically combine with some nitrogen-containing chemicals in foods (or in the body) to form nitrosamines; some nitrosamines are potent carcinogenic (cancer-causing) materials.

Sodium nitrate is readily absorbed and excreted when taken in the diet. It is, in certain circumstances, converted to nitrite (containing less oxygen in the molecule) by organisms in the digestive tract; this has been seen especially in infants less than six months old. The nitrite can react with the hemoglobin in the blood and cause a potentially fatal condition where the blood is not effective as a carrier of oxygen. The young infant has less acid production in the stomach and this favors bacterial conversion of nitrate to nitrite. Where a child has imperfect digestion (dyspepsia), even 0.05 percent of nitrate in tap water can be a real hazard, whereas healthy babies may easily tolerate 21 milligrams per kg/bw daily for a year with no adverse effect.

However, the greater concern lies with the property of nitrites to combine with certain substances (amines) to form carcinogenic nitrosamines. Some nitrosamines can cause cancer in experimental animals even at the low level of 2 parts of nitrosamine per million parts of diet (0.0002 percent, probably equivalent to 0.2 milligrams per kg/bw daily). Studies have shown that tumors in rats and mice may be formed following the simultaneous administration of an amine (present in most foods) and sodium nitrite in the diet. Commercial processing and home cooking of foods containing nitrite, before consumption, may give rise to nitrosamine formation, though this does not always occur and even if it does, the levels remain at the very minute concentration of a few parts per billion in the food. Recent studies have shown that nitrosamine formation in cured meat products, even after frying, may be greatly reduced or eliminated by an excess of ascorbic acid (vitamin C; see p. 483) or alpha-tocopherol (vitamin E; see p. 657).

In 1978, preliminary experimental findings suggested that there may be a low level of carcinogenicity from nitrite itself. However, government scientists and independent pathologists dispute this, finding on review that the higher incidence of lymphomas (malignant tumors) found in the experimental animals, compared with the controls not receiving nitrite, was the result of a rare type of cancer not likely to be caused by nitrite. With these eliminated, the occurrence of malignancies did not differ between the experimental and control animals, in effect eliminating the evidence that nitrite can cause cancer in its own right.

Long-term studies with rats showed some growth depression over a two-year period when sodium nitrate was 5 percent of the diet, but no demonstrable effect at 1 percent (500 milligrams per kg/bw daily). Fertility or reproductive performance of guinea pigs was poorer when nitrate was 0.3 percent of the drinking water, but no other adverse effects were seen. Sodium nitrite did not affect fertility, but when pregnant rats or guinea pigs consumed 0.3 to 0.5 percent of sodium nitrite in the water or diet, the offspring grew more slowly and had higher death rates, though no malformed young resulted. The dosages administered in these experiments were at least 2000 times as great as human intake of these compounds as additives.

ASSESSMENT: Sodium or potassium nitrate and sodium nitrite are used in curing meats, and some fish and poultry products. The nitrate may be converted to the nitrite, the effective chemical in preserving the meat and in controlling development of botulinus toxin. Under some circumstances, the nitrite may chemically combine with some nitrogen-containing substances during processing, cooking, or during metabolism to form cancer-causing nitrosamines. Long-term studies with experimental animals indicate that sodium nitrate and nitrite have some adverse effects on the offspring when consumed during pregnancy. It also is evident that the young infant is more susceptible than adults to problems from these additives; clearly, they should not be added to foods intended for the very young. The use of these chemicals is restricted by FDA to the minimum levels necessary for blocking development of the botulinus microorganism—a trade-off with the possibility of increasing exposure to nitrosamines.

RATING: ? for most people; X for infants and pregnant women.

SODIUM STEARYL FUMARATE

This is a synthetic chemical, the sodium salt of the acid resulting from the reactions when combining stearyl alcohol and fumaric acid (see p. 550). Both of the latter occur widely in nature and in foods, stearyl alcohol in some oils as in herring and whale, and fumaric acid in animal and plant tissues. But their combination as stearyl fumarate is not a normal food constituent. The additive is used in foods as a dough improver to give greater strength during food processing, to give better volume and improved texture to yeast-leavened baked goods, and as a stabilizing or conditioning agent that facilitates processing of dehydrated potatoes, starch-thickened foods, and dry cereals.

SAFETY: A survey of the use of sodium stearyl fumarate in 1976 by food processors showed only a minimal use of this additive; less than 0.02 milligram appeared in the average daily diet of the U.S. consumer. The calcium and sodium stearoyl lactylates (see p. 500), additives that provide similar benefits in food processing, are used far more by the industry.

The body's metabolic processes deal with the components of sodium stearyl fumarate without difficulty. The additive was rapidly metabolized by rats fed 300 milligrams of the chemical

---

MAJOR REFERENCE: The absorption and metabolism of orally administered tritium labeled sodium stearyl fumarate in the rat and dog. *Journal of Agricultural and Food Chemistry* 18:872 (1970).

per kilogram of body weight daily for 90 days. Apparently during digestion the stearyl fumarate is split into its original stearyl alcohol and fumaric acid components, and the stearyl alcohol combines with oxygen in the body to form stearic acid (see p. 479), a source of energy along with the other fatty acids. Fumaric acid is essential in tissues and, in fact, the body produces it in quantity during the metabolism of carbohydrates.

ASSESSMENT: Sodium stearyl fumarate is used as a food additive to improve the texture and handling properties of some foods, especially baked goods. The additive is split by digestive enzymes to its constituent stearyl alcohol and fumaric acid, both of which are present in a variety of foods and are easily handled in the body. It poses no hazard to the consumer.

RATING: S.

SORBIC ACID

POTASSIUM SORBATE

Sorbic acid and its salt *potassium sorbate* are made commercially by chemical synthesis. These compounds are useful preservatives which inhibit yeast and mold growth as well as some bacteria, being most effective when the food product is somewhat on the acid side. They may be used in cheeses, wine, chocolate syrups, margarine, and fruit-juice drinks.

SAFETY: The use of sorbates in foods indicated an average daily intake per person of 33 milligrams in 1975, or 0.6 milligram per kilogram of body weight (kg/bw) for a person weighing 60 kg (132 lbs.). The UN Joint FAO/WHO Expert Committee on Food Additives has estimated that even 45 times this amount would be an acceptable daily intake.

Sorbic acid is chemically related to a fatty acid, caproic acid (which is present in butter, coconut oil, and other food fats). Sorbic acid and caproic acid are metabolized and used as a source of energy by the body in the same manner. Studies with

---

MAJOR REFERENCE: Evaluation of the health aspects of sorbic acid and its salts as food ingredients. FASEB/SCOGS Report 57 (NTIS PB 262-663) 1976.

experimental animals have shown that rats can tolerate 6 grams of sorbic acid per kg/bw (10,000 times the average human intake) for four months without effect on reproductive performance but with a slight enlargement of the liver. Feeding rats sorbic acid at 4000 times human intake (adjusted for body weight) for 1000 days did not affect growth, reproduction, or survival, and there were no abnormalities in the various organs and no increase in tumors. Potassium sorbate has been shown to be negative in tests for embryo abnormalities and chromosome alteration.

ASSESSMENT: Sorbic acid and potassium sorbate are metabolized in the normal manner for fatty acids. They produce no evidence of hazard to the consumer even at levels of consumption much higher than is to be expected in the diet.

RATING: S.

SORBITAN DERIVATIVES

Polysorbate 60 (Polyoxyethylene-20-Sorbitan Monostearate); Polysorbate 65; Polysorbate 80; Sorbitan Monostearate

Sorbitan derivatives are obtained from sorbitol (see p. 638), a simple sugar-alcohol, which is modified by chemical dehydration. Extensive use is made in processed foods of four of the sorbitan derivatives. *Polysorbate 60* consists of chemically combined stearic and palmitic acids (fatty acids that are constituents of fats; see p. 479) and sorbitol, together with its dehydrated form, and condensed with 20 parts of ethylene oxide, a toxic gas frequently used as a sterilizing agent. *Polysorbate 65* differs in that it combines three molecules of stearic acid rather than a single molecule (monostearate). *Polysorbate 80* contains oleic acid (also a fatty acid) in place of stearic acid. *Sorbitan monostearate* chemically combines sorbitol with edible stearic and palmitic acid mixtures, an initial step in manufacture of polysorbate 60. Frequently the sorbitan derivatives are used in combination with mono- and diglycerides (see p. 559) and with one another, since through selection of the proper emulsifier system, significantly better aeration may be obtained, increasing volume and improving grain and texture.

Sorbitan derivatives are used in foods or in flavor composi-

MAJOR REFERENCE: Scientific literature review of sorbitan derivatives in flavor usage. FEMA Report to FDA (NTIS PB 296-016) 1979.

tions as emulsifiers or dispersing agents that blend oil and water components. The polysorbates are water-soluble, and sorbitan monostearate is oil-soluble. A wide range of products employ them: shortenings, confections, dressings, baked goods, dairy products including frozen desserts, beverages, coffee whiteners, meats and fish, and others.

SAFETY: A survey conducted by the National Academy of Sciences-National Research Council, based on 1976 usage reported by food processors, determined that the average daily diet of the U.S. adult contained 15 milligrams of polysorbate 60, 1 of polysorbate 65, 5 of polysorbate 80, and 6 of sorbitan monostearate, or 27 milligrams in total. An individual weighing 60 kg (132 lbs.) ingested on the average slightly less than 0.5 milligram daily per kilogram of body weight (kg/bw) of these sorbitan derivatives.

When ingested, the polysorbates are handled similarly by the body. The fatty acids portion is readily separated during digestion and metabolized normally, since these acids are common fat constituents. A small portion of the sorbitol also will be separated and metabolized, but most of it along with the polyethylene portion is not absorbed, and in humans is excreted.

Over the past 30 years, extensive studies on these substances, using mice, rats, hamsters, dogs, and monkeys as well as humans, have shown that at levels of 2 percent of the diet (200 to 1500 milligrams per kg/bw, depending on the species), or even more, the sorbitan derivatives will not cause harm. Such levels are many hundreds of times the average human intake of the compounds. Human infants have been given 4 grams daily of polysorbate 60 for over a month, and adults 6 grams daily, without ill effect. Humans have been fed 15 grams of polysorbate 80 daily for a month, again with no signs of any toxicity. When sorbitan monostearate was fed to infants at 1 gram per day, or to adults at 6 grams daily (approximately 100 milligrams per kg/bw), there were no adverse effects.

Studies with rats fed polysorbate 60 at 10 grams per kg/bw over most of the life span have failed to show any histological tissue changes, indicating lack of cancerous effect, and no effect on reproduction, lactation, or the offspring. Similar findings were noted with polysorbate 80 and with sorbitan monostearate.

ASSESSMENT: The sorbitan derivatives, the polysorbates and sorbitan monostearate, find important uses as emulsifying agents in foods. Dosages of these substances hundreds of times as great as are contained in the adult diet have been administered to humans without harm. They obviously are free of hazard to the consumer at their current levels in foods, or at levels far above these amounts.

RATING: S.

SORBITOL

Sorbitol is a sweet-tasting compound made by chemical addition of hydrogen to the simple sugar glucose (corn sugar). It is found as a normal constituent in many fruits and berries. Sorbitol has about half the sweetness of sucrose (cane sugar). Sorbitol also functions as a crystallization modifier to control hardening in soft sugar-based confections, and it serves to help maintain moisture, plasticity, or viscosity in certain foods. It is used in sugar-free candies and chewing gum for diabetics because its slow absorption into the bloodstream leads to little rise in blood sugar in the body. Sorbitol is not truly a sugar; it does not promote dental cavities and thus finds use in some chewing gums.

SAFETY: The 1975 estimate of the average intake of sorbitol per person was 231 milligrams, a little under 4 milligrams per kilogram body weight (kg/bw) for a person weighing 60 kg (132 lbs.).

Sorbitol is metabolized like sugars; the body converts it to fructose (fruit sugar), which is then utilized in normal energy-producing steps. Humans have consumed 10 grams (167 milligrams per kg/bw) of sorbitol daily for a month without any harm. Fifty grams acted on adults as a laxative, but not 25 grams. Infants and children consuming 500 milligrams per kg/bw have had diarrhea but no other effects. Studies throughout the life span of rats showed no evidence of adverse effects on growth, reproduction, or lactation when fed at 5 grams per kg/bw daily. There were no tissue abnormalities in the organs, suggesting lack of any carcinogenic effect, and no gene-altering response was noted. Similarly, no malformed offspring were seen when pregnant hamsters, rats, and mice received 1200 to 1600 milligrams per kg/bw daily for five to ten days.

MAJOR REFERENCE: Evaluation of the health aspects of sorbitol as a food ingredient. FASEB/SCOGS Report 9 (NTIS PB 221-951) 1973.

ASSESSMENT: Sorbitol has been used in foods by humans for many decades, with no indication that such use has had adverse effects. It can act as a laxative at levels that are considerably above normal intake. The substance is readily metabolized by the body, and as a food additive appears to pose no hazard to the consumer at current levels in the diet, or at levels that may be expected in future years.

RATING: S.

SOY PROTEIN ISOLATE (SOY PROTEIN CONCENTRATE)

Textured Vegetable Protein (TVP, Textured Soy Protein)

WHEY PROTEIN CONCENTRATE

Soybeans constitute a major agricultural crop in the U.S. that, in the past, has been grown primarily for edible oil and animal feed. Increasingly in recent years, soybean flakes, the residue after the extraction of the oil, have been further processed to separate *soy protein isolate,* a defatted protein which food manufacturers have found useful in a number of foods for human consumption. (A similar product is obtained from whey:* *whey protein concentrate.*) An alkaline extraction procedure is applied to the flakes to produce a liquor, which is then neutralized by an acid. The liquor is further processed by filtering or centrifugal force to obtain a concentrated protein isolate solution. Currently the practice is to spray-dry the solution in equipment heated by direct-fire burners. *Textured vegetable protein* (TVP) is produced by a spinning process or cooking under high pressure, which makes it fibrous, giving it a texture that better simulates meat.

Soy protein isolates are employed to provide the protein in commercially prepared milk-free formulas for the estimated 10 percent of the infants in the U.S. who cannot digest milk. As spun fiber, they represent 20 percent or more of the content of fabricated meat substitutes. A wide range of manufactured food products contain substantial amounts, among them snack

---

MAJOR REFERENCE: Evaluation of the health aspects of soy protein isolates as food ingredients. FASEB/SCOGS Report 101 (NTIS PB 300-717) 1979.

*For whey, the liquid portion of milk after separation from casein, see p. 677.

foods, gravies and sauces, seasonings and flavors, breakfast cereals, and dairy products like frozen desserts.

SAFETY: It was estimated in 1970 that the per capita daily intake in the U.S. of soy protein isolates was 150 milligrams, or 2.5 milligrams per kilogram of body weight (kg/bw) for a person weighing 60 kg (132 lbs.). Considering increasing use as replacements for animal proteins, this intake is likely to have increased materially by now. Infants who receive formulas in which the protein is supplied by these isolates may be consuming well over 1000 times the per capita amount when adjusted for body weight.

The nutritional and biologic effects of soy protein isolates differ somewhat from those of other proteins. Because of the alkali-processing procedures employed in their extraction, there is a loss of certain nutritionally essential amino acids, and the formation of lysinoalanine (LAL) as a component of the protein molecules. LAL interferes with the body's ability to metabolize certain nutrients. If these isolates were the sole source of protein in the diet, phosphorus would be utilized poorly, certain minerals (especially zinc) would be insufficiently available, and the requirements would be increased for vitamins E, K, D, and B_{12}. It is likely, however, that when the protein isolate is added to food it will be supplemented by sulfur-bearing amino acids like methionine, accompanied by the needed minerals and vitamins. The protein quality of soy isolate formulas for infants has been found to be very high and similar to milk.

A number of investigators have demonstrated that high levels of alkali-modified soy protein isolate when present as the sole source in diets of rats can cause cytomegalic inclusion disease (growth of large, partially lifeless cells) in kidneys. An LAL-induced nutritional deficiency is believed responsible. In rat diets, lesser levels of LAL than were contained in these experiments failed to produce this effect. Sensitivity to LAL appears to differ among animal species. It did not harm mice, hamsters, rabbits, quail, and rhesus monkeys, nor have these symptoms been reported in humans. LAL has been detected at low levels in samplings of soy protein isolates used in food, including the spun fiber variety employed in meat substitutes.

Soy protein isolates may contain up to 50 ppm (parts per million) of nitrite. The nitrite appears to be formed during spray drying when the product comes into contact with combustion gases. It is estimated as contributing 0.04 milligram per kg/bw in the daily diets of maximum consumers such as vegetarians who eat substitutes prepared from spun protein isolates, and 0.03 milligram per kg/bw daily to infants subsisting on formulas based on these proteins. Ingestion by others of nitrite from these protein substitutes should be materially less, since they are less likely to intentionally select them.

Nitrite can react with other nitrogen-containing compounds in foods, drugs, and other substances to produce nitrosamines, which have been shown to cause cancer in animals (see p. 631).

Soy protein isolate when fed at high levels to pregnant rats did not adversely affect their offspring. Human infants who received all their protein from soy protein isolate formulas for six months did not differ from others who subsisted on milk-based formulas. The same was true of adults who consumed their protein as soy protein isolate for 24 weeks. Tests to determine whether soy protein isolate can cause cancer have not been located.

ASSESSMENT: Consumption of unsupplemented soy protein isolate as a sole protein source in a human's diet could contribute to nutritional deficiencies and serious illness, but this is unlikely to happen. Whenever the substance is added to food, it usually is supplemented with needed amino acids, minerals and vitamins, and other and adequate proteins, and many studies with humans have confirmed it to be safe in these circumstances.

The adverse effect of lysinoalanine (LAL) on rat kidneys required far greater amounts than the levels detected in soy protein isolates that are used in food; lesser amounts, still far in excess of current human intake, failed to cause harm. Adverse effects did not occur in other animal species even at greater dosages, nor have they been reported in humans. Thus the low levels of LAL present in soy protein isolates are not regarded at present to be of concern to health.

The amount of nitrite contributed to the diet is minimal compared to the total intake from all sources of food and the nitrite that is produced in the body during digestion. About 10 percent of infants cannot tolerate milk and depend on soy protein isolate-based formulas for their protein, and the substantial benefits gained warrant whatever slight risk may be involved. Studies on the safety of whey protein concentrate have not been located, but because of close similarity to lactalbumin (p. 465) the findings for that additive will be applicable.

Since the concentrate usually is prepared by a heat treatment, there is minimal likelihood of allergy from its use.

RATING: S.

STANNOUS CHLORIDE

This compound, sometimes called tin salt, is obtained by the action of hydrochloric acid on tin. Tin is widely distributed in nature and is present in foods like fish and to a lesser extent in fresh vegetables. Humans accumulate tin in differing degrees in bone, liver, heart, stomach, muscles, and blood. It is reported to be nutritionally essential for growth in rats, but there is no direct evidence that this is true for man.

Stannous chloride is permitted in food as a chemical preservative, and also for color retention. It is used in processed vegetables such as asparagus, wax beans, and sauerkraut and also in processed fruits and nonalcoholic beverages.

SAFETY: It has been estimated that in 1976, based on industry's reported usage, 0.25 milligram of stannous chloride was added to food consumed in the diet of the average U.S. adult. An analysis of the amount of tin in a one-day institutional diet for a person determined that it contained 3.6 milligrams of tin, which provides a perspective of the relative contributions from food and from additives.

A study of anemia in rats determined that the no-adverse-effect level of tin salts was 22 milligrams per kilogram of body weight (kg/bw) if the diet contained sufficient iron. This is well over 350 times the amount found in the institutional diet reported above, when adjusted for body weight. Rats and mice receiving 5 parts per million of stannous chloride in their drinking water over their life span did not suffer any toxic effects, nor did it affect their growth rate or longevity.

The few reported observations concerning stannous chloride and elemental tin as possible causes of cancer have been negative, with the exception of one study where three malignant tumors were found in 30 mice in a feeding experiment employing a diet containing 2 percent sodium chlorostannate,

---

MAJOR REFERENCE: Evaluation of the health aspects of stannous chloride as a food ingredient. FASEB/SCOGS Report 31 (NTIS PB 254-531) 1974.

a compound containing tin. However, the investigators regarded this as probably without significance because of the small numbers involved. Tumors did not occur when the sodium chlorostannate was reduced to 1 percent. Tin, in fact, may deter cancer; fewer malignancies occurred among mice fed 0.5 percent of tin as sodium chlorostannate or sodium stannous oleate than among controls that received neither substance. Unlike some metals which caused cancerous tumors when implanted under the skin of rats, tinfoil did not.

Stannous chloride fed by tube to pregnant mice, rats, and hamsters in dosages up to 50 milligrams per kg/bw did not affect fertility or maternal or fetal survival. Tests in laboratories indicate that it is unlikely to cause defective young.

ASSESSMENT: Feeding studies of stannous chloride conducted among a number of animal species, and including observations on cancer, maternal and fetal survival and health, at dosages far greater than the amounts present in the daily diet of humans, indicate that this compound is not harmful.

RATING: S.

STARCH

GELATINIZED STARCH; MODIFIED STARCH

Starches from various sources may be used as food ingredients. They may come from tapioca, potatoes, or several of the cereals, notably corn, wheat, and rice. Starch is a carbohydrate polymer—a condensation of many glucose molecules (a sugar; see p. 518) into long chains. It is produced commercially by steeping and grinding the seed or tuber of starchy plants to produce a wet, thin mixture, or slurry. Sulfur dioxide (see p. 650) is added to help separate the protein from the starch granules and to improve the color, and the unmodified starch is settled out and separated. When starch and water are heated, the granules swell and burst; the result is a *gelatinized starch,* which is much easier to disperse in water and which readily forms a thick gel. The term *"modified starch"* does not refer

MAJOR REFERENCE: Evaluation of the health aspects of starch and modified starches as food ingredients. FASEB/SCOGS Report 115 (NTIS PB 80-128804) 1979.

to this physical change, but rather to a chemical modification, which may involve bleaching and oxidation (combining with oxygen), partial splitting of the molecules with the aid of acids, or cross-linking or bridging (combining) of the starch chains with various substances to produce starch acetate, starch succinates (see p. 648), starch phosphate, and other substances. The various starch sources and the physical and chemical modifications provide the food processor with starches with many different properties in terms of clarity, tolerance of acidic conditions, permanency of gel, and even blandness of flavor.

Starches are used in foods as thickening and gelling agents, to increase viscosity or thickness of a solution, to help keep ingredients from separating, to prevent caking of foods such as powdered sugar or baking powder, and even as dusting powders to prevent sticking of a food such as bread dough.

SAFETY: Starch is a major component of cereals and many vegetables. The average U.S. diet provides about 180 grams per person daily. By contrast, the use of unmodified starch as a food additive contributed less than 1 gram per person daily in 1971. About an equal amount of chemically modified starches was added to the food supply, a daily consumption averaging about 17 milligrams of modified starch per kilogram of body weight (kg/bw) per person. A resurvey of the food industry in 1976 indicated about the same.

The use of modified starches in infant foods has received special attention. In 1977 a food consumption survey found that somewhat less than two-thirds of infant foods contained modified starch. The survey indicated an overall daily average per infant of about 3 grams, perhaps 400 milligrams per kg/bw of modified starch. The maximum daily intake for infants up to three months of age was 3.1 grams, and at eight months was 15.8 grams.

The source of the starch and the type of modification rarely are identified on food labels, since this is not required by the FDA; occasionally the starch source is noted. It is known that most of the starch used comes from corn. Five forms of modified starch account for most of the usage in foods: bleached starch, acetylated distarch adipate, distarch phosphate, acetylated distarch phosphate, and hydroxypropyl distarch phosphate. The latter three forms are the ones commonly used in baby foods.

Starch is a food, and as such is digested, metabolized, and

used as an energy source, as are most carbohydrates. Even in infants, who have lower amounts of the digestive enzymes during the first few months, the digestibility of the different unmodified starches is complete. The starch is split by the enzymes to its constituent glucose molecules, and these are readily absorbed and metabolized. Essentially all of the starch present in the diet is gelatinized by cooking before being eaten. Some raw starches, especially ungelatinized potato starch, are more difficult for the enzymes to digest and have led to lower weight gains in experimental animals, but no other adverse effects were seen. Some humans have shown a compulsive eating of raw starch, as much as 2 pounds of laundry starch being ingested daily; they frequently are obese and anemic.

Perhaps two dozen different types of modified starches have been studied with various kinds of experiments designed to assess their safety. A few have been studied with short-term tests involving humans (showing no adverse effect when fed at 1 gram per kg/bw, some 60 times the average daily intake of modified starch in processed foods), pigs, and dogs. Most of the studies used rats and included a number of two-year feeding tests in multigeneration experiments. Some short-term tests involved diets incorporating the modified starch at 4 grams per kg/bw, but others ranged as high as 100 grams per kg/bw.

None of the animal studies showed any adverse effects of the modified starch on reproduction, the offspring, or on tumor development. With a single exception, digestibility and metabolism seemed normal. Only four effects of any significance were noted in all of these studies. Some diarrhea was observed for a very few of the modified starches; these instances were at feeding levels of 20 grams per kg/bw or higher. A few modified starches resulted in a slightly lower rate of growth when present in the diet at a level of 15 to 60 grams per kg/bw. While most of the starches were readily and completely digested, hydroxypropyl starch was less completely digested by rats. Finally, there was some evidence of greater calcium deposition in the kidney tubules when rats were fed high levels of several of the modified starches.

ASSESSMENT: Starch is present as a major carbohydrate and calorie source in the diet. All of the evidence confirms the safety of unmodified or gelatinized starch added to processed foods. Chemically modified starches come from a variety of food starch sources and are treated with many different chemicals to accentuate a desirable physical property of the additive. Some diarrhea, some slower growth, and some calcium deposition in kidney tubules have occurred with a few of these modified starches. It is uncertain whether these effects have any significant bearing on safety, and they have resulted from feeding at levels of 20 grams per kg/bw or more; the total modified starch intake in human diets would appear to average only about 17 milligrams per kg/bw daily, and this is distributed among a number of forms of modified starch rather than a single type. The poor digestibility of one modified starch, hydroxypropyl starch, causes some possible concern; however, this modified starch is not at present used in baby foods.

No modified starch that has been tested has shown any adverse effects on fertility, and none appears to have caused cancer, allergic sensitivity, abnormal offspring, or altered genes. While there are no suggestions of any adverse symptoms at the moderate levels of modified starches present in processed foods, there are some unanswered questions about whether any could pose even a slight hazard to the consumer. Several have not been specifically studied in long-term feeding tests, or for such effects as cancer or teratogenicity; the significance of the occasional calcium deposition in the kidney remains questionable.

RATING: S for unmodified or gelatinized starch; ? for modified starches (as further long-term studies are needed to resolve questions of possible hazard).

STERCULIA GUM (GUM KADAYA, GUM KARAYA, INDIA GUM, INDIAN TRAGACANTH)

Sterculia gum (also referred to by the other names above) is the dried discharge, obtained by drilling or tapping, of various species of *Sterculia,* a tree native to central and eastern India, and also found in Africa, Australia, China, and Indochina. The gum is a complex carbohydrate (a condensation of a number of simple sugars); on aging or heating it may release a high acetyl content (acetic acid such as is found in vinegar, see p. 463).

MAJOR REFERENCE: Evaluation of the health aspects of sterculia gum as a food ingredient. FASEB/SCOGS Report 5 (NTIS PB 234-905) 1973.

Sterculia gum absorbs water readily, forming viscous, semifluid mixtures at low concentrations, and is used in milk and frozen dairy products, meat products, soft candy, and nonalcoholic beverages to blend ingredients and keep them from separating.

SAFETY: The intake of sterculia gum as an additive in the daily diet of a U.S. individual, based on reported usage by the food industry in 1975, averaged less than 2 milligrams. Research and clinical experience reported on the biological effects of this substance are not extensive, but do permit a number of observations. Allergic reactions to sterculia gum have been documented in humans, both by ingestion and surface contact, with symptoms such as hay fever, asthma, dermatitis, and gastrointestinal distress, and caution has been expressed about its indiscriminate use as a laxative.

When fed to rats at 3 grams daily for a week, a dosage 100,000 times as great as human intake (adjusted for body weight), these animals appeared to be bloated and their intestinal weights were substantially greater than those of the controls. An even larger quantity, 5 grams, was fed to dogs for 30 days, and while their feces showed increased bulk and moisture, no irritating effect was observed. The laxative effect of sterculia gum was studied by oral administration of about 7 grams daily to 89 human subjects over a four-week period. Because of its water-absorbing qualities, the gum increased the bulk and moisture of the stools, but the investigators considered it to be a useful and harmless laxative.

Tests conducted with pregnant mice, rats, and hamsters fed 6000 to 34,000 times sterculia gum's presence in the human diet (adjusted for body weight) have shown that it does not cause physical defects in the embryo. However, it has been reported that the feeding of dosages equivalent to 29,000 times human intake did result in the death of a significant number of pregnant mice (but not rats or hamsters).

According to findings of laboratory tests, sterculia gum does not produce gene mutation. Studies have not been conducted to determine whether it can cause cancer.

ASSESSMENT: Since sterculia gum can cause an allergic response, people sensitive to such allergens should avoid this substance. The fatalities among pregnant mice when force-fed sterculia gum occurred at levels so far above human intake that a health hazard is unlikely.

RATING: S for most people; X for anyone who is hyperallergic.

SUCCINIC ACID

Succinic acid is present in meats, cheese, and many vegetables such as asparagus, beets, broccoli, and rhubarb (which have distinct flavors that may be related to it), since it is a key compound in the metabolic conversion of carbohydrates in plants and animals. It is used as an additive to foods to impart a distinct acid taste and is usually manufactured for this purpose by chemical addition of hydrogen to synthetic maleic or fumaric acid (see p. 550).

SAFETY: The daily intake of succinic acid added to foods in 1976 was less than 0.025 milligram per person. By contrast, up to 50 grams have been given daily in studies of diabetic patients. In normal animals or humans, succinic acid when ingested is converted to glucose for energy or stored as glycogen in the body for future use. Studies have shown that succinic acid does not adversely affect normal embryo development. It has not been investigated as a possible cause of cancer or genetic changes.

ASSESSMENT: Succinic acid occurs naturally and is present in foods, and in the body, in far greater amounts than what is added to processed foods. It is tolerated without hazard in amounts thousands of times those used in foods or likely to be used in the future.

RATING: S.

MAJOR REFERENCE: Evaluation of the health aspects of succinic acid as a food ingredient. FASEB/SCOGS Report 53 (NTIS PB 254-541) 1975.

SUCROSE (CANE OR BEET SUGAR)

Sucrose is common table sugar. A carbohydrate (like starch and cellulose), it consists of two simple sugars, glucose and fructose (see p. 518). Sugar beets and sugar cane are the sources for sucrose used by food processors, and it is found in abun-

---

MAJOR REFERENCE: Evaluation of the health aspects of sucrose as a food ingredient. FASEB/SCOGS Report 69 (NTIS PB 262-668) 1976.

dance in many other plants. As an additive, it is used as a sweetener in foods and beverages.

SAFETY: In 1975 refined sucrose accounted for approximately 89 of the 120 pounds of sweeteners used annually per person, which means an average daily consumption of this sugar of 110 grams per person, or a quarter pound.

Sucrose is split by enzymes in the digestive tract to its constituent sugars, which are then absorbed and readily metabolized and used as sources of energy. High levels of sucrose in the diet can increase the plasma triglyceride (neutral fat*) levels of human subjects, but this appears to be transitory. The more common levels of sucrose in the diet of normal individuals do not have any greater effect on blood lipids than do other carbohydrates, such as starch (see p. 643). While elevation of plasma triglycerides has been associated with increased susceptibility to coronary artery disease, the kind and amount of fat in the diet, which can raise the serum cholesterol level, is regarded as of greater significance. Carbohydrates consumed in the amounts contained in the normal diet are unlikely to be of importance here.

Sucrose is also linked by some with risk of diabetes, but the available evidence denies such a direct relationship. Since diabetes is a disorder characterized by the inability to properly metabolize glucose, carbohydrates in the diet have been suspect. However, all information at hand indicates that the cause is really calorie intake, regardless of source. Obesity can induce diabetes in genetically susceptible people, and being overweight has at times been linked with sucrose intake. But obesity occurs when more calories are consumed than expended, and sucrose in the food supply accounts for only about 15 percent of total calorie intake.

Sucrose as a replacement for starch in the diet has been found to affect the life span of some strains of animals susceptible to kidney disease. Although this may not apply to humans, these studies suggest that persons with kidney problems may take the precaution of avoiding excessive sucrose intake. Animal studies have shown that sucrose does not induce cancer or gene alteration, and has a very low toxicity.

Dental caries (cavities) is associated with most fermentable carbohydrates in the diet; it is most likely to be a problem

---

*The way in which lipids (constituents of cells and a source of body fuel) are stored, as in body fat.

when sweet foods that tend to be retained on the teeth are eaten frequently, and especially between meals. Sucrose often is used in sweet snack foods (such as sticky candies) and thus certainly contributes to dental health problems, particularly if oral hygiene (cleaning the teeth) is poor and the water consumed is not fluoridated.

ASSESSMENT: Sucrose is widely used to provide sweetness in food preparations and beverages. That this contributes to tooth decay is well documented. Too much sugar in the diet clearly can pose a hazard to diabetics, who cannot efficiently use the glucose as a source of energy. In addition, experimental evidence suggests that excessive sucrose may prove to be harmful to some individuals with kidney ailments. However, both diabetics and people with kidney ailments are likely to be under strict dietary controls and a caution in rating sucrose in food does not seem necessary.

RATING: S.*

SULFITING AGENTS

Potassium Bisulfite; Potassium Metabisulfite; Sodium Bisulfite; Sodium Metabisulfite; Sodium Sulfite; Sulfur Dioxide

Sulfites are sulfur-containing chemicals that can release sulfur dioxide. They are effective as sanitary agents for food containers and fermentation equipment. They are also used as preservatives to reduce or prevent spoilage by bacteria, to minimize browning and other discoloration of food during processing, storage, and distribution, and as inhibitors of undesir-

---

MAJOR REFERENCE: Evaluation of the health aspects of sulfiting agents as food ingredients. FASEB/SCOGS Report 15 (NTIS PB 265-508) 1977.

*Sucrose has not been rated here as hazardous because of its effect on cavities. While it does have a role in tooth decay, it does so only in combination with other factors, such as improper dental hygiene or in-between-meal snacks which contain fermentable carbohydrates or water that is not fluoridated.

The rating procedure followed here does not consider a substance hazardous if it is harmful only when it is abused by the consumer, as almost anything can be. However, the sugar content of some dry breakfast cereals and instant breakfast bars and powders is a sizable percentage of total content. Because these percentages are not readily available and because children consume much of these foods, sucrose can be regarded as of concern to health. As a caution, we have provided the percentage of sugar content in many of the brands in these categories that are listed in the inventory section of this book.

able microorganisms during fermentation. Sulfites are used in dehydration, freezing, and brining of fruits and vegetables; in fruit juices and purees, syrups, and condiments; and in winemaking.

SAFETY: Sulfur dioxide to some extent evaporates or transforms to sulfate (which is harmless) during processing and subsequent storage, and any subsequent preparation, such as cooking, at home. The amount remaining in food before it is eaten is likely to be considerably less than the quantity added originally, and when it is finally ingested, an enzyme (sulfite oxidase) transforms it rapidly to sulfate. (A rare hereditary deficiency of this enzyme in the liver has been reported in humans.)

The usual basis for calculating the margin of safety for daily human consumption of a substance is the level that caused no ill effects in animal experimentation. This level varies for sulfur dioxide, its lowest being 30 milligrams per kilogram of body weight (kg/bw). An expert panel in 1975 estimated the daily intake per person as 0.2 milligram per kg/bw for most people and not over 2.0 milligrams for excessive users of sulfited foods and beverages. This is 1/150 of the animal safety level for the bulk of consumers, but for excessive users the ratio is 1/15; note that frequently 1/100 is regarded as the desired margin of safety.

Sulfites can destroy thiamin (vitamin B_1), and they are not permitted by the FDA in meats or other foods that are known to be major sources of this vitamin. Sulfites are unlikely to cause sufficient destruction of thiamin to threaten a deficiency because they are present in only a fraction of the mixed foods eaten in the course of a day. Large dosages (70 milligrams per kg/bw) of sulfites have resulted in gastrointestinal irritation, evidenced by abdominal pains and vomiting in animals and humans. In an experiment, rats fed with sodium sulfite for over a year suffered a deficiency in vitamin E, but only when extremely large doses (500 milligrams per kg/bw) were used.

In a number of test-tube experiments sulfites caused mutations in microorganisms by altering the nucleic acids which store and transfer the genetic code. There is no evidence of this effect in studies conducted with living bodies, and it seems reasonable to assume that the rapid destruction of sulfur dioxide by the enzyme sulfite oxidase provides protection. Investigations indicate that the sulfites do not adversely affect reproduction or offspring or cause cancer.

When evaluating human exposure to sulfiting agents in food, it merits consideration that inhaling air polluted with sulfur from coal or petroleum products can add an additional burden of sulfur dioxide. It is not possible to estimate the amount of the chemical that is being inhaled.

ASSESSMENT: The sulfiting agents are rapidly transformed by oxygen to harmless sulfate in the body and are not likely to present a hazard to humans at the levels and manner in which they are currently used as additives in food and beverages, except for the very few who may be afflicted with a deficiency of the liver enzyme sulfite oxidase. In the event of a significant increase in the use of these substances, this judgment would require reassessment, particularly for individuals who are excessive in their use of foods and beverages high in sulfur content.

RATING: S.

SULFURIC ACID

AMMONIUM, CALCIUM, POTASSIUM, SODIUM SULFATE

Sulfuric acid, which becomes a sulfate when added to food, and its salts *(ammonium, calcium, potassium,* and *sodium sulfate)* are used in food in a variety of ways. Calcium sulfate, besides qualifying as a nutrient supplement (calcium is essential to life), can act as a firming agent in tomato products and canned vegetables and as a dough conditioner to give desirable texture characteristics in bakery products and flours. Sulfuric acid may also be useful in a limited number of foods as an acidifier (to provide tartness and control acid-alkali balance).

SAFETY: The daily dietary intake of sulfuric acid and its sulfates as additives approximated 140 milligrams per person in 1976, according to a survey conducted by the National Academy of Sciences-National Research Council of usage by food processors. Calcium sulfate and sulfuric acid accounted for nearly all. Additionally, most foods normally contain sulfates among their mineral constituents. And when the body metabolizes protein, sulfate is formed from the sulfur which most proteins contain. Sulfate is incorporated into mucins (the

---

MAJOR REFERENCE: Evaluation of the health aspects of sulfuric acid and sulfates as food ingredients. FASEB/SCOGS Report 33 (NTIS PB 262-652) 1975.

slime of mucous membranes) and plays a role in detoxifying some products of metabolism that might otherwise be harmful in the body or more difficult to excrete.

From various studies with animals and humans, it is clear that the sulfates do not have adverse effects until the level is very much greater than the usual exposure in the daily diet. No tumors were found in mice or rabbits administered 300 to 600 times (respectively) the daily exposure of humans (adjusted for body weight). There appear to be no investigations of specific effects such as chromosome damage, birth defects, or allergy response to sulfates.

ASSESSMENT: Since sulfates are normally found in most foods and are the final products of the body's metabolism of sulfur-containing proteins, the body is regularly called on to handle these materials. The available evidence indicates that as additives, these sulfates pose no hazard to the consumer at levels presently used or likely to be expected in the future.

RATING: S.

TANNIC ACID

Tannin is naturally present in a wide distribution of plants; extracts of it are found in wine, tea, and coffee. The tannic acids permitted in food processing differ from other tannic acids in their tannin sources and chemical characteristics and in the effects they have on the body. Only food-grade tannic acids require evaluation here. These are usually derived from a powder found in nutgalls, growths that form on twigs of species of hardwoods such as oak, usually in temperate climates of western Asia and southern Europe, or from seed pods of Peruvian, Polynesian, or Australian palms and ferns. Tannic acid adds a desirable astringent taste to butter, caramel, fruit, maple, and nut flavorings. Its ability to filter out proteins makes it useful as a clarifier in the wine and brewing industry.

SAFETY: Based on reported usage in industry, per capita consumption of added tannic acid in the U.S. in 1976 was estimated at 9 milligrams daily, less than 0.2 milligram per kilogram of body weight (kg/bw) for a person weighing 60 kg (132 lbs.). It is questionable whether anything like this amount

---

MAJOR REFERENCE: Evaluation of the health aspects of tannic acid as a food ingredient. FASEB/SCOGS Report 48 (NTIS PB 274-669) 1977.

is actually consumed. When tannic acid is used as a filtering agent, and most of the acid probably is used in this way, good manufacturing practice requires its removal.

The extent and nature of absorption of tannic acid have not been established for man. In experimental animals, before absorption the acid has caused damage to mucous membranes of the stomach and intestines, but only when large, concentrated doses are fed by tube into empty stomachs or when exposure of the membranes to tannic acid is artificially prolonged. The method of administration apparently affects the toxicity of tannic acid: 1000 milligrams per kg/bw taken orally by rats did not disturb their liver function or do damage to the intestinal tract, whereas 60 milligrams per kg/bw injected into the abdominal cavity did damage their livers. The latter dosage is still 400 times human consumption (adjusted for body weight).

Long-term feeding studies reviewed by the UN Joint FAO/WHO Expert Committee on Food Additives indicate that tannic acid taken orally will not produce tumors or cancer. One study in which this substance was administered to rats by injection reported the development of liver tumors which were suggestive of malignancy; but other studies contest this finding.

Laboratory tests indicate that tannic acid does not cause gene mutations, nor does it adversely affect maternal or fetal survival, or cause birth abnormalities.

ASSESSMENT: The total intake in the U.S. diet of tannic acid as an additive is likely to be small. While food-grade tannic acid in concentrated doses prior to absorption can cause damage to the mucous membranes of the gastrointestinal tract, this effect is unlikely from the minute amounts (100 parts per million) of this substance when present in food. The adverse effect on livers of rats, resulting from injection of doses of tannic acid 400 times human consumption, cannot be regarded as relevant to human safety because of the amount and the method of administration as well as evidence that when taken orally, a far greater dosage did not cause harm.

RATING: S.

TARTARIC ACID

POTASSIUM ACID TARTRATE (CREAM OF TARTAR; POTASSIUM BITARTRATE)

Tartaric acid is a normal constituent of grapes and wines, fruits, coffee, and even sugar cane juice. *Potassium acid tartrate* is one of its salts. For commercial purposes, these are obtained from the waste products of wine manufacture. Both substances are added to augment fruit flavors in beverages and candies. Tartaric acid is used as a stabilizing agent in some foods to prevent discoloration or flavor changes that occur from rancidity. As an ingredient in baking powders, potassium acid tartrate (cream of tartar) prevents fermentation in baked goods.

SAFETY: As food additives, the average consumption of tartaric acid and potassium acid tartrate combined has been estimated at 2 milligrams per person in the U.S. in 1976, or 0.03 milligram per kilogram of body weight (kg/bw) for a person weighing 60 kg (132 lbs.), according to a survey conducted by the National Academy of Sciences-National Research Council of usage by food processors.

There were no ill effects in rabbits fed a daily diet containing 2300 milligrams per kg/bw for 60 to 150 days, nor did rats suffer measurable adverse changes after a two-year feeding of a diet containing 1200 milligrams per kg/bw daily. Humans have been fed dosages ranging from 3500 to 11,500 milligrams, the former amount for as long as nine consecutive days, without injury. When tartrates were ingested by humans, it was found that they can be absorbed but may be only minimally metabolized by the body; they are either excreted in the urine or destroyed by bacteria in the intestinal tract.

Experiments on a variety of animal species failed to disclose any indication that tartaric acid could cause birth defects, nor did laboratory tests provide evidence of gene mutation.

ASSESSMENT: Many studies with humans and with experimental animals have demonstrated that tartaric acid and potassium acid tartrate pose no hazard to the public when used at current levels or amounts that may be expected in future years.

RATING: S.

MAJOR REFERENCE: Evaluation of the health aspects of tartaric acid and tartrates as food ingredients. FASEB/SCOGS Report 107 (NTIS PB 301-403) 1979.

TBHQ (TERTIARY BUTYLHYDROQUINONE)

TBHQ is a chemical used as an antioxidant to prevent uptake of oxygen which can cause deteriorating flavors and odors (rancidity) in oils, fatty foods, or even low-fat products incorporating some unsaturated fat. It may be used in combination with BHA or BHT (see p. 489), and its chemical structure resembles that of BHA. If a combination is used, the total antioxidant is still limited to no more than 0.02 percent of the food product. TBHQ has proved to be very effective in polyunsaturated vegetable oils such as safflower, soybean, and cottonseed oils.

SAFETY: The 1976 survey of the food industry usage conducted by the National Academy of Sciences-National Research Council indicated that the average daily intake per person of TBHQ in foods was 0.5 milligram, less than 0.01 milligram per kilogram of body weight (kg/bw) for a person weighing 60 kg (132 lbs.). Heavy users of foods containing it may consume five times this amount.

A number of short- and long-term studies of the safety of TBHQ have been made on rats and dogs. When dogs were fed for two years on diets containing 0.5 percent of TBHQ (200 milligrams per kg/bw), all evidence indicated no effect on behavior, appearance, growth, biochemical studies of blood and urine, organ weights, gross pathology, or microscopic structure of the cells. The only effect of this high dosage was a slight change in some of the blood picture, principally a slightly lower red-blood-cell count. There was no toxicological effect at the lower level of 75 milligrams per kg/bw, equivalent to thousands of times the presence of TBHQ in the daily diet of U.S. humans.

A 20-month study with rats, again with a diet of 0.5 percent of TBHQ (250 milligrams per kg/bw, or 25,000 times the average intake by humans) disclosed no effects of the compound. There was no evidence of any significant storage of TBHQ in the tissues of the body. The animals were carried through three successive generations, and very detailed studies were made. Reproduction efficiency and litter size were normal, but some increase in early mortality of the offspring was noted, though

---

MAJOR REFERENCE: Toxicological evaluation of some food colours, thickening agents, and certain other substances. WHO Food Additives Series No. 8 (Geneva, 1975).

there was no increase in any abnormalities. TBHQ did not appear to cause birth defects.

ASSESSMENT: TBHQ is effective as an antioxidant preventing flavor changes in food fats. Feeding studies with dogs and rats have demonstrated its safety when TBHQ is included at levels thousands of times as great as its presence in the human diet. The UN Joint FAO/WHO Expert Committee on Food Additives considers 0.5 milligram per kg/bw safe and acceptable as a daily intake for humans (50 times the amount actually in the U.S. diet). There appears to be no hazard to the consumer from use of TBHQ as a food-preserving chemical, even at levels that may be expected in the future.

RATING: S.

TOCOPHEROLS (VITAMIN E)

Alpha-Tocopherol; Alpha-Tocopherol Acetate

The tocopherols are present naturally in a wide range of plant and animal tissue; they are either produced synthetically or isolated from cereal and vegetable oils for use in food. *Alpha-tocopherol* contains the most potent type of vitamin E, which is needed for normal muscular development and organic processes. It also prevents destruction in the body by oxygen of some essential fatty substances (required for cell repair and replacement and for immediate energy), including vitamin A and certain hormones. As additives in food, the tocopherols act as antioxidants, preventing or retarding flavor deterioration and rancidity caused by oxygen. Because vitamin E is an essential nutrient, they also are used in food fortification and as a vitamin supplement. Synthetic *alpha-tocopherol acetate* is the usual form of vitamin E used.

SAFETY: The daily intake of tocopherols contained in food in the diet of U.S. individuals was estimated as averaging 7 to 9 milligrams, much of it present naturally. About 3.5 milligrams were added to food in 1975. Not all the vitamin E is available for absorption in the body at the time it is eaten, as some is

---

MAJOR REFERENCE: Evaluation of the health aspects of the tocopherols and α-tocopherol acetate as food ingredients. FASEB/SCOGS Report 36 (NTIS PB 262-653) 1976.

destroyed in cooking and even more during commercial processing and storage. The UN Joint FAO/WHO Expert Committee on Food Additives, basing its judgment on human clinical experience, concluded that an acceptable daily intake (ADI) of alpha-tocopherol for man should not be more than 2 milligrams per kilogram of body weight. For an adult weighing 60 kilograms (132 lbs.), this would be 120 milligrams, or many times as great as the daily consumption reported above.

Investigators working with animals document adverse effects on growth, development, and metabolism caused by the administration of vitamin E orally or ingested in the diet. But this happened only when the levels tested far exceeded human intake, and the deviations were eliminated when dosing with this vitamin was stopped. A variety of animal species can tolerate oral dosages 100 times the FAO/WHO ADI for humans (adjusted for body weight). Hypervitaminosis E (an excess of this vitamin), which can affect the liver, heart, and reproductive cycle, is only a possibility when over 400 milligrams per day are ingested by an individual over an extended period of time, and the evidence is not convincing that it will occur even then. But neither have benefits been demonstrated, as some claim, from the addition of large amounts of vitamin E to the diet. Research on the possibility of tocopherols causing birth defects, gene mutations, or cancer has not revealed such alterations.

ASSESSMENT: Alpha-tocopherol is a prime source of vitamin E, an essential constituent for maintenance of a number of the body's normal biological and organic functions. As an additive, it contributes possibly a tenth of the total tocopherol content ingested in the daily diet, for it is also present naturally in cereals, vegetables, and other foods. Experimental and clinical evidence indicates that in quantities present in the diet, both naturally and added, alpha-tocopherol is not harmful and actually is needed. Any hazard from it appears possible only in the event of excessive doses taken over long periods.

RATING: S.

VANILLIN

ETHYL VANILLIN

Vanillin, the component in the vanilla bean that provides the aroma and flavor, is naturally present in many foods. Other than in vanilla bean extracts, a synthetic chemical substitute, *ethyl vanillin,* may be used, since its flavor is many times stronger than that of vanillin. When natural vanilla extracts are fortified with additional vanillin or ethyl vanillin, they are labeled artificial or imitation. Vanillin and ethyl vanillin are used as flavoring agents in a wide range of products, which include ice cream, candy, soft drinks, gelatin desserts, toppings and frostings, butter, margarine, and chocolate products.

SAFETY: Use of the two flavor ingredients by the U.S. food industry in 1970 indicated an average daily intake per person of 11 milligrams of vanillin and a little under 3 milligrams of ethyl vanillin. Together, they total less than 0.2 milligram daily per kilogram of body weight (kg/bw) in the diet of a person weighing 60 kg (132 lbs.).

Humans and experimental animals normally metabolize vanillin by converting most of it to vanillic acid, by combining it with oxygen or reducing it to an alcohol and eliminating both in the urine. One human subject consumed 100 milligrams of vanillin, and during the next 24 hours excreted 94 percent of it as vanillic acid. Metabolic changes of ethyl vanillin do not appear to have been reported.

Vanillin added to the diet of rats for three months at 500 milligrams per kg/bw was reported by one research group to produce some "mild toxic symptoms." This level is comparable to over 2000 times the usual human intake; lower levels of 150 milligrams per kg/bw did not produce these effects at all. Other studies have reported feeding rats dosages of vanillin at 14,000 times human intake (adjusted for body weight) for one to two years without evidence of harm.

Again, while one group of investigators found that ethyl vanillin in the diet of rats at 64 milligrams per kg/bw caused some growth retardation and tissue changes after two months (still well over 1000 times human intake), others have fed ethyl vanillin to rats for one to two years at levels up to 60,000 times

---

MAJOR REFERENCE: Scientific literature review of vanillin and derivatives in flavor usage. FEMA Report to FDA (NTIS PB 285-495) 1978.

the average per capita consumption of this chemical by humans and reported no adverse effects.

Studies do not appear to have been conducted that evaluate the effects of these two additives on reproduction, or the genes or chromosomes, or as a possible cause of cancer.

ASSESSMENT: Vanillin and its synthetic substitute ethyl vanillin are popular flavor ingredients used in a variety of foods. Available data indicate that dietary intakes far above usual human consumption do not cause adverse effects in experimental animals. Current information suggests no hazard to the consumer from vanillin or ethyl vanillin as food additives.

RATING: S.

VEGETABLE OILS

Hydrogenated, Partially Hydrogenated, Unhydrogenated Coconut, Corn, Cottonseed, Olive, Palm, Palm Kernel, Peanut, Rapeseed, Safflower (High Linoleic, High Oleic), Sheanut, Soybean, Sunflower Oils

Many of the vegetable oils, which are fats, are used directly in foods such as margarine, shortening, various cooking and salad oils, and mayonnaise. Most are highly unsaturated fats, initially in liquid rather than semisolid form. Coconut and palm kernel oils are exceptions in that they contain mainly saturated fatty acids, yet retain the character of oils. A vegetable oil can be converted, or hardened, by the chemical addition of hydrogen to the oil, and called partially hydrogenated or hydrogenated. When fully hydrogenated, the oil is converted to a semisolid fat, making it more stable and less subject to rancidity or other flavor changes. Coconut or palm kernel oil when hydrogenated can produce fats that substitute for cocoa butter in some foods.

The most common of the partially hydrogenated vegetable oils used in foods, soybean oil, is widely used in salad dress-

---

MAJOR REFERENCES: Evaluation of the health aspects of coconut oil, peanut oil, and oleic acid as they may migrate to food from packaging materials, and linoleic acid as a food ingredient. FASEB/SCOGS Report 65 (NTIS PB 274-475) 1977; Evaluation of the health aspects of hydrogenated soybean oil as a food ingredient. FASEB/SCOGS Report 70 (NTIS PB 266-280) 1977.

ings, shortenings, and margarines. The oil is usually extracted from soybeans with hexane, a solvent from petroleum. It is then purified, and hydrogen is chemically incorporated into some (but not all) of the polyunsaturated fatty acid molecules;* this improves the flavor and stability of the oil and produces some texture changes such as viscosity. Hydrogenation changes these fatty acids, increasing saturation and altering some of their components into "unnatural" arrangements, which may affect their nutritional value and biological effect.

Soybean oil usually is only partially hydrogenated, the extent depending on its intended use. Less liquid and more plastic compositions of soybean oil needed for shortenings, for example, can be obtained by hydrogenating to a greater degree (which increases saturation); and the hardened product may also be blended with liquid vegetable oils to get a desired consistency and fatty acid makeup, especially when a higher content of polyunsaturated fatty acids is sought.

Other processed or treated oils may include any of the common vegetable oils, notably coconut, corn, cottonseed, palm, palm kernel, peanut, rape, safflower, and sunflower oils. They may be found as additives in food products such as crackers, cookies, margarines, frozen yogurt pies, and potato chips. A less common oil is sheanut oil. It comes from a tropical African tree seed; its fatty acid makeup is largely equal parts of oleic and stearic acids,† and only small amounts of polyunsaturated fatty acids, a composition of fatty acids resembling some shortenings.

Safflower oil is somewhat different in that plant breeders have separated the plant into two varieties: one with an oil high in polyunsaturated linoleic acid, about three-fourths of the fatty acids being linoleic; and another that is high in oleic acid, about three-fourths of the fatty acids being oleic acid, and only 14 percent being linoleic acid. In composition, high-oleic safflower oil resembles olive oil (rather than the polyunsaturated vegetable oils), which contains approximately 73 percent oleic acid and 8 percent polyunsaturated fatty acids and is relatively low in saturated fatty acids. Both types of safflower oil are used as cooking and salad oils. The high-linoleic oil also is used

---

*A fat or oil is made up of various fatty acids chemically linked with glycerin (see p. 563).

†Stearic acid is fully saturated—a fully hydrogenated fatty acid. Oleic acid has some unsaturated (monosaturated) linkages, and will convert to stearic acid if hydrogenated.

in polyunsaturated margarines. Olive oil is usually not hydrogenated before its use in foods.

Rapeseed oil as a food additive is fully hydrogenated and bleached. It can serve as a stabilizer and thickener or as an emulsifier in shortenings.

SAFETY: Food use of soybean oil has increased from an average of about 19 pounds per person in 1964 to 29 pounds in 1974. About half of this was probably partially hydrogenated soybean oil, which thus would account for a daily consumption of 19 grams per person. Coconut oil consumption in 1970 amounted to an average daily intake of slightly over 2 grams per person; palm kernel oil intake came to 0.5 gram per person, and peanut oil to about 1 gram.

Surveys of populations have demonstrated a relationship between intake of saturated fat and blood cholesterol level, which is regarded as a risk factor in coronary heart disease. Saturated fats in the diet may raise the cholesterol level in the blood. While partial hydrogenation does cause saturation of some of the polyunsaturated fat in vegetable oils, it represents only a fraction of the total. The saturated fat content in commercial shortening made with partially hydrogenated soybean oil ranges from 15 to 40 percent, and is 13 to 20 percent in soybean oil margarines. By contrast, butter contains over 65 percent saturated fatty acids.

Polyunsaturated fats, substituting for saturated fats in the diet, can lower elevated blood cholesterol levels, and in this way may reduce the risk of coronary heart attack. It has been shown that partially hydrogenated soybean oil, when substituted for substances with higher saturated fat content, also can lower these levels because of the unsaturated fatty acids they retain. While this is true for the majority of the vegetable oils, some, such as coconut oil and palm kernel oil, do not have appreciable amounts of polyunsaturated fatty acids and thus will not have their protective effect.

There are studies in humans showing that coconut oil in the diet may raise the serum cholesterol level, which could add to the risk of heart disease if the diet included a high intake of such an oil for a prolonged period. Olive oil appears to have little effect on cholesterol levels, probably because it has few of the cholesterol-elevating saturated fatty acids, but also few of the protective polyunsaturated fatty acids.

Many of the biological studies have focused on the "unnatural" forms of the fatty acids produced by the partial hydrogenation of unsaturated vegetable oil. These acids have somewhat different physical properties, but they are absorbed and readily digested and will appear in the fat stores in the body in a similar manner. The natural polyunsaturated linoleic acid is an essential fatty acid, one which the body cannot make itself and must obtain from the diet. It is needed for healthy tissues and cell membranes, regulation of cholesterol metabolism, and production of certain hormones.

The bulk of the linoleic acid present in the diet is in the natural form, which retains these biological capabilities. However, some forms (isomers) of linoleic acid produced through hydrogenation lose this essential fatty acid property, but when metabolized still retain their energy value for the body. When incorporated into the lipid (fats, waxes, etc.) component of the membranes in various cells they may somewhat alter the membrane properties, but the significance of this remains uncertain.

Feeding studies with rats receiving 54 percent of the diet as partially hydrogenated soybean oil for several months showed no abnormal effects when a variety of physical and metabolic measures were examined. This was also true in long-term studies followed through 60 generations. The tests showed no adverse effect on fertility and reproduction, nor did they show abnormal embryo growth, or alteration of the chromosomes, or cancer development.

One study on pregnant rats receiving 20 percent high-linoleic safflower oil in their diet found fewer completed gestations (completions of pregnancy), poorer lactation, and smaller brains in the offspring when compared with rats on a commercial ration. The study needs to be confirmed.

Rapeseed sometimes contains erucic acid, an unusual fatty acid that is toxic. Selective breeding has eliminated this compound in some sources of rapeseed oil; however, the oil used as a food additive is always completely hydrogenated to chemically eliminate any residue of erucic acid and to ensure its safety in foods.

Coconut oil has been studied; it does not induce allergies and does not enhance the ability of a known carcinogen to produce cancer. The refining procedures used in preparing the oil for food uses should remove any cancer-causing contaminants that may arise from smoke-drying of coconut meats.

Besides energy value and essential fatty acids, the other nutrient commonly found in a vegetable oil is alpha-tocopherol, or vitamin E (see p. 657). Coconut and palm kernel oils are low in this vitamin, but the other common vegetable oils contain from 11 to 45 milligrams per 100 grams. Most of the vitamin remains after the hydrogenation processing and incorporation into food products.

ASSESSMENT: The practice of partially hydrogenating a vegetable oil for various food uses converts some of the fatty acids to forms not normally present. In all but coconut and palm kernel oils, the polyunsaturated "essential fatty acid" content remains high, however, and the caloric value is unchanged. The polyunsaturated fatty acids that are present have a cholesterol-lowering effect when fed as a replacement for saturated fats or carbohydrates in the diet. No hazard to the consumer is posed by use of partially hydrogenated vegetable oils in the diet, other than coconut, palm kernel, and rapeseed oils, or if the vegetable oils are fully hydrogenated. These may pose some problems. Evidence indicates that a high degree of saturation of fat or oil is related to adverse effects on blood clotting, cholesterol deposits in arteries, and somewhat poorer fat digestibility.

The potential health problem that a saturated fat may pose to a consumer will depend on the amount consumed in the diet, not merely its presence in the food. The ingredient label can alert one to the presence of a highly saturated fat, but quantities are not given. However, since the ingredients are listed in descending order of their presence in the food, it is possible to estimate when the amount of a given oil may be of concern. Accordingly, a caution is raised in the Inventory of this volume about the safety of a food where one of the first three listed ingredients (other than water), alone or as part of a shortening blend containing other vegetable oils, is coconut oil, palm kernel oil, rapeseed oil, or when "hydrogenated" rather than "partially hydrogenated" precedes the named fat or oil ingredient; these are the highly saturated vegetable oils used in foods.*

---

*The accuracy of this procedure for identifying highly saturated ingredients was tested with laboratory analyses of fatty acids in over 200 purchased foods of all types contained in a 1980 thesis by Mary Gertrude Enig, conducted in collaboration with the USDA. In only five of the 200 foods would this rating procedure have been in error. Four foods having appreciable amounts of saturated fats or oils would have been missed, and one food would have been cited incorrectly as having an excess of highly saturated fats. This figure does not include two foods that were mislabeled, or peanut butter, which has been excluded from the procedure as it naturally contains a considerable amount of unsaturated oil.

RATING: S for all vegetable oils except coconut oil, palm kernel oil, rapeseed oil, hydrogenated oils; for these three oils, and for any oil labeled "hydrogenated" rather than "partially hydrogenated," ? if it appears as part of one of the first three ingredients on a food label.

VITAMIN A

Vitamin A Acetate; Vitamin A Palmitate

Vitamin A is essential for bone growth, normal eyesight, and protection against impaired cell membranes. It is a complex substance called retinol, and its chemical combination with acetic acid *(vitamin A acetate)* or palmitic acid *(vitamin A palmitate)* provides the compounds used as additives in food fortification. The active form of the chemical has been synthesized, and this is the product used in foods or dietary supplements.

Vitamin A, as distinct from its "provitamin," carotene (see p. 505), from which the body is able to form the vitamin, is found naturally in animal products such as cheese, eggs, cream, butter, oysters, and particularly in liver. It may be added as a nutritional supplement in breakfast cereals, milk, margarine, baby formulas, beverages, and other foods.

SAFETY: Vitamin A commonly has been expressed in units of activity rather than weight. One milligram of vitamin A palmitate equals 1820 International Units (IU) of activity; 1 milligram of vitamin A acetate equals 2940 IU. A survey of the vitamin A added to fortify foods in the diet from 1971 to 1974 indicated that these additives would provide an average daily intake per person of about 800 IU. Based on nationwide food consumption studies, the total dietary intake of vitamin A from all sources is probably 5000 IU, just meeting the Recommended Dietary Allowance to maintain good nutritional health for an adult.

Any excess of vitamin A is stored in the liver. Studies in the U.S. and Canada have indicated that human liver stores of vitamin A are minimal; but where dietary intake is appreciable, as it is in some foods such as the polar bear liver, which

---

MAJOR REFERENCE: Evaluation of the health aspects of vitamin A, vitamin A acetate, and vitamin A palmitate as food ingredients. FASEB/SCOGS Report 118 (NTIS PB 80-178650) 1980.

contains as much as 1 gram of vitamin A palmitate in 4 ounces, acute poisoning has occurred in humans. There is also danger from excessive intake of vitamin A, and even 50,000 IU daily in an adult, and less for a child, can lead to severe headaches, nausea, irritability, weakness, pain, hair loss, and other toxicity symptoms. The highest daily safe level appears to be about 700 IU per kilogram of body weight, or 42,000 IU for a person weighing 60 kg (132 lbs.). This intake is highly unlikely from the vitamin A present naturally or as an additive in most foods.

High levels of vitamin A (usually over 100 times the average human intake from food) have produced some adverse effects on reproduction among laboratory animals, and have caused congenital defects in offspring. The vitamin does not cause chromosome damage. There have been no studies to determine whether vitamin A can cause cancer; however, there is some evidence that it may reduce the risk of cancer deliberately induced by some chemical agents.

ASSESSMENT: Vitamin A and its acetate and palmitate compounds provide an important vitamin essential for humans. This vitamin is toxic at high levels, but these amounts are many times as high as would be supplied by foods. The fortification of foods with vitamin A derivatives appears to pose no hazard to the consumer at current levels or levels likely to be encountered in the future.

RATING: S.

VITAMIN B_1 (THIAMIN)

Thiamin Hydrochloride; Thiamin Mononitrate

Vitamin B_1, or thiamin, is widely distributed in plant and animal tissues, especially in pork products, liver, cereal grains, and legumes. It is essential for tissue growth and maintenance, and for many of the metabolic reactions in the body, in particular for converting carbohydrates to energy. Practically all the thiamin that is added to food is produced synthetically in two forms: *thiamin hydrochloride,* which is very soluble in water, and *thiamin mononitrate,* which is only moderately soluble in

---

MAJOR REFERENCE: Evaluation of the health aspects of thiamin, thiamin hydrochloride and thiamin mononitrate as food ingredients. FASEB/SCOGS Report 109 (NTIS PB 288-674) 1978.

water but is more stable than the hydrochloride and is preferred by the food industry for the enrichment of flour mixes. These compounds are added to various foods such as baked goods, cereals, and pasta products as nutritive and dietary supplements, since the milling of wheat (which lightens it and produces a finer flour texture) removes fractions of the wheat which contain a good deal of this vitamin.

SAFETY: The daily intake per person in the U.S. of added vitamin B_1 was estimated at 2 milligrams in 1975, based on reported usage by the food industry; 0.33 milligram of this was as thiamin hydrochloride. The Food and Nutrition Board of the National Academy of Sciences-National Research Council has suggested a Recommended Dietary Allowance (RDA) of 0.5 milligram of vitamin B_1 per 1000 calories of food intake (up to 1.5 milligrams for an adult male), but not less than 1 milligram for anyone consuming less than 2000 calories. A survey in 1971-74 determined that the mean intake in the U.S. was 0.64 milligram per 1000 calories, and that all population groups, according to age, sex, race, and income, exceeded the RDA.

Extreme deficiency of thiamin leads to the severe nerve and muscular disease beriberi. No comparable clinical condition in humans is associated with an excess of the vitamin, since it is readily excreted, and oral doses of several grams daily would be required for toxic symptoms to develop. There are infrequent cases of allergy to thiamin due to exposure during pharmaceutical handling and packaging of the vitamin, or to its use as a drug.

Thiamin hydrochloride or thiamin mononitrate fed to rats or mice at levels near 1000 times normal human intake (adjusted for body weight differences) over several generations failed to affect fertility, lactation, growth, or reproductive performance. Conflicting findings have been reported on the effect of thiamin on cancer development induced by chemical carcinogens; in some cases it retarded growth, in others accelerated it, and in another had no discernible effect. It did not cause malformations of the fetus in experiments with mice and rats, and had no adverse effects on chromosomes or genes in laboratory tests.

ASSESSMENT: Vitamin B_1 is essential for good nutrition, and is added to foods to assure an adequate intake. It is free of adverse effects at levels hundreds of times normal intake. This vitamin used as a food additive poses no health hazard to consumers even at levels that might be expected in future use.

RATING: S.

VITAMIN B_2

Riboflavin; Riboflavin-5′-Phosphate; Sodium Riboflavin Phosphate (Disodium Riboflavin Phosphate)

Riboflavin (vitamin B_2) is found in minute amounts in practically all animal and plant tissues and cells. Its phosphate ester, *riboflavin-5′-phosphate,* is more soluble in water, and is the form in which the vitamin is found in nature and stored in the body. Vitamin B_2 is a component of a group of important enzymes (catalysts) that enable utilization of oxygen in cell respiration and in the metabolism of foods into usable forms. An adequate supply is necessary for healthy skin and hair and good vision. Milk, liver and other organ meats, leafy vegetables, eggs, and yeast are the best sources, but little of the vitamin is present in many foods, and supplementation may be desirable by means of the synthetic riboflavin phosphate and its sodium salt *(sodium riboflavin phosphate),* produced for commercial use as a nutrient. It is added for this reason to such products as macaroni, noodles, bread, and beverages. The vitamin may be lost when exposed to light, or when cooking water is discarded. As riboflavin is yellow to orange-yellow in color, it may be used as a food coloring as well as an enrichment or fortification.

SAFETY: The daily intake of vitamin B_2 added to foods in 1975 has been estimated at a little over 1 milligram per person, almost wholly in the form of riboflavin. Possibly four times as much is consumed in vitamin preparations. The Recommended Dietary Allowance suggested by the National Academy of Sciences-National Research Council is 0.6 milligram per 1000 calories, or up to 1.7 milligrams for young adult men; women need additional intakes during pregnancy and lactation. A survey that attempted to assess the nutritional status of the U.S. population found that this amount was

---

MAJOR REFERENCE: Evaluation of the health aspects of riboflavin and riboflavin-5′-phosphate as food ingredients. FASEB/SCOGS Report 114 (NTIS PB 301-406) 1979.

exceeded in the diet of every age, sex, race, and income group; the average was 0.96 milligram per 1000 calories or, translated to individuals, a daily intake of almost 2 milligrams of vitamin B_2 from all food sources, including fortified foods.

A deficiency of vitamin B_2 leads to metabolic impairments and clinical changes, including a type of skin eczema and inflammation, defects in the cornea of the eye, and subnormal growth. Large doses of riboflavin can be tolerated by the body. Even 10 grams per kilogram of body weight taken orally produced no toxic effects in rats. Humans have been given riboflavin in doses of 4 grams daily for various medical conditions, again with no adverse effects.

Studies with rats followed over several generations indicate that riboflavin in substantial amounts has no adverse effect on fertility or reproductive performance. On the other hand, a deficiency is known to cause severe malformation of offspring. The influence of this vitamin on experimentally induced cancer in mice and rats appears uncertain: a deficiency of riboflavin in the diet in two instances retarded cancerous growth, in another it accelerated it. In one case of deficiency where the growth had been retarded, supplementation reversed the effect. This variability may be related to the chemical that initially caused the cancer. Riboflavin does not alter genes.

ASSESSMENT: Vitamin B_2 (riboflavin) is essential in the human diet. It is relatively nontoxic, and doses thousands of times normal intake have proved harmless to humans and experimental animals. No health hazard is indicated from the addition of this vitamin to foods at current levels or any that might be expected in the future.

RATING: S.

VITAMIN B_6

Pyridoxine; Pyridoxine Hydrochloride

Pyridoxine and two derivatives, pyridoxal and pyridoxamine, are collectively called vitamin B_6, which is essential to humans and must be obtained from the diet. It is present naturally in foods mostly as pyridoxal, and less so as pyridoxamine. Good

---

MAJOR REFERENCE: Evaluation of the health aspects of pyridoxine (vitamin B_6) and pyridoxine hydrochloride as food ingredients. FASEB/SCOGS Report 100 (NTIS PB 275-340) 1977.

sources for it include eggs, yeast, liver, kidney, meats, poultry, fish, whole grains, and legumes. When vitamin B_6 is added to food, it is in the form of a synthetic compound, *pyridoxine hydrochloride*. Vitamin B_6 plays a vital role as a coenzyme in combination with more than 60 enzymes (see p. 527) in the metabolic processes that break down ingested food into usable forms in the body, and it is particularly important in the metabolism of amino acids and proteins. It is needed in the formation of hemoglobin in the red blood cells. Pyridoxine hydrochloride is added as a nutrient or diet supplement in prepared baby foods, breakfast cereals, baked goods and mixes, milk and milk products, and dairy product analogs.

SAFETY: The Recommended Dietary Allowance (RDA) for vitamin B_6 suggested by the Food and Nutrition Board of the National Academy of Sciences-National Research Council is 2.0 to 2.2 milligrams per day for normal adults, and 2.5 to 2.6 milligrams during pregnancy or lactation; higher amounts are not considered necessary for women taking contraceptives. Intakes recommended for infants and adolescents range up to 2 milligrams daily. The estimated daily intake of vitamin B_6 as a food additive in 1975, based on usage reported by the food industry, was 0.6 milligram per person, about a fourth of the RDA. A U.S. Department of Agriculture survey conducted in 1977-78 estimated that the total dietary intake of vitamin B_6 for females was only 60 to 65 percent of the RDA; the mean dietary intake for men older than 65 years was 78 percent of the RDA.

Humans have ingested as much as 100 milligrams of pyridoxine hydrochloride, equivalent to some 170 times the estimated daily intake as a food additive, without adverse effects. Patients suffering from inherited disorders known to require greater amounts of vitamin B_6 have been given large doses (100 to 1500 milligrams daily) for periods of up to three or four years; these intakes did not cause harmful reactions.

Repeated intramuscular administration of a large amount of pyridoxine hydrochloride to a pregnant woman resulted in a pyridoxine dependency in the infant, manifested by convulsions and the death of the infant. Other cases have not been reported, but studies have been conducted with pregnant rats to determine the effect on the newborn of large doses during pregnancy. None of these have produced such an effect, nor did pyridoxine or pyridoxine hydrochloride affect infant growth or development, or reproductive performance by mothers. In one experiment the dosage employed was the equivalent of over 3000 times human intake of added pyridoxine (adjusted for body weight).

Laboratory tests of this vitamin indicate that it does not cause birth abnormalities. Studies to determine whether it has cancer-producing properties have not been conducted.

ASSESSMENT: Pyridoxine is an essential vitamin and is naturally present in a wide variety of foods. With a recommended dietary intake of 2 milligrams per day for adults, the daily average intake of pyridoxine hydrochloride resulting from its addition to foods may amount to no more than 10 percent of total intake, which frequently is less than desired. The evidence indicates that vitamin B_6 (pyridoxine or pyridoxine hydrochloride) poses no hazard to the public when used as a food additive at levels now current or that might reasonably be expected in the future.

RATING: S.

VITAMIN B_{12}

Cobalamin; Cyanocobalamin

Vitamin B_{12}, which is found in organ meats (liver, brain, kidney), as well as in oysters, clams, and egg yolk in relatively substantial amounts, is the active factor that is clinically effective in the treatment of pernicious anemia. All forms contain the trace mineral cobalt. The form in which it is usually present in these natural sources (all are animal products), *cobalamin,* has an unstable linkage with the cobalt; when vitamin B_{12} is added to processed foods, another more stable form, *cyanocobalamin,* in which cyanide is linked with the cobalt, is produced from the fermentation of microorganisms. Cobalamin is the vitamin minus the cyanide group. Vitamin B_{12} is added to a limited number of food products as a nutrient or dietary supplement. Among them are breakfast cereals, baby-food baked products and prepared formulas, rice and pasta dishes, milk products, and some snack foods.

MAJOR REFERENCE: Evaluation of the health aspects of vitamin B_{12} as a food ingredient. FASEB/SCOGS Report 104 (NTIS PB 289-922) 1978.

SAFETY: Based on use of vitamin B_{12} for fortification by food processors, the daily intake per person in 1975 was estimated as 39 micrograms (millionths of a gram). Vitamin B_{12} is essential in the physiologic effectiveness of folic acid (see p. 532) and in cell functioning, especially in the metabolic conversions related to nerve tissue, bone marrow, and red-blood-cell formation. It should be noted that this vitamin is one of the most biologically active known; the Food and Nutrition Board of the National Academy of Sciences-National Research Council recommends only 3 micrograms as being needed in the daily diet to meet all needs of the normal, healthy adult to prevent the anemia caused by a vitamin B_{12} deficiency.

Vitamin B_{12} in huge doses, as much as 100 milligrams, has been taken orally by patients with pernicious anemia, without harmful effects. There are no reports of any sensitivity to taking vitamin B_{12} orally, although allergy may arise from injections. Large doses of the vitamin fed to female rats did not affect reproduction and had no ill effects on the offspring, nor did they cause any abnormalities. Various studies have reported that developing cancers arising from other causes may be enhanced by the vitamin, but no studies have been conducted to determine whether the vitamin can itself cause cancer or alter genes.

ASSESSMENT: Vitamin B_{12} is essential in the metabolic processes of the body related to cell functioning; an insufficiency in the diet or failure to absorb it can result in pernicious anemia. Very large doses administered to humans have proved to be free of hazard. Thus the addition of vitamin B_{12} to foods is considered safe at present levels, or even at higher levels in the future.

RATING: S.

VITAMIN D (CALCIFEROL)

Vitamin D_2 (Ergocalciferol); Vitamin D_3 (Cholecalciferol)

Vitamin D is needed for calcium absorption and normal bone growth; its absence will lead to rickets. It occurs naturally in some foods, such as liver, eggs, and fish oils (like codfish oil). However, the limited food sources have led to fortification of a few products, particularly those used by growing children; they include milk, infant formulas, margarines, and cereals.

Two forms of vitamin D are used as additives in fortifying foods. *Vitamin D_2 (ergocalciferol)* is produced by ultraviolet irradiation of a compound of very similar chemical structure obtained from yeast. *Vitamin D_3 (cholecalciferol)* is the form of the vitamin produced through activation of the compound 7-dehydrocholesterol by sunlight, which normally occurs in humans through skin exposed to sun.

SAFETY: The quantity of vitamin D is reported as International Units (IU).* The recommended intake per day is 400 IU for children, adolescents, and pregnant and lactating women (since these individuals are not likely to meet their vitamin D needs through exposure to sunlight). Dietary requirement for normal healthy adults is less as some vitamin D is supplied by sunlight radiation. Estimates have been made on vitamin D intake from supplements and foods. They indicate that a baby may be receiving 800 IU daily from food and vitamin supplements and, in total, not more than 1600 IU (160 to 300 IU per kilogram of body weight, kg/bw). An eight-year-old's intake may be 800 IU from food together with supplements and up to 2900 IU in total (30 to 100 IU per kg/bw). The data on which the estimates are based are questionable and need improvement.

For its use in calcium metabolism, vitamin D must first be transformed by the liver and the kidneys into potent hormonelike regulatory products. Sometimes kidney failure may prevent this and thus lead to bone diseases.

While too little vitamin D can cause rickets or other disturbances in calcium metabolism, there is a real danger from

---

MAJOR REFERENCE: Evaluation of the health aspects of vitamin D, vitamin D_2 and vitamin D_3 as food ingredients. FASEB/SCOGS Report 95 (NTIS PB 293-099) 1978.

*An IU is 1/40 of a microgram, which is a millionth of a gram.

prolonged and excessive intake of this vitamin. Reduced kidney function along with calcium deposits in soft tissues is frequently seen. Behavioral disturbances may result from vitamin D intakes of even 1000 to 3000 IU per kg/bw if this is continued for very long periods of time. There are some suggestions that prolonged intake of more than 1250 IU daily per person may lead to greater risk of heart disease.

Vitamin D fed to pregnant rats and rabbits in dosages some thousands of times as great as likely human consumption may have provided a clue to the role an overdose of this substance may play in infantile hypercalcemia, a disorder which in mild form can slightly retard growth, but which when severe will prevent a baby from thriving, impair kidney function, and possibly cause mental retardation. The findings from these animal experiments indicate that large doses of vitamin D taken during pregnancy pass through the placenta barrier to the fetus, with effects on offspring that are comparable with infantile hypercalcemia. An accompanying symptom of this ailment in humans is a serious vascular disorder affecting the aorta (the main artery), and this too appeared in infant rabbits.

Anticonvulsant drugs taken regularly can inhibit the benefits of vitamin D, producing effects similar to vitamin D deficiency. Children can develop rickets; adults may experience below-normal blood calcium levels or bone softening.

The evidence indicates that large amounts of vitamin D have not caused cancer in experimental animals. Research relating to its effects on chromosome or gene alteration appears not to have been conducted.

ASSESSMENT: Vitamin D is an essential nutrient; its use in fortified foods has largely eliminated rickets in the U.S. The few foods with added vitamin D will assure that the daily requirement is met, and current levels of fortification make it unlikely that excessive intakes that can be harmful will come from the diet. However, many vitamin supplements and fish liver oils contain this vitamin, and undesirably high intakes can occur if these substances are used indiscriminately. Since the margin of safety is small between recommended intakes and toxic effects associated with higher intake levels, caution is advised on use of supplements containing this vitamin. Anyone taking anticonvulsant drugs should be alert to the possibility of an interaction between some of these and vitamin D, which can cause a vitamin deficiency.

Present food fortification practices will not pose any hazard to the consumer from dietary sources.

RATING: S.

VITAMIN K

Menadione (Vitamin K_3); Menaquinone (Vitamin K_2); Phylloquinone (Vitamin K_1, Phytonadione)

Vitamin K consists of a group of substances that are needed in the blood-clotting systems of higher animals. Three types are relevant to the dietary needs of humans: *vitamin* K_1 is a natural component of green plants, such as carrot tops and leafy vegetables (spinach, kale, broccoli, cabbage); *vitamin* K_2 is produced by various bacteria, including those in the intestine, and is present in beef liver and various tissues of chickens, rats, and other animals; and *vitamin* K_3, a simpler form which is produced synthetically, can be made into water-soluble forms which offer the advantage of being absorbable in the intestine without using bile, which vitamins K_1 and K_2 require. Vitamin K's function as a food additive at present is as a nutritional supplement in prepared foods for small infants. It is used medicinally for a number of illnesses related to blood conditions and side effects caused in their treatment.

SAFETY: The Food and Nutrition Board of the National Academy of Sciences-National Research Council in its 1980 revision of Recommended Dietary Allowances included an "estimated safe and adequate daily dietary intake" for vitamin K. For infants up to six months of age, it is 12 micrograms (millionths of a gram), gradually rising with age to an intake of 70 to 140 micrograms for adults.

Estimates of the average daily intake of vitamin K from natural sources are 300 to 500 micrograms. Vitamin K deficiency is rarely found. However, supplementation may be required for the newborn infant who often has low levels and no appreciable stores of blood-clotting factors, for which this vitamin is needed. Until the intestinal bacteria to synthesize it in the body are established, an infant may not be able to

---

MAJOR REFERENCE: The vitamins, volume 3. R. S. Harris and W. H. Sebrell, Jr., eds. (New York: Academic Press, 1971).

produce the prothrombin and other coagulation factors needed to prevent hemorrhage (abnormal bleeding).

In the form of menadione, vitamin K given to premature infants and pregnant women can easily rise to excessive levels, causing a toxicity characterized by anemia, flushing, and sweating. Phylloquinone does not have this undesirable property, and is the preferred form of vitamin K, unless there is evidence of malabsorption. There are studies showing that 2 milligrams of vitamin K_1 may be adequate and safe given orally to newborn infants, and may be adequate to prevent hemorrhagic disease in the newborn.

In people taking antibiotic drugs, bacterial growth in the digestive tract may be suppressed to the point where more vitamin K may be needed. Similarly, where there is liver disease or poor absorption because of inadequate bile secretion, a deficiency may occur even on an adequate diet.

Patients may take anticoagulant drugs to prevent blood clots that can cause heart attacks. Many of the anticoagulants are vitamin K antagonists, and vitamin K_1 may be used to maintain a balance against too little capacity of the blood to coagulate because of an excess of the anticoagulant drug. Doses of as much as 72 milligrams of menadione have been given intravenously following overdosages of oral anticoagulants; but far smaller dosages are usual, even in emergency cases, and vitamin K_1 is the form used.

ASSESSMENT: Usual dietary habits, supplemented by bacterial synthesis in the intestinal tract, assure an adequate intake of vitamin K for nearly everyone. Newborn infants, those with poor absorption or inadequate bile secretions, and people receiving antibiotics or anticoagulants may be exceptions. Vitamin K as a food additive is aimed at the needs of infants, and there is no reason to believe that it will be a hazard to these consumers from its presently limited use as a nutritional supplement in special foods.

RATING: S.

WHEY

DELACTOSED WHEY; DEMINERALIZED WHEY

When the principal protein of milk, casein, is separated from it (as in the manufacture of cheese), the remaining liquid portion is called *whey*. Some whey is processed to obtain lactose, or milk sugar (see p. 579); it also is a source of the valuable protein lactalbumin (see p. 465), which is rich in animo acids essential for the building, replacement, and repair of tissues. Dried whey solids are nearly three-fourths lactose, contain about 13 percent lactalbumin, and are rich in mineral elements such as potassium, sodium, phosphorus, and calcium.

Whey solids are used in more than 100 processed foods, which include ice cream, baked goods, confections, imitation dairy products, milk products such as eggnog, and breakfast cereals. Whey solids can serve as a binder in sausage products and meat loaves. Sometimes the whey may be treated to remove either the mineral salts *(demineralized whey)* or the lactose sugar *(delactosed whey)*.

SAFETY: A survey of food processors in 1977 indicated that the average daily diet included 2.2 grams of whey per person. Whey, a normal part of milk (an important food in the diet) contributes to the energy requirements as well as providing a very nutritious protein and many essential minerals for the body's needs. The National Academy of Sciences-National Research Council recommends a daily intake of 44 to 56 grams of protein for adults; whey as a food additive will contribute only a few percentage points of this.

The only question about the safety of whey as an additive in the diet relates to its extremely high content of the sugar lactose. Many people have an intolerance or inability to digest this sugar because they do not have the necessary enzymes to break it down and make possible its absorption and use by the body (see p. 579).

ASSESSMENT: Whey is a nutritious food and contributes needed protein and energy for the body. For those who are lactose-intolerant, whey may pose some problems of discomfort and diarrhea if excessive amounts are included in the diet; the average daily intake of whey as an additive would contribute an amount of lactose equivalent to a little over 1 ounce of skim milk. No other hazards are likely from including whey in processed foods.

RATING: S for delactosed whey; S for whey for most individuals; X for those allergic to milk and those with recognized lactose intolerance.

XANTHAN GUM

This additive is a complex carbohydrate-containing gum composed of sugar and sugar-acid units. It is produced from a microorganism by controlled fermentation of dextrose (see p. 518). When dissolved in water it results in a very viscous or gummy solution, even at low concentrations. This property makes it useful in a wide variety of foods as a stabilizer or thickener, as an emulsifier to keep water and oily components from separating, and as a suspending agent. It is used as a food additive in low-calorie products to simulate the viscosity and texture of sugar or oil, and to replace the starch normally used in such foods as puddings and pie fillings.

SAFETY: A survey by the National Academy of Sciences-National Research Council, based on usage reported by food manufacturers in 1976, determined that the daily diet of the average person in the U.S. contained 8 milligrams of xanthan gum, or 0.14 milligram per kilogram of body weight (kg/bw) for an individual weighing 60 kg (132 lbs.).

As is common in vegetable and microbial gums used as food additives, xanthan gum is not digested by the body and is mostly excreted unchanged. A long-term study using both rats and dogs, covering two years of feeding up to 1 gram of xanthan gum per kg/bw, well over 1000 times human consumption, established that this substance is harmless. Compared with the control animals whose diets did not contain this gum, there were no differences in survival, weight gain, organ weight, blood components, blood pressure or heart rate, microscopic tissue structure, or tumor incidence. The rats were studied over three generations, and the high level of the gum in the diet had no effect on parental or offspring survival, repro-

---

MAJOR REFERENCE: Toxicological evaluation of some food colours, enzymes, flavour enhancers, thickening agents, and certain other food additives. WHO Food Additive Series No. 6 (Geneva, 1975).

ductive performance, litter size or condition, or birth weight of the young.

Xanthan gum did not produce sensitization or irritation in tests for possible allergic effect. With rats, a diet containing 15 percent of the gum did not cause diarrhea; with dogs, some diarrhea was seen when the animals received 1 gram per kg/bw daily, and softer stools with 0.5 gram per kg/bw.

ASSESSMENT: Xanthan gum is used as a thickener and suspending agent in foods. This gum is essentially inert and poses no hazard even at very high levels in the diet. Like other gums that hold considerable amounts of water, it may cause diarrhea, but only at high dosages. The UN Joint FAO/WHO Expert Committee on Food Additives considers up to 10 milligrams per kg/bw as a safe and acceptable daily intake for humans, which is 75 times its presence in the U.S. diet.

RATING: S.

XYLITOL

Xylitol is not a true sugar, but rather a carbohydrate alcohol that is used as a synthetic sweetener in place of sucrose (cane or beet sugar; see p. 648) in certain dietary foods and "sugar-free" chewing gums. Although it has the same caloric value as sugar, xylitol is metabolized differently and thus has been used in diets for diabetics. Xylitol occurs naturally in many berries, fruits, and mushrooms. Commercially it is produced from "wood sugar," or xylose, which is abundant in wood and in plants. Finland has been the major producer of xylitol for use as a food additive.

SAFETY: Xylitol is one of the normal products in carbohydrate metabolism; the body itself produces 5 to 15 grams daily as an intermediate to aid in the conversion of carbohydrates into energy and its storage form, glycogen. Xylitol has been useful for diabetics because, although it does not increase blood sugar levels as much as do normal sugars in the diet, it is well metabolized and can be used as a source of energy by humans, even when a few hundred grams are consumed daily. It also has been found useful in reducing dental caries (cavi-

MAJOR REFERENCE: Dietary sugars in health and disease. II. Xylitol. FASEB/LSRO Report to FDA (NTIS PB 285-494) 1978.

ties) because the caries-producing bacteria in the mouth do not grow on and ferment xylitol to produce undesirable acids, as they do with sugar.

When xylitol was administered orally to rats in daily amounts of as much as 30 percent of the diet (equivalent to 30 grams per kilogram of body weight, kg/bw) for a period of 12 weeks, it did not adversely affect growth, reproduction, or the function or microscopic structure of major organs. Long-term studies in which three generations of rats were given 100 milligrams of xylitol per kg/bw daily during a period of two years showed that this additive did not adversely affect reproduction or fertility; nor did it cause cancers or pathological changes in any tissue. Preliminary findings are available from long-term animal feeding trials under way in England, in which significantly greater quantities of xylitol were administered than in the preceding study. Male mice fed daily diets for two years, 10 to 20 percent of which consisted of xylitol (approximately 8 to 17 grams per kg/bw), suffered bladder damage associated with abnormal concentration of mineral salts. Female mice, and rats and dogs fed similar diets for extended periods did not experience this effect. However, at the 20 percent level rats (but not dogs) developed adrenal tumors, while dogs at both the 10 and 20 percent levels exhibited an increase in liver weight caused by cell enlargement.

Evidence from animal tests and laboratory microbial studies indicates that xylitol neither causes birth defects nor alters genes.

Because animal experiments have shown that cataracts can be induced by feeding high-xylose diets (xylitol is formed from xylose), a question arises whether xylitol can produce a similar effect. There is no such evidence; in fact, it is unlikely that it can, as absorbed xylitol is removed from the blood as it is metabolized in the liver; and even if it weren't and high levels remained, the cells of the eye are known to be resistant to the diffusion that would be necessary to cause this effect.

Humans have taken xylitol at levels in the diet of up to 220 grams daily—3.7 grams per kg/bw for a person weighing 60 kg (132 lbs.)—in a three-week study of tolerance of this chemical; no significant adverse effects were observed, though loose stools were noted. Male and female volunteers have ingested 53 grams of xylitol per day over a two-year period without harm. In addition, normal children were born to these subjects during the period.

Xylitol has been fed intravenously to people with such

symptoms as kidney failure. In a few instances, where dosage levels were quite high, adverse changes were noted (including some kidney, liver, and brain disturbances). In Australia some deaths occurred in patients receiving this treatment. The most common effect of high doses of xylitol (several grams per kg/bw) is a transient diarrhea. Xylitol is known to be more slowly absorbed than sugars and thus tends to hold water and produce a watery stool. This has been seen in rats, monkeys, and humans.

ASSESSMENT: Xylitol provides calories but does not behave the way most carbohydrates do in raising blood sugar levels. In addition to its use in dietary foods, it has been used in sugar-free gum, which appears to be effective in reducing cavities in the teeth. Until recently, research indicated that xylitol in the diet had no adverse effect other than a possible mild diarrhea. However, the discovery that xylitol has caused tumors and organ injury in some animals administered high dosages in long-term feeding studies has caused concern. There are also the disturbing hazardous effects observed in humans receiving intravenous feeding of high levels of xylitol as an energy source. Xylitol's status as a permissible additive in food is currently under review by the FDA, and manufacturers in the U.S. have voluntarily ceased using it for the present. Until additional research can assure its safety in foods, uncertainties remain regarding whether this food additive may pose a hazard to the consumer.

RATING: ?

YEASTS*

Baker's Yeast; Brewer's Yeast; Dried Yeast; Smoked Yeast; Torula Yeast

Yeast, a type of fungus, is produced or grown by the fermentation of carbohydrates. The yeast used in food may be *baker's yeast* (a strain of *Saccharomyces cerevisiae* used in breadmaking and producing the leavening effect of copious amounts of gaseous carbon dioxide); *brewer's yeast* (a different strain

MAJOR REFERENCE: Single Cell Protein, II. S. R. Tannenbaum and D. I. Wang, eds. (Cambridge, Mass.: M.I.T. Press, 1975).

*For autolyzed yeast, a hydrolyzed brewer's yeast, see p. 613.

which produces greater amounts of alcohol in fermenting sugar, but is not effective in leavening), which is obtained as a by-product from the fermentation of beer made from cereal and hops (after removal of the bitter material derived from hops); or *torula yeast* (*Candida* species), which is obtained from cultures grown on molasses, the carbohydrate residues of papermaking from wood pulp, or more recently, petroleum. *Dried yeast* consists of the dry cells of any suitable yeast fungi, usually from brewer's yeast. It is high in protein (45 percent), and is rich in many of the B vitamins. It is also high in nucleic acids, and this has limited use of yeast as a major protein source. *Smoked yeast* is used as a flavoring agent in soups, cheese spreads, crackers, and snack foods; it is prepared by exposing dried yeast to wood smoke.

Yeasts are useful in foods as dough conditioners and leavening agents in baked goods; as a fermenting aid, particularly for alcoholic beverages; in formulating flavors in soup mixes, gravies, and other foods; and in providing nutrients.

SAFETY: In 1975 yeast used in food processing averaged 545 milligrams per person in the daily diet. Dried yeast approximated a tenth of the total. Smoked yeast flavoring had an average daily consumption of 4 milligrams per person in 1978.

In earlier years, yeast was used as a dietary source of vitamins. Today pure vitamins are available at much lower cost, so the use of yeast as a nutrient is primarily for its protein value. A high-lysine baker's yeast has been suggested as a protein supplement to improve the nutritional quality of cereal foods, which tend to be limited in lysin (see p. 475). Yeast also has enhanced the nutritional benefit to humans of several kinds of formulation of vegetable protein mixtures. The usefulness of yeast as a diet supplement has been demonstrated many times during the past decades; at levels up to 10 percent of yeast in the diet, weight gain has increased and the nutritive value of the dietary protein has improved. Many thousands of tons of yeast were used as meat substitutes and to extend meat, and in army rations in Germany, Russia, and Japan during World War II.

In humans, the nucleic acids in yeast are converted to uric acid when metabolized in the body. A large excess of uric acid can cause gout, a painful inflammation of the toes and joints. A safe intake of nucleic acid is about 2 grams per day. Since the daily intake of yeasts in the diet totals less than 0.5 gram, a harmful excess of uric acid from this source is unlikely unless yeast is consumed as a major source of protein in the diet, perhaps 20 grams or more, and this is not the way yeast is used as a food additive.

Clinical studies with human subjects indicate that an intake of 20 grams of yeast may result in nausea and diarrhea. At these high levels of consumption, there can be a sensitization to yeast.

In the 1970s there was interest in growing torula yeast on petroleum rather than using carbohydrate sources. The safety of this practice has been examined primarily because petroleum products may contain small amounts of cancer-inducing chemicals. Yeast grown on petroleum hydrocarbons has been dried and fed to rats to provide 30 percent of the protein; in 90-day studies, there were no significant effects of these yeasts on appearance, behavior, growth, food intake, blood components (including blood uric acid, though rats can metabolize uric acid and degrade it further, in contrast to humans), or on various pathological measures (including microscopic examination of the tissues and organs for precancerous changes) as compared with animals on a casein (see p. 507) diet. Proteins prepared from such yeasts have been fed to rats as the sole source of protein (20 percent of the diet). During the 100-day study there was no effect on deaths of the animals or on their general condition and behavior, but there was some occurrence of calcium deposits in the kidney. The level of feeding in this study would be equivalent to well over a thousand times the average human intake, adjusted for body weight.

A study has been conducted in which mice were given an injection of some 30,000 cancer cells. The mice were then tested for effects of feeding a yeast preparation as a food supplement. The tumor growth over the next four weeks was reduced, apparently because the yeast in the diet antagonized the establishment and early growth of the cancer. Yeasts grown on petroleum fractions have also been tested to see what effect they might have on tumor growth. In one test, rats were treated with a cancer-inducing chemical and fed yeast at a level of 17 to 27 percent of the diet (up to 80 percent of the protein) for seven months. The yeast did not influence growth or food consumption, nor did it affect the cancer development or incidence in the treated rats.

ASSESSMENT: Yeasts are useful nutrient supplements. They have been used for centuries and are indispensable for certain fermentation processes, such as making bread or brewing. No safety problem appears to come from growing yeast on either carbohydrate by-products or on petroleum. The use of yeast as a food additive poses no hazard to the consumer at levels now used or likely to be used in foods in the future. However, smoked yeast has not been adequately tested for safety, and there are reasons for concern about possible health hazards from the wood smoking (see p. 625).

RATING: S for all yeasts except smoked yeast; ? for smoked yeast.

ZINC SALTS

Zinc Acetate; Zinc Carbonate; Zinc Chloride; Zinc Gluconate; Zinc Hydrosulfite; Zinc Oxide; Zinc Stearate; Zinc Sulfate

Zinc is an essential element required in the diet of man; it is present in every cell, and is a component of enzymes, the specialized proteins in cells that act on substances in the body which initiate the chemical changes involved in metabolism. It is believed to be vital also for the transport of vitamin A from the liver, where it is mostly stored. Manifestation of inadequate intake of zinc in humans includes stunted growth and delayed sexual maturation.

In food, zinc is found chiefly in meats and cereals; liver, oysters, and eggs are good sources; vegetables, fruits, and milk less so. A number of zinc salts are permitted in food; 75 percent of the amount used is zinc oxide. The salts are used chiefly as nutritional supplements and are found for the most part in infant foods and ready-to-eat cereals.

SAFETY: Reported use of zinc salts in 1976 as a food additive in the U.S. diet averaged slightly more than 1 milligram per person daily; this represents only a fraction of the total intake because of the zinc content present naturally in foods. It has been estimated that the consumption of zinc in the daily diet of an adult is 5 to 22 milligrams. The Food and Nutrition Board advises a daily allowance of 3 milligrams for infants, 5 for children, 15 for adults, 20 for pregnant women, and 25 during lactation.

A wide margin exists between present human intake levels of zinc salts and levels that can produce adverse effects. Feed-

---

MAJOR REFERENCE: Evaluation of the health aspects of certain zinc salts as food ingredients. FASEB/SCOGS Report 21 (NTIS PB 266-879) 1977.

ing tests with a number of experimental animal species have shown that the salts caused no harm below 100 milligrams per kilogram of body weight, 250 times the maximum estimate of average human consumption adjusted for body weight. A few investigations of zinc sulfate have been conducted with man, without evidence of toxicity at dosages of up to 660 milligrams daily for as long as three months. One human fatality attributed to zinc sulfate has been reported following the accidental consumption of 30 grams (30,000 milligrams).

The most important effect of ingesting excess zinc is the appearance of a type of anemia in which there is a decrease in hemoglobin (the oxygen-carrying pigment in red blood cells), probably caused by interference with the body's utilization of iron and copper. Supplementation of these minerals in the diet can reverse the condition.

Studies performed through several generations of mice employing a variety of zinc salts did not show adverse effects on fertility, fetus health and development, or maternal or fetal enzymatic activities. Nor did oral administration of zinc sulfate, in quantities well in excess of human consumption, to three species of animals cause discernible harm to maternal survival or result in defective offspring.

Experiments with rats fed several zinc salts over three generations failed to produce evidence of cancer. In the early 1960s two studies of mice given zinc chloride in drinking water did report occurrence of cancer. However, control animals (mice exposed to the same conditions except for the zinc chloride) were not employed, nor were other relevant data made available to enable assessment of the validity of the findings. Another study with mice, using zinc sulfate in the same manner, did not show this result. Experienced investigators and laboratories specializing in experimental cancer have concluded, after reviewing the scientific literature, that zinc salts taken orally are not a cancer hazard.

ASSESSMENT: Zinc is an essential component for human life which plays a vital role in the body's metabolic processes. The Food and Nutrition Board of the National Academy of Sciences-National Research Council has suggested the fortification of cereal grain products at a level of 10 milligrams of zinc per pound, to help assure the adequacy of this mineral in the diet. A review of the evidence indicates that taken orally, zinc salts are not a cause of cancer.

RATING: S.

ADDITIVES INDEX

IDENTIFYING LEGEND FOR ADDITIVES-OF-CONCERN

**	Additives-of-concern to everyone
*	Additives-of-concern to some people only, and further defined for easy reference in the following manner:
*A	Anyone hyperallergic
*A (c)	Anyone allergic to corn, wheat, sugar beets
*A (e)	Anyone allergic to eggs
*A (g)	Anyone sensitive to gluten
*A (l)	Anyone intolerant of lactose
*A (m)	Anyone allergic to milk
*C (h)	Children, especially hyperactive ones
*C (y)	Young children
*CV	Anyone with a cardiovascular ailment
*E	Anyone with an eye ailment
*GI	Anyone with a gastrointestinal ailment
*H	Anyone with high blood pressure (hypertension)
*I	Infants
*N	Nursing mothers
*P	Pregnant women
*W	Anyone on a weight-reducing diet

APPENDIX 1

GENERAL TERMS IN LISTS OF INGREDIENTS WHICH MAY OR MAY NOT INCLUDE ADDITIVES-OF-CONCERN

These general terms will be found in the additives-of-concern column alongside an item in the Inventory of Brands when they appear in the list of ingredients on its label. It is not possible to determine whether these contain additives regarded in this volume as warranting a caution.

There are 13 of these general terms contained in the accompanying table. In each case, listed below it, are the ingredients they can be referring to, divided into those of concern and those that are not. All are reviewed elsewhere in this book.

ADDITIVES-OF-CONCERN	NOT ADDITIVES-OF-CONCERN
ANIMAL FATS	
Beef fat (tallow); butter (butter fat); lard (pork fat); mutton (fat); stearic acid; calcium stearate	Marine (fish) oil; poultry fat and skin
CELLULOSE DERIVATIVES	
Hydroxypropyl cellulose; methyl ethyl cellulose	Cellulose derivatives *other than* hydroxypropyl cellulose: methyl ethyl cellulose.
COLORING, FOOD COLORS	
Artificial color; certified color; FD&C colors; cochineal and carmine; paprika, turmeric, and their oleoresins; saffron	Natural color *other than* cochineal and carmine, paprika, turmeric, and their oleoresins; saffron.

NATURAL COLORS, VEGETABLE COLORS	
Cochineal and carmine; paprika, turmeric, and their oleoresins; saffron	Natural color *other than* cochineal and carmine; paprika, turmeric and their oleoresins; saffron
SEAWEED	
Alginates; carrageenan and furcelleran; dulse and kelp	Agar-agar
SHORTENING, VEGETABLE OILS	
Coconut oil; hydrogenated vegetable oils; palm kernel oil; rapeseed oil	Non- and partially hydrogenated vegetable oils *other than* coconut, palm kernel, and rapeseed oils
SOFTENERS, STABILIZERS, THICKENERS, VEGETABLE GUMS	
Alginates; carrageenan and furcelleran; dulse and kelp; glycerol esters of wood rosin; gum arabic; gum tragacanth; guar gum; hydroxypropyl cellulose; locust bean gum; methyl ethyl cellulose; modified starch	Agar-agar; cellulose derivatives *other than* hydroxypropyl cellulose, methyl ethyl cellulose; unmodified or gelatinized starch; xanthan gum

SUBSTANCES DERIVED FROM MILK* AND FLOUR THAT MAY CONTAIN INGREDIENTS REGARDED AS ADDITIVES-OF-CONCERN

A number of substances prepared from milk may contain ingredients which, if specified as such on a food label, would be viewed here as additives-of-concern.

	Ingredients of Concern
Buttermilk Solids Milk Derivatives Milk Protein Milk Solids Sour Milk Solids Sour Cream Solids	All may contain casein, lactalbumin, or whey.

A similar caution applies to High Protein Flour, which may contain added albumin, casein, or gluten.

These products prepared from milk and flour, when they appear on a food item's ingredients list, have been placed alongside the item in the additives-of-concern column. The ingredients of concern listed above are reviewed elsewhere in this volume.

*Milk itself is considered a food, not an additive. It is not included as an additive-of-concern in this book, although some people may have a milk intolerance.

APPENDIX 2

A COMPARISON IN RATINGS OF SAFETY OF FOOD ADDITIVES BETWEEN SCOGS AND THE FOOD ADDITIVES BOOK

For those food additives which were separately rated by both the Select Committee on GRAS Substances (SCOGS) and the authors, the data used were identical. They were contained in the reports issued by SCOGS.

The SCOGS final rating of additives had to conform to the requirements of the 1958 Food Additives Amendment of the Food, Drug, and Cosmetic Act. This stipulated that "credible evidence of, or reasonable grounds to suspect, adverse biological effects had to be present in whatever information was available before the pronouncement of a potential health hazard was to be advanced."* Upon completion of the initial information base to a given GRAS substance or a group of related substances, and later when tentative conclusions had been made, these were made available to the public by announcements in the *Federal Register,* and public hearings were scheduled for anyone desiring an opportunity to provide information or to express an opinion.† The effect, of these requirements and procedures on the SCOGS' final judgments imposed the same standards as would be required of the FDA if it were to challenge the safety of a GRAS substance as an ingredient in food.

The Food Additives Book (TFAB), which is directed to the food consumer, often found it more suitable for this purpose to base its judgments on information supplied in the SCOGS reports even when the data did not fully satisfy the standards of evidence demanded by federal legislation to support an action to ban a substance.

Because the information that was available often was far less complete than normally would be desirable, the Select Committee found it necessary to establish five ratings, two for degrees of safety, two for hazard, and one when data were insufficient or totally absent:

---

*Evaluation of health aspects of GRAS food ingredients. *Federation Proceedings,* volume 36, p. 2534 (1977)

†Ibid, p. 2531.

I There is no evidence at hand to suspect a hazard to the public when used at current levels, or levels that might reasonably be expected in the future.

II Identical with "I" except that evidence of safety was not sufficient in the event of a significant increase in current usage.

III Uncertainties exist about safety that require additional studies for their resolution.

IV Evidence of adverse effects exists.

V Data at hand are insufficient to evaluate safety.

By contrast, *TFAB's* rating system made provision for three categories of safety:

S Safe for everyone.

? Uncertain about safety.

X Unsafe for everyone.

It also employed combinations of these ratings, such as S for some people, ? and/or X for others (who would be identified). These disparities in categories of safety, as a consequence, make for difficulty in comparing the degree of correspondence between SCOGS and *TFAB* in their judgments of safety of additives. Despite this, a comparison has been attempted by means of combining some of the ratings in order to achieve greater comparability.

SCOGS Rating Categories		*TFAB's Rating Categories*
I	equivalent to	S
II	equivalent to	All combinations that include S, S/?, S/X, S/?/X
III & IV	equivalent to	?, X, ?/X

V has been omitted as safety ratings were not possible for these additives because of insufficient information.

In these rating categories, SCOGS I and *TFAB* S are essentially identical; additives allocated to these groups have been judged to be safe as ingredients in food without qualification. In the ratings suggesting hazard, SCOGS III and IV are regarded here as similar to *TFAB*'s ? and X and ?/X combinations.

The relationship of SCOGS II and the various combinations of *TFAB*'s ratings that include S is not as clear cut. What is similar is that SCOGS II rating expresses less certainty of the safety of these additives than those in its I category; and the same is true of *TFAB*'s S combinations compared with its unqualified S ratings. The S combinations caution some consumers of the possibility of a hazard.

The table which follows contains a comparison of 279 additives rated for safety by both SCOGS* and *TFAB*, based on the joined categories outlined above.

TFAB Categories	*SCOGS Categories*			
	I	II	III & IV	Total
S	195	32	0	227
S/?, S/X, S/?/X	3	16	2	21
?, ?/X, X	6	8	17	31
Total	204	56	19	279

Given the different requirements and objectives of SCOGS and *TFAB*, it is not surprising that differences exist in their ratings of the same additives, although it should be noted that there is considerable correspondence, as the table below reveals. Also of interest is that when SCOGS and *TFAB* are in disagreement, SCOGS arrives at a more severe rating in twice the number of instances as *TFAB*.

Extent of Correspondence
and
Direction of Severity of Rating
in Instances of Disagreement

	TFAB less severe	*TFAB* & SCOGS in Correspondence	*TFAB* more severe
S	32	195	0
S Combinations	2	16	3
?, X, ?/X	0	17	14
Total	34	228	17

*SCOGS ratings were obtained from Final Report FDA 223-75-2004. *Evaluation of GRAS Monographs,* FASEB Life Sciences Research Office, April 30, 1980.

ABOUT THE AUTHORS

WILLIS A. GORTNER is a biochemist with degrees from the University of Minnesota and the University of Rochester. He is now retired after a long and distinguished career in education, government, and private industry. Among the positions he has held are: Associate Professor of Biochemistry, School of Nutrition, Cornell University (1943-48); Head of the Department of Chemistry, The Pineapple Research Institute of Hawaii (1948-64); Director of the Human Nutrition Research Division of the Agricultural Research Service of the U.S. Department of Agriculture (1964-72); and Executive Officer, American Institute of Nutrition (1976-78).
Gortner has been honored by his fellow professionals on numerous occasions and has served often and with distinction on many committees within the food science community. He has been widely published in the nutrition and biochemical sciences and brings to this popular reference work an extraordinary degree of scientific expertise. He and his family live in Portola Valley, California.

NICHOLAS FREYDBERG has had a career in publishing—in newspapers *(New York World-Telegram, New York Herald Tribune),* in magazines *(Liberty, Newsweek)* and in books as co-founder of Basic Books. He has a doctorate in psychology and has participated in epidemiological research on the stresses experienced by humans in everyday life. He and his wife live on Martha's Vineyard, Massachusetts.